Compendium of Organic Synthetic Methods

Compendium of Organic Synthetic Methods

Volume 5

LEROY G. WADE, Jr.

DEPARTMENT OF CHEMISTRY
COLORADO STATE UNIVERSITY
FORT COLLINS, COLORADO

A Wiley-Interscience Publication

JOHN WILEY & SONS
New York • Chichester • Brisbane • Toronto • Singapore

Library of Congress Catalog Card Number: 71–162800
ISBN 0-471-86728-4

Printed in the United States of America

10 9 8 7 6 5 4 3 2 1

PREFACE

By their compilation of Volumes 1 and 2 of this *Compendium,* Ian and Shuyen Harrison filled one of the greatest needs of the synthetic community: a method for rapidly retrieving needed information from the literature by reaction type rather than by the author's name or publication date.

Compendium of Organic Synthetic Methods, Volume 5, presents the functional group transformations and difunctional compound preparations of 1980, 1981, and 1982. We have attempted to follow as closely as possible the classification schemes of the first three volumes; the experienced user of the *Compendium* will require no additional instructions on the use of this volume.

Perhaps it is fitting here to echo the Harrison's request stated in Volume 2 of the *Compendium:* The synthetic literature would become easily accessible and more useful if chemists could write well-organized, concise papers with charts and diagrams that allow the reader to assess quickly and easily the scope of the published research. In addition, the reporting of actual, isolated yields and detailed experimental conditions will save a great deal of wasted effort on the part of other chemists hoping to apply the reported reactions to their own synthetic problems.

I wish to express my gratitude to the many people who helped to bring this book to completion: To Mrs. Rosalie Jaramillo for her patience and dedication in the preparation of the camera-ready copy; to Dr. James McKearin, Dr. Forrest Sheffy, and Ron Wilde for proofreading the manuscript with great care and offering hundreds of helpful suggestions; and to my wife Betsy for her patience, help and moral support throughout the preparation of this Compendium.

LEROY G. WADE, JR.

Fort Collins, Colorado
November 1983

CONTENTS

ABBREVIATIONS   ix

INDEX, MONOFUNCTIONAL COMPOUNDS   xi

INDEX, DIFUNCTIONAL COMPOUNDS   xii

INTRODUCTION   xiii

 1 PREPARATION OF ACETYLENES   1

 2 PREPARATION OF CARBOXYLIC ACIDS, ACID HALIDES,
       AND ANHYDRIDES   9

 3 PREPARATION OF ALCOHOLS AND PHENOLS   24

 4 PREPARATION OF ALDEHYDES   92

 5 PREPARATION OF ALKYLS, METHYLENES, AND ARYLS   124

 6 PREPARATION OF AMIDES   168

 7 PREPARATION OF AMINES   191

 8 PREPARATION OF ESTERS   224

 9 PREPARATION OF ETHERS AND EPOXIDES   256

10 PREPARATION OF HALIDES AND SULFONATES   272

11 PREPARATION OF HYDRIDES   289

12 PREPARATION OF KETONES   304

13 PREPARATION OF NITRILES   353

14 PREPARATION OF OLEFINS   366

15 PREPARATION OF DIFUNCTIONAL COMPOUNDS   400

ABBREVIATIONS

An attempt has been made to use only abbreviations whose meaning will be readily apparent to the reader. Some of those more commonly used are the following:

Ac	Acetyl
AIBN	Azobisisobutyronitrile
Am	Amyl
Ar	Aryl
9-BBN	9-borabicyclo[3.31]nonane
BOC (t-Boc)	t-Butyloxycarbonyl
Bu	Butyl
Bz	Benzyl
Cp	Cyclopentadienyl
DBU	1,5-diazabicyclo[5.4.0]undecene-5
DCC	Dicyclohexylcarbodiimide
DDQ	2,3-Dichloro-5,6-dicyanobenzoquinone
DEAD	Diethyl azodicarboxylate
DIBAL (DIBAH)	Diisobutylaluminum hydride
DMAD	Dimethyl acetylenedicarboxylate
DME	1,2-Dimethoxyethane
DMF	Dimethylformamide
DMSO	Dimethyl sulfoxide
ee	Enantiomeric excess
Et	Ethyl
Hex	Hexyl
HMPA, HMPT	Hexamethylphosphoramide (hexamethylphosphoric triamide)
$h\nu$	Irradiation with light
L	Triphenylphosphine ligand (if not specified)
LAH	Lithium aluminum hydride
LDA	Lithium diisopropylamide
MCPBA	meta-Chloroperbenzoic acid
Me	Methyl
MEM	β-Methoxyethoxymethyl
Ms	Methanesulfonyl
MTM	Methylthiomethyl
MVK	Methyl vinyl ketone
NBS	N-bromosuccinimide

Ni	Raney nickel
Ⓟ	Polymeric backbone
PCC	Pyridinium chlorochromate
Ph	Phenyl
PPA	Polyphosphoric acid
PPE	Polyphosphate ester
Pr	Propyl
PTC	Phase-transfer catalysis
Py, Pyr	Pyridine
RT	Room temperature
Sia	*secondary*-isoamyl
Tf	Trifluoromethane sulfonate
TFA	Trifluoroacetic acid
TFAA	Trifluoroacetic anhydride
THF	Tetrahydrofuran
THP	Tetrahydropyranyl
TMEDA	Tetramethylethylenediamine
TMP	2,2,6,6-Tetramethylpiperidine
TMS	Trimethylsilyl
Tol	Tolyl
Ts	*p*-Toluenesulfonyl
Z	Benzyloxycarbonyl
Δ	Heat

INDEX, MONOFUNCTIONAL COMPOUNDS

Sections – heavy type
Pages – light type

PREPARATION OF → / FROM ↓

FROM \ PREPARATION OF	Acetylenes	Carboxylic acids, acid halides, anhydrides	Alcohols, phenols	Aldehydes	Alkyls, methylenes, aryls	Amides	Amines	Esters	Ethers, epoxides	Halides, sulfonates	Hydrides (RH)	Ketones	Nitriles	Olefins
Acetylenes	**1** 1	**16** 9	**31** 24	**46** 92	**61** 124			**106** 224		**136** 272		**166** 304		**196** 366
Carboxylic acids, acid halides, anhydrides	**2** 2	**17** 9	**32** 24	**47** 93	**62** 125	**77** 168		**107** 225	**122** 256	**137** 272	**152** 289	**167** 306	**182** 353	**197** 368
Alcohols, phenols		**18** 11	**33** 25	**48** 96	**63** 125	**78** 175	**93** 191	**108** 233	**123** 256	**138** 273	**153** 290	**168** 310	**183** 354	**198** 369
Aldehydes	**4** 2	**19** 12	**34** 25	**49** 103	**64** 126	**79** 176	**94** 192	**109** 237	**124** 260			**169** 318	**184** 354	**199** 371
Alkyls, methylenes, aryls				**50** 104	**65** 127					**140** 275	**155** 292	**170** 319		**200** 375
Amides		**21** 14	**36** 37			**81** 177	**96** 195	**111** 239					**186** 358	**201** 378
Amines		**22** 15	**37** 37	**52** 105		**82** 180	**97** 196			**142** 276	**157** 293	**172** 321	**187** 358	**202** 379
Esters		**23** 15	**38** 38		**68** 128	**83** 183		**113** 240	**128** 261	**143** 278	**158** 293	**173** 322		**203** 379
Ethers, epoxides			**39** 41	**54** 108	**69** 129	**84** 184		**114** 243	**129** 263	**144** 278	**159** 294	**174** 324		**204** 380
Halides, sulfonates, sulfates	**10** 4	**25** 17	**40** 48	**55** 109	**70** 130	**85** 184	**100** 200	**115** 245	**130** 263	**145** 279	**160** 295	**175** 325	**190** 359	**205** 382
Hydrides (RH)		**26** 18	**41** 49	**56** 112	**71** 141	**86** 185	**101** 203	**116** 247	**131** 264	**146** 280		**176** 327	**191** 361	**206** 388
Ketones	**12** 6	**27** 19	**42** 52	**57** 114	**72** 145	**87** 186	**102** 204	**117** 251	**132** 265	**147** 284		**177** 329	**192** 362	**207** 389
Nitriles		**28** 19		**58** 116	**73** 149	**88** 187	**103** 208	**118** 253			**163** 301	**178** 337	**193** 363	
Olefins	**14** 8	**29** 19	**44** 75	**59** 117	**74** 149	**89** 188	**104** 209	**119** 254	**134** 266	**149** 285		**179** 339	**194** 364	**209** 394
Miscellaneous compounds	**15** 8	**30** 20		**60** 118	**75** 166	**90** 189	**105** 212		**135** 271	**150** 287	**165** 302	**180** 341		**210** 395

PROTECTION

	Sect.	Pg.
Carboxylic acids	30A	20
Alcohols, phenols	45A	79
Aldehydes	60A	119
Amides	90A	190
Amines	105A	216
Esters	120A	255
Ketones	180A	343
Olefins	210A	398

Blanks in the table correspond to sections for which no additional examples were found in the literature.

INDEX, DIFUNCTIONAL COMPOUNDS

Sections — heavy type

Pages — light type

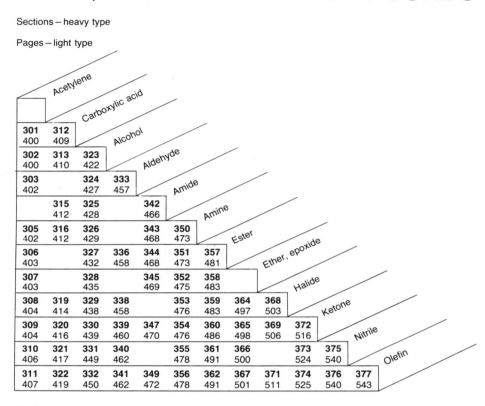

Acetylene	Carboxylic acid	Alcohol	Aldehyde	Amide	Amine	Ester	Ether, epoxide	Halide	Ketone	Nitrile	Olefin
301 400	**312** 409										
302 400	**313** 410	**323** 422									
303 402		**324** 427	**333** 457								
	315 412	**325** 428		**342** 466							
305 402	**316** 412	**326** 429		**343** 468	**350** 473						
306 403		**327** 432	**336** 458	**344** 468	**351** 473	**357** 481					
307 403		**328** 435		**345** 469	**352** 475	**358** 483					
308 404	**319** 414	**329** 438	**338** 458		**353** 476	**359** 483	**364** 497	**368** 503			
309 404	**320** 416	**330** 439	**339** 460	**347** 470	**354** 476	**360** 486	**365** 498	**369** 506	**372** 516		
310 406	**321** 417	**331** 449	**340** 462		**355** 478	**361** 491	**366** 500		**373** 524	**375** 540	
311 407	**322** 419	**332** 450	**341** 462	**349** 472	**356** 478	**362** 491	**367** 501	**371** 511	**374** 525	**376** 540	**377** 543

Blanks in the table correspond to sections for which no additional examples were found in the literature.

INTRODUCTION

Relationship between Volume 5 and Previous Volumes. *Compendium of Organic Synthetic Methods, Volume 5* presents over 1000 examples of published methods for the preparation of monofunctional compounds, updating the 6000 in Volumes 1 through 4. In addition, Volume 5 contains over 1000 additional examples of preparations of difunctional compounds and various functional groups, updating the sections introduced in Volume 2. The same systems of section and chapter numbering are used in all five volumes.

Classification and Organization of Reactions Forming Monofunctional Compounds. Chemical transformations are classified according to the reacting functional group of the starting material and the functional group formed. Those reactions that give products with the same functional group form a chapter. The reactions in each chapter are further classified into sections on the basis of the functional group of the starting material. Within each section, reactions are listed in a somewhat arbitrary order, although an effort has been made to put similar reactions together when possible.

The classification is unaffected by allylic, vinylic, or acetylenic unsaturation appearing in both starting material and product, or by increases or decreases in the length of carbon chains; for example, the reactions t-BuOH $\rightarrow$ t-BuCOOH, $PhCH_2OH$ $\rightarrow$ PhCOOH and $PhCH=CHCH_2OH$ $\rightarrow$ $PhCH=CHCOOH$ would all be considered as preparations of carboxylic acids from alcohols. Conjugate reduction and alkylation of unsaturated ketones, aldehydes, esters, acids, and nitriles have been placed in category 74, Alkyls from Olefins.

The terms hydrides, alkyls, and aryls classify compounds containing reacting hydrogens, alkyl groups, and aryl groups, respectively; for example, RCH_2-H $\rightarrow$ RCH_2COOH (carboxylic acids from hydrides), RMe $\rightarrow$ RCOOH (carboxylic acids from alkyls), RPh $\rightarrow$ RCOOH (carboxylic acids from aryls). Note the distinction between R_2CO $\rightarrow$ R_2CH_2 (methylenes from ketones) and RCOR' $\rightarrow$ RH (hydrides from ketones). Alkylations involving additions across double bonds are found in section 74, Alkyls from Olefins.

The following examples illustrate the classification of some potentially confusing cases:

$RCH=CHCOOH \rightarrow RCH=CH_2$	(hydrides from carboxylic acids)
$RCH=CH_2 \rightarrow RCH=CHCOOH$	(carboxylic acids from hydrides)
$ArH \rightarrow ArCOOH$	(carboxylic acids from hydrides)
$ArH \rightarrow ArOAc$	(esters from hydrides)
$RCHO \rightarrow RH$	(hydrides from aldehydes)
$RCH=CHCHO \rightarrow RCH=CH_2$	(hydrides from aldehydes)
$RCHO \rightarrow RCH_3$	(alkyls from aldehydes)
$R_2CH_2 \rightarrow R_2CO$	(ketones from methylenes)
$RCH_2COR \rightarrow R_2CHCOR$	(ketones from ketones)
$RCH=CH_2 \rightarrow RCH_2CH_3$	(alkyls from olefins)
$RBr+RC\equiv CH \rightarrow RC\equiv CR$	(acetylenes from halides; also (acetylenes from acetylenes)
$ROH+RCOOH \rightarrow RCOOR$	(esters from alcohols; also esters from carboxylic acids)
$RCH=CHCHO \rightarrow R_2CHCH_2CHO$	(alkyls from olefins)
$RCH=CHCN \rightarrow RCH_2CH_2CN$	(alkyls from olefins)

How to Use the Book to Locate Examples of the Preparation or Protection of Monofunctional Compounds. Examples of the preparation of one functional group from another are located via the monofunctional index on p. xi, which lists the corresponding section and page. Thus Section 1 contains examples of the preparation of acetylenes from other acetylenes; Section 2, acetylenes from carboxylic acids; and so forth.

Sections that contain examples of the reactions of a functional group are found in the horizontal rows of the index. Thus Section 1 gives examples of the reactions of acetylenes that form other acetylenes; Section 16, reactions of acetylenes that form carboxylic acids; and Section 31, reactions of acetylenes that form alcohols.

Examples of alkylation, dealkylation, homologation, isomerization, and transposition are found in Sections 1, 17, 33, and so on, lying close to a diagonal of the index. These sections correspond to such topics as the preparation of acetylenes from acetylenes, carboxylic acids from carboxylic acids, and alcohols and phenols from alcohols and phenols. Alkylations which involve conjugate additions across a double bond are found in section 74, Alkyls from Olefins.

Examples of name reactions can be found by first considering the nature of the starting material and product. The Wittig reaction, for instance, is in Section 199 on olefins from aldehydes and Section 207 on olefins from ketones.

Examples of the protection of acetylenes, carboxylic acids, alcohols, phenols, aldehydes, amides, amines, esters, ketones, and olefins are also indexed on p. xi.

The pairs of functional groups alcohol, ester, carboxylic acid, ester; amine, amide; carboxylic acid, amide can be interconverted by simple reactions. When a member of these groups is the desired product or starting material, the other member should, of course, also be consulted on the text.

The original literature must be used to determine the generality of reactions. A reaction given in this book for a primary aliphatic substrate may also be applicable to tertiary or aromatic compounds. This book does not attempt to provide experimental conditions or precautions, under the assumption that the reader will study the original literature before attempting a reaction. Not to do so would be hazardous, as well as foolish. The original papers usually yield a further set of references to previous work. Subsequent publications can be found by consulting the Science Citation Index.

Classification and Organization of Reactions forming Difunctional Compounds. This chapter considers all possible difunctional compounds formed from the groups acetylene, carboxylic acid, alcohol, aldehyde, amide, amine, ester, ether, epoxide, halide, ketone, nitrile, and olefin. Reactions that form difunctional compounds are classified into sections on the basis of the two functional groups of the product. The relative positions of the groups do not affect the classification. Thus preparations of 1,2-aminoalcohols, 1,3-aminoalcohols and 1,4-aminoalcohols are included in a single section. The following examples illustrate the application of this classification system:

Difunctional Product	*Section Title*
RC≡C-C≡CR	Acetylene – Acetylene
RCH(OH)COOH	Carboxylic Acid – Alcohol
RCH=CHOMe	Ether – Olefin
$RCHF_2$	Halide – Halide
$RCH(Br)CH_2F$	Halide – Halide
$RCH(OAc)CH_2OH$	Alcohol – Ester
RCH(OH)COOMe	Alcohol – Ester
$RCH=CHCH_2COOMe$	Ester – Olefin
RCH=CHOAc	Ester – Olefin

How to Use the Book to Locate Examples of the Preparation of Difunctional Compounds. The difunctional index on p. xii gives the section and page corresponding to each difunctional product. Thus Section 327 (Alcohol – Ester) contains examples of the preparation of hydroxyesters;

Section 323 (Alcohol — Alcohol) contains examples of the preparation of diols.

Some preparations of olefinic and acetylenic compounds from olefinic and acetylenic starting materials can, in principle, be classified in either the monofunctional or difunctional sections; for example, RCH=CHBr $\rightarrow$ RCH=CHCOOH, Carboxylic acids from Halides (monofunctional sections) or Carboxylic acid — Olefin (difunctional sections). In such cases both sections should be consulted.

Reactions applicable to both aldehyde and ketone starting materials are in many cases illustrated by an example that uses only one of them.

Many literature preparations of difunctional compounds are extensions of the methods applicable to monofunctional compounds. Thus the reaction RCl $\rightarrow$ ROH can clearly be extended to the preparation of diols by using the corresponding dichloro compound as a starting material. Such methods are not fully covered in the difunctional sections.

The user should bear in mind that the pairs of functional groups alcohol, ester; carboxylic acids, ester; amine, amide; and carboxylic acid, amide can be interconverted by simple reactions. Compounds of the type RCH(OAc)CH$_2$OAc (Ester — Ester) would thus be of interest to anyone preparing the diol RCH(OH)CH$_2$OH (Alcohol — Alcohol).

CHAPTER 1
PREPARATION OF ACETYLENES

Section 1 <u>Acetylenes from Acetylenes</u>

$$C_5H_{11}-C{\equiv}CH \xrightarrow[\text{NaOAc, MeOH}]{\text{PdCl}_2,\ \text{CuCl}_2,\ \text{CO}} C_5H_{11}-C{\equiv}C-CO_2Me \quad 74\%$$

Tetr Lett, <u>21</u>, 849 (1980)

$$\begin{array}{c} HC{\equiv}C-\overset{\oplus}{\underset{\underset{\displaystyle Co_2(CO)_6}{|}}{C}}(CH_3)_2 \\ \\ + \\ \\ TMS- \end{array}$$

$$\xrightarrow{\text{2) Fe(NO}_3)\cdot 9H_2O}$$

$$HC{\equiv}C-\overset{CH_3}{\underset{CH_3}{\overset{|}{\underset{|}{C}}}}-$$

~90%

Tetr Lett, <u>21</u>, 1595 (1980)

Section 2 Acetylenes from Acid Derivatives

Ph-C≡CH

+

$\underset{\text{toluene}}{\xrightarrow{\text{CuI, Et}_3\text{N}}}$

Ph-C≡C-C(=O)X   81%

Bull. Acad. USSR Chem. **30**, 918 (1981)

Section 3 Acetylenes from Alcohols

No additional examples

Section 4 Acetylenes from Aldehydes

$\underset{2\ \underline{t}\text{-BuOK}}{\xrightarrow{\text{Ph}_3\overset{\oplus}{\text{P}}\text{CH}_2\text{Br}}}$

79%

Tetr Lett, **21**, 4021 (1980)

1) DBU
2) Pb(OAc)$_4$

40%

Tetr Lett, **23**, 4607 (1982)

Hx-CHO $\xrightarrow[\text{2) BuLi}]{\text{1) PPh}_3\text{, CBr}_4}$ Hx-C≡C-Li

$$\text{Hx-C≡C-(CH}_2)_4\text{-OTHP}$$

$\underset{\overset{|}{\text{OTHP}}}{\text{CH}_2\text{CH}_2\text{-CHO}}$ $\xrightarrow[\text{2) BH}_3\cdot\text{THF}]{\text{1) Ph}_3\text{P=CH}_2}$ $\left[\underset{\overset{|}{\text{OTHP}}}{\text{CH}_2\text{(CH}_2)_3}\right]_3$ —B

71%

Chem Ber, <u>115</u>, 828 (1982)

Section 5 <u>Acetylenes from Alkyls, Methylenes, and Aryls</u>

No examples

Section 6 <u>Acetylenes from Amides</u>

No additional examples

Section 7 <u>Acetylenes from Amines</u>

No additional examples

Section 8 <u>Acetylenes from Esters</u>

No additional examples

Section 9 <u>Acetylenes from Halides</u>

No examples

Section 10 <u>Acetylenes from Halides</u>

83%

Liebigs Ann. Chem, 1 (1980)

$$Hx-\underset{\underset{Br}{|}}{CH}-\underset{\underset{Br}{|}}{CH_2} \xrightarrow[\text{P.T.C.}]{\text{KOH}} Hx-C{\equiv}CH \qquad 95\%$$

Tetrahedron, <u>37</u>, 1653 (1981)

$$\underset{\underset{Br}{|}}{CH_2}-\underset{\underset{Br}{|}}{CH}-CH(OEt)_2 \xrightarrow[\text{H}_2\text{O, pentane}]{Bu_4N^+ \ \bar{O}H} H-C{\equiv}C-CH(OEt)_2 \quad 67\%$$

Org Syn, <u>59</u>, 10 (1980)

$$Hx-CH{=}CF_2 \xrightarrow[\text{2) LDA}]{\text{1) BuLi}} Hx-C{\equiv}C-Bu \qquad 68\%$$

Chem Lett, 935 (1980)

$C_5H_{11}-C{\equiv}CLi$

+

$\xrightarrow[\text{LiI}]{\text{THF}}$ $C_5H_{11}-C{\equiv}C$ 75%

Chem Lett, 669 (1980)

$I(CH_2)_9COOH$

+ $\xrightarrow{\text{HMPA}}$ $HC{\equiv}C(CH_2)_9COOH$ 96%

$LiC{\equiv}CH{\cdot}EDA$

Synth Comm, <u>10</u>, 653 (1980)

$Bu-C{\equiv}C-Li$

+ $\xrightarrow{\qquad\qquad}$ $Bu-C{\equiv}C-(CH_2)_8OH$ 55%

$Br(CH_2)_8O-THP$ 2) $H_3O^{\oplus}$

JOC (USSR), <u>16</u>, 1728 (1980)

$PhC{\equiv}C-SnMe_3$

+ $\xrightarrow{\qquad\qquad}$

94%

JOC (USSR), <u>17</u>, 18 (1981)

1) 2 $HC{\equiv}C-TMS$
   L_2PdCl_2, CuI

2) ^-OH, H_2O

~80%

Synthesis, 627 (1980)

$$\text{Me}_3\text{Si-C≡C-SiMe}_3 \xrightarrow[\text{2) } \underline{\text{t}}\text{-BuCl, AlCl}_3]{\text{1) PhBr, AlCl}_3} \text{Ph-C≡C-}\underline{\text{t}}\text{-Bu} \quad 86\%$$

JCS Chem Comm, 959 (1982)

Section 11 Acetylenes from Hydrides

no examples

For examples of the reaction RC≡CH → RC≡C-C≡CR' see section 300 (Acetylene - Acetylene)

Section 12 Acetylenes from Ketones

1) LDA
2) ClPO(OEt)$_2$
3) LDA
4) H$_3$O$^\oplus$

85%

JOC, 45, 2526 (1980)

$$\text{Ph-}\overset{O}{\underset{}{\text{C}}}\text{-CH}_2\text{-}\overset{O}{\underset{}{\text{C}}}\text{-CH}_3 \xrightarrow{\text{KF, Et}_2\text{NCF}_2\text{CHFCF}_3} \text{Ph-C≡C-}\overset{O}{\underset{}{\text{C}}}\text{-CH}_3 \quad 72\%$$

Chem Lett, 1327 (1980)

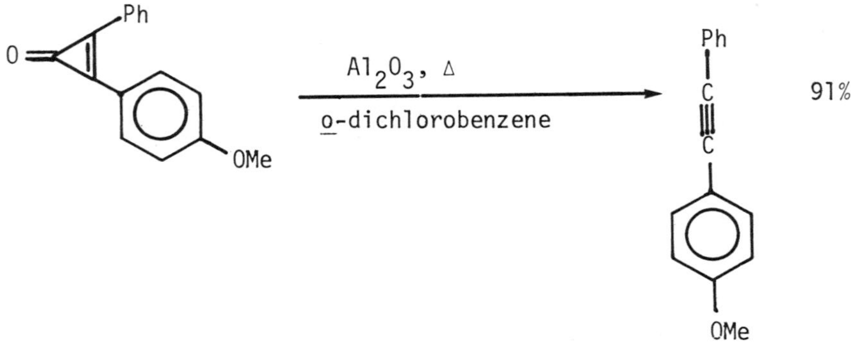
Ph
O
OMe
Al_2O_3, Δ
o-dichlorobenzene
Ph
C
C
OMe
91%

Synthesis, 285 (1981)

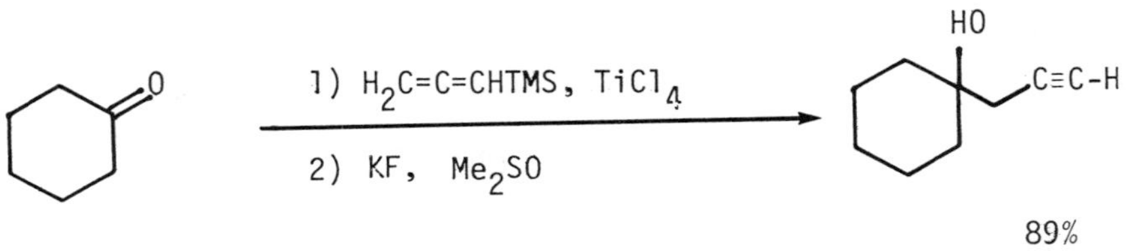
O
1) $H_2C{=}C{=}CHTMS$, $TiCl_4$
2) KF, Me_2SO
HO
C≡C-H
89%

JOC, 45, 3925 (1980)

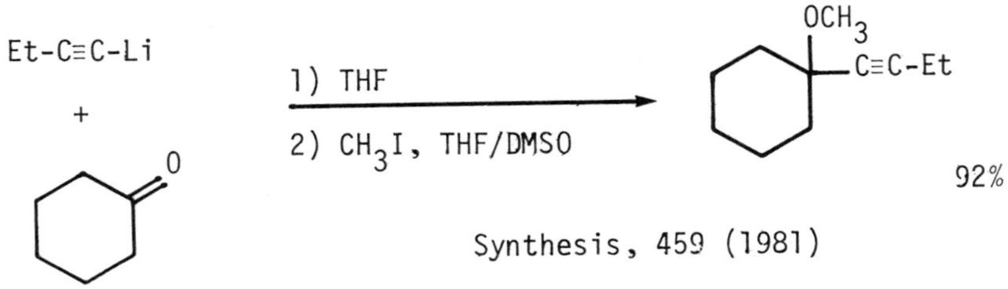
Et-C≡C-Li
+
O
1) THF
2) CH_3I, THF/DMSO
OCH_3
C≡C-Et
92%

Synthesis, 459 (1981)

Section 13 Acetylenes from Nitriles

no examples

Section 14 Acetylenes from Olefins

2   [structure: OMe-substituted phenyl vinyl NO$_2$] $\xrightarrow[\text{Et}_3\text{N}]{\text{H}_2\text{O}_2}$ [structure: bis(4-methoxyphenyl)acetylene] $\sim 30\%$

Tetr Lett, 22, 2301 (1981)

Section 15 Acetylenes from Miscellaneous Compounds

Review: "Heterocyclic Rearrangements: New Cumulenes and
         Acetylenes"

Bull Soc Chim Belges, 91, 997 (1982)

Section 15A Protection of Acetylenes

no additional examples

CHAPTER 2
PREPARATION OF CARBOXYLIC ACIDS, ACID HALIDES, AND ANHYDRIDES

Section 16 <u>Carboxylic Acids from Acetylenes</u>

$$CH_3(CH_2)_5-C\equiv CH \xrightarrow[\text{RuCl}_2\text{L}_3]{\text{PhIO}} CH_3(CH_2)_5-COOH \qquad 81\%$$

Helv Chim Acta, <u>64</u>, 2531 (1981)

$$H_2N-CH_2-C\equiv CH \xrightarrow{\qquad\qquad\qquad}$$

H ,,CH$_2$OMe

N-CH(OMe)$_2$

2) similar, using BzBr

3) CH$_3$I

4) several steps

Bz ,,,CH$_3$

H$_2$N   COOH

63%
84%ee

Angew Chem Int Ed, <u>19</u>, 725 (1980)

Section 17 <u>Carboxylic Acids from Acid Halides from</u>

<u>Carboxylic Acids</u>

$$PhCH_2CH_2COOH \xrightarrow[\text{CH}_2\text{Cl}_2]{\text{Ph}_3\text{PBr}_2} PhCH_2CH_2\overset{O}{\overset{\|}{C}}-Br \qquad 84\%$$

Synthesis, 684 (1982)

$$CH_3-\underset{\underset{O}{\|}}{C}-\underset{\underset{O}{\|}}{C}-OH \xrightarrow{CH_3OCHCl_2} CH_3-\underset{\underset{O}{\|}}{C}-\underset{\underset{O}{\|}}{C}-Cl \qquad 50\%$$

Org Syn, <u>61</u>, 1 (1983)

$$\underline{n}-C_7H_{15}\underset{\underset{O}{\|}}{C}-Cl \xrightarrow{Me_3SiBr} \underline{n}-C_7H_{15}\underset{\underset{O}{\|}}{C}-Br \qquad 93\%$$

Synthesis, 216 (1981)

$$CH_3-\underset{\underset{O}{\|}}{C}-Br \xrightarrow[glyme]{(CF_3)_2Cd} CH_3-\underset{\underset{O}{\|}}{C}-F \qquad 95\%$$

JCS Chem Comm, 670 (1980)

$$Me_2C\underset{\underset{O}{\|}}{\overset{\overset{O}{\|}}{\begin{matrix}C-Cl\\C-Cl\end{matrix}}} \xrightarrow[CH_3CN]{2NaI} Me_2C\underset{\underset{O}{\|}}{\overset{\overset{O}{\|}}{\begin{matrix}C-I\\C-I\end{matrix}}} \qquad 90\%$$

Synthesis, 237 (1982)

75%

Synthesis, 715 (1981)

$$2 \ C_{11}H_{23}COOH \quad \xrightarrow[\text{PhNH}]{\text{PhO}\overset{O}{\underset{}{\|}}\text{P-Cl}} \quad \left(C_{11}H_{23}-\overset{O}{\underset{}{\|}}C \right)_2 O \qquad 90\%$$

Synthesis, 218 (1981)

1) ClSO$_2$NCO, Et$_3$N

2) ^~~COOH

$$\left(\overset{O}{\underset{}{\|}}C \right)_2 O \qquad 86\%$$

Synthesis, 506 (1982)

1) Meldrum's acid, DMAP
2) NaBH$_3$CN, HOAc/THF
3) HCl, heat

$$Cl-\overset{O}{\underset{}{\|}}C-(CH_2)_n-\overset{O}{\underset{}{\|}}C-Cl \quad \longrightarrow \quad HO-\overset{O}{\underset{}{\|}}C-(CH_2)_{n+4}-\overset{O}{\underset{}{\|}}C-OH$$

~50% overall

Synth Comm, 12, 19 (1982)

Section 18 Carboxylic Acids from Alcohols

1-octanol $\xrightarrow{\text{solid NaMnO}_4}$ octanoic acid   67%

Tetr Lett, 22, 1655 (1981)

$$C_9H_{19}\text{-}CH_2OH \xrightarrow[CH_2Cl_2]{Cu(MnO_4)_2} C_9H_{19}COOH$$

81%

JOC, 47, 2790 (1982)

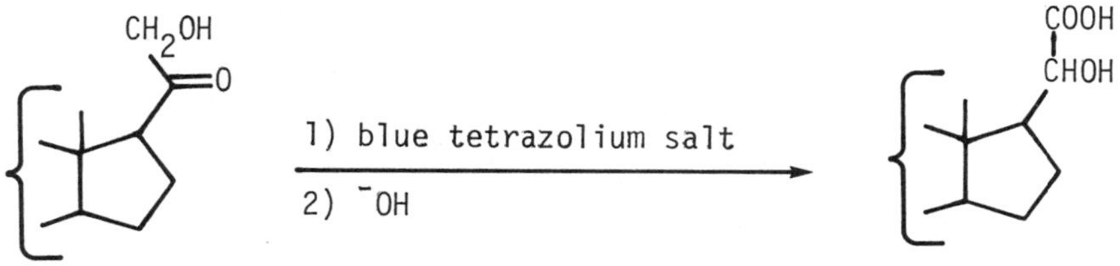
CH2OH
=O
1) blue tetrazolium salt
2) ⁻OH
COOH
CHOH

74%

Synthesis, 739 (1980)

Section 19 Carboxylic Acids from Aldehydes

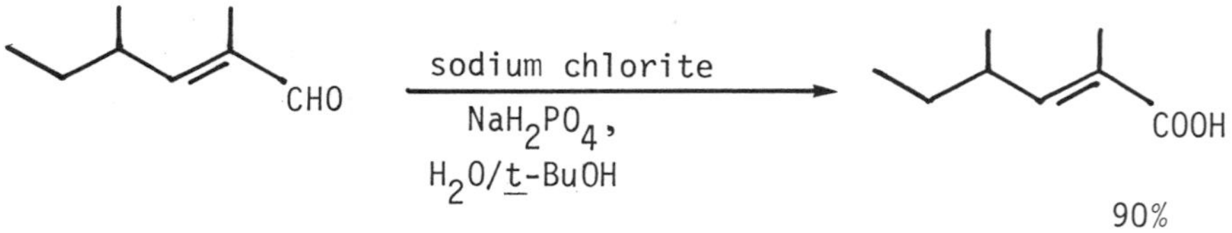
CHO
sodium chlorite
NaH2PO4,
H2O/t-BuOH
COOH

90%

Tetrahedron, 37, 2091 (1981)

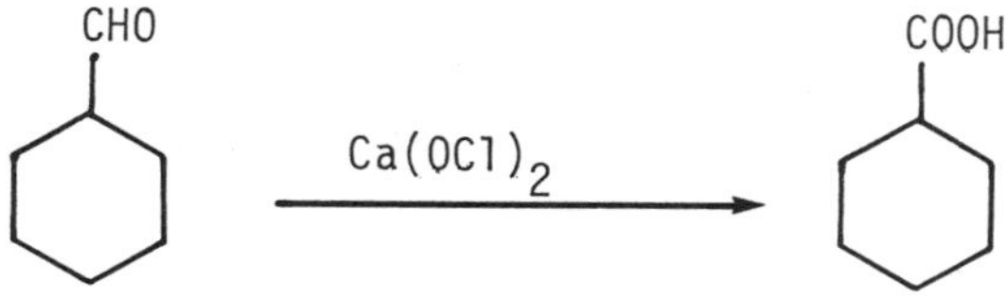
CHO
Ca(OCl)2
COOH

77%

Tetr Lett, 23, 3131 (1982)

Syn Comm, 10, 951 (1980)

91%

Tetr Lett, 21, 685 (1980)

solid NaMnO$_4$

octanal ———————————→ octanoic acid   77%

Tetr Lett, 22, 1655 (1981)

$$(MeO)_2\overset{\underset{\displaystyle \|}{O}}{P}H, \; NaOMe$$

Ph-CHO ———————————————→   89%

ClCH$_2$COOH, MeOH

JOC, 46, 2514 (1981)

Related methods: Carboxylic Acids from Ketones (Section 27).

Also via: Esters - Section 109).

Section 20 Carboxylic Acids from Alkyls

no additional examples

Section 21 Carboxylic Acids from Amides

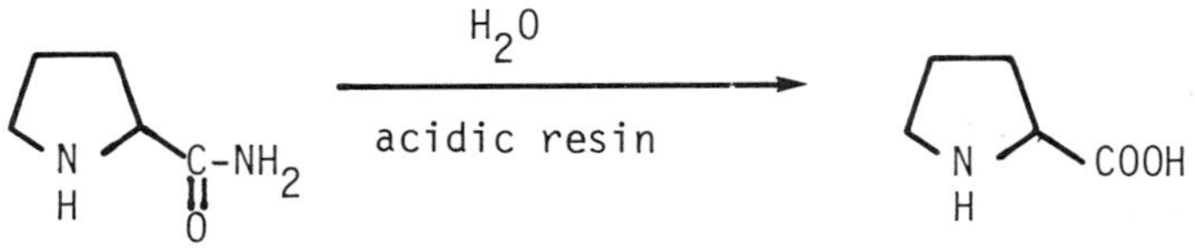

90%

JOC, 46, 5351 (1981)

98%

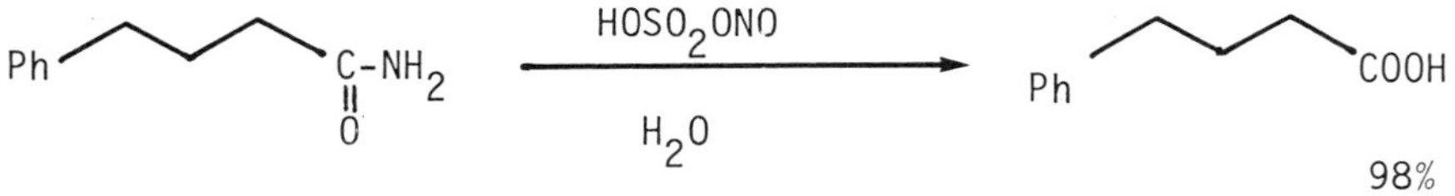

Org Prep Proc Int, 14, 357 (1982)

$CH_3(CH_2)_6$-C(=O)-$NHNH_2$ $\xrightarrow[H_2O]{Cu^{++}}$ $CH_3(CH_2)_6COOH$ 98%

Tetrahedron, 36, 1311 (1980)

85%
(R)

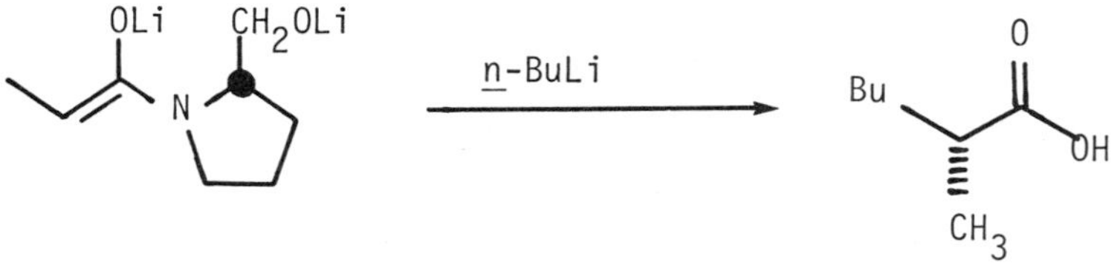

Tetrahedron Lett, 21, 4233 (1980)

1) BuLi
2) MeI
3) H_3O^+

56%

Synth Comm, <u>10</u>, 837 (1980)

Section 22 <u>Carboxylic Acids from Amines</u>

$$Ph-N_2^{\oplus} \; {}^{\ominus}BF_4 \xrightarrow[\text{NaOAc, } CH_3CN]{\text{CO, } Pd(OAc)_2} Ph-COOH$$

85%

JOC, <u>45</u>, 2365 (1980)

Section 23 <u>Carboxylic Acids from Esters</u>

$$\underline{n}\text{-}C_{10}H_{21}\text{-}O\text{-}\overset{\overset{\displaystyle O}{\|}}{C}\text{-}Ph \xrightarrow[\text{2) } H^+]{\text{1) KOH, } Al_2O_3}$$

Ph-C-OH   89%

+

$\underline{n}\text{-}C_{10}H_{21}OH$ 91%

Synth Comm, <u>11</u>, 413 (1981)

HBr

PTC*   +   94%

*hexadecyltributylphosphonium bromide

JOC, <u>47</u>, 154 (1982)

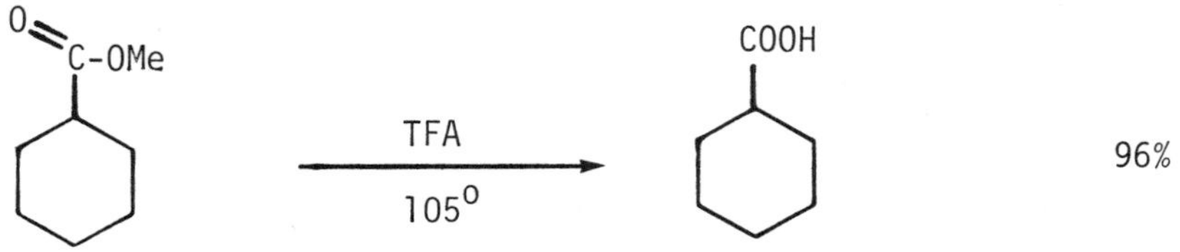

Synth Comm, <u>12</u>, 855 (1982)

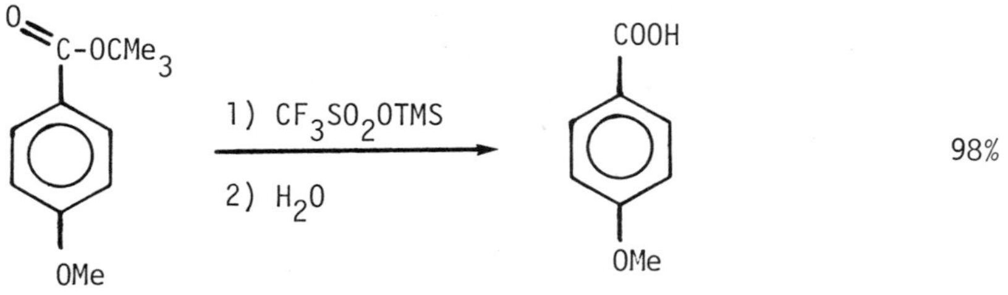

Synthesis, 545 (1980)

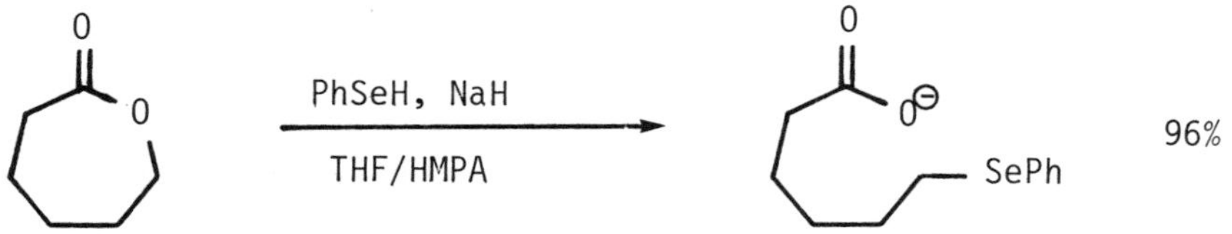

JOC, <u>46</u>, 2605 (1981)

Other reactions useful for the hydrolysis of esters may be found in Section 30A (Protection of Carboxylic Acids).

Section 24 <u>Carboxylic Acids from Ethers</u>

No additional examples

Section 25 <u>Carboxylic Acids from Alkyl Halides</u>

$$\xrightarrow[\text{Co}_2(\text{CO})_8,\ \text{P.T.C.}]{h\nu,\ \text{CO},\ \text{NaOH}}$$

96%

Tetr Lett, <u>22</u>, 1013 (1981)

$$\xrightarrow[\text{2) H}_2\text{O, H}^\oplus]{\text{1) } \underline{i}\text{-PrMgBr, CuCl}}$$

78%

Bull Chem Soc Japan, <u>55</u>, 3555 (1982)
Chem Lett, 571 (1980)

+

Bu$_2$CuMgBr

$$\xrightarrow[\text{2) H}_3\text{O}^\oplus]{}$$

Bu-CH$_2$CH$_2$COOH   92%

Tetr Lett, <u>21</u>, 935 (1980)
Tetr Lett, <u>21</u>, 2181 (1980)

$$\text{Ph}-\underset{\text{CH}_3}{\overset{}{\diagup}}\text{C}=\text{CF}_2 \xrightarrow{\text{H}_2\text{SO}_4} \underset{\overset{|}{\text{CH}_3}}{\text{Ph-CH- COOH}} \qquad 94\%$$

Chem Lett, **651** (1980)

$$\underset{\overset{|}{\text{N}}}{\underset{\text{CH}_2\text{Ph}}{}}\text{---CH}_3 \xrightarrow[\begin{array}{l}\text{3) NaOH, H}_2\text{O}\\\text{4) H}_3\text{O}^+\end{array}]{\begin{array}{l}\text{1) BuLi}\\\text{2) }\underline{n}\text{-C}_8\text{H}_{17}\text{Br}\end{array}} \underline{n}\text{-C}_8\text{H}_{17}\text{COOH} \qquad 66\%$$

Tetr Lett, <u>**22**</u>, 261 (1981)

Also via: Esters - Section 115

Section 26 <u>Carboxylic Acids from Hydrides</u>

+

Tol-S-CH$_2$—$\overset{\overset{\text{O}}{\|}}{\underset{\text{Tol}}{\text{S}}}$⁚

several steps

∿20%

38% ee

Synthesis, 74 (1981)

Section 27 <u>Carboxylic Acids from Ketones</u>

$$\text{Tetr Lett, } \underline{22}, 2595 \text{ (1981)}$$

86%

$$\text{Synthesis, } 563 \text{ (1980)}$$

75%

Also via: Esters - Section 117.

Section 28 <u>Carboxylic Acids from Nitriles</u>

$$CH_3(CH_2)_6CN \quad \xrightarrow[\text{2) } H^+]{\begin{array}{c}\text{1) } \underline{t}\text{-BuOK, } O_2 \\ \text{THF, 18-crown-6}\end{array}} \quad CH_3(CH_2)_5COOH \qquad 89\%$$

$$\text{JOC, } \underline{45}, 3630 \text{ (1980)}$$

Section 29 <u>Carboxylic Acids from Olefins</u>

$$\text{Chem Lett, } 651 \text{ (1980)}$$

94%

Section 30 <u>Carboxylic Acids from Miscellaneous Compounds</u>

Use of polymer-bound oxazolines for the synthesis of chiral carboxylic acids:

1) BuLi
2) BzCl
3) H_2SO_4
   EtOH/THF

45%
56% ee

JOC, <u>46</u>, 3097 (1981)

Section 30A <u>Protection of Carboxylic Acids</u>

$$R-\overset{O}{\underset{\|}{C}}-Cl \xrightarrow{(CH_3\overset{O}{\underset{\|}{S}}CH_2\overset{\ominus}{B}Bu_3)Li^{\oplus}} R-\overset{O}{\underset{\|}{C}}-OCH_2SCH_3 \qquad 29\text{-}72\%$$

R = alkyl, subst. Ph, -CH$_2$OPh

Tetr Lett, <u>23</u>, 4539 (1982)

$$R-\overset{O}{\underset{\|}{C}}-OBz \xrightarrow[10\% \text{ Pd/C}]{\overset{\oplus}{N}H_4 \ H\overset{\ominus}{C}OO} R-COOH \qquad >90\%$$

Synthesis, 929 (1980)

$H_2/K_3[Co(CN)_5]$ removes benzyl ester protecting groups from amino acids and peptides in 83-94% yields.

Z. Chem, 188 (1981)

$$R-\overset{\overset{\text{O}}{\|}}{C}-O-CH_2-\underset{}{\bigcirc}-NO_2 \xrightarrow[\text{H}_2\text{O/MeCN}]{\text{Na}_2\text{S}_2\text{O}_4} R-COOH \qquad 85\text{-}95\%$$

Synth Comm, <u>12</u>, 219 (1982)

$$Br-CH_2-\underset{}{\bigcirc}-COOMe$$

$$R-\overset{\overset{\text{O}}{\|}}{C}-ONa \qquad R-\overset{\overset{\text{O}}{\|}}{C}-OCH_2-\underset{}{\bigcirc}-COOMe$$

$\xleftarrow{\text{electrochemical reduction}}$ stable to hydro-
$\qquad\qquad\text{DMF}$ genation

JCS Chem Comm, 1083 (1980)

$$R-\overset{\overset{\text{O}}{\|}}{C}-O-\underset{}{\bigcirc}-OMe \xrightarrow{[\text{Ar}_3\text{N}]^{+\cdot}} R-COOH$$

Angew Chem Int Ed, <u>21</u>, 780 (1982)

$$R-\overset{\overset{\text{O}}{\|}}{C}-OH \xrightarrow{CH_2=CH-CH_2SiMe_3} R-\overset{\overset{\text{O}}{\|}}{C}-OSiMe_3 \qquad \sim 90\%$$

Tetr Lett, <u>21</u>, 835 (1980)

$$R\text{-}COOH \xrightarrow[\text{ClCH}_2\text{CH}_2\text{Cl}]{\text{Me}_3\text{SiCl}} R\overset{\overset{\displaystyle O}{\|}}{-}\text{C-OSiMe}_3 \qquad \sim70\text{-}90\%$$

R = alkyl, vinyl, allyl, acetylenic, etc.

Synthesis, 626 (1980)

$$R\overset{\overset{\displaystyle O}{\|}}{-}\text{C-OH} \xrightarrow[\text{H}_2\text{SO}_4]{(\text{Me}_3\text{Si})_2\text{O}} R\overset{\overset{\displaystyle O}{\|}}{-}\text{C-OSiMe}_3 \qquad \sim70\text{-}90\%$$

Chem Lett, 1475 (1980)

proton sponge, DMAP

ceric ammonium nitrate

$$R\overset{\overset{\displaystyle O}{\|}}{-}\text{C-OH}$$

Chem Lett, 1551 (1981)

The 5,6-dihydrophenanthridine protecting group may be activated
and removed by oxidation:

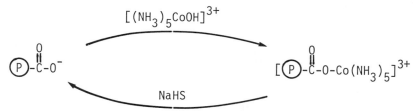

$$\xrightarrow{\text{Ce}^{+4}, \text{H}_2\text{O}} \text{R-COOH}$$

$$\xrightarrow[\text{R'NH}_2]{\text{Ce}^{+4}, \text{CuO}} \overset{\overset{\text{O}}{\|}}{\text{R-C-NHR'}}$$

Chem Lett, 991 (1982)

$$[(\text{NH}_3)_5\text{CoOH}]^{3+}$$

$$\textcircled{P}\text{-}\overset{\overset{\text{O}}{\|}}{\text{C}}\text{-O}^{-} \underset{\text{NaHS}}{\rightleftarrows} [\textcircled{P}\text{-}\overset{\overset{\text{O}}{\|}}{\text{C}}\text{-O-Co(NH}_3)_5]^{3+}$$

Used as a C-terminal protecting group in peptide synthesis.

JACS, 104, 3910 (1982)

Review: "Recent developments in Methods for the Esterification
and Protection of the Carboxyl Group"

Tetrahedron, 36, 2409 (1980)

Other reactions useful for the protection of carboxylic acids are
included in Section 107 (Esters from Carboxylic Acids and Acid
Halides) and Section 23 (Carboxylic Acids from Esters).

CHAPTER 3

PREPARATION OF ALCOHOLS
AND PHENOLS

Section 31 <u>Alcohols from Acetylenes</u>

$$Ph-C{\equiv}C-CH_3 \xrightarrow[\text{3) NaOH, } H_2O_2]{\begin{array}{l}\text{1) 2 BuLi}\\ \text{2) } BH_3 \cdot THF\end{array}}$$

Ph ⟍ ⟋ ⟍ ⟋

    OH   OH

70%

Tetrahedron, <u>36</u>, 299 (1980)

Section 32 <u>Alcohols from Carboxylic Acids</u>

$$\underline{n}\text{-}C_8H_{17}COOH \xrightarrow[\text{TiCl}_4]{\text{NaBH}_4} \underline{n}\text{-}C_8H_{17}CH_2OH$$

93%

Synthesis, 695 (1980)

$$C_5H_{11}COONa \xrightarrow[\text{2) } H_3O^{\oplus}]{\text{1) 2 } BH_3 \cdot THF} C_5H_{11}CH_2OH$$

100%

Tetr Lett, <u>23</u>, 2475 (1982)

$$CH_3(CH_2)_8COOH \longrightarrow \longrightarrow CH_3(CH_2)_8 \overset{O}{\underset{}{C}}-N \overset{S}{\underset{S}{\bigcirc}} \xrightarrow{\text{NaBH}_4} CH_3(CH_2)_8CH_2OH$$

JCS Perkin I, 2470 (1980)   98%

Section 33 <u>Phenols from Phenols</u>

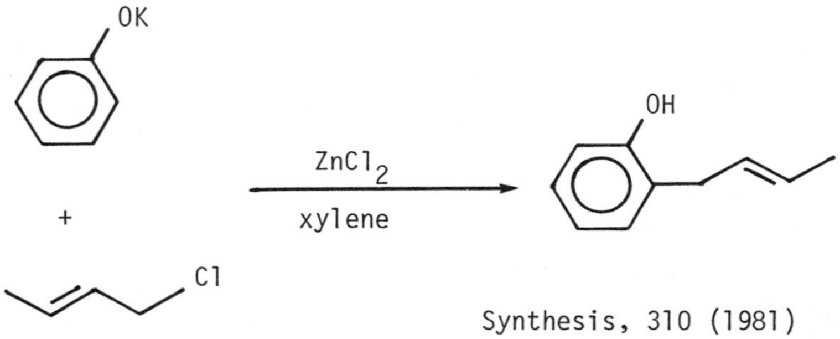

$$\xrightarrow[\text{xylene}]{\text{ZnCl}_2}$$

63%

Synthesis, 310 (1981)

Section 34 <u>Alcohols from Aldehydes</u>

The following reaction types are included in this section:

A. Reductions of aldehydes to alcohols.

B. Nucleophilic additions to aldehydes, forming alcohols.

C. Coupling of aldehydes to give diols.

Section 34A: <u>Reductions of Aldehydes to Alcohols</u>

$$\xrightarrow[\text{H}_2\text{O/dioxane/DMF}]{\text{Na}_2\text{S}_2\text{O}_4}$$

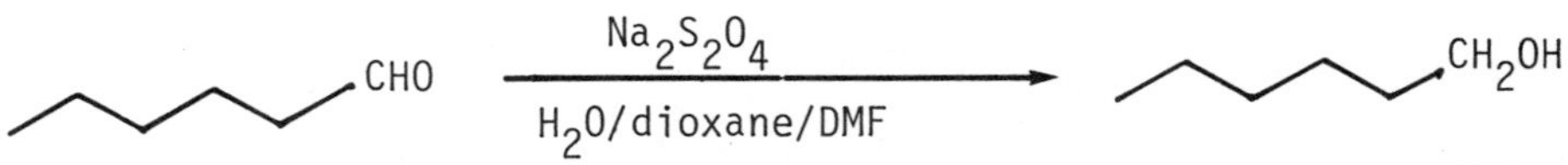

63%

JOC, <u>45</u>, 4126 (1980)

$$\text{(pentyl)}-CHO \xrightarrow{\text{LiAlH(OCEt}_3)_3} \text{(pentyl)}-CH_2OH$$

100%

JOC, <u>46</u>, 4628 (1981)

$$\text{(cyclohexyl)}-CHO \xrightarrow[\text{Et}_2O]{\underline{t}\text{-BuNH}_2\cdot BH_3} \text{(cyclohexyl)}-CH_2OH \quad 95\%$$

Tetr Lett, <u>21</u>, 693 and 697 (1980)

$$\text{(heptyl)}-CHO \xrightarrow[\text{THF}]{\text{poly(2-vinylpyridine)}-BH_3}$$

$$\text{(heptyl)}-CH_2OH$$

87%

JOC, <u>45</u>, 2724 (1980)

$$C_8H_{17}-CHO \xrightarrow[\text{2) } H_2O_2, \ ^\ominus OH]{\text{1) } Et_4\overset{\oplus}{N} \ \overset{\ominus}{B}H_4, \ CH_2Cl_2} C_8H_{17}CH_2OH \quad 75\%$$

Tetr Lett, <u>21</u>, 3963 (1980)

$$C_9H_{19}CHO \xrightarrow{\text{(TFA)}_2\text{BH·THF}} C_9H_{19}CH_2OH \qquad 96\%$$

JOC, <u>46</u>, 355 (1981)

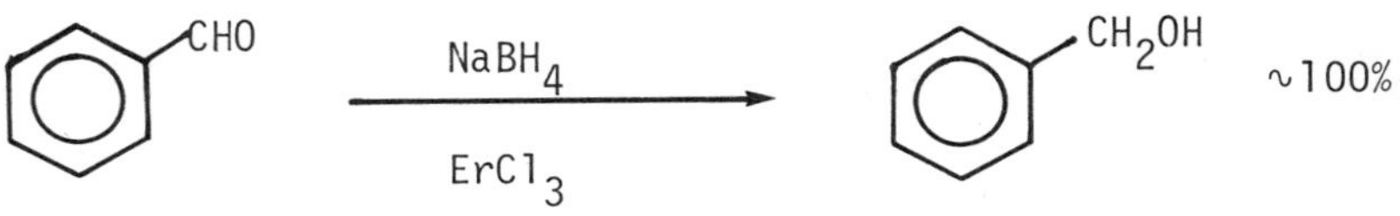

∼100%

Reduces conjugated aldehydes in the presence of non-conjugated ones.

Tetr Lett, <u>22</u>, 4077 (1981)

Aldehydes may be reduced in the presence of ketones:

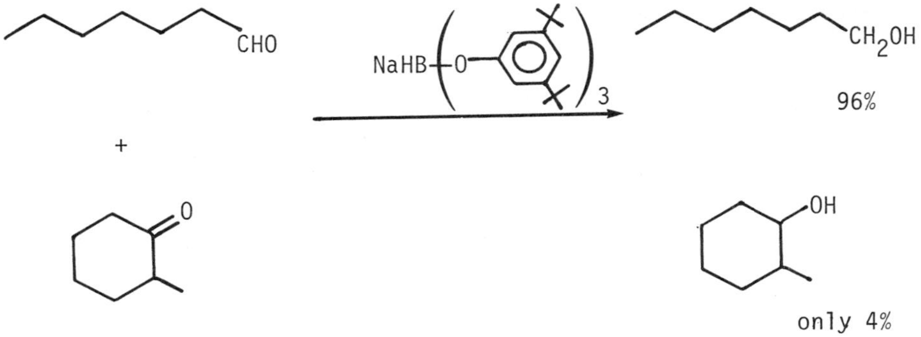

96%

only 4%

Chem Lett, 461 (1981)

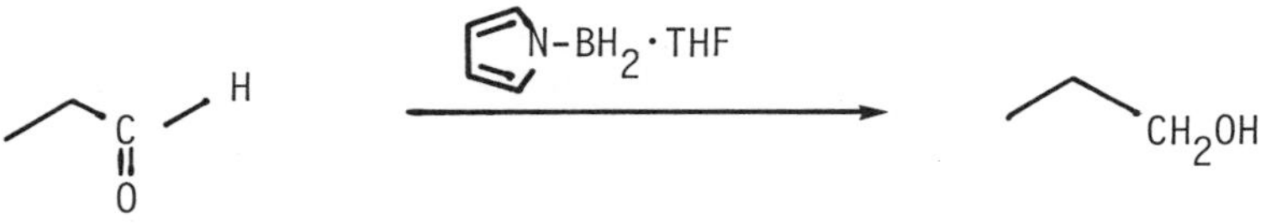

Synthesis, 214 (1981)

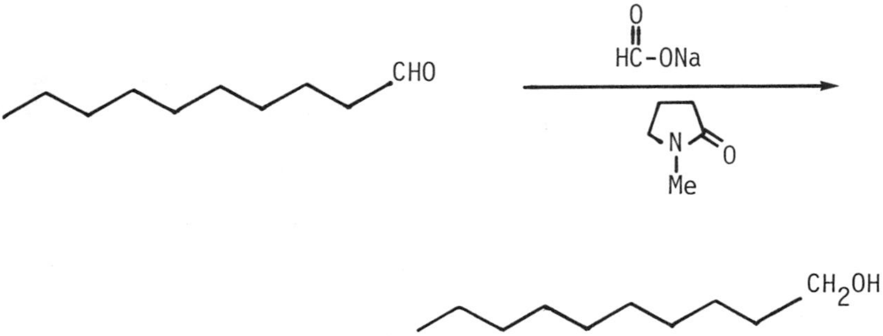

JOC, 46, 3367 (1981)

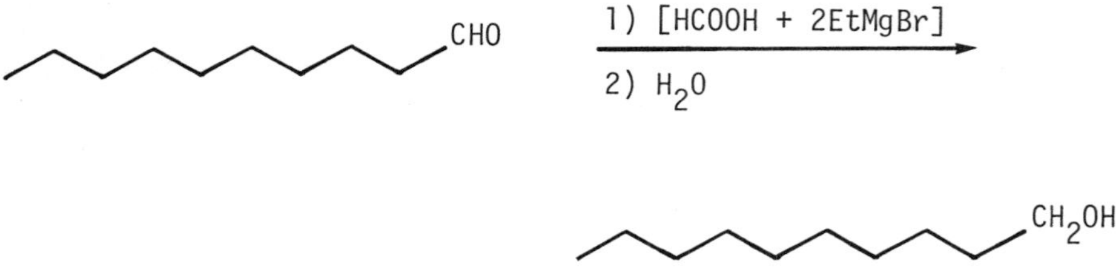

Tetr Lett, 22, 621 (1981)

$$Ph-CH=CH-CHO \xrightarrow[\text{KF}]{\text{(EtO)}_2\text{SiHMe}} Ph-CH=CH-CH_2OH \quad 95\%$$

JCS Chem Comm, 121 (1981)

$$Ph-CH=CH-CHO \xrightarrow{\text{(Ph}_3\text{P)}_2\text{CuBH}_4} Ph-CH=CH-CH_2OH \quad 99\%$$

Tetr Lett, **22**, 675 (1981)

$$Ph-CHO \xrightarrow{\text{Al}_2\text{Te}_3, \text{H}_2\text{O}} Ph-CH_2OH \quad 100\%$$

Angew Chem Int Ed, **19**, 1008 and
1009 (1980)

34B: Nucleophilic Additions to Aldehydes, Forming Alcohols

Aldol reactions are listed in:

Section 324 (Aldehyde-Alcohol) and
Section 330 (Ketone-Alcohol)

CHO $\xrightarrow{\text{MeTiCl}_3}$ (cyclohexyl)CH(OH)CH$_3$ >90%

Angew Chem Int Ed, <u>19</u>, 1011 (1980)

CHO $\xrightarrow[\text{2) Et}_2\text{Zn}]{\text{1) TiCl}_4}$ (cyclohexyl)CH(OH)Et

Synth Comm, <u>11</u>, 261 (1981)

$C_6H_{13}CHO$

+ $\longrightarrow$ $\longrightarrow$ $C_6H_{13}\overset{H}{\underset{Pr}{C}}$-OH   73%

$Pr-CrCl_2(THF)_3$

Ketones are relatively unreactive.

Angew Chem Int Ed, <u>21</u>, 144 (1982)

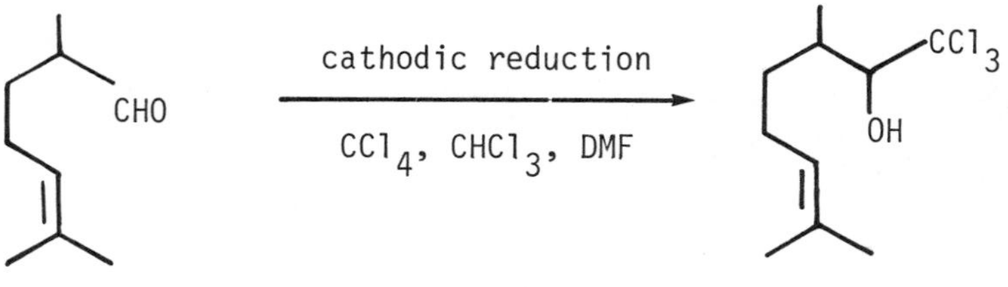
CHO
cathodic reduction
CCl_4, $CHCl_3$, DMF
CCl_3
OH

62%

Tetr Lett, 23, 1609 (1982)

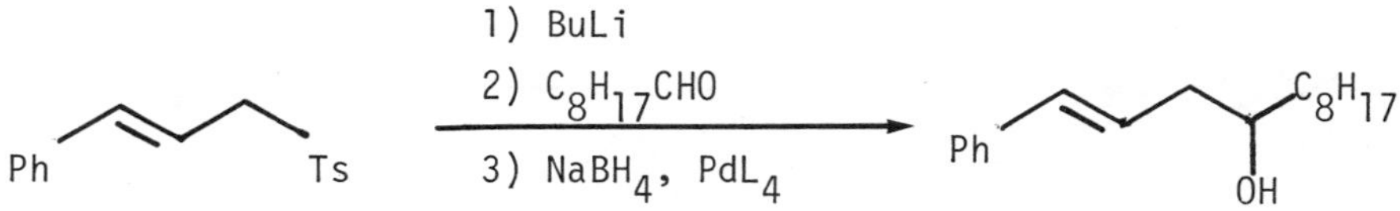
Ph
Ts
1) BuLi
2) $C_8H_{17}CHO$
3) $NaBH_4$, PdL_4
Ph
C_8H_{17}
OH

73%

Chem Lett, 1331 (1982)

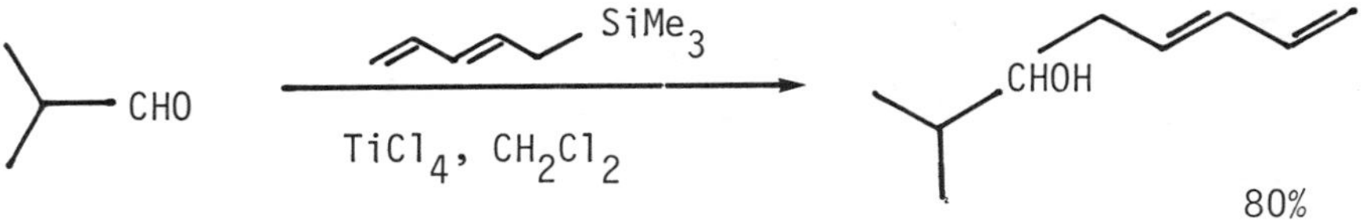
CHO
$SiMe_3$
$TiCl_4$, CH_2Cl_2
CHOH

80%

Organometallics, 1, 1651 (1982)

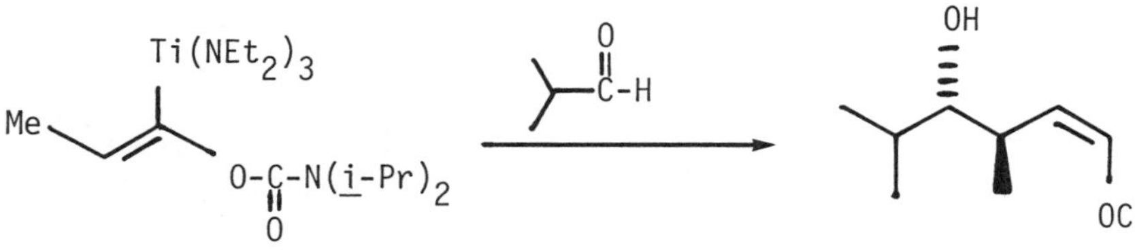

Angew Chem Int Ed, 21, 372 (1982)

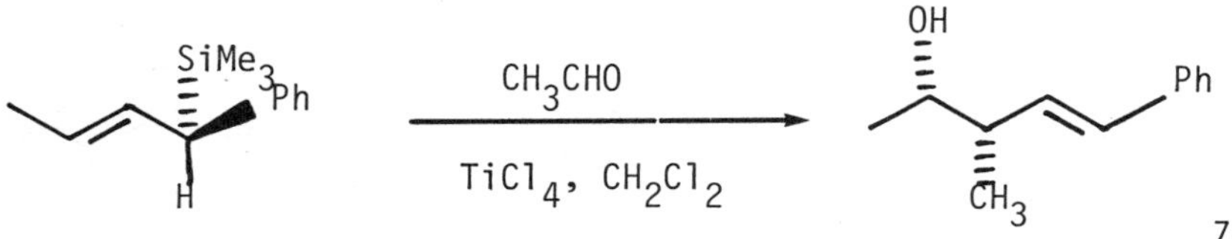

JACS, 104, 4963 (1982)

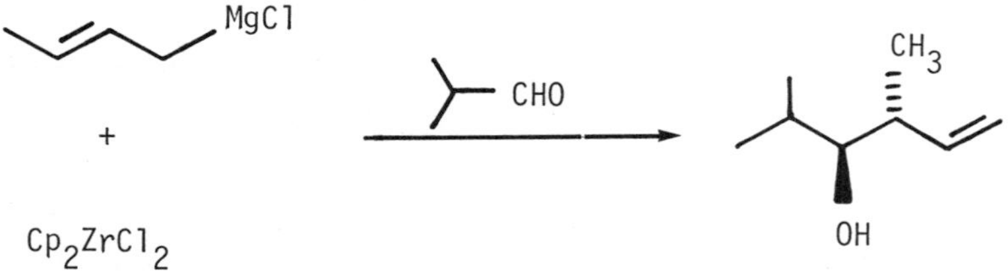

Tetr Lett, 22, 2895 (1981)

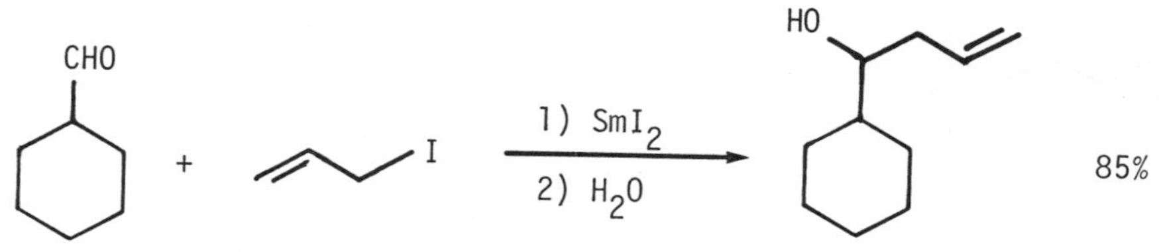

Tetr Lett, <u>23</u>, 3497 (1982)

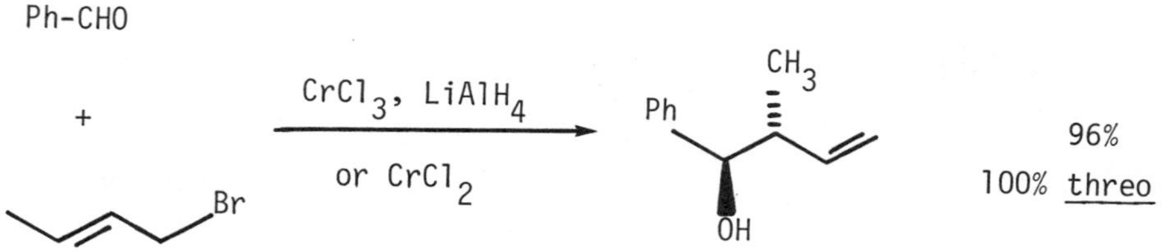

Ph-CHO

96%

100% <u>threo</u>

Tetr Lett, <u>22</u>, 1037 (1981)

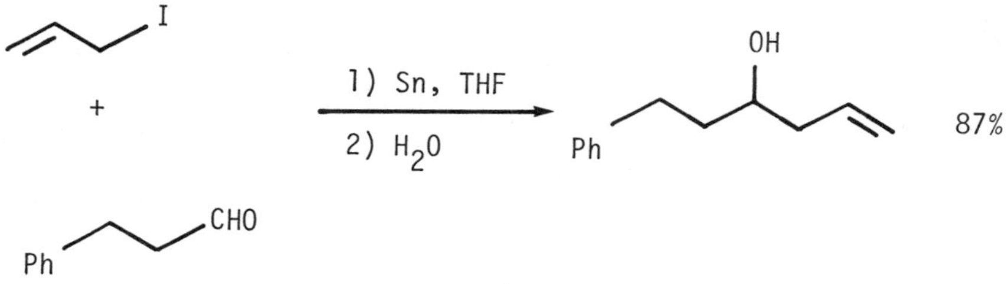

Chem Lett, 1527 (1981)

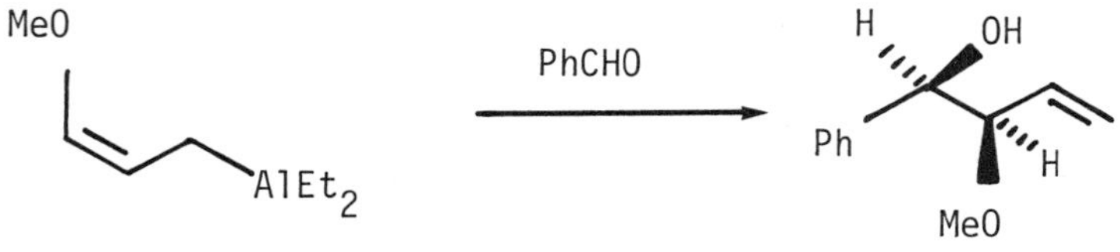

JCS Chem Comm, 845 (1982)

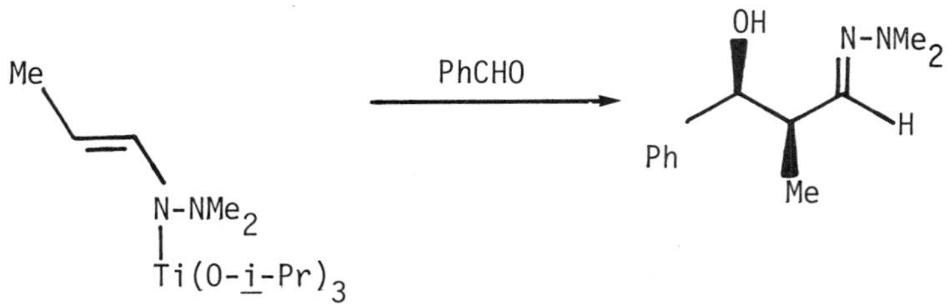

Angew Chem Int Ed, 21, 864 (1982)

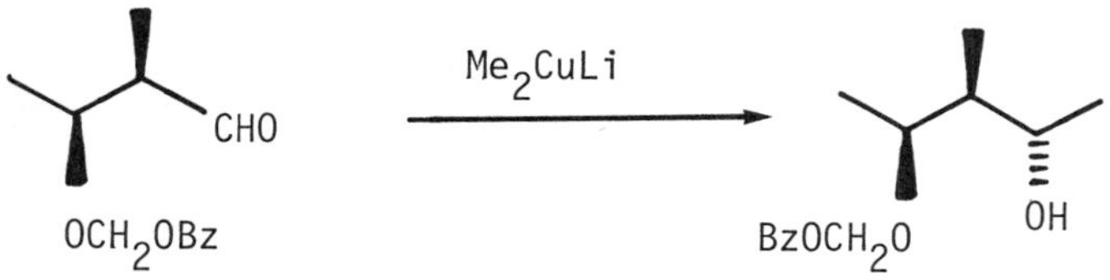

>20:1 diastereoselectivity

Tetr Lett, 21, 1031 and 1035 (1980)

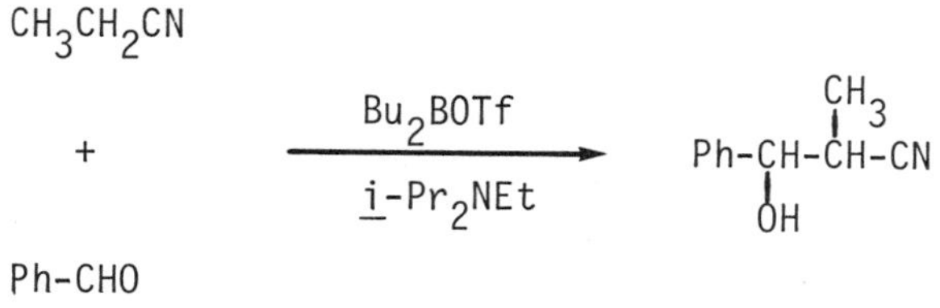

CH_3CH_2CN

+ $\xrightarrow[\underline{i}\text{-Pr}_2\text{NEt}]{\text{Bu}_2\text{BOTf}}$ Ph-CH-CH-CN

Ph-CHO

90%

Chem Lett, 1401 (1982)

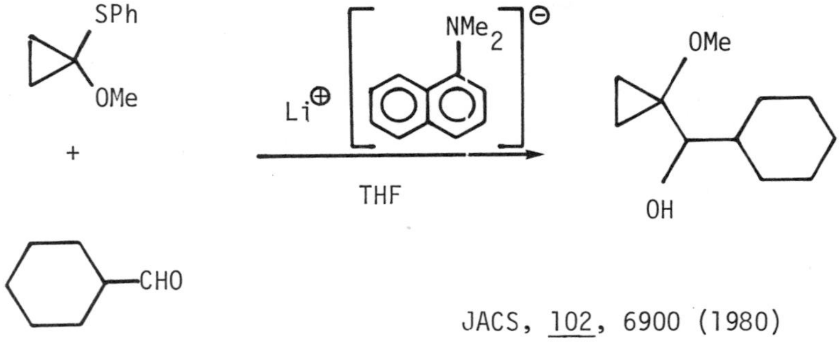
SPh
OMe
+
CHO
Li⊕
NMe2
⊖
THF
OMe
OH

90%

JACS, 102, 6900 (1980)

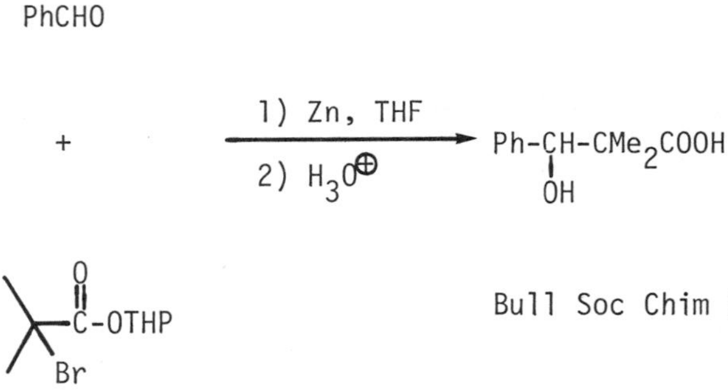
PhCHO
+
1) Zn, THF
2) H3O⊕
Ph-CH-CMe2COOH
OH
C-OTHP
Br

88%

Bull Soc Chim France II, 145 (1980)

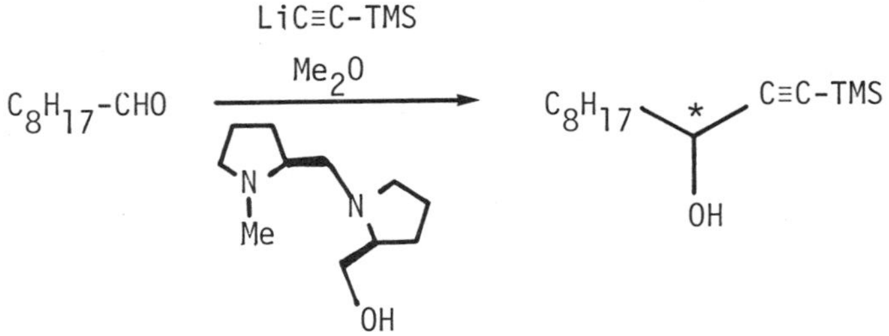
LiC≡C-TMS
Me2O
C8H17-CHO
N
Me
N
OH
C8H17
*
C≡C-TMS
OH

83%

80% ee

Chem Lett, 255 (1980)

60%

Tetr Lett, <u>22</u>, 3269 (1981)

34C: Coupling of Aldehydes to Give Diols

2 Ph-CHO   1) Fe(CO)$_5$, pyridine / 2) H$_3$O$^\oplus$ →   Ph-CH-CH-Ph   88%
                                                                 | |
                                                                OH   OH

Chem Lett, 1141 (1980)

2 Ph-CHO   TiCl$_3$ / NaOH   →   Ph-C——C-Ph
                                       | |
                                      OH   OH
                                      (H) (H)

Tetr Lett, <u>23</u>, 3517 (1982)

Related methods: Alcohols from Ketones (Section 42)

Section 35 Alcohols and Phenols from Alkyls, Methylenes and Aryls

No examples of the reaction RR' → ROH (R' = alkyl, aryl, etc.)
occur in the literature. For reactions of the type RH → ROH
(R = alkyl or aryl) see Section 41 (Alcohols and Phenols from
Hydrides).

Section 36 Alcohols from Amides

$$\xrightarrow{\text{LiEt}_3\text{BH}}$$

95%

JOC, 45, 1 (1980)

1) PhMgBr

2) MeMgCl

3) H_3O^+

$$Ph-\overset{\overset{\displaystyle OH}{|}}{\underset{\underset{\displaystyle H}{|}}{C}}-Me$$

82%

Tetr Lett, 22, 1085 (1981)

Section 37 Alcohols from Amines

$\underline{n}\text{-C}_{16}\text{H}_{33}\text{NH}_2$

1)

2)

P.T.C.

$\underline{n}\text{-C}_{16}\text{H}_{33}\text{OH}$

∿50%

J.C.S. Perkin I, 1492 (1981)

Section 38 Alcohols from Esters

$$C_5H_{11}\overset{\overset{O}{\|}}{C}\text{-OMe} \xrightarrow[\substack{\text{polyethylene} \\ \text{glycol}}]{NaBH_4} C_5H_{11}CH_2OH \qquad 80\%$$

JOC, <u>46</u>, 4584 (1981)

$$CH_3(CH_2)_{16}\overset{\overset{O}{\|}}{C}\text{-OMe} \xrightarrow[\text{t-BuOH, MeOH}]{NaBH_4} CH_3(CH_2)_{16}\text{-}CH_2OH \quad 79\%$$

Synth Comm, <u>12</u>, 463 (1982)

$$\xrightarrow[\text{EtOH}]{NaBH_4} \qquad 93\%$$

Synth Comm, <u>11</u>, 599 (1981)

$$\xrightarrow[(MeO)_3B]{LiBH_4} \qquad 81\%$$

JOC, <u>47</u>, 1604 (1982)

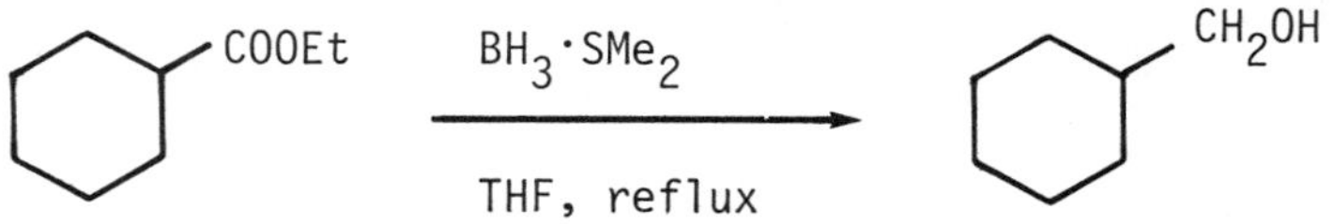

89%

Synthesis, 439 (1981)
JOC, 47, 3153 (1982)

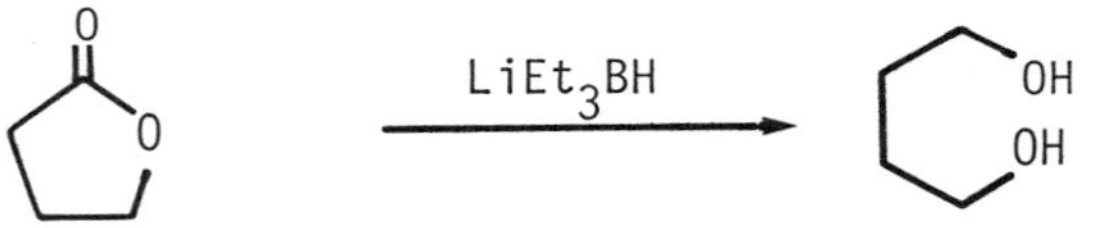

94%

JOC, 45, 1 (1980)

1) HSi(OEt)$_3$
   CsF

H$_2$C=CH(CH$_2$)$_8$-COOMe ⟶ H$_2$C=CH(CH$_2$)$_8$CH$_2$OH   70%

2) H$_2$O

Synthesis, 558 (1981)

94%

Tetr Lett, 21, 2171 and 2175 (1980)

EtCOOCH$_3$

+

PrMgBr

$\xrightarrow{\text{Cp}_2\text{TiCl}_2}$

Et-$\overset{\overset{\displaystyle OH}{|}}{\underset{\underset{\displaystyle H}{|}}{C}}$-Pr

83%

Tetr Lett, 21, 2171 and 2175 (1980)

72%

JOC, 45, 1828 (1980)

$N\equiv C-CH_2-\overset{\overset{\displaystyle O}{\|}}{C}-SBz$

1) NaH, i-BuX
2) NaH, PhCH$_2$X
3) NaBH$_4$

$N\equiv C-\overset{\overset{\displaystyle \text{i-Bu}}{|}}{\underset{\underset{\displaystyle CH_2Ph}{|}}{C}}-CH_2OH$

76%

Tetr Lett, 23, 3151 (1982)

Related Methods: Carboxylic Acids from Esters - Section 23,
                    Protection of Alcohols - Section 45A
                    Hydrolysis of Esters is covered in Section 23

Section 39 <u>Alcohols and Phenols from Ethers and Epoxides</u>

$$BBr_3, NaI$$

$$CH_2Cl_2, \text{15-crown-5}$$

100%

Tetr Lett, <u>22</u>, 4239 (1981)

$$BBr_3 \cdot SMe_2$$

~80%

Tetr Lett, <u>21</u>, 3731 (1980)

$$MeSiCl_3$$

$$NaI$$

85%

Angew Int Ed, <u>20</u>, 690 (1981)

OMe
(cyclohexane ring)

1) Me₃SiI, pyridine
$\longrightarrow$
2) CH₃OH

OH
(cyclohexane ring)

89%

Org Syn, <u>59</u>, 35 (1980)

R-OMe
or
R-OBz

MeSSiMe₃
$\longrightarrow$
or PhSSiMe₃

R-OH

~90%

Tetr Lett, <u>21</u>, 2305 (1980)

R-OBz

(cyclohexene)   EtOH
$\longrightarrow$
20% Pd(OH)₂/C

R-OH

~98%

Synthesis, 396 (1981)

Ph-OMe

SiCl₄
$\longrightarrow$
NaI

Ph-OH

90%

Synthesis, 1048 (1982)

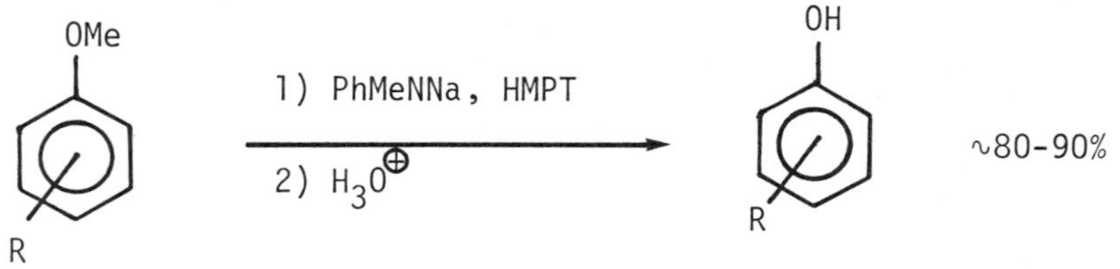

Synthesis, 638 (1980)

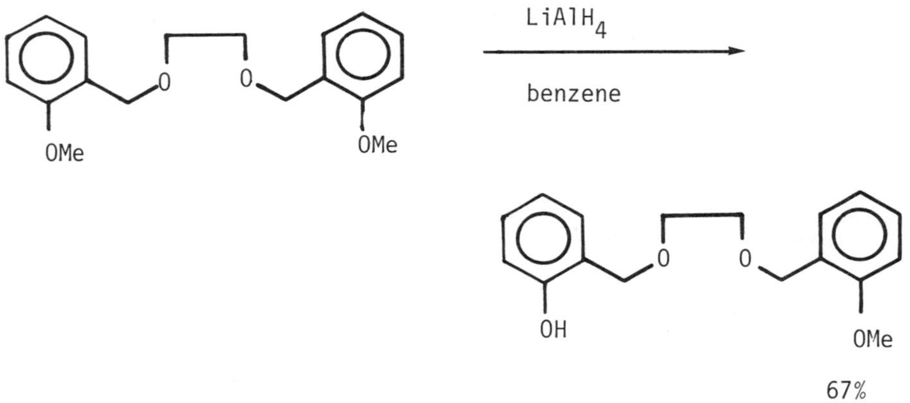

67%

JCS Chem Comm, 507 (1980)

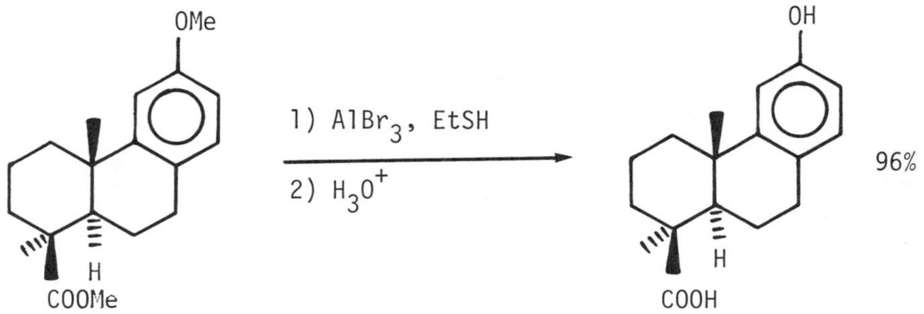

96%

JOC, 46, 1991 (1981)

Additional examples of ether cleavages may be found in Section 45A (Protection of Alcohols and Phenols).

Synthesis, 280 (1981)

Tetr Lett, <u>23</u>, 4541 (1982)

Tetr Lett, <u>23</u>, 2719 (1982)

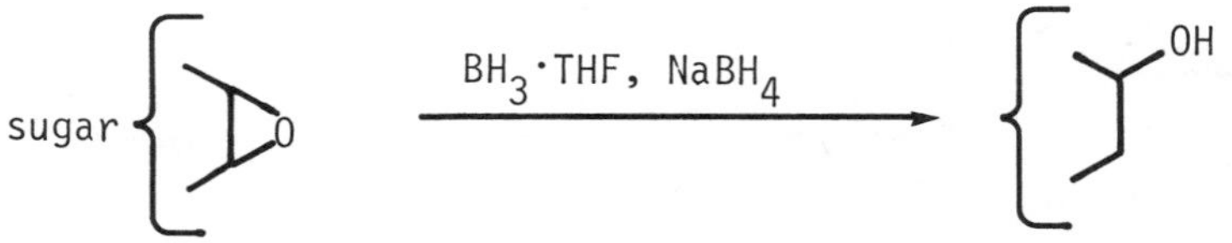

$$\text{BH}_3\cdot\text{THF, NaBH}_4$$

Can be accomplished in the presence of OTs, ketal, and acetal groups.

JOC, <u>45</u>, 3836 (1980)

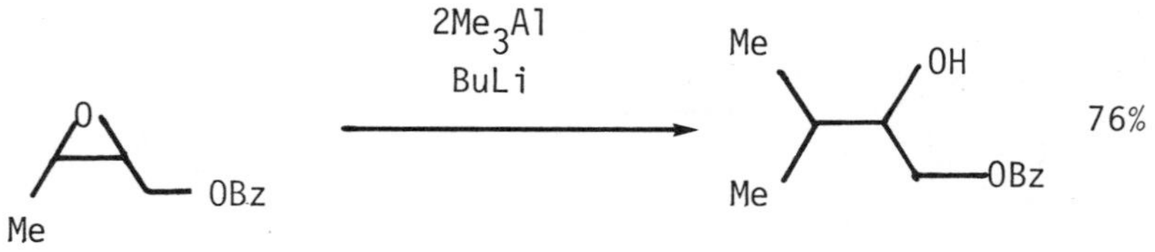

$$2\text{Me}_3\text{Al}$$
$$\text{BuLi}$$

76%

Angew Chem Int Ed, <u>21</u>, 71 (1982)

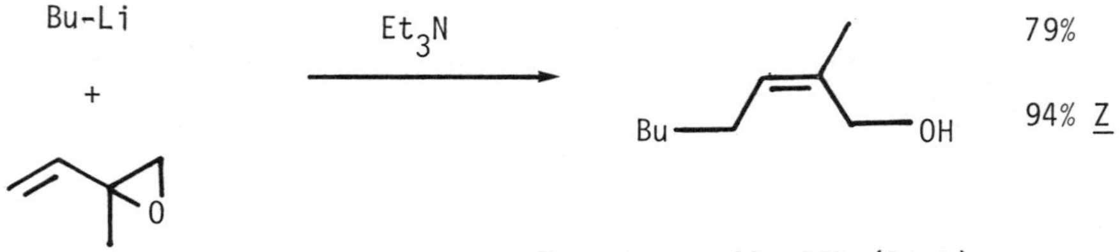

Bu-Li

+

$$\text{Et}_3\text{N}$$

79%

94% <u>Z</u>

Tetr Lett, <u>22</u>, 577 (1981)

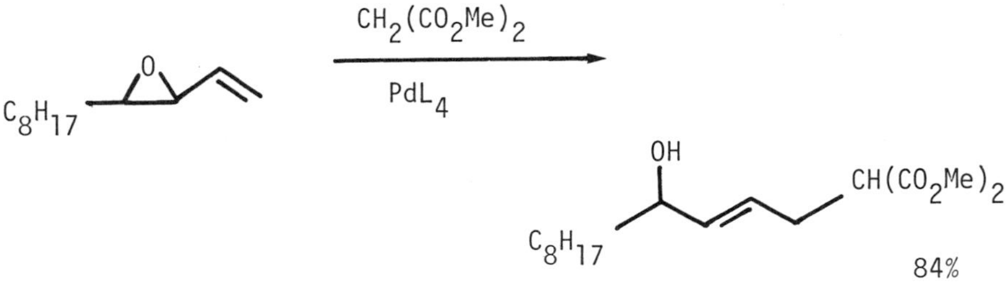

$$CH_2(CO_2Me)_2$$

$$PdL_4$$

C$_8$H$_{17}$ —

OH

C$_8$H$_{17}$ — CH(CO$_2$Me)$_2$

84%

Tetr Lett, 22, 2575 (1981)

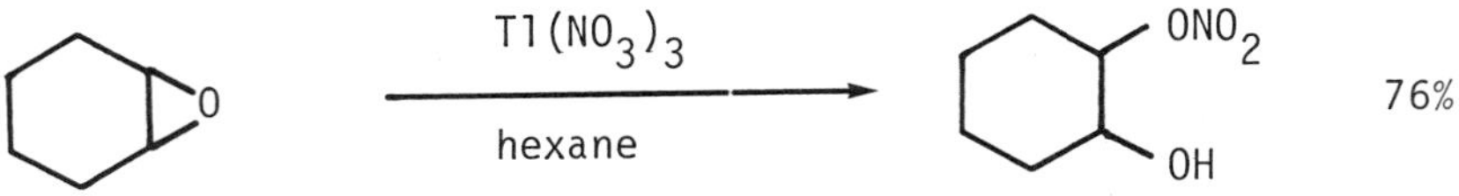

$$Tl(NO_3)_3$$

hexane

ONO$_2$

OH

76%

Tetr Lett, 21, 1149 (1980)

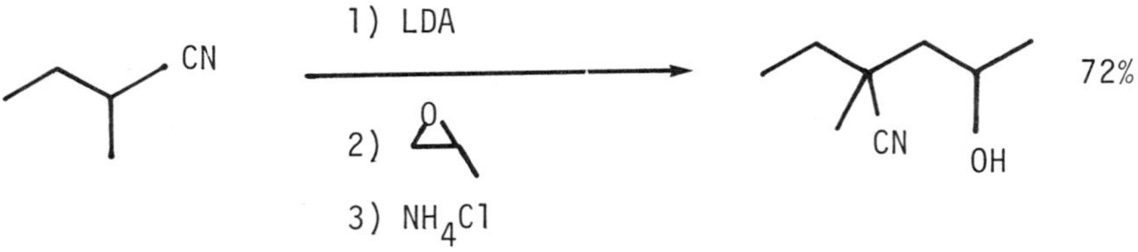

CN

1) LDA

2)

3) NH$_4$Cl

CN   OH

72%

Synth Comm, 10, 49 (1980)

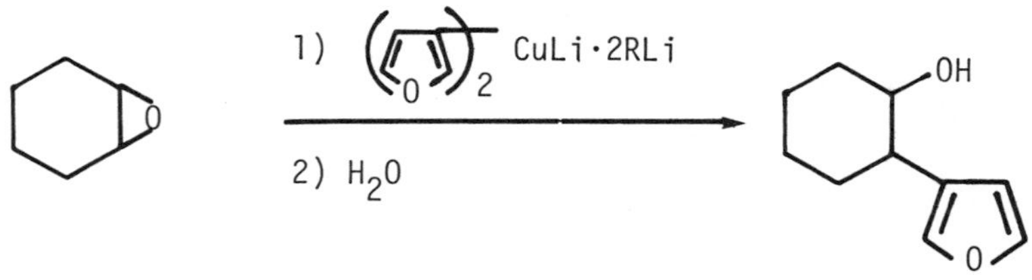
1)
CuLi·2RLi
2
2) H2O
OH
O

90%

Tetr Lett, 21, 4365 (1980)

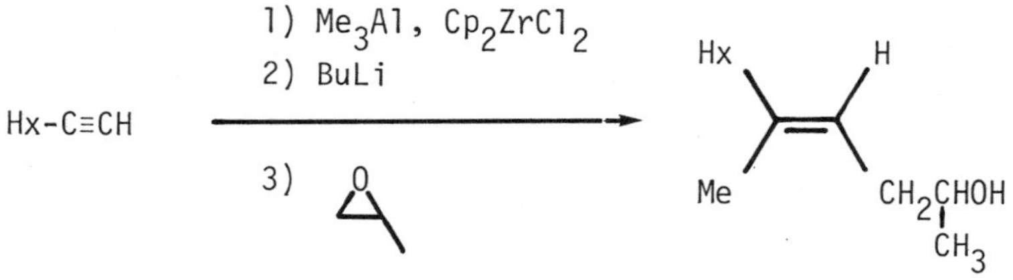
1) Me3Al, Cp2ZrCl2
2) BuLi
Hx-C≡CH
3)
O
Hx
H
Me
CH2CHOH
CH3

87%

Synthesis, 1034 (1980)

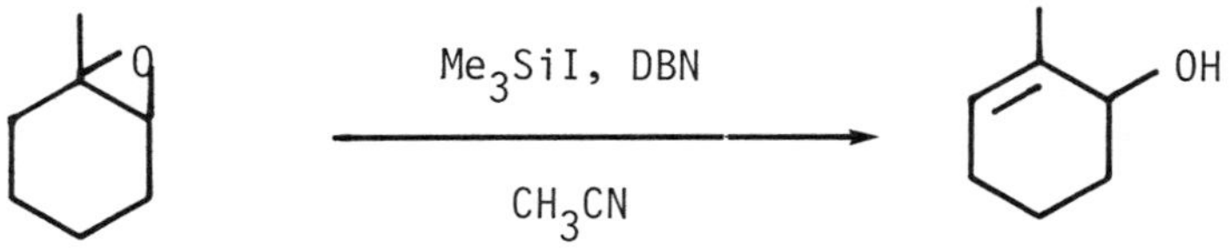
Me3SiI, DBN
CH3CN
OH

68%

JOC, 45, 2579 (1980)

Section 40 <u>Alcohols from Halides</u>

$$\underline{n}\text{-}C_8H_{17}Cl \xrightarrow[\text{benzene}]{\text{polymer-bound carbonate}} \underline{n}\text{-}C_8H_{17}OH \quad 90\%$$

Synthesis, 793 (1981)

NaOH

DMF/H$_2$O

95%

JOC, <u>47</u>, 4024 (1982)

K$_2$CO$_3$/acetone

or Bu$_4$NOH/DMF

85-90%

Steroids, <u>39</u>, 345 (1982)

$$\text{BuBr} \xrightarrow{\text{PhLi, CO}} \underset{\underset{Bu}{|}}{Ph_2C\text{-}OH} \quad 80\%$$

JOC, <u>46</u>, 4625 (1981)

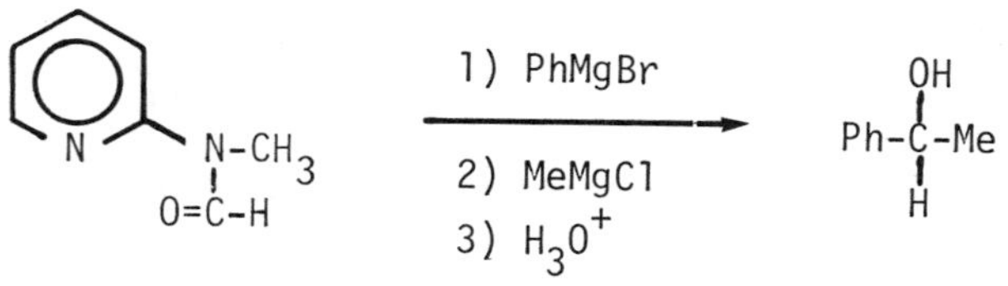
N
N-CH3
O=C-H
1) PhMgBr
2) MeMgCl
3) H3O+
OH
Ph-C-Me
H

82%

Tetr Lett, 22, 1085 (1981)

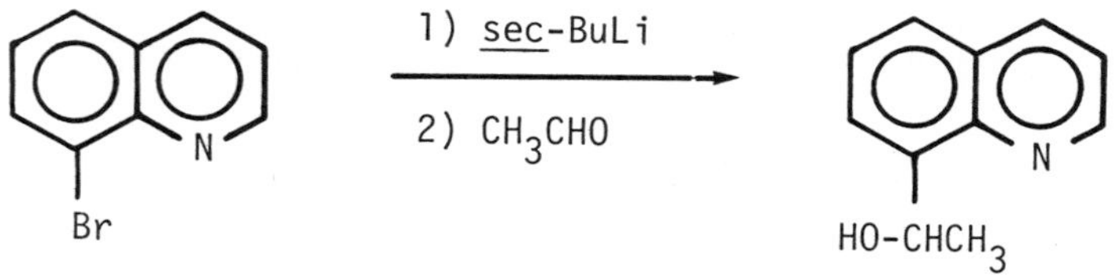
N
Br
1) sec-BuLi
2) CH3CHO
N
HO-CHCH3

64%

JOC, 45, 1514 (1980)

Section 41 Alcohols and Phenols from Hydrides

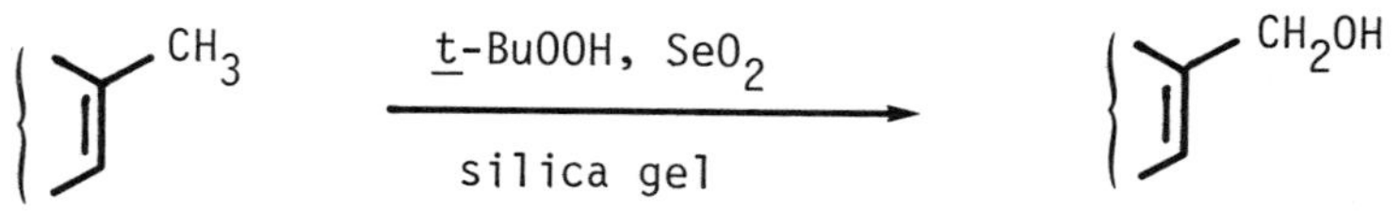
CH3
t-BuOOH, SeO2
silica gel
CH2OH

100%

Chem Lett, 1703 (1981)

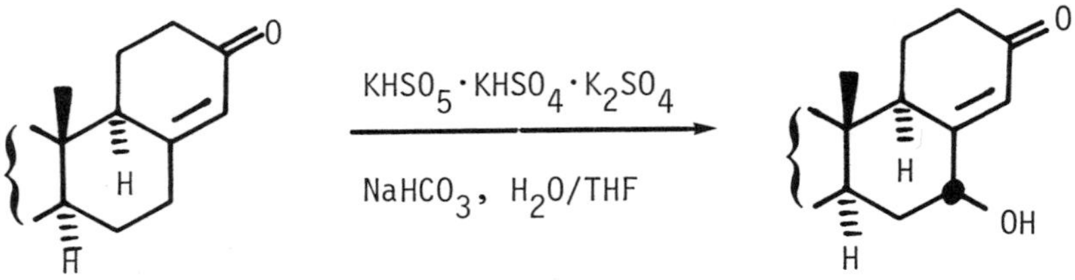

KHSO$_5$·KHSO$_4$·K$_2$SO$_4$

NaHCO$_3$, H$_2$O/THF

Tetr Lett, <u>22</u>, 4201 (1981)

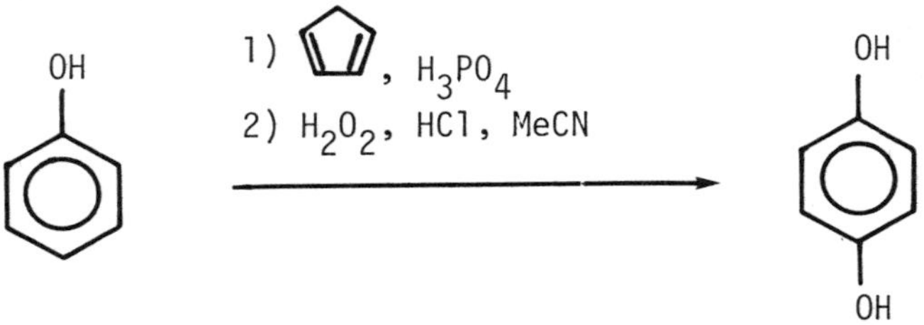

1) ⬠, H$_3$PO$_4$
2) H$_2$O$_2$, HCl, MeCN

~75%

Tetr Lett, <u>22</u>, 2337 (1981)

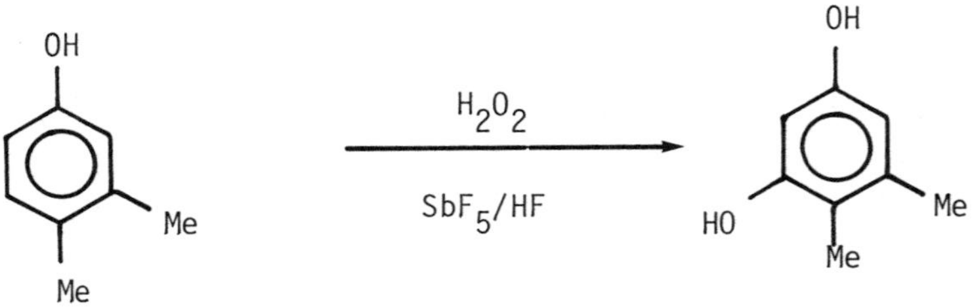

H$_2$O$_2$

SbF$_5$/HF

88%

JCS Chem Comm, 1128 (1980)

JACS, <u>103</u>, 6263 (1981)

Tetr Lett, <u>23</u>, 1717 (1982)

Tetr Lett, <u>22</u>, 1283 (1981)

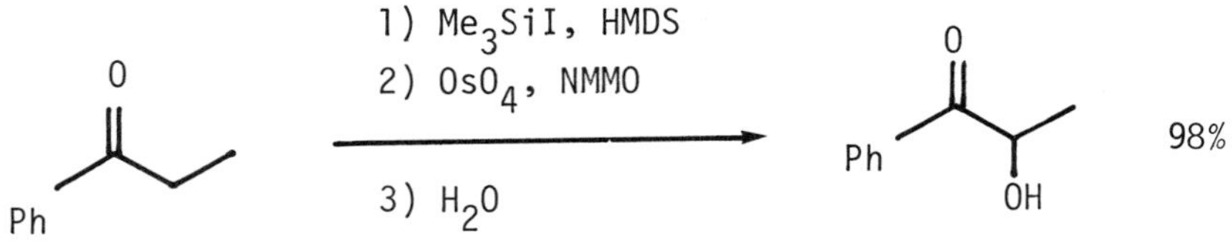

1) Me$_3$SiI, HMDS

2) OsO$_4$, NMMO

3) H$_2$O

98%

Tetr Lett, 22, 607 (1981)

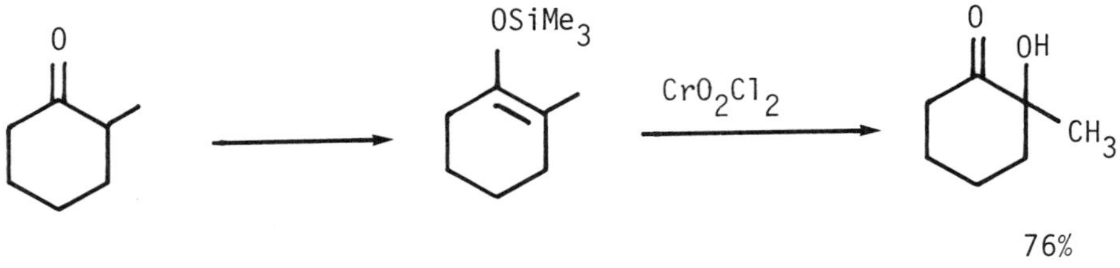

OSiMe$_3$

CrO$_2$Cl$_2$

76%

Tetr Lett, 23, 2917 (1982)

1) PhI(OAc)$_2$

Ph-CH$_2$COOMe   →   Ph-CH-COOH   50%
                            |
2) KOH, H$_2$O   OH

3) H$^+$

Tetr Lett, 22, 2747 (1981)

Section 42 Alcohols from Ketones

The following reaction types are included in this section:

A. Reductions of ketones to alcohols.

B. Nucleophilic additions to ketones, forming alcohols.

C. Coupling of ketones to give diols.

42A: Reductions of Ketones to Alcohols

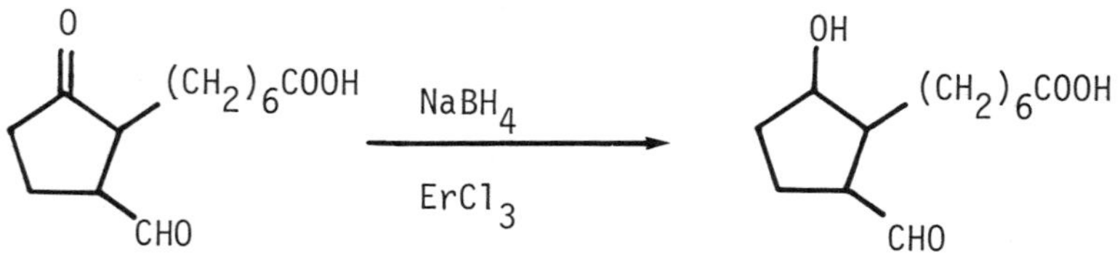

Tetr Lett, <u>22</u>, 4077 (1981)

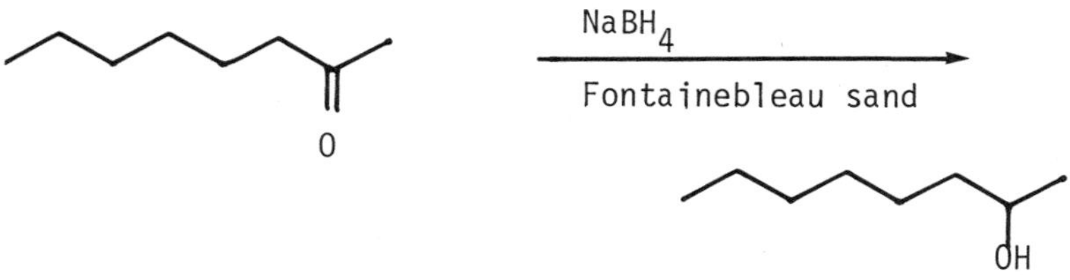

JCS Chem Comm, 1066 (1981)

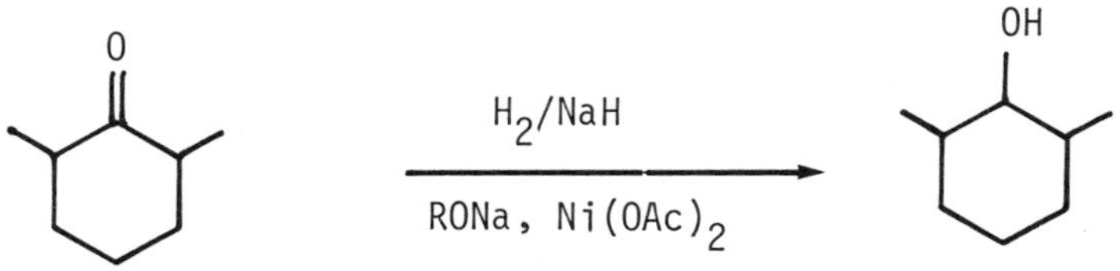

JOC, <u>45</u>, 1946 (1980)

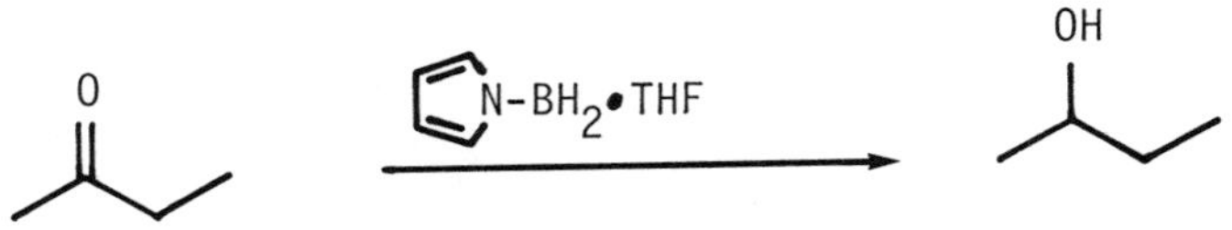

Synthesis, 214 (1981)

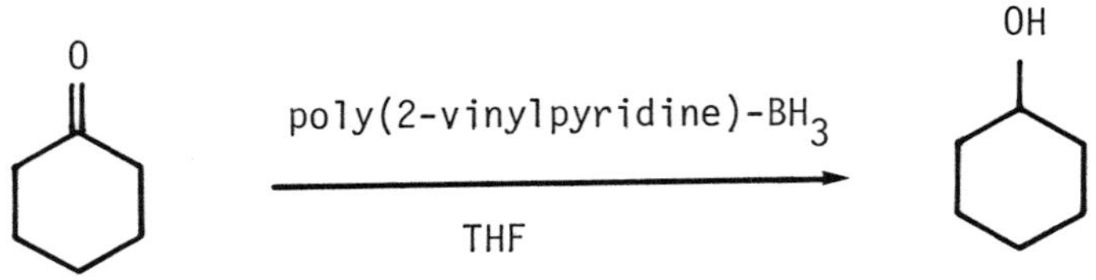

JOC, <u>45</u>, 2724 (1980)

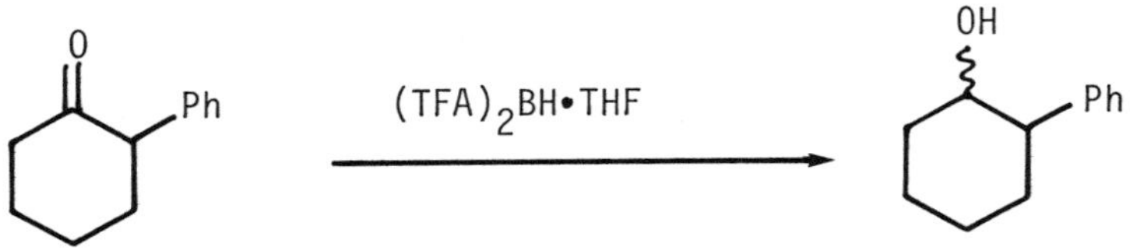

JOC, <u>46</u>, 355 (1981)

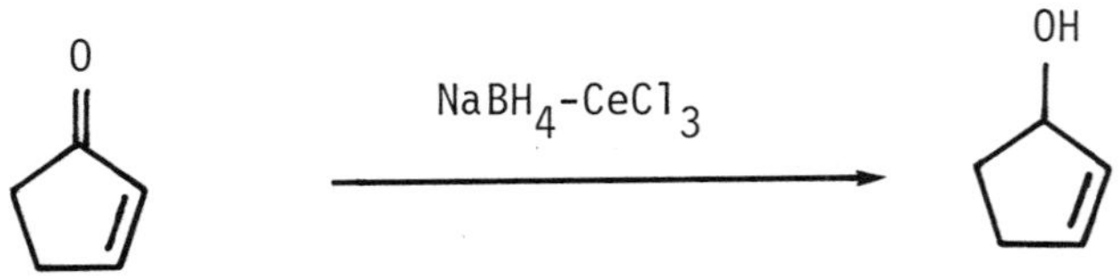

JACS, <u>103</u>, 5454 (1981)

$$\underset{Et_2O}{\xrightarrow{\underline{t}\text{-BuNH}_2 \cdot BH_3}}$$

87%

Tetr Lett, 21, 693 and 697 (1980)

$$\underset{\text{toluene/hexane}}{\xrightarrow{\text{LiBuBH}_3,\ -78^0}}$$

96%

JOC, 47, 3311 (1982)

$$CH_3-\overset{O}{\underset{\parallel}{C}}-(CH_2)_3-\overset{O}{\underset{\parallel}{C}}-OMe \quad \underset{\text{ether}}{\xrightarrow{\text{LiAlH}_4\text{-SiO}_2}} \quad CH_3-\overset{OH}{\underset{|}{C}}H(CH_2)_3-\overset{O}{\underset{\parallel}{C}}-OMe \quad 84\%$$

Tetr Lett, 23, 4585 (1982)

$$\underset{2)\ NH_4Cl,\ H_2O}{\overset{1)\ (\underline{i}\text{-Bu})_3Al}{\xrightarrow{\hspace{2cm}}}}$$

60%

JOC, 47, 4640 (1982)

$$\text{cycloheptanone} \xrightarrow[\text{H}_2\text{O/dioxane/DMF}]{\text{Na}_2\text{S}_2\text{O}_4} \text{cycloheptanol} \quad 97\%$$

JOC, <u>45</u>, 4126 (1980)

$$\text{cyclohexanone} \xrightarrow[\text{RuCl}_2\text{L}_3]{\text{HCOOH}} \text{cyclohexanol} \quad 78\%$$

Bull Chem Soc Jpn, <u>55</u>, 2441 (1982)

$$\text{2-methylcyclohexanone} \xrightarrow{\text{H-C-ONa, N-Me-pyrrolidinone}} \text{2-methylcyclohexanol} \quad 76\%$$

JOC, <u>46</u>, 3367 (1981)

$$\text{cyclohexanone} \xrightarrow{\text{H}_2\text{Se, } h\nu} \text{cyclohexanol} \quad 93\%$$

Angew Chem Int Ed, <u>19</u>, 1008 and 1009 (1980)

1) Ph$_2$SiH$_2$
   L$_3$RhCl

2) CH$_3$OH

97%

Organometallics, <u>1</u>, 1390 (1982)

$$Ph-\overset{O}{\overset{\|}{C}}-\underset{\underset{Br}{|}}{CH}-CH_3$$

(EtO)$_2$SiHMe

KF

$$Ph-\overset{OH}{\overset{|}{CH}}-\underset{\underset{Br}{|}}{CH}-CH_3$$

70%

JCS Chem Comm, 121 (1981)

1) (Ph$_3$P)$_3$RuCl$_2$
   Et$_3$SiH

2) H$_3$O$^+$, MeOH

95% equatorial

JOC, <u>47</u>, 2469 (1982)

NaBH$_4$, Ce^{3+}

MeOH

85%
98% trans

Synth Comm, <u>10</u>, 623 (1980)

Ind J. Chem, <u>21B</u>, 212 (1982)

Tetr Lett, <u>22</u>, 675 (1981)

Synth Comm, <u>12</u>, 723 (1982)

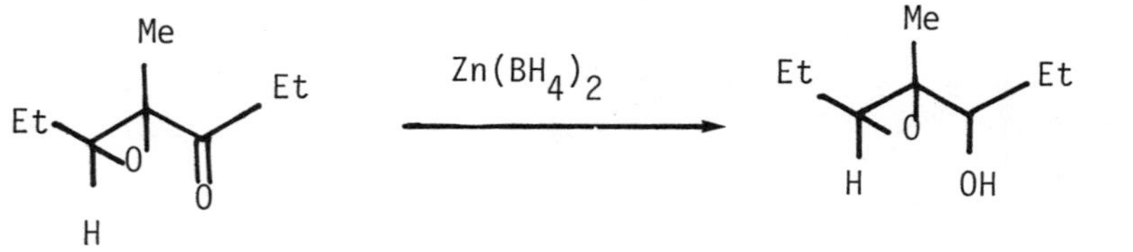

83%

99% erythro

Tetr Lett, 22, 4723 (1981)

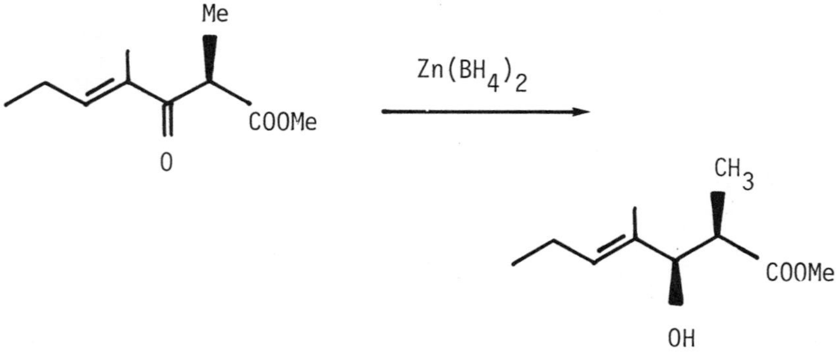

85%

>99% erythro

Tetr Lett, 21, 1641 (1980)

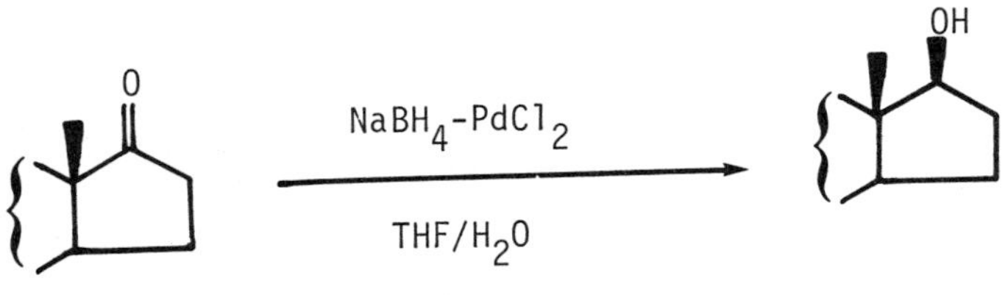
$NaBH_4$-$PdCl_2$
THF/H_2O
OH
O

96%

Chem Lett, 1029 (1981)

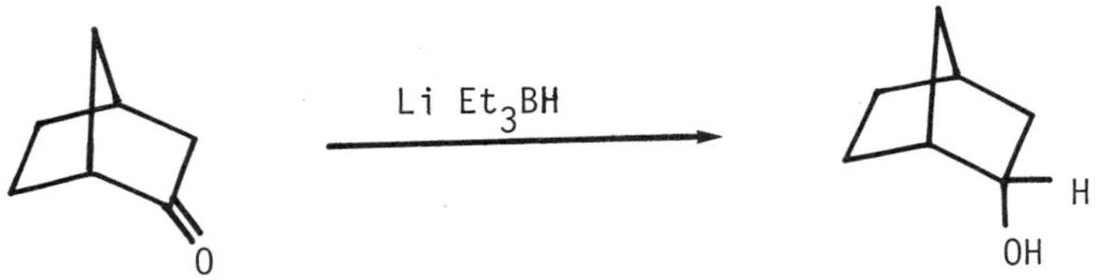
Li Et_3BH
O
H
OH

99%

JOC, 45, 1 (1980)

Prostaglandin

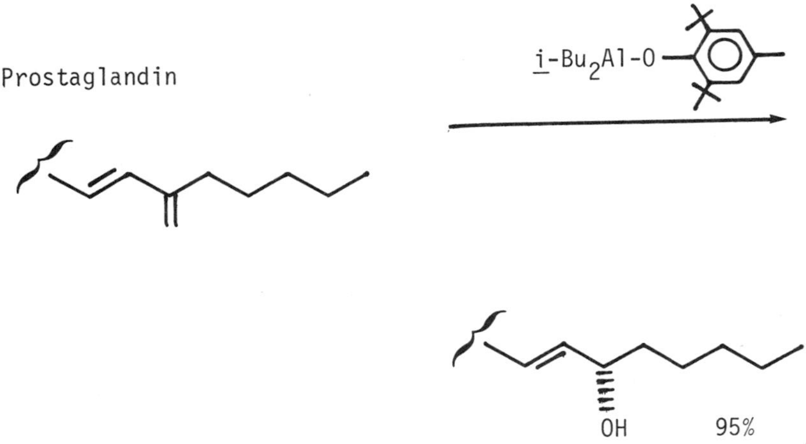
i-Bu_2Al-O
OH

95%

92% stereo-selectivity

Bull Chem Soc Jpn, 54, 3033 (1981)

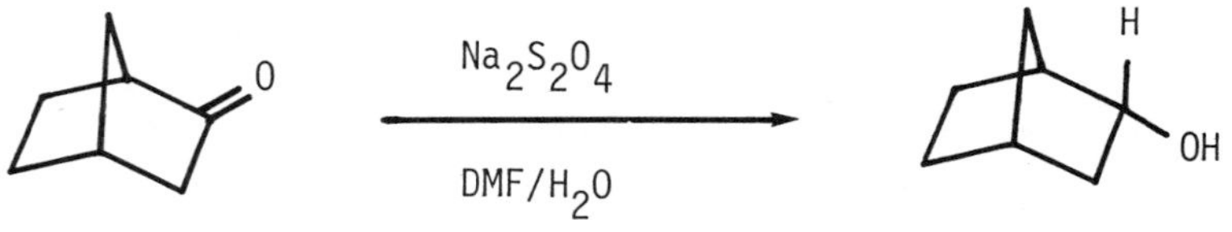

75% yield

85% <u>endo</u>

Tetr Lett, <u>22</u>, 179 (1981)

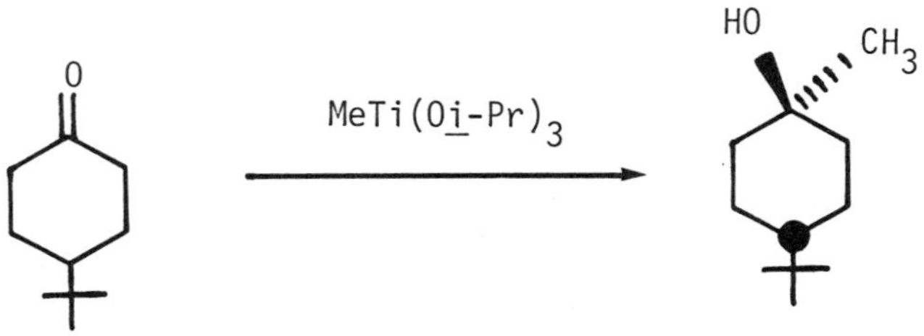

87%

Helv Chim Acta, <u>63</u>, 2451 (1980)

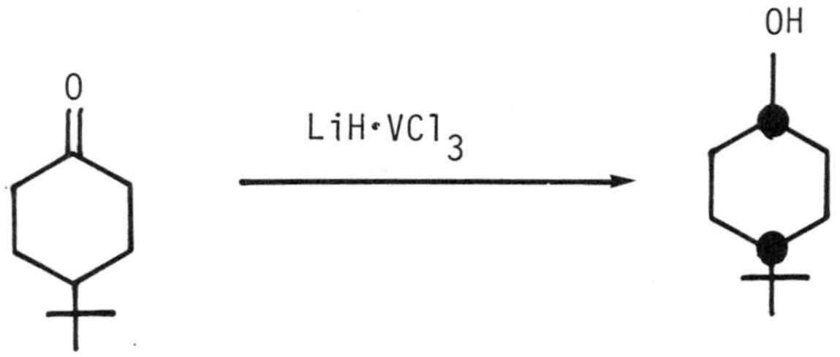

86%
82% <u>cis</u>

JOC, <u>45</u>, 1041 (1980)

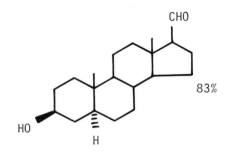

1) $\underline{t}$-BuNH$_2$
   (protects the aldehyde)
2) Li($\underline{t}$-BuO)$_3$AlH

3) H$_2$O
4) basic alumina

83%

Tetrahedron, <u>38</u>, 1827 (1982)

Potassium tri(R,S-$\underline{s}$-butyl)borohydride reduces 3-oxo steroids to the axial alcohols, without affecting the 17- and 20-ketone groups.

Use of a chiral hydrosilane-rhodium phosphine reagent allows greater stereoselectivity of 17α-alcohol formation than with other methods.

JCS Chem Comm, 1238 and 1239 (1982)

$$Ph-\overset{\overset{\displaystyle O}{\|}}{C}-Et \xrightarrow[\text{chiral glucose derivatives}]{NaBH_4}$$

H⟍ ⟋OH
 C
Ph⟋ * ⟍Et

100%

51% ee

JOC, <u>45</u>, 4229 (1980)

JOC, <u>45</u>, 4231 (1980)

Bull Chem Soc Jpn, <u>54</u>, 1424 (1981)

2-octanone ⟶ (S)-2-octanol

79% ee

$$C_5H_{11}-\overset{\overset{\displaystyle O}{\|}}{C}-C\equiv C-CH_3 \xrightarrow{\text{similar conditions}}$$

HO⟍ H
    C
C₅H₁₁⟋ ⟍C≡C-CH₃

79%

91% ee

JOC, <u>47</u>, 2496 (1982)

JOC, <u>47</u>, 2814 (1982)

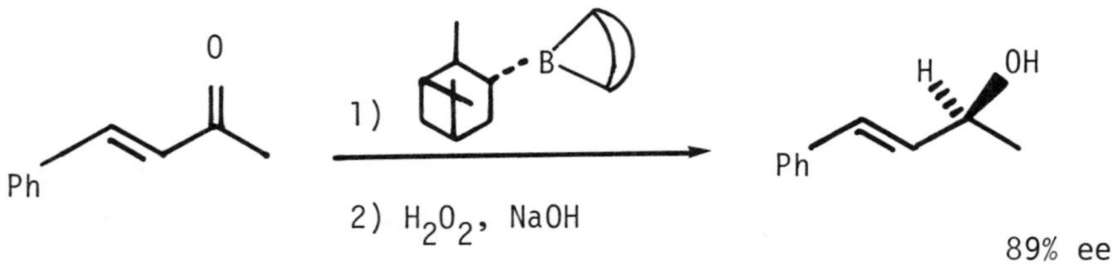

JOC, <u>47</u>, 1606 (1982)

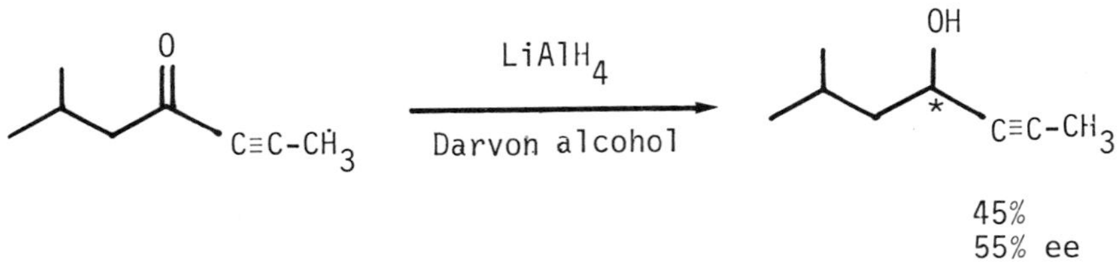

45%
55% ee

JOC, <u>45</u>, 582 (1980)

92%
78% ee

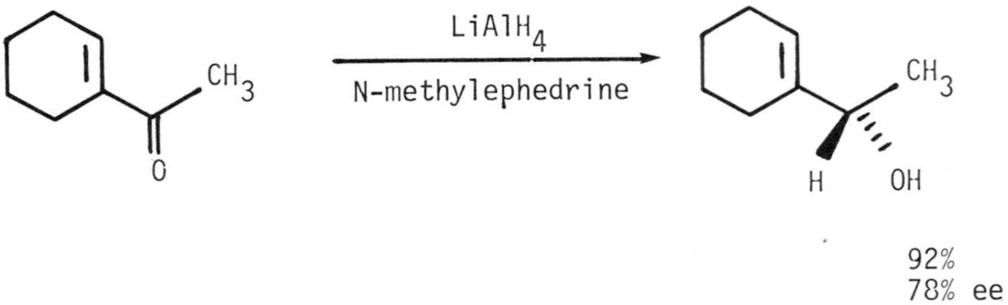

JCS Chem Comm, 1026 (1980)
Chem Lett, 981 (1980)

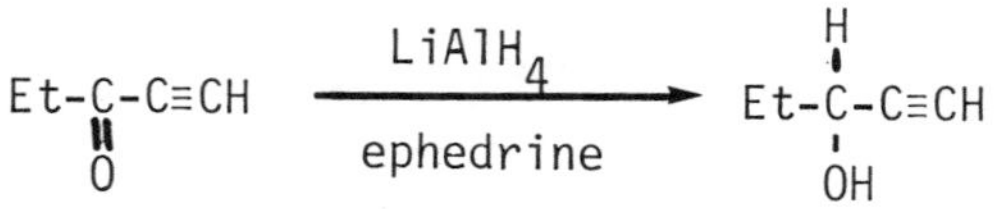

$$Et-\underset{\underset{O}{\|}}{C}-C\equiv CH \xrightarrow[\text{ephedrine}]{\text{LiAlH}_4} Et-\underset{\underset{OH}{|}}{\overset{\overset{H}{|}}{C}}-C\equiv CH$$

60-80%
84% ee

Tetr Lett, <u>21</u>, 1735 and 1739 (1980)

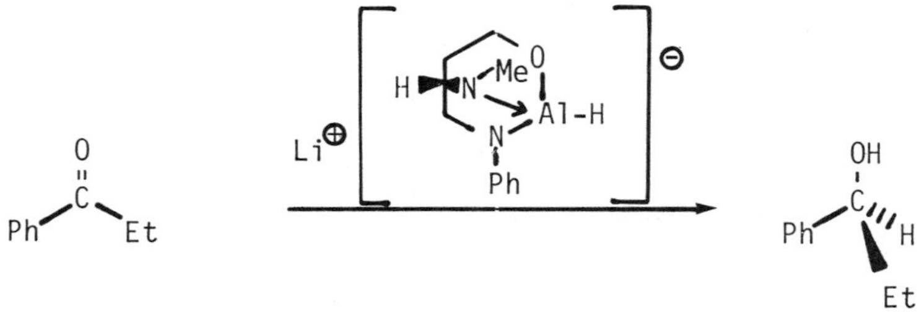

93%
68% ee

Tetr Lett, <u>23</u>, 4111 (1982)

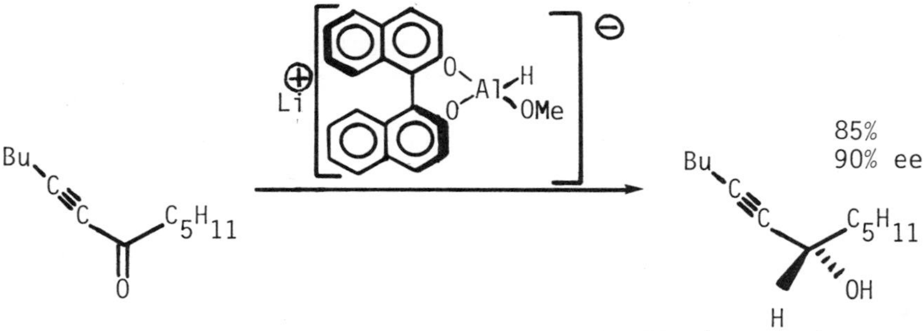

85%
90% ee

Tetr Lett, <u>22</u>, 247 (1981)

$$\underset{\substack{\text{O}\\ \text{R-C-R'}}}{} \xrightarrow{\quad \text{Li} \left[\text{binaphthyl} \overset{\text{O}}{\underset{\text{O}}{\diagdown}} \text{Al} \overset{\text{H}}{\underset{\text{OMe}}{\diagup}} \right] } \underset{\substack{\text{R}\quad\text{R'}}}{\overset{\text{HO}\quad\text{H}}{\diagup \text{C} \diagdown}}$$

R = alkyl

R' = vinyl, Ph, acetylenic

Pure and Appl Chem, <u>53</u>, 2315 (1981)

$$\underset{\text{}}{\text{(pentane-2,4-dione)}} \xrightarrow[\text{RaNi, tartaric acid}]{2H_2, \text{ NaBr}} \underset{\text{OH}\quad\text{OH}}{\text{}}$$

65% (R,R)

$[\alpha] = -54^{o}$

Bull Chem Soc Japan, <u>53</u>, 3367 (1980)

$$CH_3(CH_2)_6\overset{\text{O}}{\underset{}{C}}-CH_2COOMe \xrightarrow{\text{chiral Ni complex}} CH_3(CH_2)_6\underset{*}{\overset{\text{OH}}{CH}}-CH_2COOMe$$

76%

87% ee

Bull Chem Soc Japan, <u>55</u>, 2186 (1982)

O
||
Ph-C-COOEt

$\xrightarrow[\text{Mg}^{++}]{\text{chiral dihydropyridine}}$

OH
|
Ph-CH-COOEt
    *

80%
83% ee

JACS, 103, 2091 (1981)

JACS, 103, 4613 (1981)

O
||
Ph-C-CH$_2$Cl

$\xrightarrow[\text{RhCl Diop}]{\text{ArPhSiH}_2}$

H   OH
  \   /
   C
  /   \
Ph   CH$_2$Cl

63% ee

J Chem Research(S), 320 (1980)

O
||
Ph-C-CH$_2$Cl

$\xrightarrow{\text{microbial reduction}}$

H   OH
  \   /
   C
  /   \
Ph   CH$_2$Cl

80%
100% ee

JOC, 45, 3352 (1980)

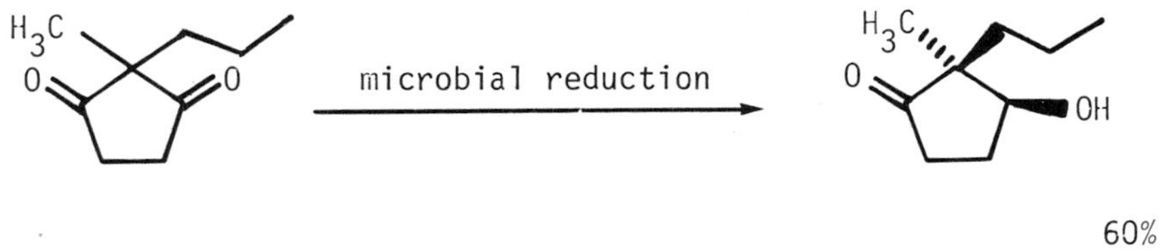

60%

JOC, <u>47</u>, 2820 (1982)

Use of LiAlH$_4$-aminodiol complexes to reduce ketones asymmetrically. The chiral aminodiols are synthesized from tartaric acid.

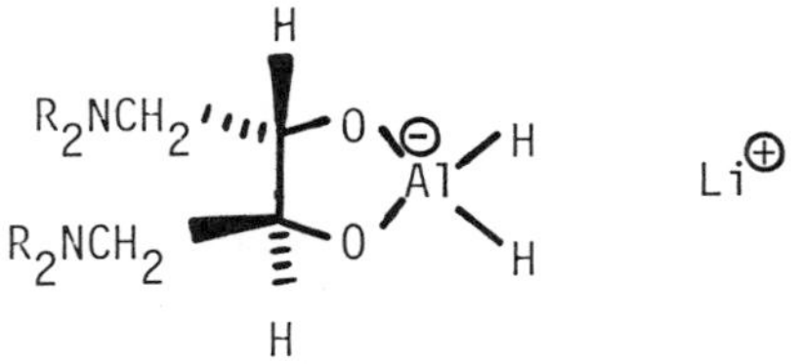

Chem Ber, <u>113</u>, 1691 (1980)

42B: Nucleophilic Additions to Ketones, Forming Alcohols

Aldol reactions are listed in Section 330 (Ketone-Alcohol).

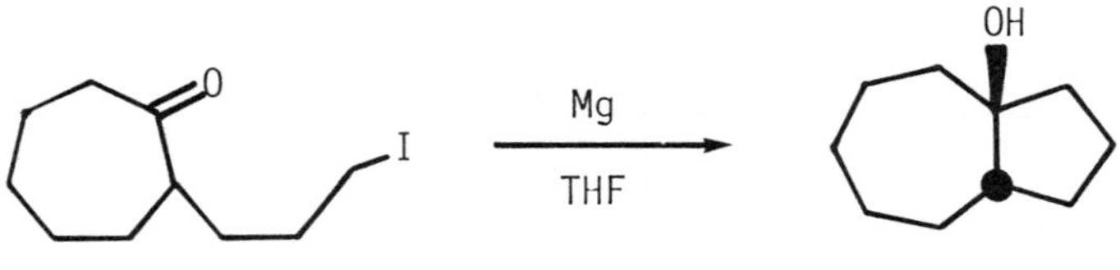

(85% <u>cis</u>)

54%

JOC, <u>47</u>, 5368 (1982)

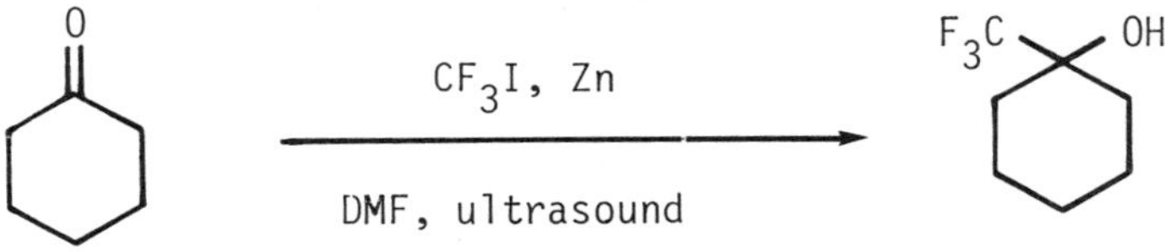

48%

Chem Lett, 1679 (1981)

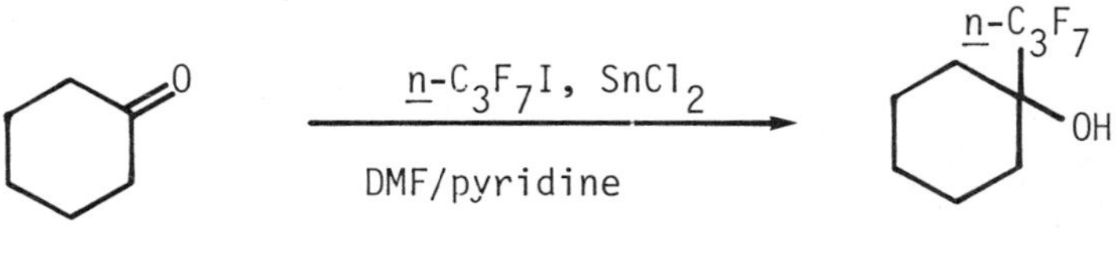

57%

Chem Lett, 1337 (1981)

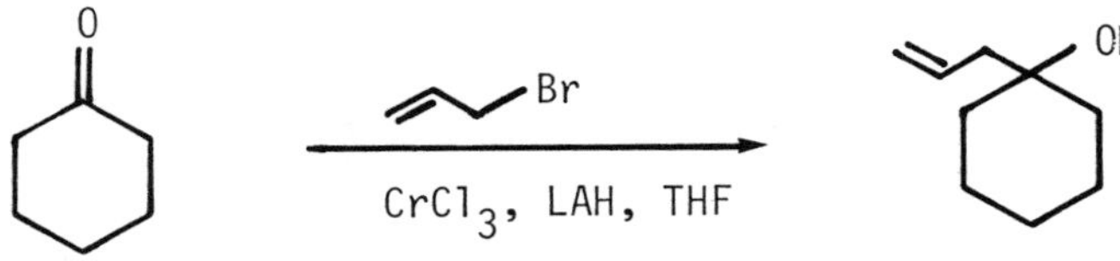

73%

BCS Japan, 55, 561 (1982)

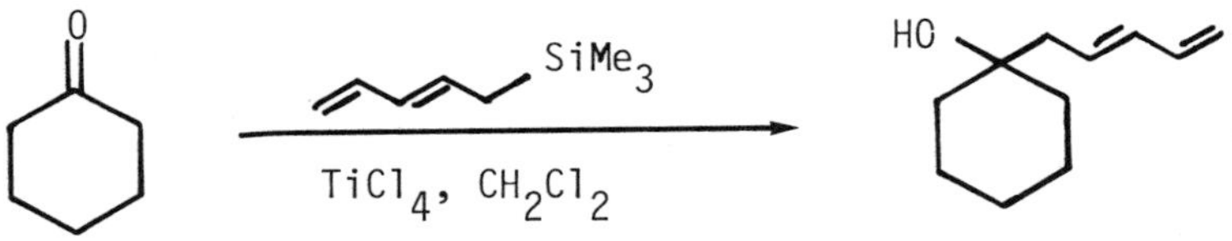

64%

Organometallics, 1, 1651 (1982)

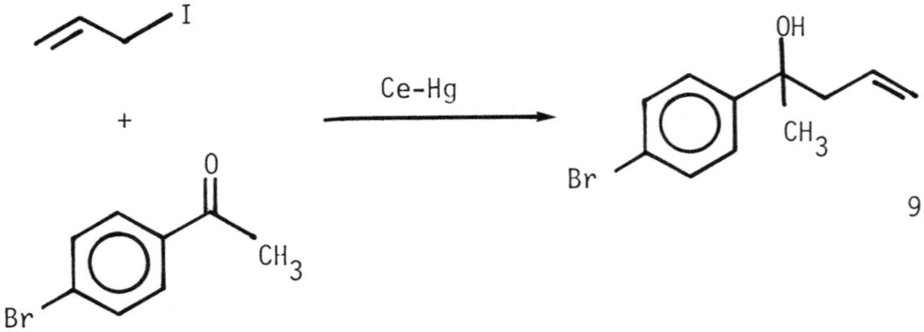

95%

Tetr Lett, 22, 4985 (1981)

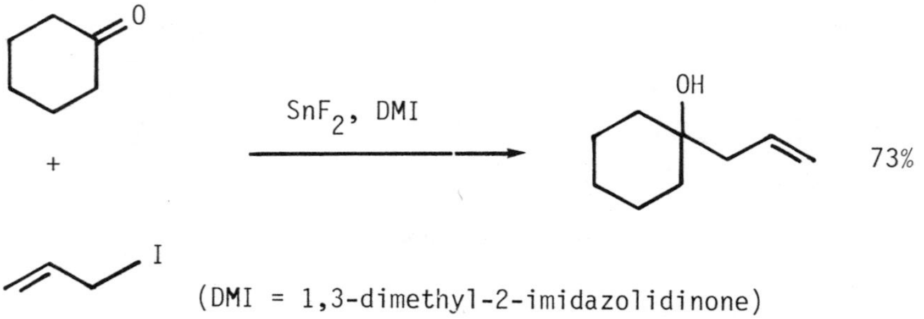

(DMI = 1,3-dimethyl-2-imidazolidinone)

Chem Lett, 1507 (1980)

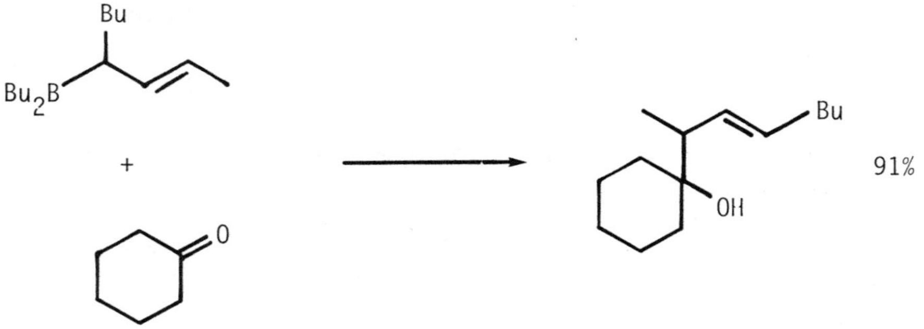

Chem Lett, 993 (1980)

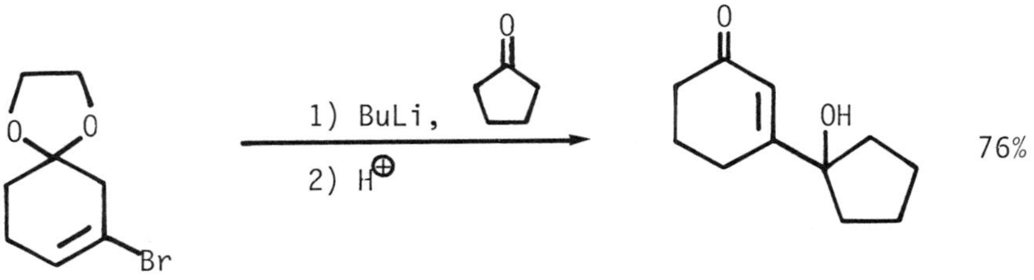

JOC, _47_, 2825 (1982)

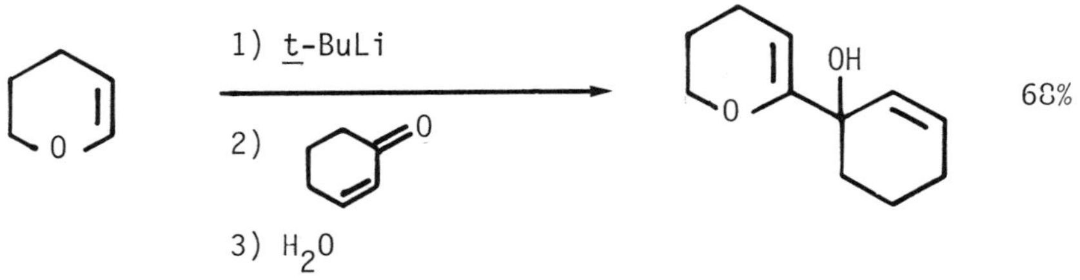

Tetrahedron, _37_, 3997 (1981)

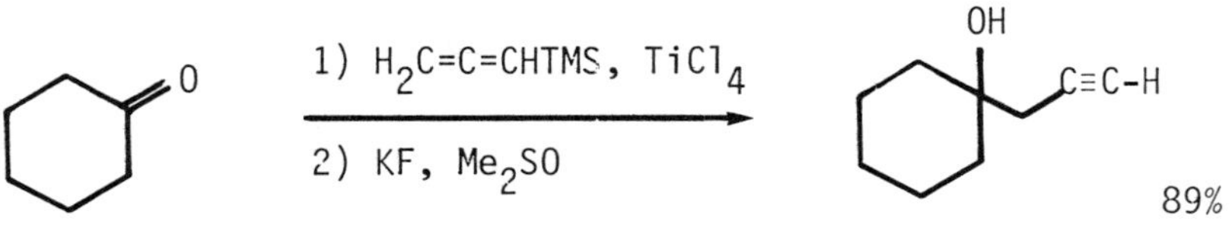

JOC, _45_, 3925 (1980)

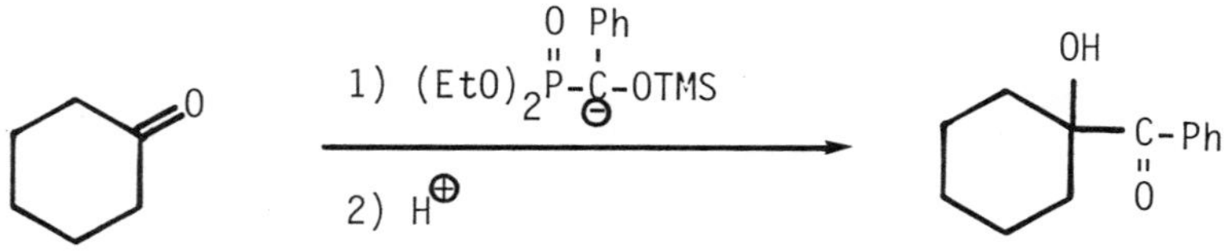

1) (EtO)$_2$P-C-OTMS

2) H$^{\oplus}$

83%

Tetr Lett, 21, 1017 (1980)

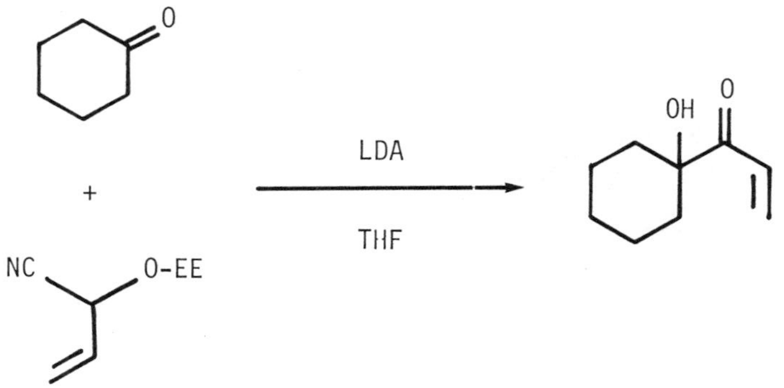

LDA

THF

72%

(EE = ethoxyethyl)   JOC, 45, 395 (1980)

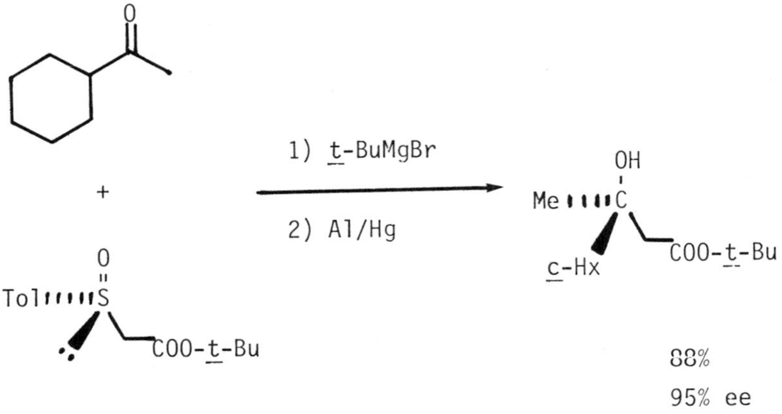

Tetrahedron, <u>36</u>, 227 (1980)

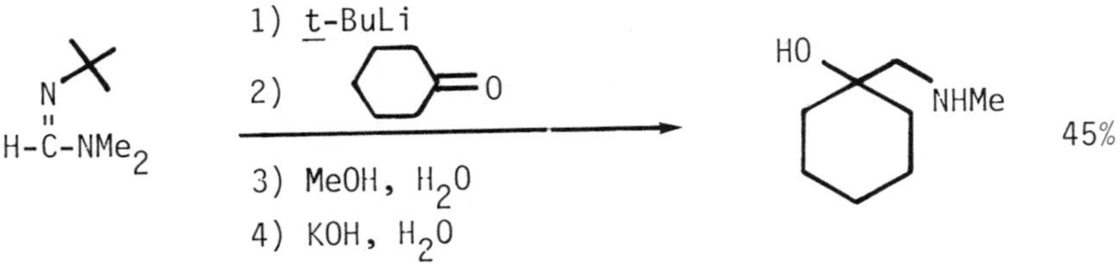

JACS, <u>102</u>, 7125 (1980)

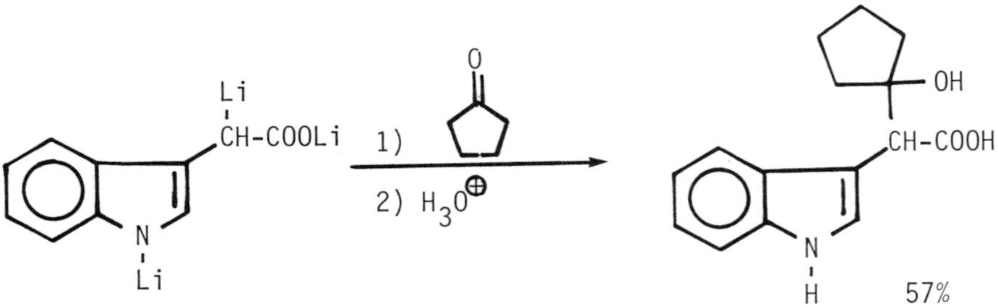

JCC, <u>45</u>, 447 (1980)

42C: Coupling of Ketones to Give Diols

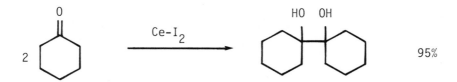

$$2 \xrightarrow{\text{Ce-I}_2}$$

95%

Tetr Lett, 23, 1353 (1982)

Related methods: Alcohols from Aldehydes (Section 34)

Section 43 Alcohols and Phenols from Nitriles

No Additional Examples.

Section 44 Alcohols from Olefins

For the preparation of diols from olefins see Section 323
(Alcohol-Alcohol)

$$\xrightarrow[\text{DME}]{\text{NaBH}_4, \text{TiCl}_4}$$

71%

JCS Chem Comm, 414 (1980)

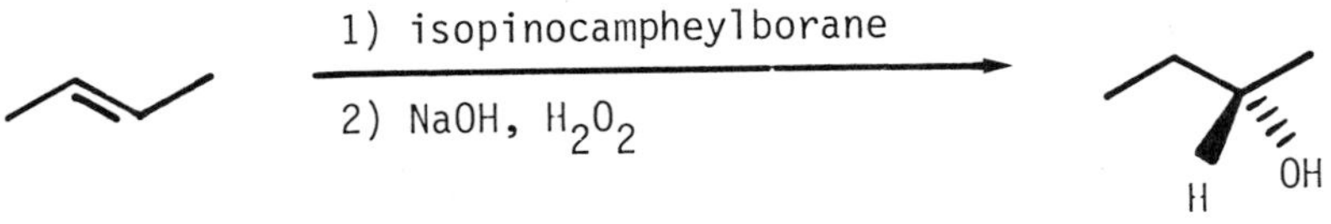

1) isopinocampheylborane

2) NaOH, H_2O_2

73%

73%ee

JOC, <u>46</u>, 2988 (1981)

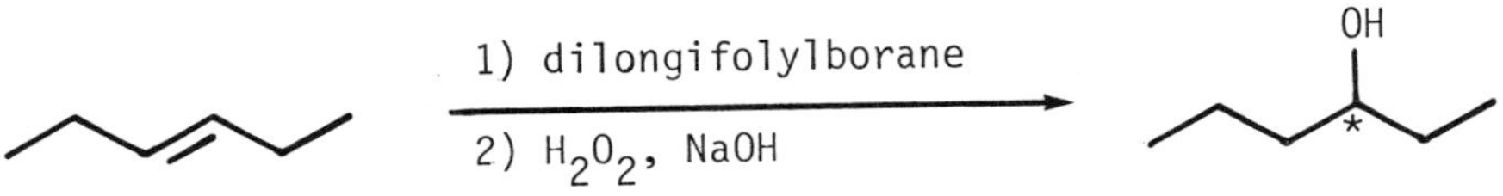

1) dilongifolylborane

2) H_2O_2, NaOH

81%

71% ee

JOC, <u>46</u>, 2988 (1981)

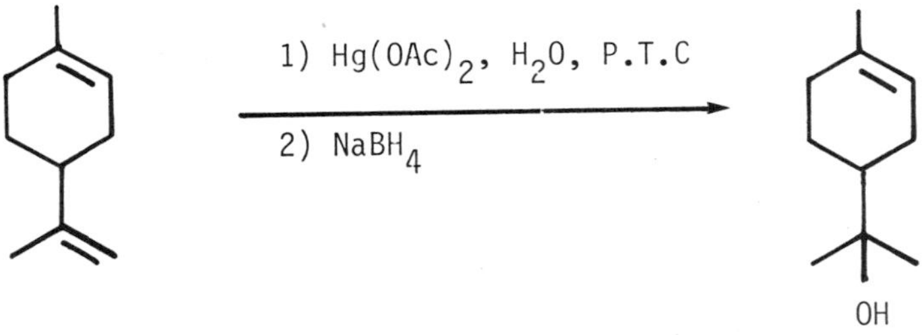

1) Hg(OAc)$_2$, H_2O, P.T.C

2) NaBH$_4$

97%

JACS, <u>102</u>, 7798 (1980)

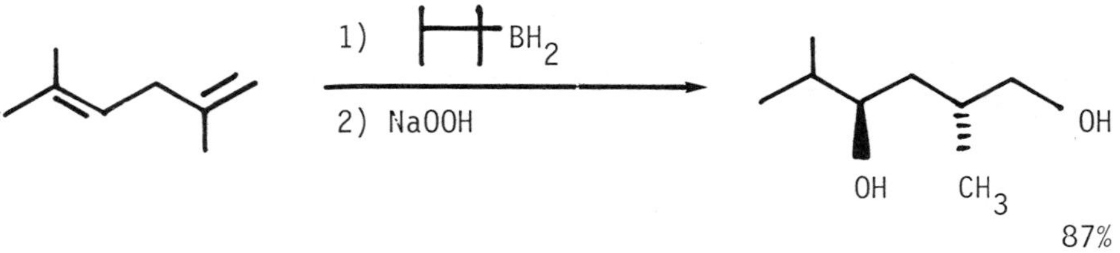
1)
BH2
2) NaOOH
OH
OH
CH3
87%

JACS, 102, 7385 (1980)

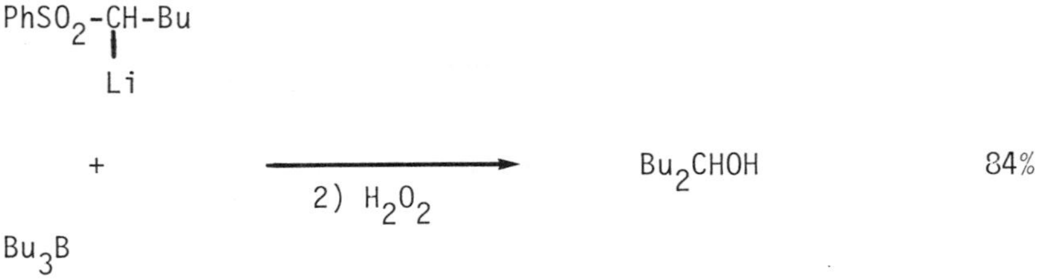
PhSO2-CH-Bu
Li
+
Bu3B
2) H2O2
Bu2CHOH
84%

Bull Soc Chim France II, 99 (1981)

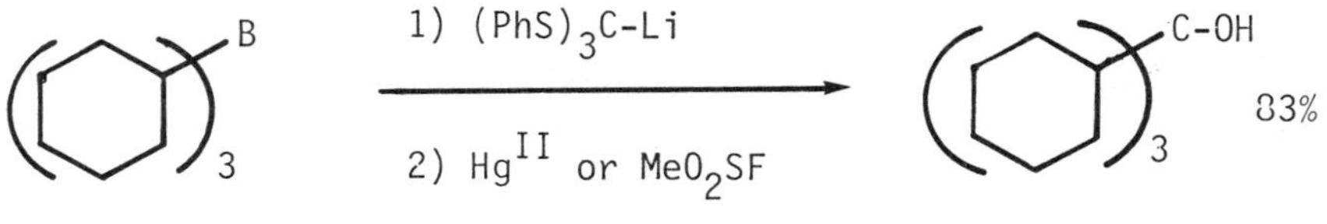
B
3
1) (PhS)3C-Li
2) HgII or MeO2SF
C-OH
3
83%

JCS Chem Comm, 1149 (1981)

$$CH_3(CH_2)_9CH=CH_2 \xrightarrow[\text{BF}_3 \cdot \text{OEt}_2]{\text{BH}_3\text{CN}^\ominus} CH_3(CH_2)_9CH_2CH_2OH \qquad 83\%$$

JOC, <u>46</u>, 5214 (1981)

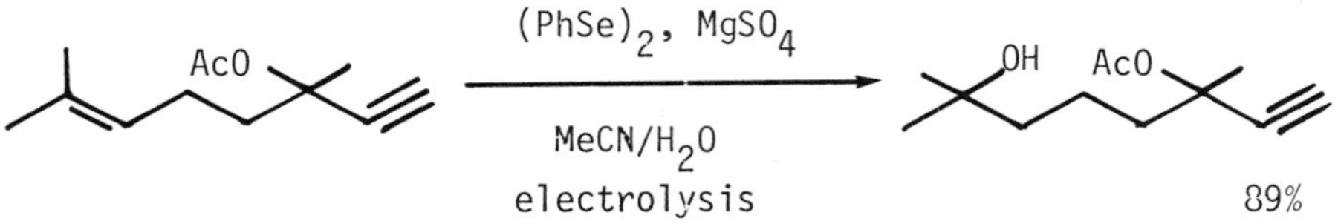

(PhSe)$_2$, MgSO$_4$

MeCN/H$_2$O
electrolysis

89%

JACS, <u>103</u>, 4606 (1981)

Section 45 Alcohols from Miscellaneous Compounds

No Additional Examples

For conversions of boranes to alcohols, see Section 44

Section 45A <u>Protection of Alcohols and Phenols</u>

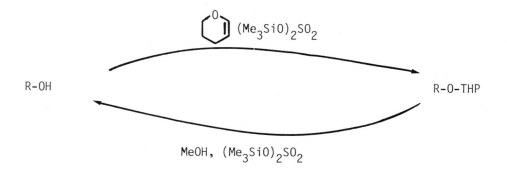

R-OH   R-O-THP

MeOH, $(Me_3SiO)_2SO_2$

Synthesis, 899 (1981)

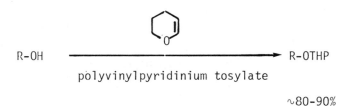

R-OH ————————————————————→ R-OTHP

polyvinylpyridinium tosylate

~80-90%

R = 1^O, 2^O, 3° alkyl, benzyl, etc.

JOC, <u>46</u>, 5044 (1981)

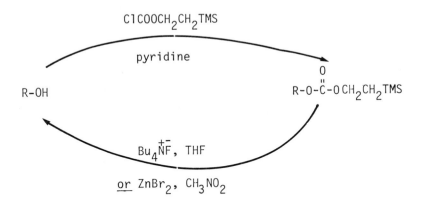

Tetr Lett, <u>22</u>, 969 (1981)

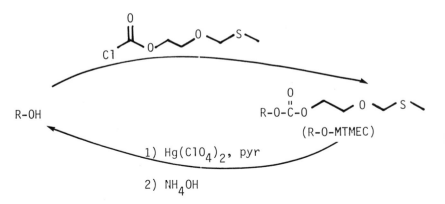

Tetr Lett, <u>22</u>, 1933 (1981)

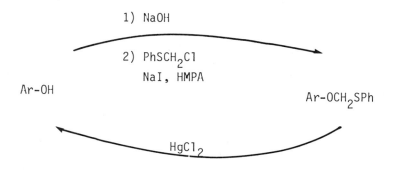

1) NaOH

2) PhSCH$_2$Cl
   NaI, HMPA

Ar-OH   Ar-OCH$_2$SPh

HgCl$_2$

Synth Comm, <u>10</u>, 911 (1980)

R-OH $\dfrac{CH_2(OMe)_2}{Nafion-H, \Delta}$ R-O-CH$_2$OMe   57-96%

Synthesis, 471 (1981)

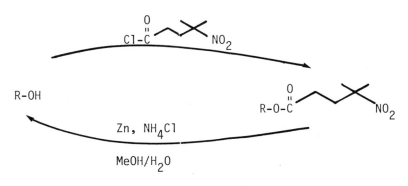

R-OH

Zn, NH$_4$Cl

MeOH/H$_2$O

Synth Comm, <u>10</u>, 469 (1980)

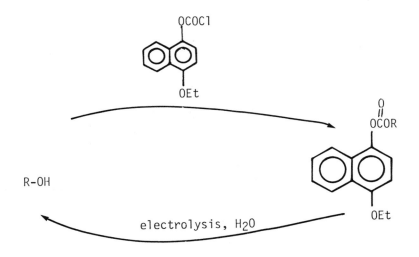

R-OH

electrolysis, H₂O

Tetr Lett, 22, 3719 (1981)

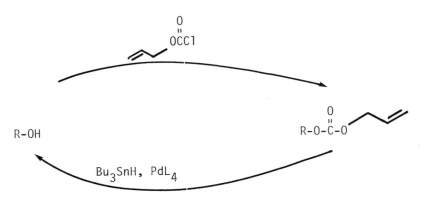

R-OH

Bu₃SnH, PdL₄

Tetr Lett, 22, 3591 (1981)

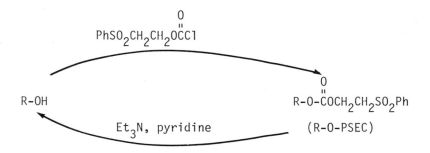

Tetr Lett, <u>22</u>, 3667 (1981)

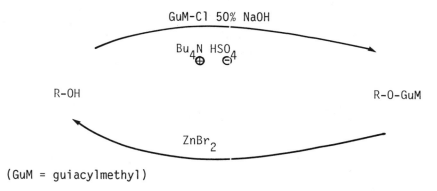

(GuM = guiacylmethyl)

Tetr Lett, <u>22</u>, 1973 (1981)

R-O-CH$_2$—⬡—OMe $\xrightarrow[\text{CH}_2\text{Cl}_2/\text{H}_2\text{O}]{\text{DDQ}}$ R-OH

Isopropylidene, THP, MEM, TMS, Bz, etc. groups are unaffected.

Tetr Lett, <u>23</u>, 885 and 889 (1982)

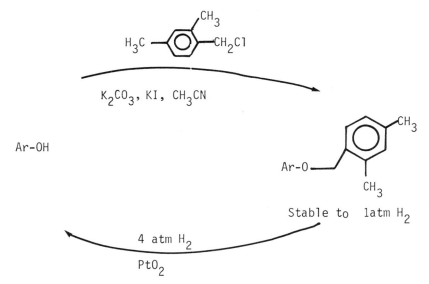

Ar-OH

Stable to 1atm H_2

K_2CO_3, KI, CH_3CN

4 atm H_2

PtO_2

Ar = subst. Ph

Synthesis, 987 (1982)

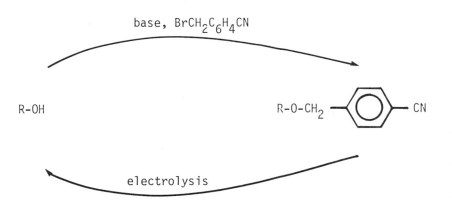

base, $BrCH_2C_6H_4CN$

R-OH

electrolysis

Chem Ber, 114, 946 (1981)

R-O-Bz $\xrightarrow[\text{Me}_2\text{S}]{\text{BF}_3 \cdot \text{Et}_2\text{O}}$ R-OH

Chem Pharm Bull, <u>28</u>, 3662 (1980)

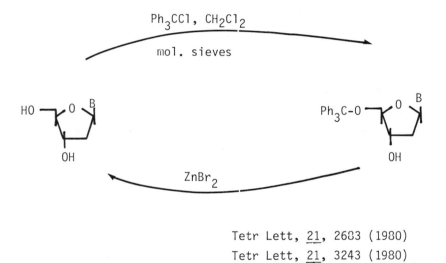

Tetr Lett, <u>21</u>, 2683 (1980)
Tetr Lett, <u>21</u>, 3243 (1980)

R-OH $\xrightarrow[\text{I}_2]{\text{Me}_3\text{SiCH}_2\text{CH}=\text{CH}_2}$ R-O-SiMe$_3$ >90%

Chem Lett, 85 (1981)

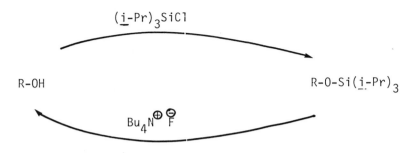

(More stable to H$^+$ than t-butyldimethylsilyl)

JOC, <u>45</u>, 4797 (1980)

HC≡C, OH → HC≡C OSiMe$_3$

ClSiMe$_3$, Et$_3$N

cat. DMSO or HMPA

>80%

Rec Trav Chim Pays-Bas, <u>99</u>, 70 (1980)

R-OH $\xrightarrow[\text{CH}_3\text{CN}]{\text{CH}_2=\text{CH-CH}_2-\text{SiMe}_3}$ R-OSiMe$_3$ ∿90%

R-OH $\xrightarrow[\text{CH}_3\text{CN}]{\text{CH}_2=\text{CH-CH}_2\text{SiMe}_2\text{t-Bu}}$ R-OSiMe$_2$t-Bu

Tetr Lett, <u>21</u>, 835 (1980)

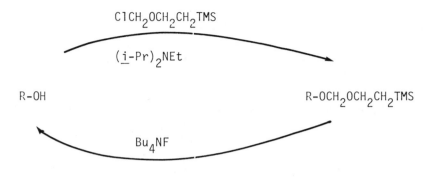

$$\text{R-OH} \xrightarrow[\text{($\underline{i}$-Pr)}_2\text{NEt}]{\text{ClCH}_2\text{OCH}_2\text{CH}_2\text{TMS}} \text{R-OCH}_2\text{OCH}_2\text{CH}_2\text{TMS}$$

$$\text{R-OCH}_2\text{OCH}_2\text{CH}_2\text{TMS} \xrightarrow{\text{Bu}_4\text{NF}} \text{R-OH}$$

Tetr Lett, $\underline{21}$, 3343 (1980)

$$\text{R-OH} \xrightarrow[\text{H}_2\text{C=CH-CH}_2\text{SiMe}_3]{\text{F}_3\text{C-SO}_3\text{H, CH}_2\text{Cl}_2} \text{R-O-SiMe}_3 \qquad \text{92-96\%}$$

Synthesis, 745 (1981)

AcCH=C-O-X (Me above C)

85%

X = SiMe$_2$-$\underline{t}$-Bu

Tetr Lett, $\underline{22}$, 1299 and 1303 (1981)

$$\text{R-OH} \xrightarrow{\underset{\substack{\\ }}{CF_3-\overset{\overset{\displaystyle O}{\|}}{C}-\overset{\overset{\displaystyle CH_3}{|}}{N}-\underset{\underset{\displaystyle CH_3}{\diagdown}}{\overset{\overset{\displaystyle CH_3}{\diagup}}{Si}}-C(CH_3)_3}} \text{R-O-Si-}\overset{\overset{\displaystyle CH_3}{|}}{\underset{\underset{\displaystyle CH_3}{|}}{C}}\text{-C(CH}_3)_3 \qquad 96\text{-}100\%$$

JOC, <u>47</u>, 3336 (1982)

$$\text{R-OSiMe}_2\underline{t}\text{-Bu} \xrightarrow[\text{DMSO}]{\text{NBS}} \text{R-OH} \qquad \sim 80\%$$

Synthesis, 234 (1980)

$$\text{R-O-SiMe}_2\underline{t}\text{-Bu} \xrightarrow{\text{LiBF}_4} \text{R-OH}$$

Tetr Lett, <u>21</u>, 35 (1980)

$$\underset{\underset{\displaystyle R'}{|}}{\text{R-CH-OTs}} \xrightarrow[\text{diglyme}]{\text{K, crown ether}} \underset{\underset{\displaystyle R'}{|}}{\text{R-CHOH}}$$

Chem Pharm Bull, <u>30</u>, 3178 (1982)

Use of the levulinoyl group as an OH-protecting group in the synthesis of oligosaccharides. Removed by hydrazine.

Rec Trav Chim Pays-Bas, 100, 65 (1981)

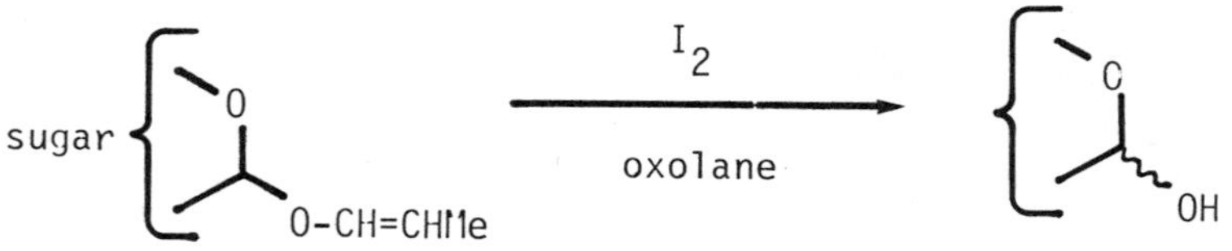
sugar
O
O-CH=CHMe
I2
oxolane
O
OH

JCS Chem Comm, 1274 (1982)

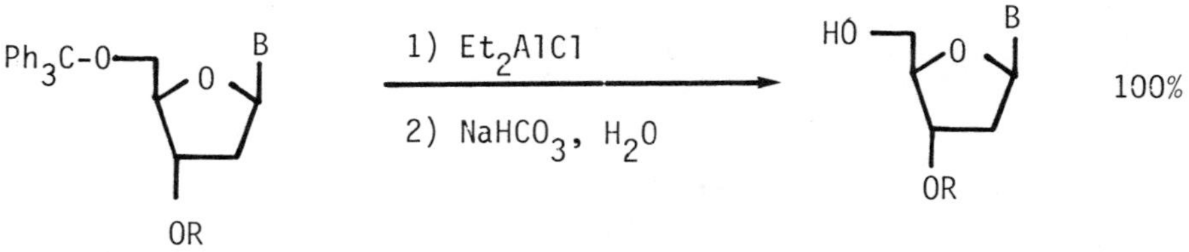
Ph3C-O
O
B
OR
1) Et2AlCl
2) NaHCO3, H2O
HO
O
B
OR
100%

Tetr Lett, 23, 2641 (1982)

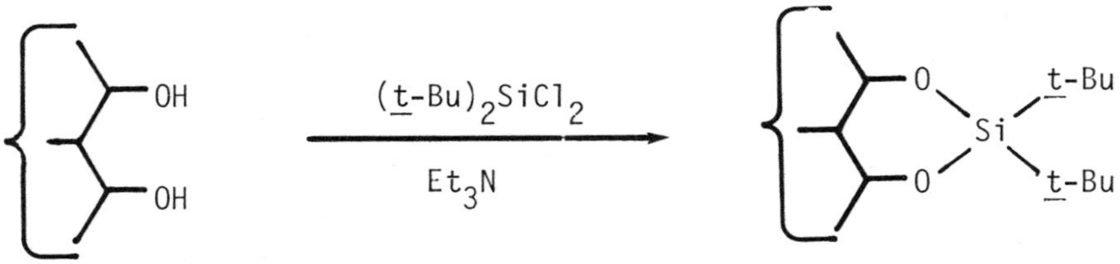

Removed using pyridinium hydrofluoride.   ~50-80%

Tetr Lett, 22, 4999 (1981)

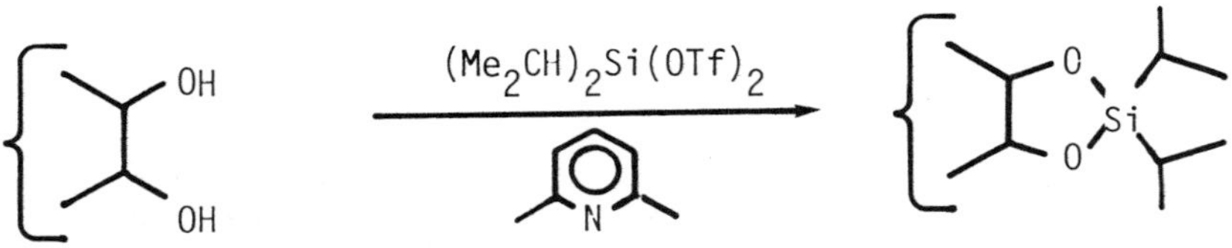

Protects 1,2-, 1,3-, and 1,4-diols.
Deprotected by 48% aqueous HF in CH_3CN.

Tetr Lett, 23, 4871 (1982)

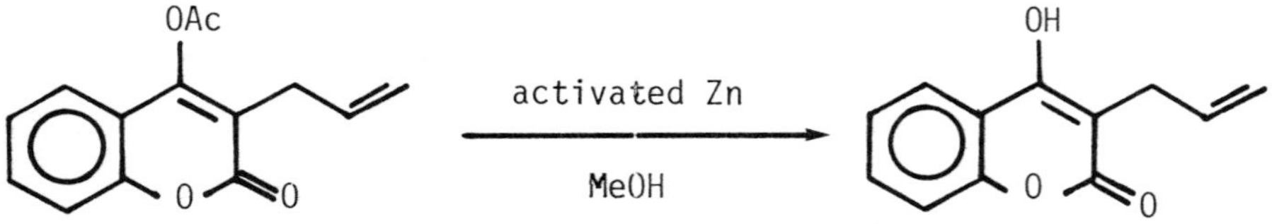

100%

Tetr Lett, 22, 335 (1981)

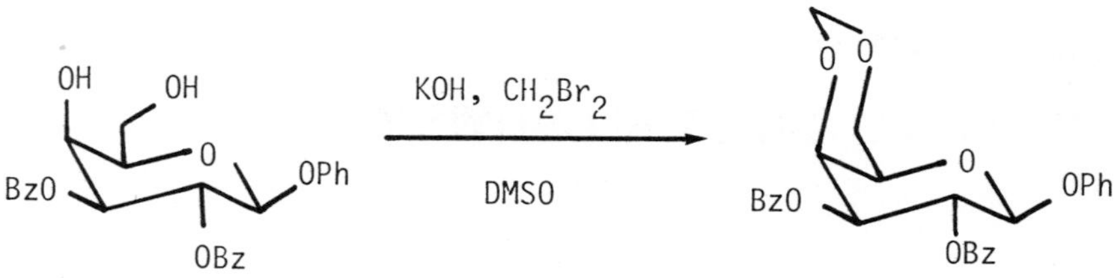

Synthesis, 421 (1982)

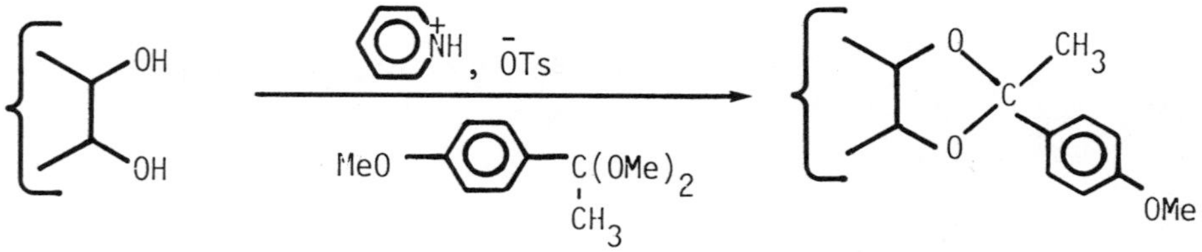

Removed using $SnCl_4$, followed by ^-OH.

JOC, 46, 2419 (1981)

Related Methods:

    Ethers from Alcohols - Section 123

    Alcohols from Ethers - Section 39

    Esters from Alcohols - Section 108

    Alcohols from Esters - Section 38

CHAPTER 4
PREPARATION OF ALDEHYDES

Section 46 <u>Aldehydes from Acetylenes</u>

$$Ph-C{\equiv}C-Ph \xrightarrow{\text{(BiPy)H}_2\text{CrOCl}_5} 2\ PhCHO \qquad 96\%$$

Org Prep Proc Int, <u>14</u>, 362 (1982)

HO⟍⟋C≡C-TMS

1) PhSCl, Et$_3$N
2) H$_2$O

3) AgNO$_3$, H$_3$O$^{\oplus}$/MeCN

⟶ (product with CHO) 62%

Tetr Lett, <u>22</u>, 2021 (1981)

HO⟍⟋C≡CH

1) I$_2$

2) pyridium dichromate

⟶ (product with I, CHO) 66%

Tetr Lett, <u>22</u>, 1041 (1981)

Section 47 <u>Aldehydes from Carboxylic Acids and Acid Halides</u>

Synthesis, 871 (1981)

$$Et\text{-}\overset{O}{\overset{\|}{C}}\text{-}OH \xrightarrow{\substack{1) \text{ PPA,} \\ 2) \text{ NaOEt, EtOH} \\ 3) \text{ MeI} \\ 4) \text{ LiAlH}_4 \\ 5) \text{ H}_3\text{O}^{\oplus}}} Et\text{-}CHO \qquad \sim 45\%$$

Synthesis, 303 (1981)

$$CH_3(CH_2)_8COOH \longrightarrow \longrightarrow CH_3(CH_2)_8\text{-}\overset{O}{\overset{\|}{C}}\text{-}N \xrightarrow{\text{DIBAH}} CH_3(CH_2)_8\text{-}CHO$$

72%

JCS Perkin I, 2470 (1980)

$$\text{(CH}_2)_8\text{-}\overset{\text{O}}{\underset{\text{}}{\text{C}}}\text{-Cl} \xrightarrow[\text{DMF, THF, -70}^{\text{o}}]{\text{NaBH}_4\text{, pyridine}} \text{(CH}_2)_8\text{-CHO}$$

Synth Comm, _12_, 839 (1982)

$$\text{(CH}_2)_8\text{-}\overset{\text{O}}{\underset{\text{}}{\text{C}}}\text{-Cl} \xrightarrow[\text{EtOCH=CH}_2]{\begin{array}{l}\text{1) NaBH}_4\\ \text{2) EtCOOH, HCl}\end{array}} \text{(CH}_2)_8\text{-CHO} \quad \sim58\%$$

Tetr Lett, _22_, 11 (1981)

$$\text{CH}_3\text{(CH}_2)_5\text{-}\overset{\text{O}}{\underset{\text{}}{\text{C}}}\text{-Cl} \xrightarrow[\text{PdL}_4]{\text{Bu}_3\text{SnH}} \text{CH}_3\text{(CH}_2)_5\text{-CHO} \quad 81\%$$

JCS Chem Comm, 432 (1980)
JOC, _46_, 4439 (1981)

$$\text{CH}_3\text{(CH}_2)_7\text{-}\overset{\text{O}}{\underset{\text{}}{\text{C}}}\text{-Cl} \xrightarrow{\text{L}_2\text{CuBH}_4} \text{CH}_3\text{(CH}_2)_7\text{-CHO} \quad 76\%$$

JOC, _45_, 3449 (1980)

$$Ph-\overset{O}{\overset{\|}{C}}-TMS \xrightarrow[\text{H}_2\text{O/THF}]{\overset{+-}{Bu_4NF}} Ph-CHO \qquad 75\%$$

Tetr Lett, <u>22</u>, 1881 (1981)

$$Ph-CH_2-\overset{OH}{\overset{|}{CH}}-COOH \xrightarrow{\text{(N-I succinimide)}} Ph-CH_2-CHO \qquad 97\text{-}101\%$$

JOC, <u>47</u>, 3006 (1982)

$$Bu-\overset{OH}{\overset{|}{CH}}-COOH \xrightarrow[\text{CHCl}_3 \text{ or dioxane}]{Bu_4\overset{\oplus}{N} IO_4^{\ominus}} Bu-CHO \qquad 90\%$$

Synthesis, 563 (1980)
Tetr Lett, <u>21</u>, 2655 (1980)

$$CH_3(CH_2)_{16}\overset{O}{\underset{}{\overset{\|}{C}}}-COOH \xrightarrow[\text{2) } H_3O^{\oplus}]{\overset{\text{1) } \ce{[NH ring]} \text{ TsOH}}{\text{benzene}}} CH_3(CH_2)_{16}\overset{O}{\underset{}{\overset{\|}{C}}}-H \qquad 100\%$$

Tetr Lett, <u>23</u>, 459 (1982)

$$\xrightarrow[\text{CHCl}_3 \text{ or dioxane}]{Bu_4\overset{\oplus}{N} \ IO_4^{\ominus}} \qquad 70\%$$

Synthesis, 563 (1980)
Tetr Lett, <u>21</u>, 2655 (1980)

Section 48 <u>Aldehydes from Alcohols</u>

$$\xrightarrow[\text{(on Alumina)}]{\ce{N \cdot CrO_3 \cdot HCl}}$$

87%

Synthesis, 223 (1980)

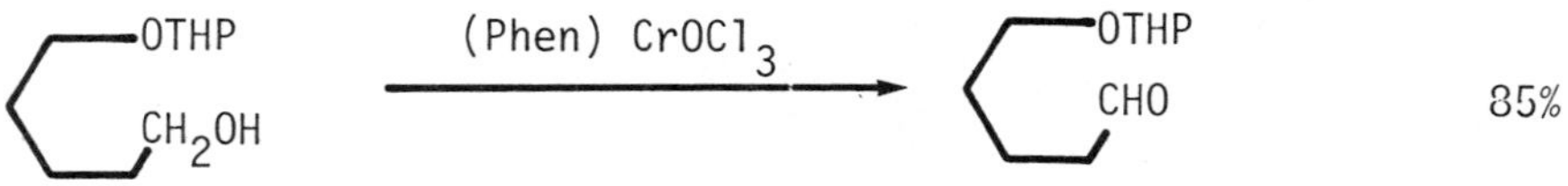

Tetr Lett, 21, 1583 (1980)

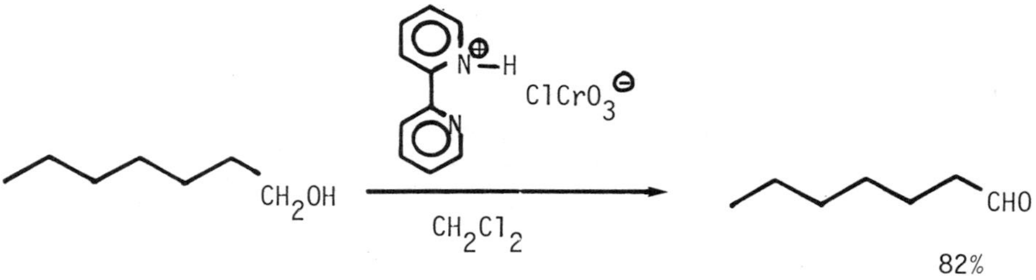

Synthesis, 691 (1980)

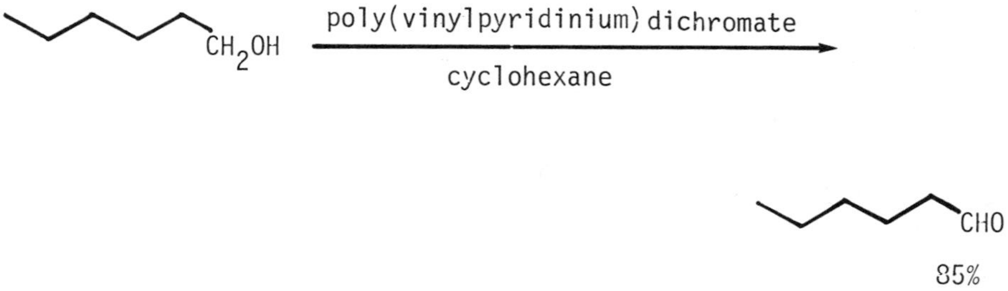

JOC, 46, 1728 (1981)

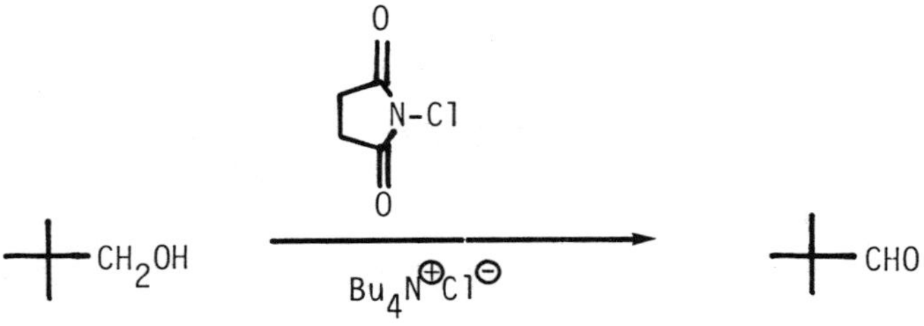

98%

Synthesis, 394 (1981)

$$\underline{n}\text{-}C_7H_{15}\text{-}CH_2OH \xrightarrow[\text{2) } Ph_3P]{\text{1) DEAD}} \underline{n}\text{-}C_7H_{15}CHO$$

62%

Tetr Lett, 22, 2295 (1981)

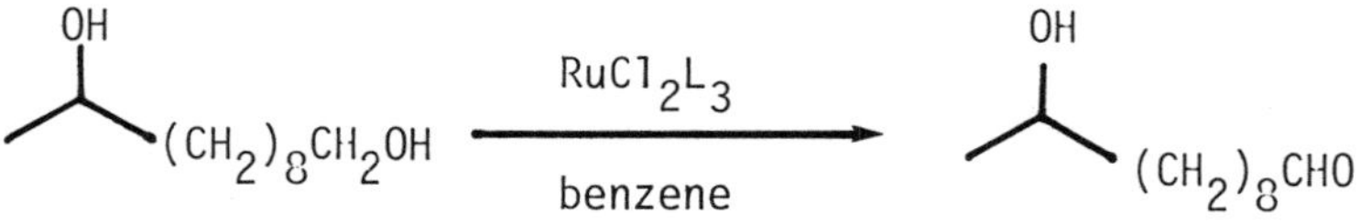

89%

Tetr Lett, 22, 1605 (1981)

$$\text{1-octanol} \xrightarrow[\text{PhIO or PhI(OAc)}_2]{RuCl_2L_3} \text{octanal}$$

97%

Tetr Lett, 22, 2361 (1981)

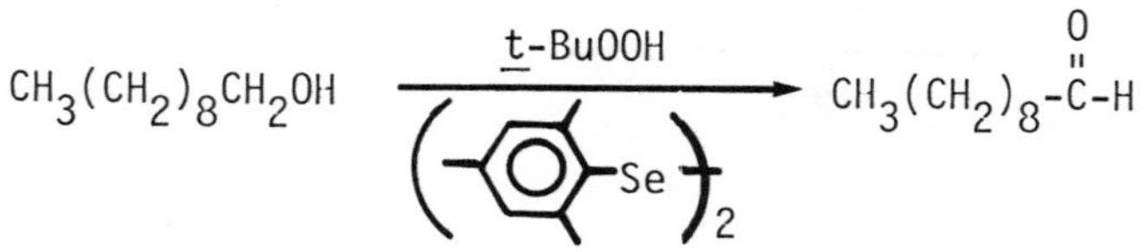

$$CH_3(CH_2)_8CH_2OH \xrightarrow[\left(\text{—}\underset{2}{\overset{}{\text{Se}}}\right)]{\underline{t}\text{-BuOOH}} CH_3(CH_2)_8\text{-}\overset{\overset{\text{O}}{\|}}{C}\text{-H} \qquad 92\%$$

JOC, <u>47</u>, 837 (1982)

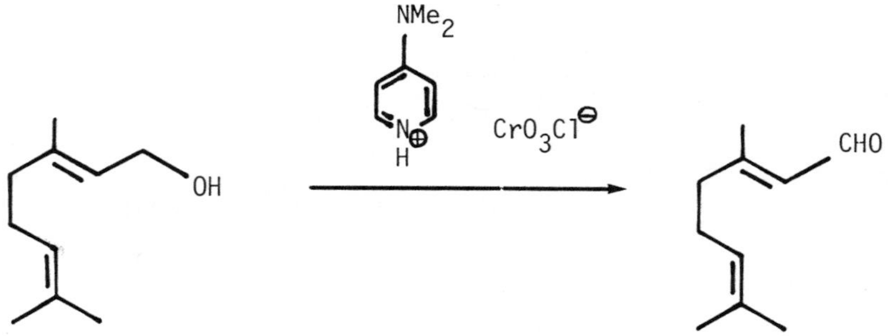

JOC, <u>47</u>, 1787 (1982)

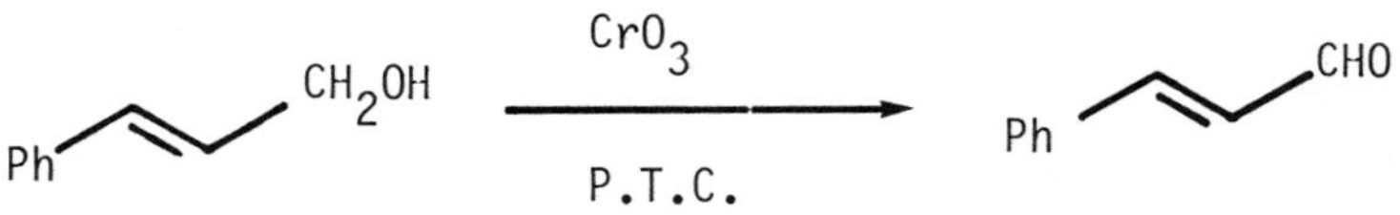

Tetr Lett, <u>21</u>, 4653 (1980)

$$Ph\text{-}CH=CH\text{-}CH_2OH \xrightarrow[\text{DMSO}]{K_2Cr_2O_7} Ph\text{-}CH=CH\text{-}CHO \quad 88\%$$

Synthesis, 646 (1980)

$$\underline{n}\text{-Hx}\diagdown C=CH\text{-}CH_2OH \xrightarrow[\text{CH}_2\text{Cl}_2]{(Bu_4N)_2Cr_2O_7} \underline{n}\text{-Hx}\diagdown C=CH\text{-}CHO \quad 78\%$$

Synth Comm, 10, 75 (1980)

JOC, 47, 1787 (1982)

$$Ph\text{-}CH_2OH \xrightarrow[\text{HMPT}]{[BzNEt_3]_2 \overset{\oplus}{} Cr_2O_7^{\ominus}} Ph\text{-}\overset{O}{\overset{\|}{C}}\text{-}H \quad 90\%$$

Synthesis, 1091 (1982)

CH$_2$OH / OMe / OMe → PCC* or PDC → CHO / OMe / OMe   80%

*PCC = pyridinium chlorochromate
PDC = pyridinium dichromate

JCS Perkin I, 1967 (1982)

$$\text{Ph-CH}_2\text{OH} \xrightarrow[\text{bentonite clay}]{\text{Fe(NO}_3)_3,\ \text{H}^+} \text{Ph-CHO}$$ 81%

Synthesis, 849 (1980)

OH / OMe / CH$_2$OH → DDQ / dioxane → OH / OMe / CHO   85%

JOC, **45**, 1596 (1980)

$$Ph\text{-}CH_2OH \xrightarrow[\text{DMSO}]{ClSO_2NCO} Ph\text{-}CHO \qquad 90\%$$

Synthesis, 141 (1980)

$$Ph\text{-}CH_2OH \xrightarrow[Br_2]{Ni(OBz)_2} Ph\text{-}CHO \qquad 92\%$$

Synth Comm, 10, 881 (1980)

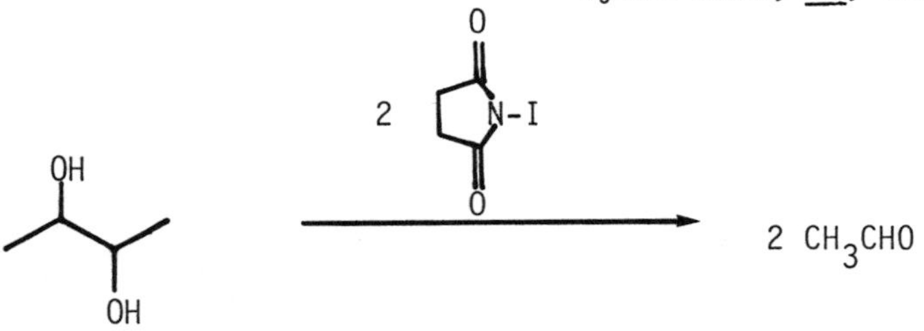
OH
OH
2
O
N-I
O
2 CH3CHO

83%

JOC, 46, 1927 (1981)

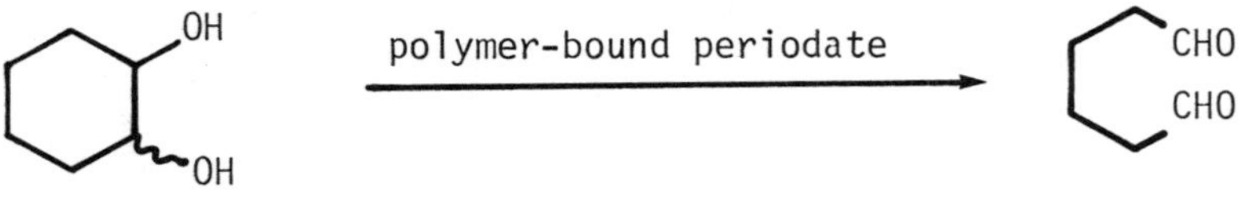
OH
OH
polymer-bound periodate
CHO
CHO

90%

JCS Perkin I, 509 (1982)

Ph-I

+

$\diagup\!\!\diagdown$—CH$_2$OH

$\xrightarrow[\text{Et}_3\text{N}]{\text{Pd(OAc)}_2}$

CHO   82%

Ph

Org Syn, <u>61</u>, 82 (1983)

Related methods: Ketones from Alcohols and Phenols (Section 168)

Section 49 <u>Aldehydes from Aldehydes</u>

Conjugate reductions and Michael alkylations of conjugated aldehydes are listed in Section 74 (Alkyls from Olefins).

CHO

Cl

1) Me-N   N-Li

2) BuLi

3) CH$_3$I
4) H$_3$O$^+$

CHO

CH$_3$

Cl   61%

Tetr Lett, <u>23</u>, 3979 (1982)

Synthesis, 677 (1980)

Related methods: Aldehydes from Ketones (Section 57).   Ketones from Ketones (Section 177).   Also via: Olefinic aldehydes (Section 341).

Section 50 Aldehydes from Alkyls

Org Prep Proc Int, 12, 201 (1980)

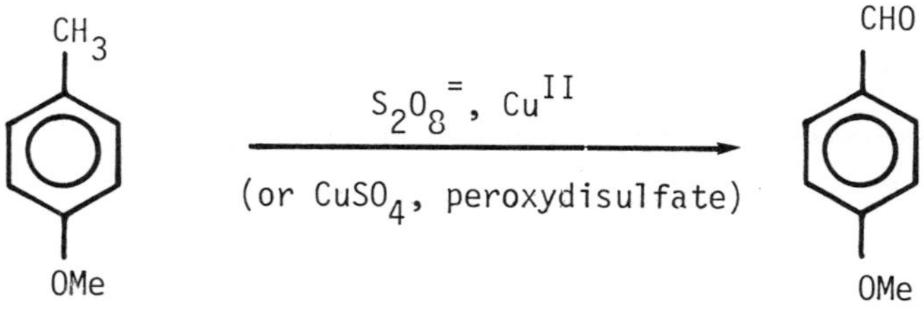

70%

Indian J Chem, 20B, 153 (1981)

Section 51 Aldehydes from Amides

No Additional Examples.

Section 52 Aldehydes from Amines

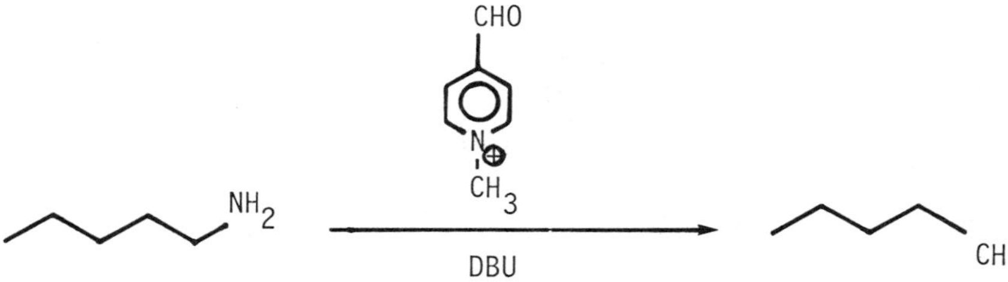

84%

JACS, 104, 4446 (1982)

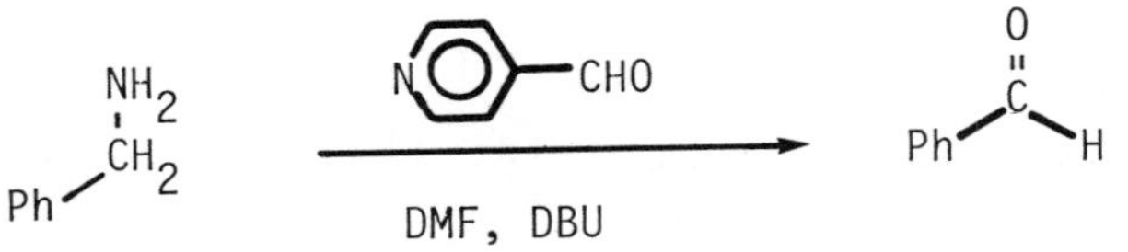

Synthesis, 756 (1982)

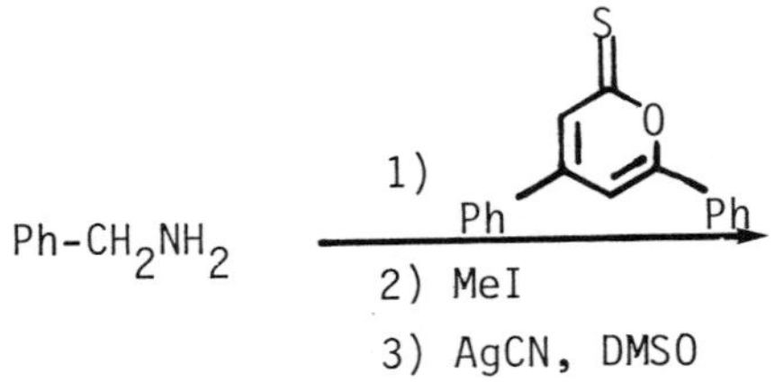

Synthesis, 711 (1981)

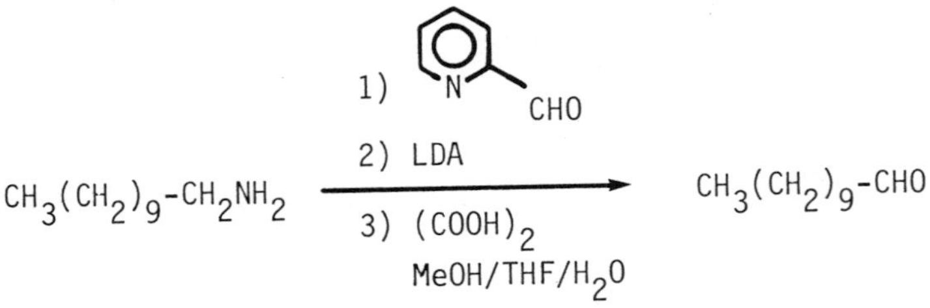

JOC, 46, 1937 (1981)

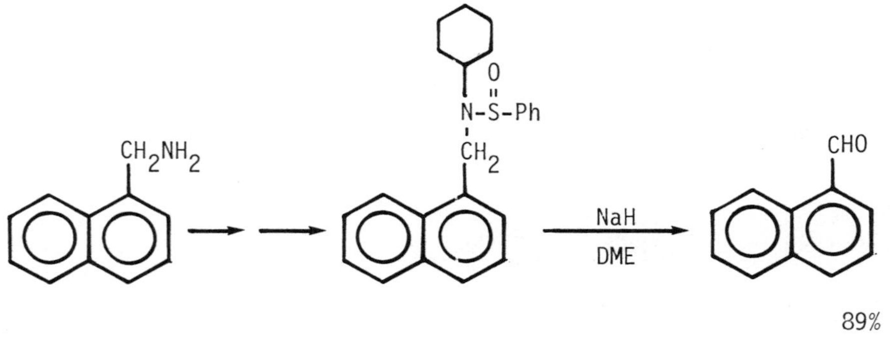

JOC, <u>46</u>, 4617 (1981)

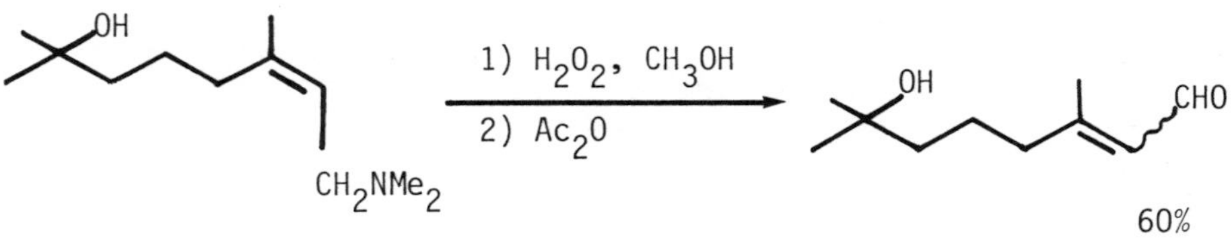

Chem Lett, 1987 (1982)

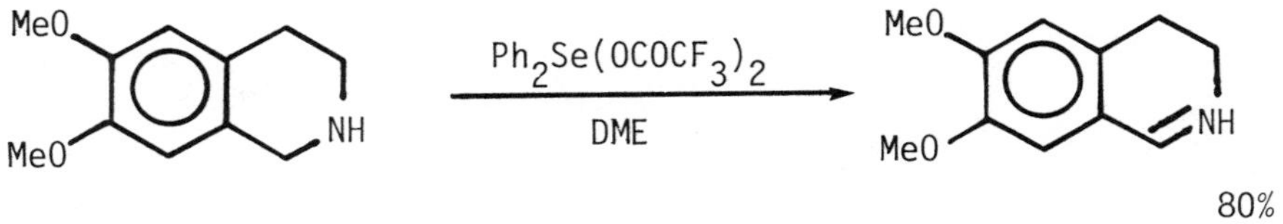

JACS, <u>103</u>, 4642 (1981)

Related methods: Section 172 (Ketones from Amines).

Section 53 <u>Aldehydes from Esters</u>

No Additional Examples.

Section 54 <u>Aldehydes from Ethers</u>

$$\text{Bu-CH}_2\text{-OBu} \quad \xrightarrow[\text{2) H}_3\text{O}^+]{\text{1)} \quad h\nu} \quad \text{Bu-CHO} \qquad \sim 30\%$$

Aust J Chem, <u>32</u>, 2787 (1979)

$$\xrightarrow[\text{DMF}]{\text{Pd(OAc)}_2} \quad \text{Pr} \diagup\!\!\!\diagdown\!\!\!\diagup \text{CHO} \qquad 98\%$$

Chem Lett, 1997 (1982)

Related methods: Section 174 (Ketones from Ethers and Epoxides).

Section 55 <u>Aldehydes from Halides</u>

$$\begin{array}{c} O \\ \parallel \\ H\text{-}C\text{-}OMgBr \end{array}$$

+

$\xrightarrow[\text{2) } H_2O]{}$ Hx-CHO   75%

Hx-MgBr

Tetr Lett, <u>21</u>, 2869 (1980)

$C_6H_{13}MgBr$ $\xrightarrow[\text{2) HOAc}]{\text{1) } Fe(CO)_5}$ $C_6H_{13}CHO$ 91%

Bull Chem Soc Japan, <u>55</u>, 1663 (1982)

1) [piperidine]N-CHO

2) $H_3O^{\oplus}$

→ cyclopentyl-CHO   72%

Angew Int Ed, <u>20</u>, 878 (1981)

$\underline{n}$-C$_9$H$_{19}$MgBr

1) [structure: benzene ring fused with S-C(+)-D dithiolium]

2) HgO, HBF$_4$
   THF/H$_2$O

$\underline{n}$-C$_9$H$_{19}$-$\overset{O}{\overset{\|}{C}}$-D   68%

Tetr Lett, $\underline{22}$, 1821 (1981)

$\underline{n}$-C$_8$H$_{17}$Cl

1) Me$_2$N—[pyridine ring]—N→O

2) DBU

$\underline{n}$-C$_7$H$_{15}$CHO   88%

Bull Chem Soc Japan, $\underline{54}$, 2221 (1981)

n-Hx-Br

1) PhSĊHSiMe$_3$
2) MCPBA, Δ
3) H$_3$O$^{\oplus}$

n-Hx-CHO   65%
overall

Tetr Lett, $\underline{21}$, 1677 (1980)

$$\text{Bu-CH}_2\text{Br} \xrightarrow{\begin{array}{l}\text{1) Ph}\overset{\overset{\displaystyle O}{\|}}{\diagdown}\text{COOMe}\\ \text{2) NaOH, MeOH}\\ \text{3) NaBH}_4\text{, EtOH}\\ \text{4) Pd(OAc)}_2\text{, I}_2\text{, benzene}\end{array}} \begin{array}{l}\text{Bu-CH}_2\text{CH}_2\text{CHO}\\ \text{(heptanal)}\\ \quad\sim 50\%\end{array}$$

Indian J Chem, <u>21B</u>, 408 (1982)

$$\begin{array}{l}\text{1) (EtOCH=CH})_3\text{-B}\\ \quad\text{NaOH, PdL}_4\\ \text{2) H}_3\text{O}^+\end{array}$$

96%

JOC, <u>47</u>, 2117 (1982)

$$\underline{n}\text{-C}_{10}\text{H}_{21}\text{CH=CHF} \xrightarrow[\text{H}_2\text{O, NaHCO}_3]{\text{Hg(OAc)}_2\text{, TFA}} \underline{n}\text{-C}_{10}\text{H}_{21}\text{CH}_2\text{CHO} \qquad 95\%$$

Chem Lett, 651 (1980)

Bu-CBr$_2$Li

+ $\xrightarrow{\hspace{3cm}}$ Bu-CBr$_2$-CHO   78%

                    2) H$_3$O$^{\oplus}$

O
‖
H-C-OMe   may also be used to form α-monohalo aldehydes

Synthesis, 644 (1980)

Ph-I

+ $\xrightarrow[\text{Et}_3\text{N}]{\text{Pd(OAc)}_2}$

CH$_2$OH   CHO   82%

                                  Ph

Org Syn, <u>61</u>, 82 (1983)

Section 56 <u>Aldehydes from Hydrides</u>

$\xrightarrow[\text{AlCl}_3]{\overset{\text{OH}}{\underset{}{\text{Me}_2\text{C-CN}}}}$ CHO   50%

Synth Comm, <u>12</u>, 485 (1982)

Review: "The Reimer-Tiemann Reaction:

Org React, 28, 1 (1982)

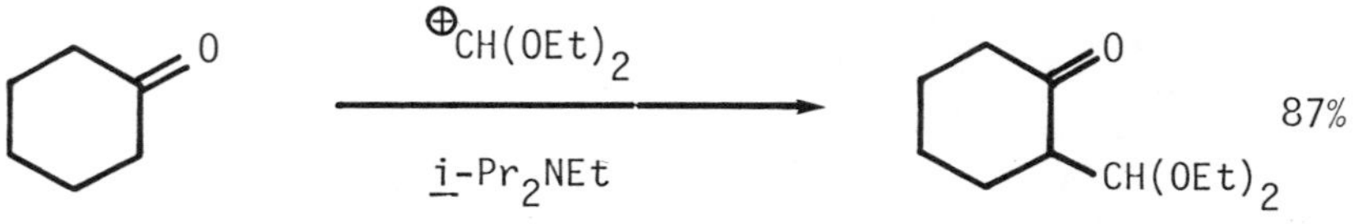
O
⊕CH(OEt)2
i-Pr2NEt
O
CH(OEt)2
87%

JOC, 46, 2557 (1981)

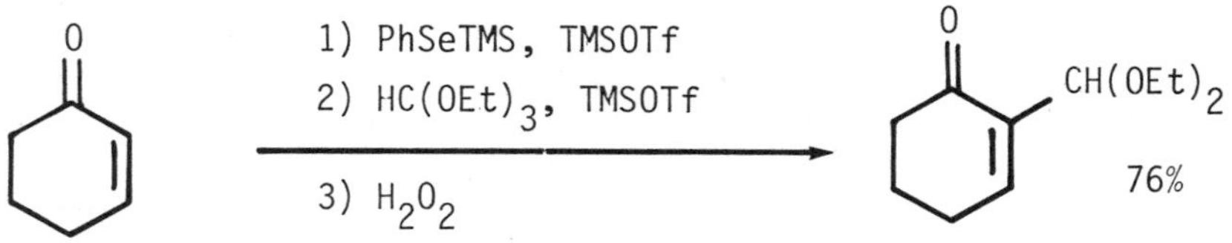
O
1) PhSeTMS, TMSOTf
2) HC(OEt)3, TMSOTf
3) H2O2
O
CH(OEt)2
76%

Tetr Lett, 22, 1809 (1981)

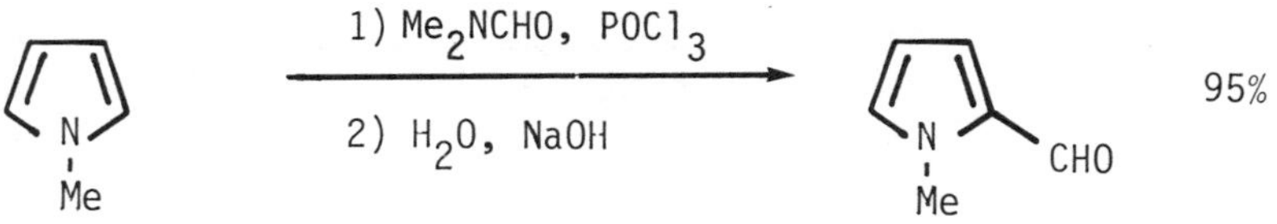
N
Me
1) Me2NCHO, POCl3
2) H2O, NaOH
N
Me
CHO
95%

Org Prep Proc Int, 13, 97 (1981)

Section 57 <u>Aldehydes from Ketones</u>

JOC, <u>45</u>, 1091 (1980)

Tetr Lett, <u>21</u>, 3535 (1980)

Helv Chim Acta, <u>63</u>, 1665 (1980)

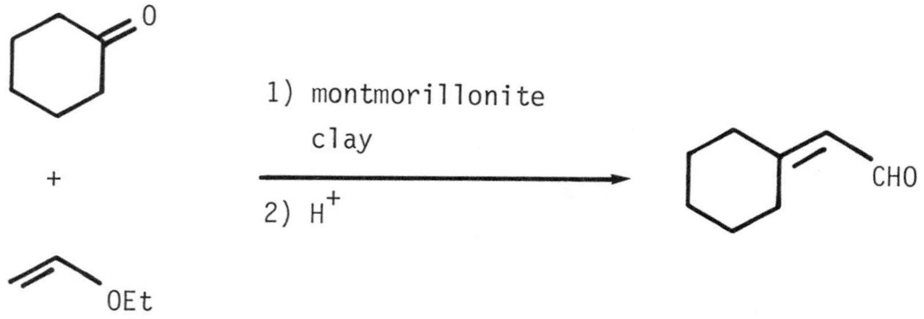

1) montmorillonite
   clay

2) H$^+$

50%

Synthesis, 137 (1981)

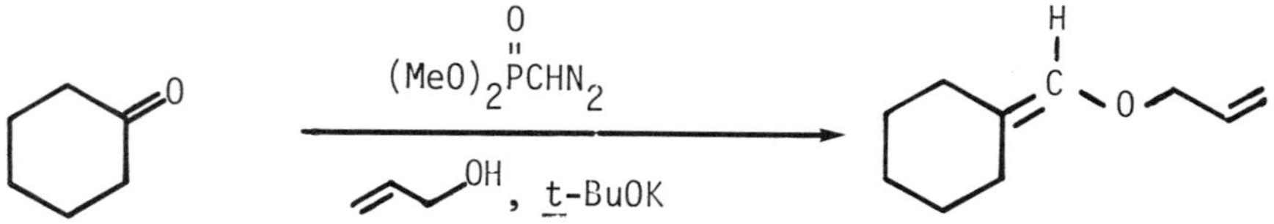

(MeO)$_2$$\overset{\overset{O}{\|}}{P}CHN_2$

$\diagup\!\!\!\diagdown$OH, $\underline{t}$-BuOK

60%

Tetr Lett, 21, 5003 (1980)

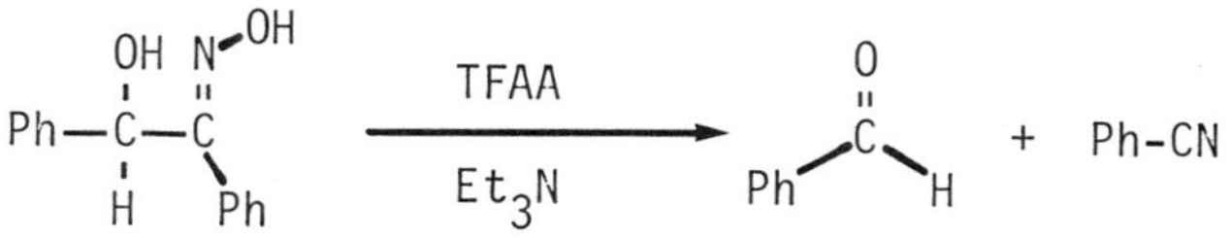

Ph—$\overset{OH}{\underset{H}{C}}$—$\overset{N{\sim}OH}{\underset{Ph}{C}}$ TFAA

Et$_3$N

Ph$\overset{\overset{O}{\|}}{C}$H   +   Ph-CN   78%

Synthesis, 45 (1980)

Section 58 <u>Aldehydes from Nitriles</u>

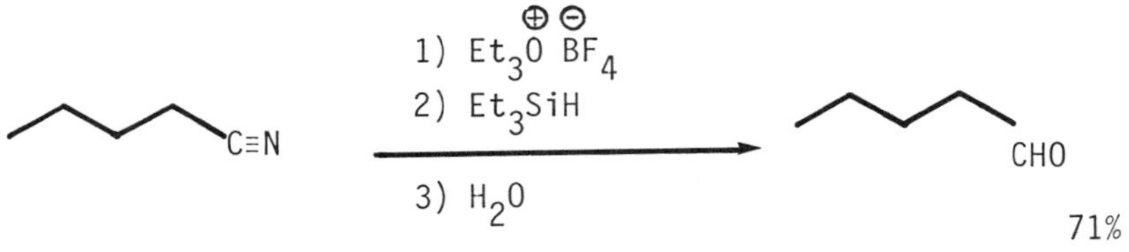

1) $Et_3\overset{\oplus}{O} \overset{\ominus}{BF_4}$
2) Et_3SiH
3) H_2O

71%

JOC, <u>46</u>, 602 (1981)

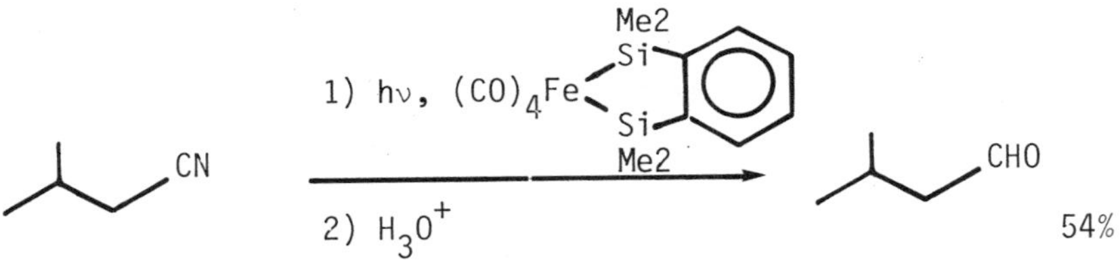

1) $h\nu$, $(CO)_4Fe$⟨Si(Me2)C₆H₄Si(Me2)⟩

2) H_3O^+

54%

JOC, <u>46</u>, 3372 (1981)

Bu-CH₂CN

1) DIBAL-H
2) LDA, HMPA
3) $C_5H_{11}Br$

Bu-CH-CHO
   |
   C_5H_{11}

85%

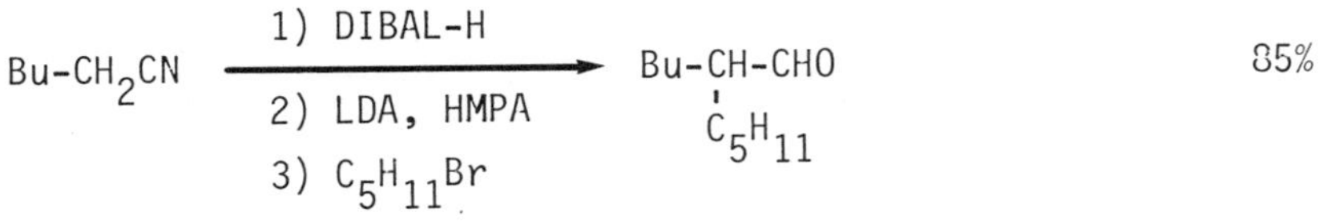

JOC, <u>46</u>, 5250 (1981)

Section 59 Aldehydes from Olefins

1) Sia$_2$BH

2) PCC

CHO   67%

Synthesis, 151 (1980)

Ph

Ph

(BiPy)H$_2$CrOCl$_5$

2 Ph-CHO   96%

Org Prop Proc Int, 14, 362 (1982)

Review: "Ozonolysis -- A Modern Method in the Chemistry of
          Olefins"

                                Russ Chem Rev, 50, 636 (1981)

Review: "Asymmetric Hydroformylation"

                                Topics in Current Chem, 105,77 (1982)

Related methods: Section 179 (Ketones from Olefins)

Section 60 <u>Aldehydes from Miscellaneous Compounds</u>

$$CH_3(CH_2)_8-CH_2NO_2 \xrightarrow[\text{2) KMnO}_4]{\text{1) NaH}} CH_3(CH_2)_8-\overset{\overset{\displaystyle O}{\|}}{C}-H \qquad 92\%$$

JOC, <u>47</u>, 4534 (1982)

$$CH_3(CH_2)_4-CH_2NO_2 \xrightarrow[\text{MeOH/H}_2O]{H_2O_2, \ K_2CO_3} CH_3(CH_2)_4-CHO \qquad 80\%$$

Synthesis, 44 and 662 (1980)

1) NaH, <u>t</u>-BuOH
2) KMnO$_4$
3) Na$_2$S$_2$O$_5$, H$_2$SO$_4$

81%

JOC, <u>47</u>, 4534 (1982)

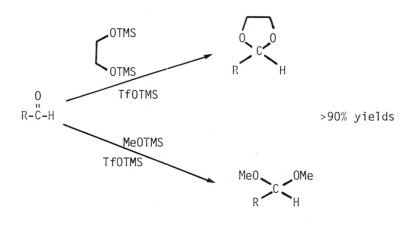

JOC, <u>46</u>, 1037 (1981)

Section 60A <u>Protection of Aldehydes</u>

>90% yields

Tetr Lett, <u>21</u>, 1357 (1980)

$$R\text{-CHO} \xrightarrow[\text{Nafion-H}]{\text{Ac}_2\text{O}} \begin{array}{c} R \\ \diagdown \\ H \end{array} C \begin{array}{c} OAc \\ \diagup \\ OAc \end{array}$$

50-99%

R = alkyl, aryl, heterocyclic

Synthesis, 962 (1982)

$$\underset{R}{\overset{O}{\underset{\|}{C}}}\underset{H}{} \xrightarrow[\text{AlCl}_3]{2R''SH} \begin{array}{c} R''S \\ \diagdown \\ R \end{array} C \begin{array}{c} SR'' \\ \diagup \\ H \end{array}$$

∿100%

R" = alkyl, dithiol

Tetr Lett, <u>21</u>, 4225 (1980)

$$\begin{array}{c} S \diagdown \diagup S \\ C \\ R \diagup \diagdown H \end{array} \xrightarrow{(\text{PhSeO})_2\text{O}} \underset{R}{\overset{O}{\underset{\|}{C}}}\underset{H}{}$$

JCS Perkin I, 1654 (1980)

$$\begin{array}{c} S \diagdown \diagup S \\ C \\ R \diagup \diagdown H \end{array} \quad \xrightarrow[\text{DMSO}]{\text{Me}_3\text{SiI(or Br)}} \quad R-\overset{\overset{O}{\|}}{C}-H \qquad 65\text{-}99\%$$

Synthesis, 965 (1982)

$$\begin{array}{c} S \diagdown \diagup S \\ C \\ R \diagup \diagdown H \end{array} \quad \xrightarrow[\underset{\oplus}{\text{or MeS(SMe)}_2 \ {}^{\ominus}\text{SbCl}_6}]{\text{HCl, H}_2\text{O/DMSO/dioxane}} \quad R \overset{\overset{O}{\|}}{C} H \qquad >95\%$$

Synthesis, 679 (1982)

$$\begin{array}{c} S \diagdown \diagup S \\ C \\ R \diagup \diagdown H \end{array} \quad \xrightarrow[\text{(p-Tol)}_3\text{N}]{\text{electrochemical oxidation}} \quad R \overset{\overset{O}{\|}}{C} H \qquad 85\text{-}97\%$$

Tetr Lett, 21, 511 (1980)

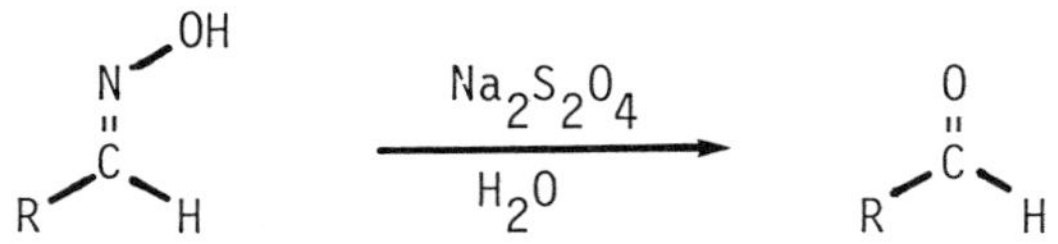

Aust J Chem, $\underline{32}$, 201 (1979)

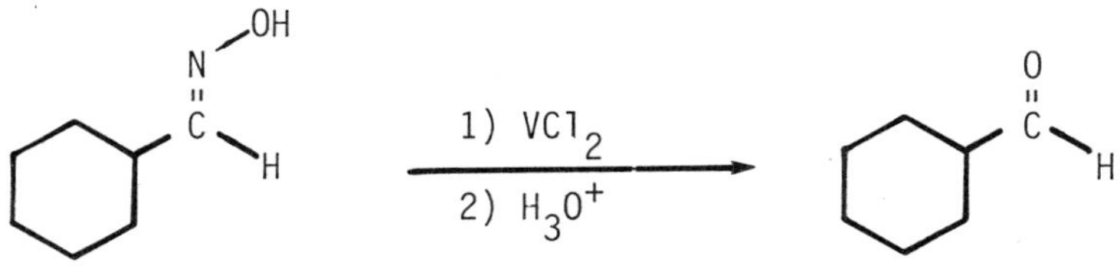

77%

Synthesis, 220 (1980)

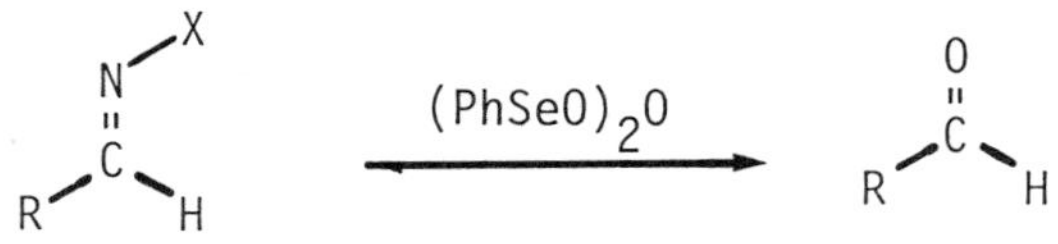

$\sim$70-90%

X = OH, NHPh

JCS Perkin I, 1212 (1980)

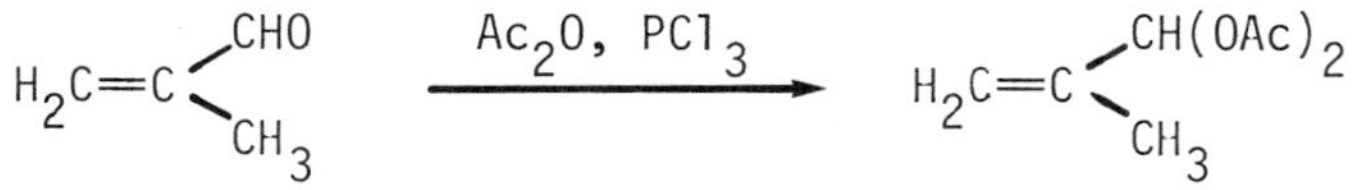

Synthesis, 824 (1981)

Stable to R'Li and R'MgX. Deprotection occurs during the normal workup.

Tetr Lett, <u>22</u>, 4213 (1981)

See Section 367 (Ether - Olefin) for the formation of enol ethers. Many of the methods in Section 180A (Protection of Ketones) are also applicable to aldehydes.

CHAPTER 5

PREPARATION OF ALKYLS,
METHYLENES, AND ARYLS

This chapter lists the conversion of functional groups into Me, Et..., CH$_2$, Ph, etc.

Section 61 Aryls from Acetylenes

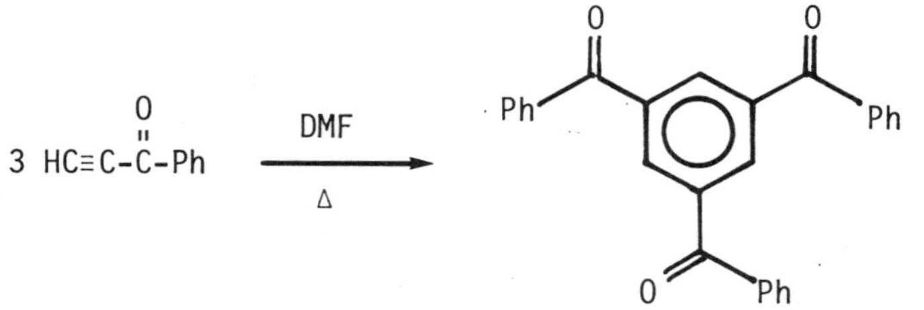

Synthesis, 29 (1980)

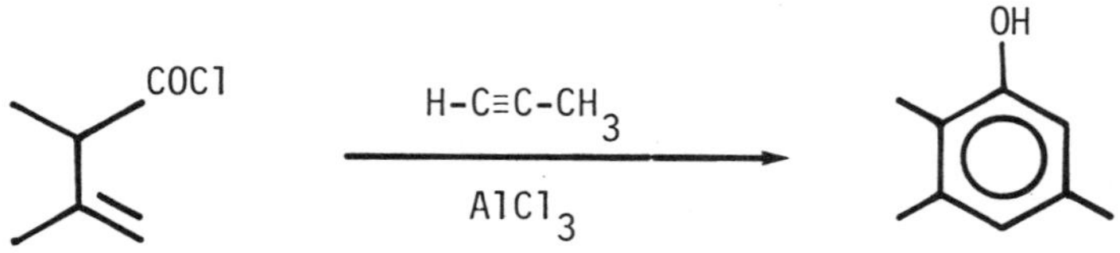

Tetr Lett, 23, 4923 (1982)

Section 62 <u>Alkyls and Methylenes from Carboxylic Acids</u>

1) Na, NH$_3$
2) C$_7$H$_{15}$Br
3) H$_3$O$^+$

~50%

Org Syn, <u>61</u>, 59 (1983)

$$2 \ MeO\text{-}\overset{O}{\overset{\|}{C}}\text{-}(CH_2)_4COOH \quad \xrightarrow[\text{MeOH, NaOMe}]{\text{anodic oxidation}} \quad MeO\text{-}\overset{O}{\overset{\|}{C}}\text{-}(CH_2)_8\text{-}\overset{O}{\overset{\|}{C}}\text{-}OMe \qquad 71\%$$

Org Syn, <u>60</u>, 1 (1981)

Section 63 <u>Alkyls from Alcohols</u>

OH
Ph-CH-Ph

SiMe$_3$, BF$_3$

Ph-CH-Ph

100%

JOC, <u>47</u>, 2125 (1982)

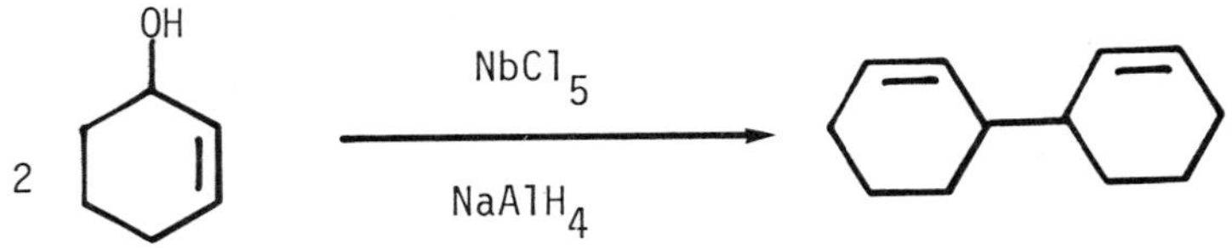

75%

Chem Lett, 157 (1982)

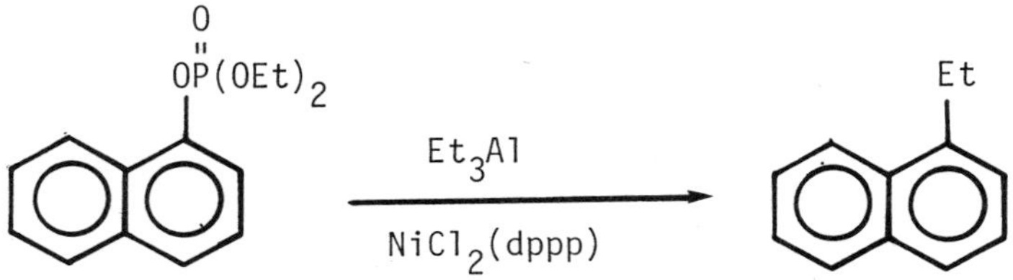

95%

Tetr Lett, 22, 4449 (1981)

Section 64 Alkyls from Aldehydes

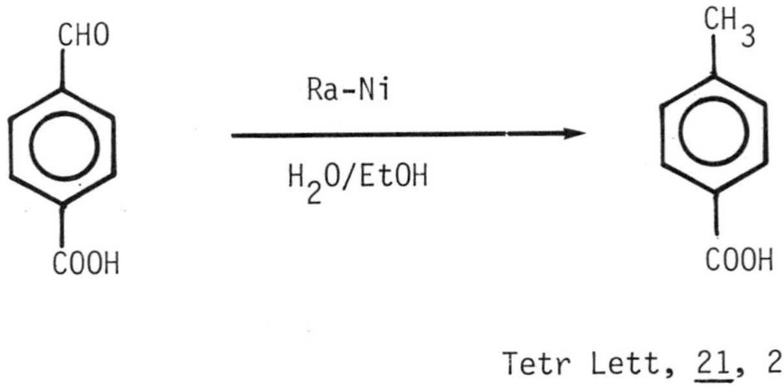

89%

Tetr Lett, 21, 2637 (1980)

Related methods: Alkyls, Methylenes, and Aryls from Ketones
(Section 72)

Section 65 Alkyls and Aryls from Alkyls and Aryls

JACS, 102, 6519 (1980)

Section 66 Alkyls, Methylenes, and Aryls from Amides

No Additional Examples.

Section 67 Alkyls, Methylenes, and Aryls from Amines

No Additional Examples.

Section 68 Alkyls, Methylenes, and Aryls from Esters

Tetr Lett, <u>21</u>, 3237 (1980)

JCS Chem Comm, 30 (1981)

Tetr Lett, <u>22</u>, 5097 (1981)

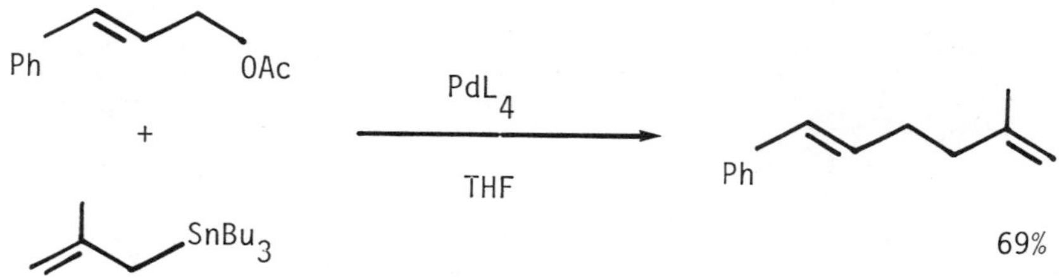

Tetr Lett, <u>21</u>, 2591, 2595, and 2599

(1980)

Section 69 <u>Alkyls and Aryls from Ethers</u>

The conversion ROR → RR' (R' = alkyl, aryl) is included in this
section.

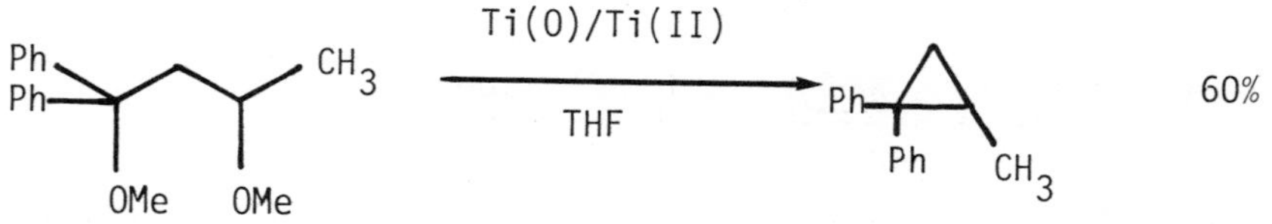

JACS, <u>102</u>, 426 (1980)

97%

Synthesis, 143 (1982)

$$\text{t-BuO}\overset{O}{\overset{\|}{C}}\text{-CH}_2\text{MgCl}$$

+

$$\text{BuO-CH-NMe}_2$$
$$\text{Ph}$$

$$\longrightarrow \quad \text{t-BuO}\overset{O}{\overset{\|}{C}}\text{-CH}_2\text{-CH-NMe}_2 \quad 76\%$$
$$\text{Ph}$$

Bull Soc Chim France II, 395 (1982)

Section 70 Alkyls and Aryls from Halides

The replacement of halogen by alkyl or aryl groups is included in this section. For the conversion RX → RH (X = halo) see Section 160 (Hydrides from Halides and Sulfonates).

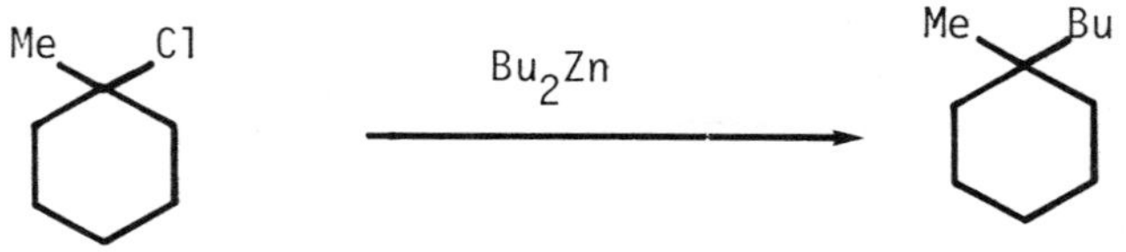

$$Bu_2Zn$$

70%

JCS Chem Comm, 1202 (1980)

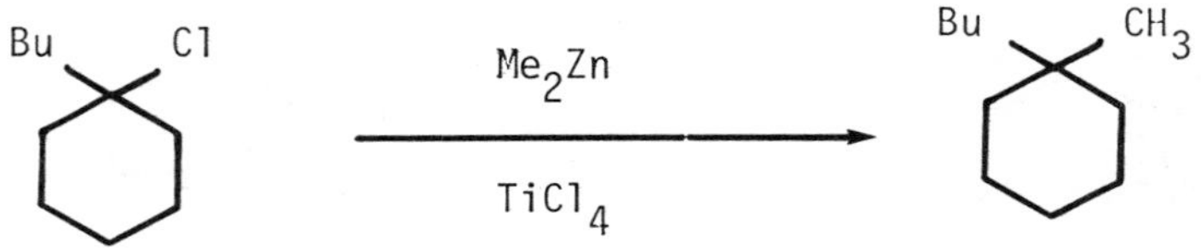

$$Me_2Zn$$

$$TiCl_4$$

95%

Angew Chem Int Ed, 19, 900 and 901 (1980)

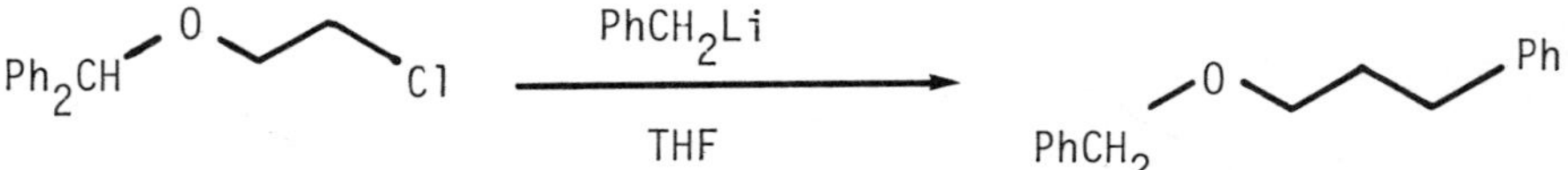

$$PhCH_2Li$$

THF

80%

JOC, 47, 3008 (1982)

1) LiCH$_2$TePh

2) SO$_2$Cl$_2$, Br$_2$, or I$_2$

R-X $\xrightarrow{\hspace{3cm}}$ R-CH$_2$X   59-95%

3) 100°, DMF

R = 1° alkyl

Chem Lett, 1081 (1982)

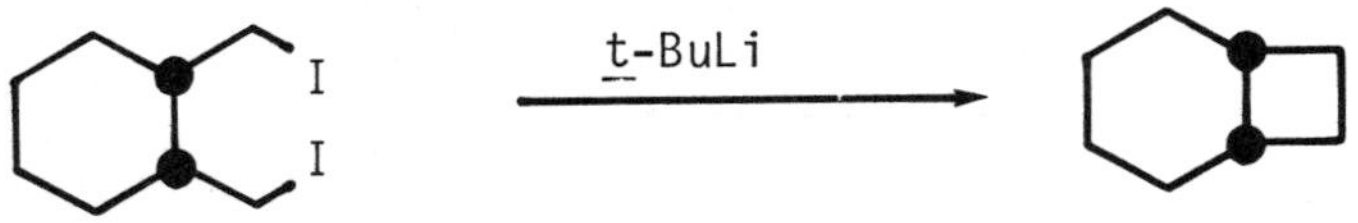

t-BuLi

97%

Tetr Lett, 23, 5123 (1982)

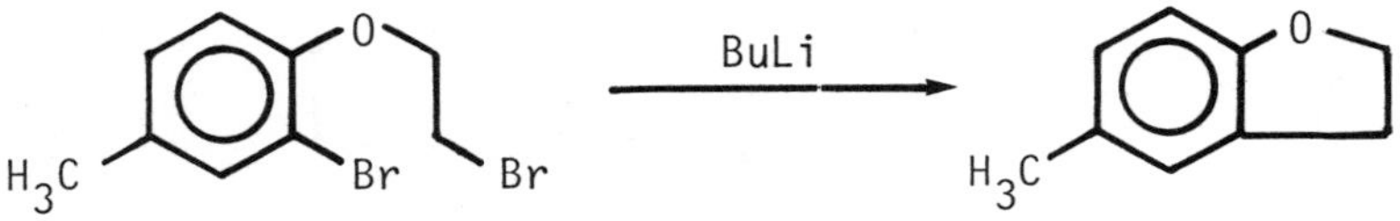

BuLi

73%

JOC, 46, 1384 (1981)

$\dfrac{CoClL_3}{benzene}$

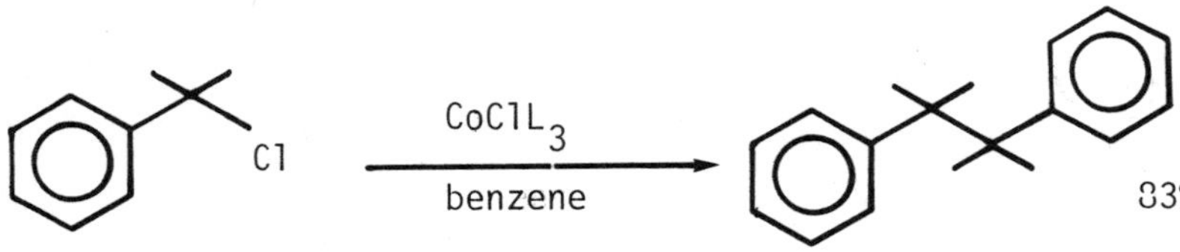

83%

(Also works with 1° and 2° benzyl chlorides.)

Chem Lett, 1277 (1981)

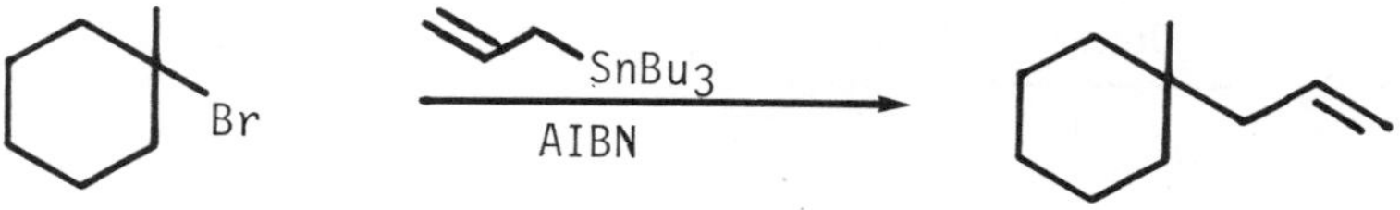

JACS, <u>104</u>, 5829 (1982)

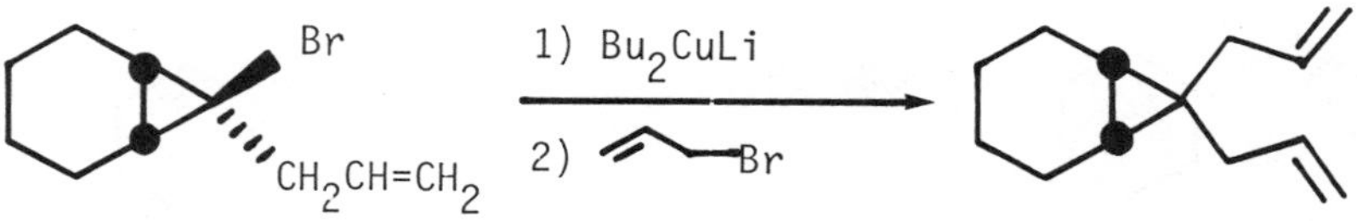

Bull Chem Soc Japan, <u>52</u>, 3632 (1979)

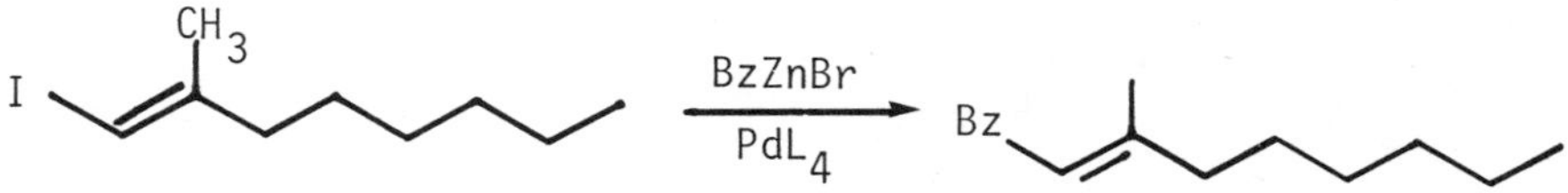

96%

78%

Tetr Lett, <u>22</u>, 2715 (1981)

Hx-CH=CF$_2$ $\xrightarrow[\text{Et}_2\text{O}]{\text{BuLi}}$ HxCH=CHBu   76%

Chem Lett, 935 (1980)

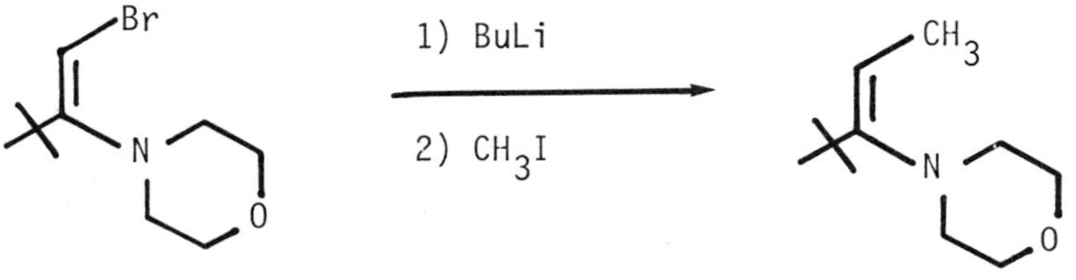

$\xrightarrow[\text{2) CH}_3\text{I}]{\text{1) BuLi}}$ 75%

Bull Soc Chim France II, 297 (1982)

2Ph-Br $\xrightarrow[\text{DMF}]{\text{Ni(COD)}_2}$ Ph-Ph   82%

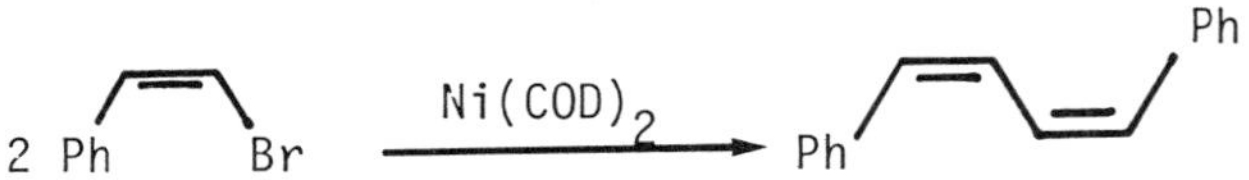

$\xrightarrow{\text{Ni(COD)}_2}$ 70%

JACS, <u>103</u>, 6460 (1981)

Tetr Lett, $\underline{23}$, 2469 (1982)

JCS Chem Comm, 434 (1980)

Organometallics, $\underline{1}$, 259 (1982)

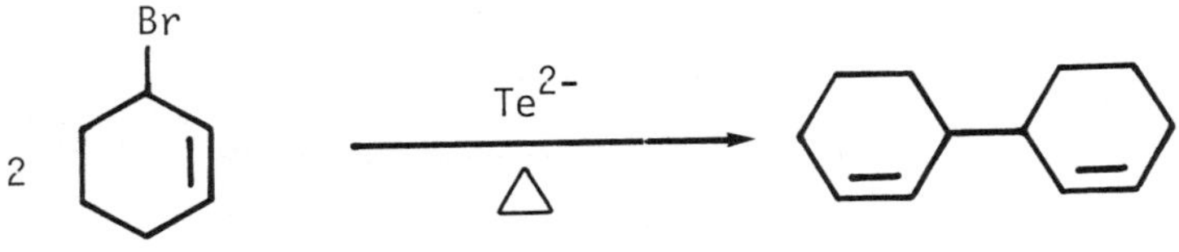

JOC, <u>47</u>, 1641 (1982)

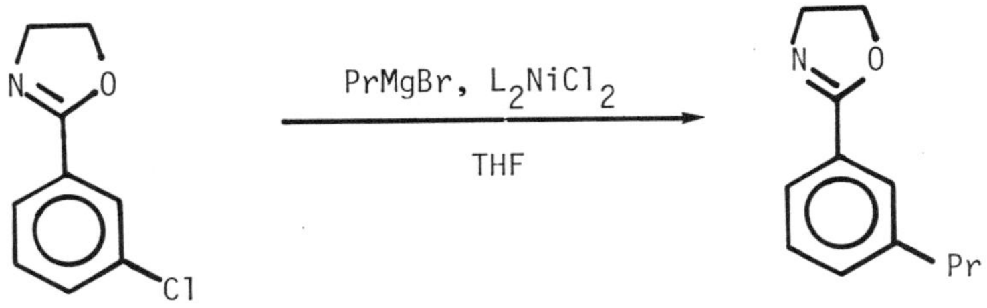

JOC, <u>47</u>, 4319 (1982)

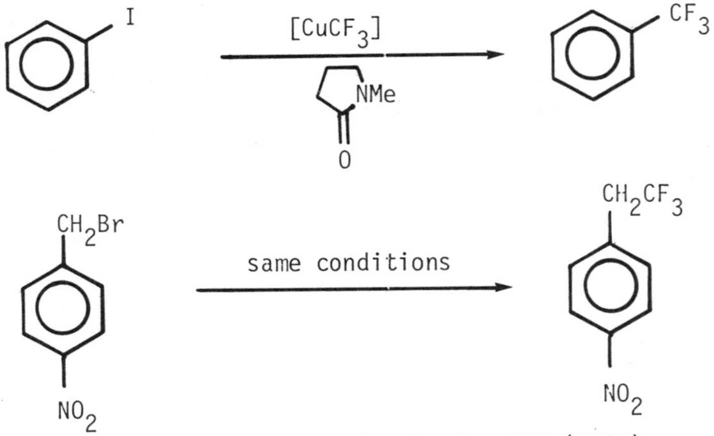

Synthesis, 932 (1980)

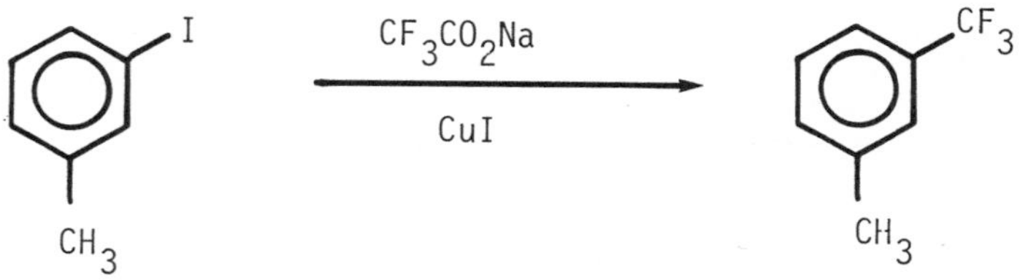

Chem Lett, 1719 (1981)

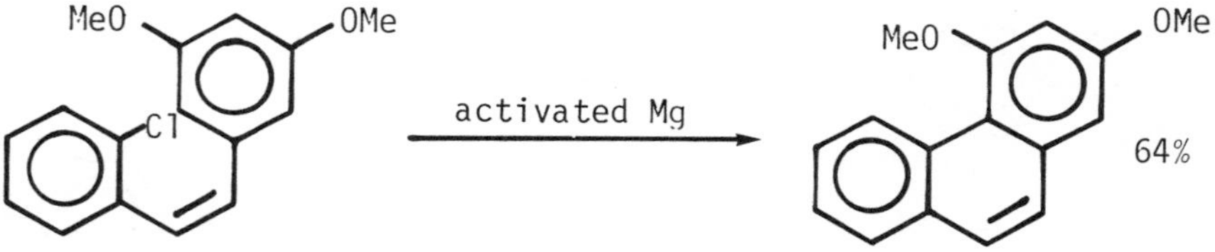

JCS Perkin I, 3007 (1982)

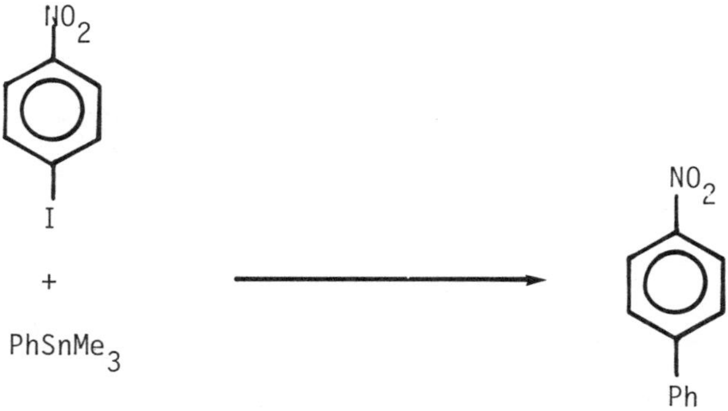

JOC (USSR), <u>17</u>, 18 (1981)

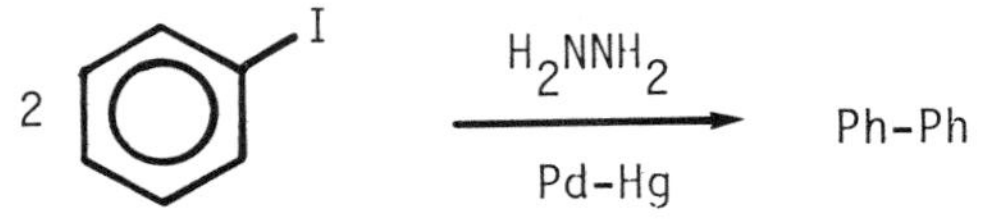

Bull Chem Soc Japan, 53, 1767 (1980)

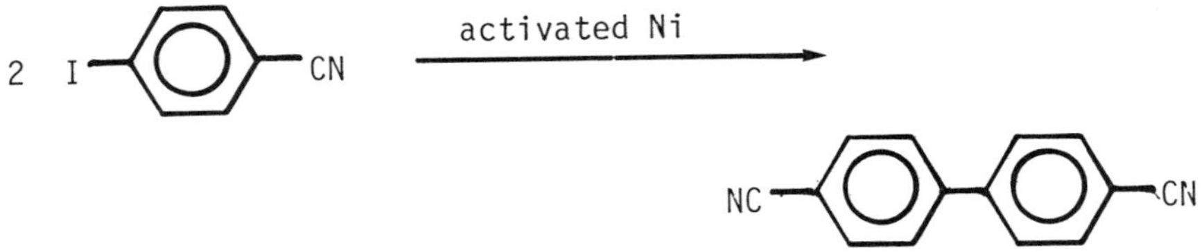

85%

Tetr Lett, 23, 4215 (1982)

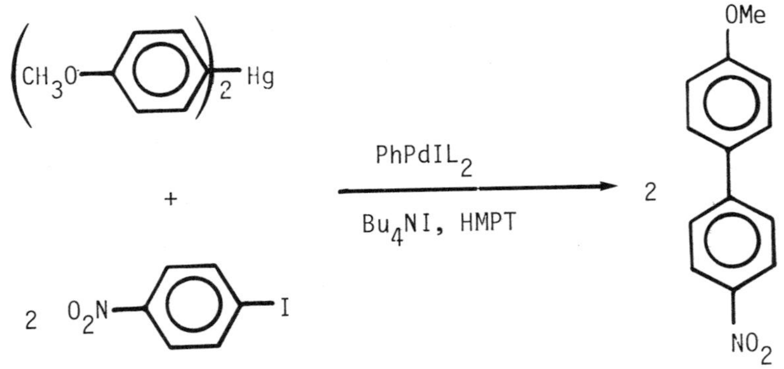

97%

Bull Acad USSR Chem, 30, 1993 (1982)

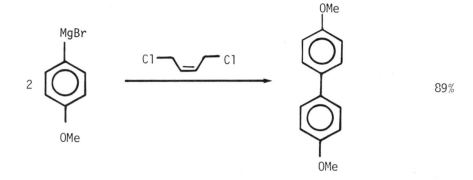

JOC, <u>46</u>, 2194 (1981)

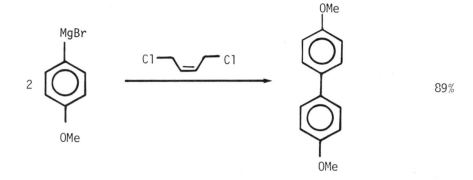

Tetr Lett, <u>23</u>, 3115 (1982)

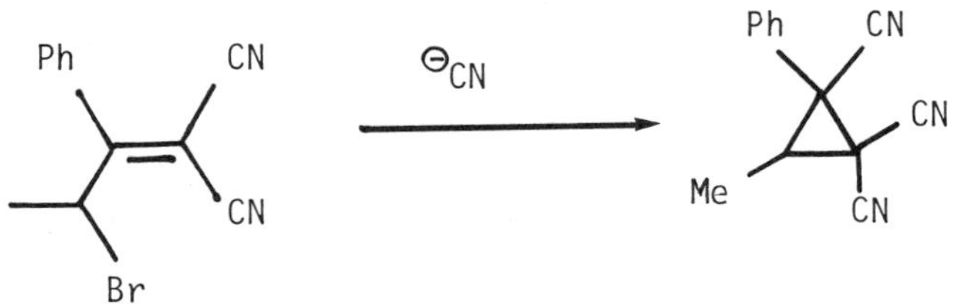

92%

Acta Chem Scand B, 34, 289 (1980)

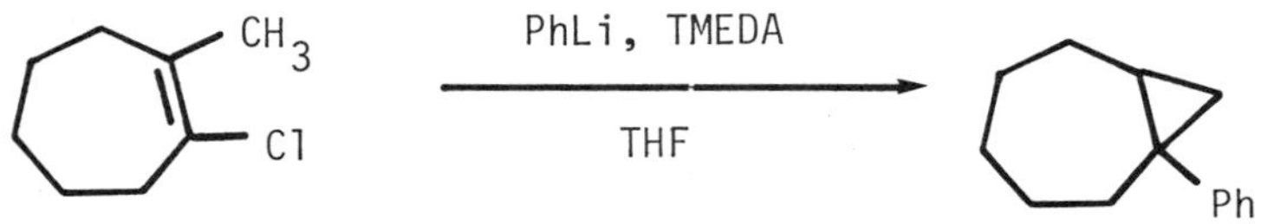

66%

JACS, 102, 6519 (1980)

Review: "The synthesis of substituted cyclopropanes and cyclo-
         propenes by the reductive cyclization of polychloro-
         alkanes."

Russ Chem Rev, 51, 368 (1982)

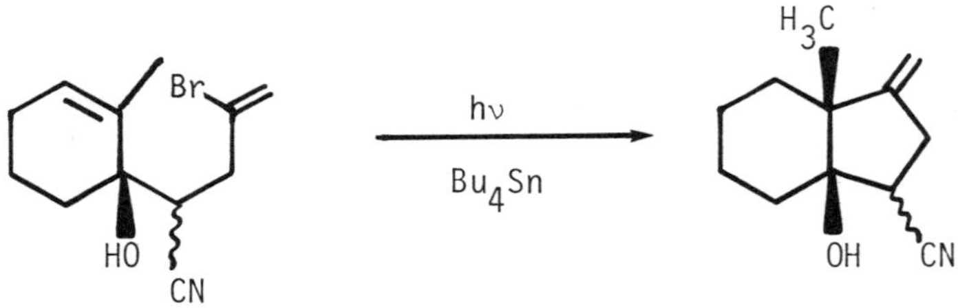

70%

JACS, 104, 2321 (1982)

Section 71 <u>Alkyls and Aryls from Hydrides</u>

This section lists examples of the reaction RH → RR' (R, R' =
alkyl or aryl). For the reaction C=CH →C=CR (R = alkyl or aryl)
see Section 209 (Olefins from Olefins). For alkylations of ketones
and esters, see Section 177 (Ketones from Ketones) and Section 113
(Esters from Esters).

$$Ph-NH_2, \underline{t}-BuONO$$
$$Pd(dba)_2$$

78%

JOC, <u>46</u>, 4885 (1981)

Pb(OAc)$_3$

DMSO

71%

Tetr Lett, <u>22</u>, 783 (1981)

2 Ar-H $\xrightarrow[\text{TFA}]{\text{Tl(TFA)}_3}$ Ar-Ar

Ar = subst. Ph   widely varying yields

JACS, <u>102</u>, 6504 (1980)

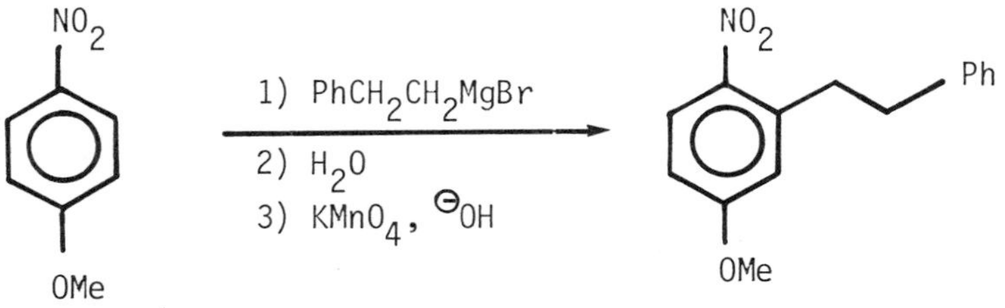

NO2
1) PhCH2CH2MgBr
2) H2O
3) KMnO4, ⊖OH
OMe
NO2
Ph
OMe

62%

JOC, 45, 522 (1980)

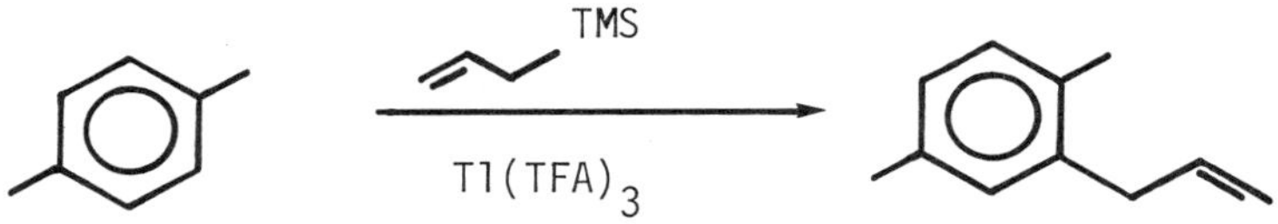

TMS
Tl(TFA)3

67%

Tetr Lett, 22, 4491 (1981)

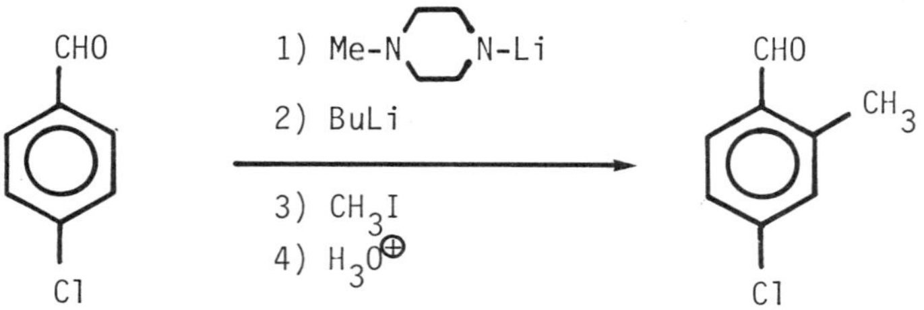

CHO
Cl
1) Me-N N-Li
2) BuLi
3) CH3I
4) H3O⊕
CHO
CH3
Cl

61%

Tetr Lett, 23, 3979 (1982)

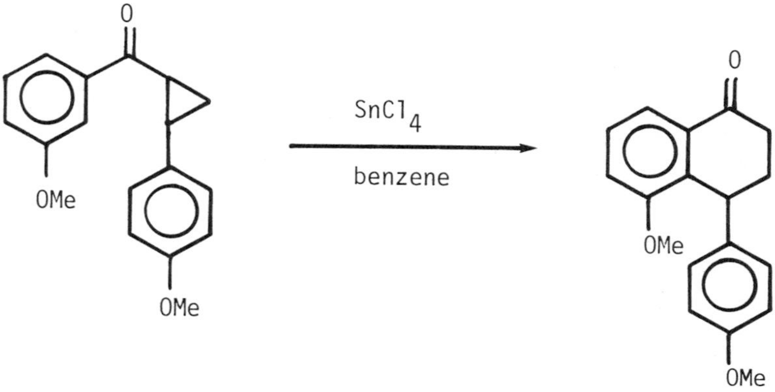
O
SnCl4
benzene
OMe
OMe
O
OMe
OMe

80%

Tetr Lett, 21, 1887 (1980)

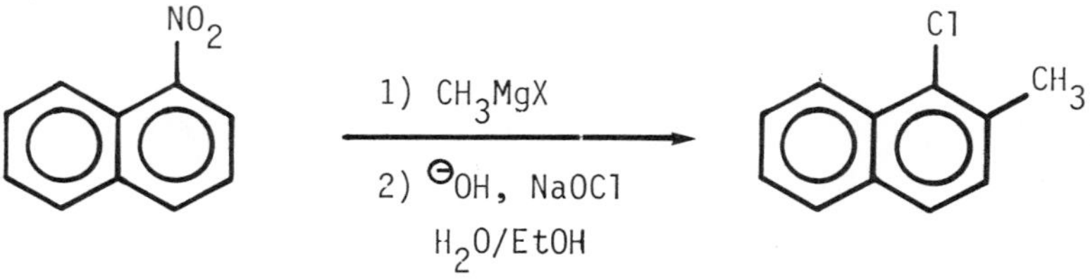
NO2
1) CH3MgX
2) ⊖OH, NaOCl
H2O/EtOH
Cl
CH3

60%

Synthesis, 616 (1980)

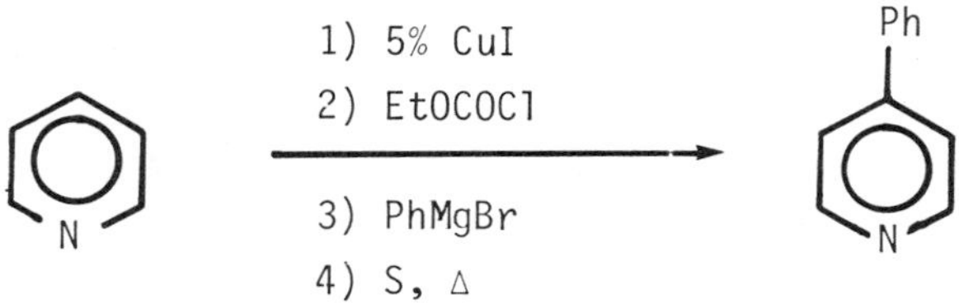
1) 5% CuI
2) EtOCOCl
3) PhMgBr
4) S, Δ
N
Ph
N

77%

JOC, 47, 4315 (1982)

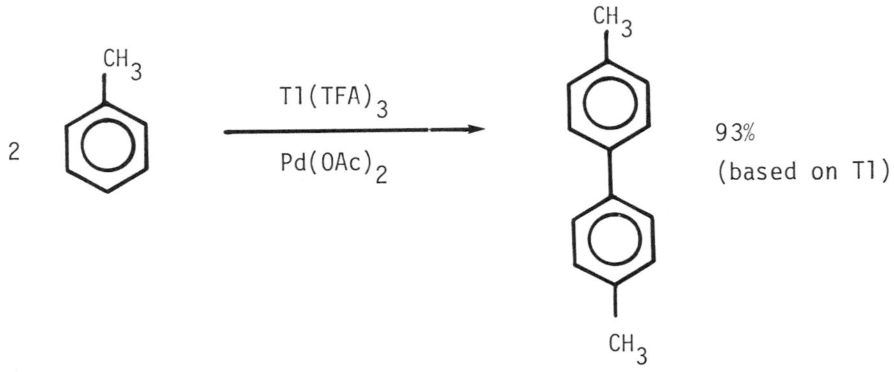

Tetr Lett, 22, 3793 (1981)

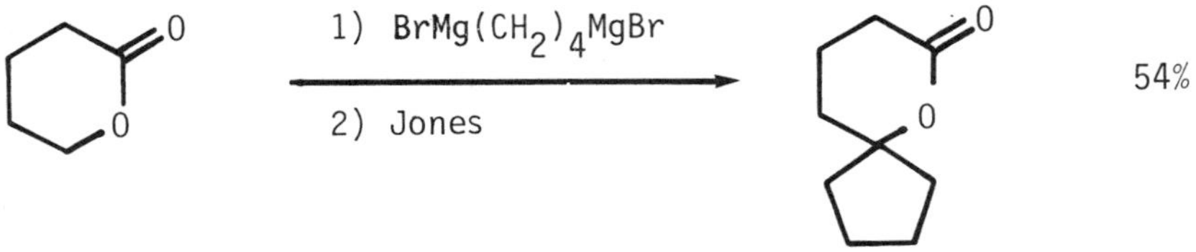

JOC, 45, 1828 (1980)

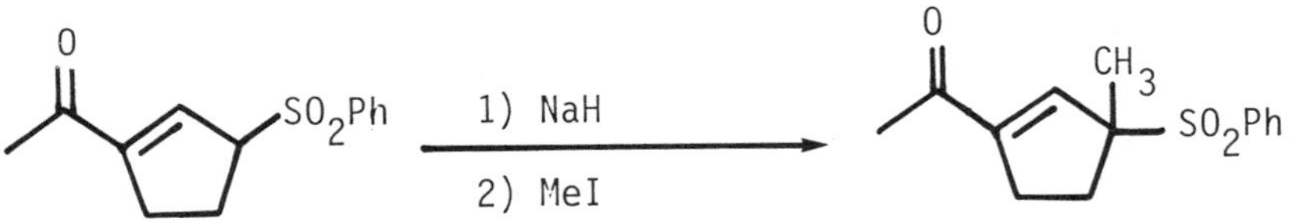

JACS, 102, 1602 (1980)

Section 72 <u>Alkyls, Methylenes, and Aryls from Ketones</u>

The conversions $R_2CO \rightarrow RR$, R_2CH_2, R_2CHR', etc. are listed in this section.

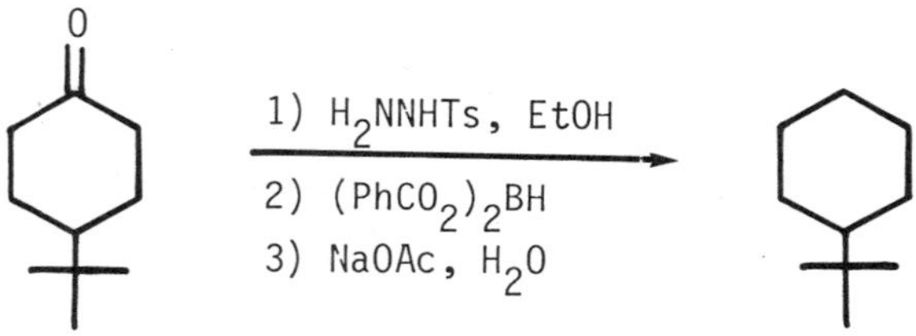

1) H_2NNHTs, EtOH

2) $(PhCO_2)_2BH$

3) NaOAc, H_2O

82%

JOC, <u>46</u>, 1217 (1981)

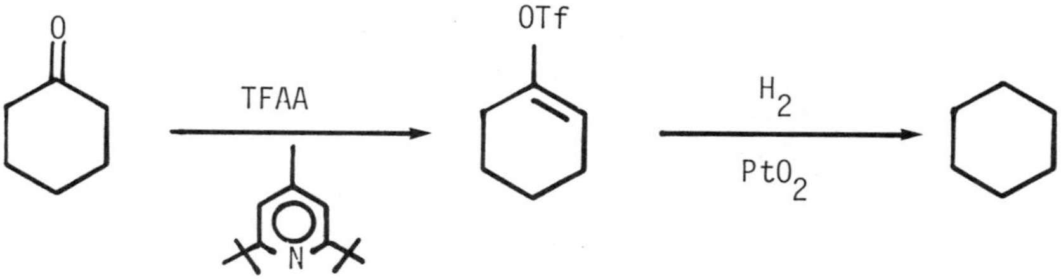

TFAA

OTf

H_2

PtO_2

90%

Tetr Lett, <u>23</u>, 117 (1982)

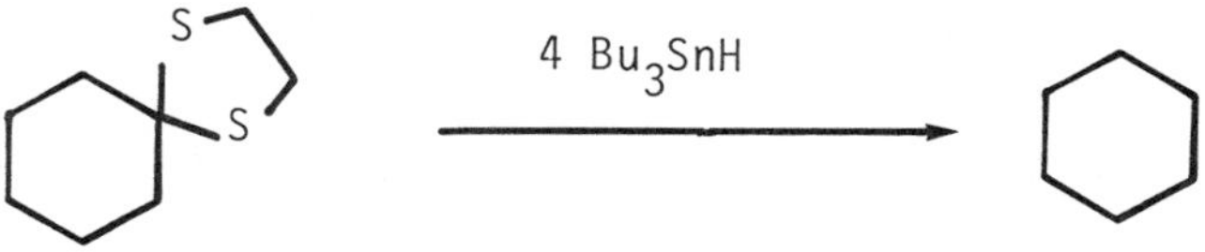

4 Bu_3SnH

76%

JOC, <u>45</u>, 3393 (1980)

$$\underset{\substack{\text{O} \\ \|}}{\text{Ph-C-CH}_3} \xrightarrow[\text{H}_2\text{O/EtOH}]{\text{Ra-Ni}} \text{PhCH}_2\text{CH}_3 \qquad 94\%$$

Tetr Lett, <u>21</u>, 2637 (1980)

$$\xrightarrow[\text{CH}_2\text{Cl}_2]{\text{Et}_3\text{SiH, BF}_3} \qquad 92\%$$

Org Syn, <u>60</u>, 108 (1981)

$$\text{Ph}_2\text{C=O} \xrightarrow[\text{MeOH}]{\text{NaBH}_4\text{-PdCl}_2} \text{Ph-CH}_2\text{-Ph} \qquad 83\%$$

Chem Lett, 1029 (1981)

$$\xrightarrow[\text{CH}_2\text{Cl}_2]{\text{Me}_2\text{TiCl}_2} \qquad 70\%$$

JCS Chem Comm, 237 (1981)

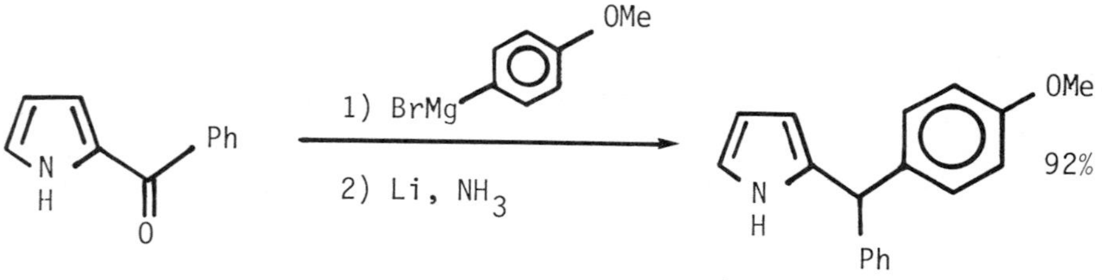

JOC, 46, 5060 (1981)

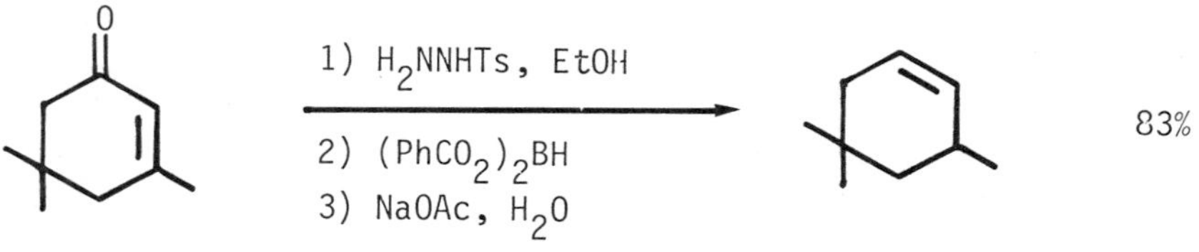

JOC, 46, 1217 (1981)

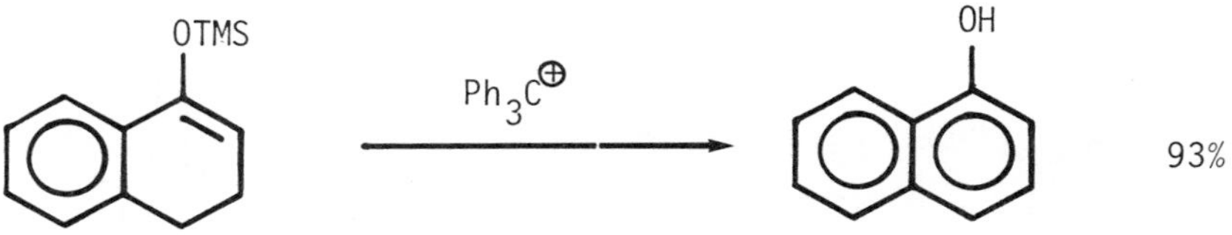

Liebigs Ann Chem, 533 (1980)

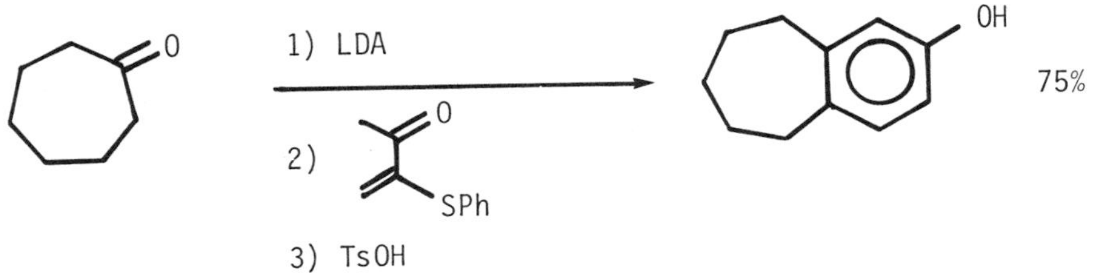

JOC, <u>47</u>, 1200 (1982)

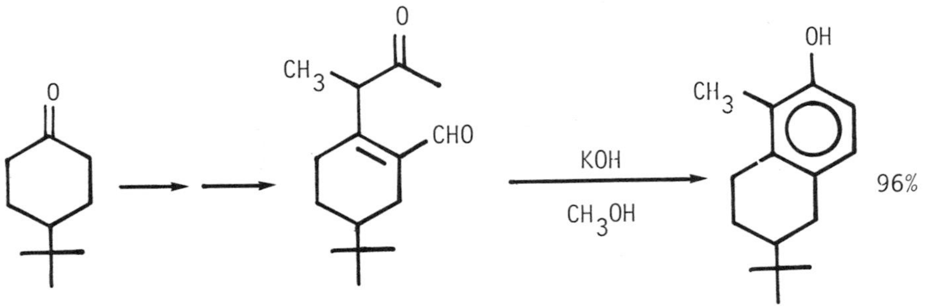

Tetr Lett, <u>23</u>, 2823 (1982)

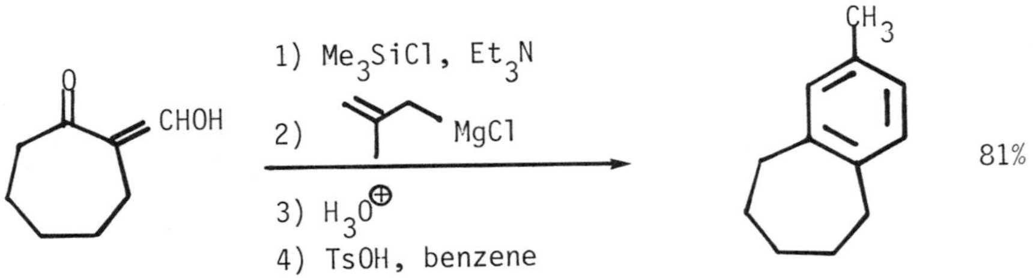

JOC, <u>47</u>, 3163 (1982)

Section 73 Alkyls, Methylenes, and Aryls from Nitriles

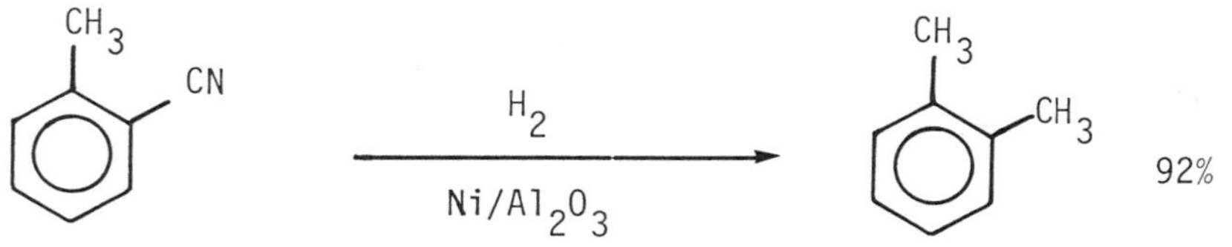

$$\xrightarrow[\text{Ni/Al}_2\text{O}_3]{\text{H}_2}$$

92%

Synthesis, 802 (1980)

$$\xrightarrow[\text{HCO}_2^{\ominus} \ {}^{\oplus}\text{NH}_4]{\text{10\% Pd-C}}$$

100%

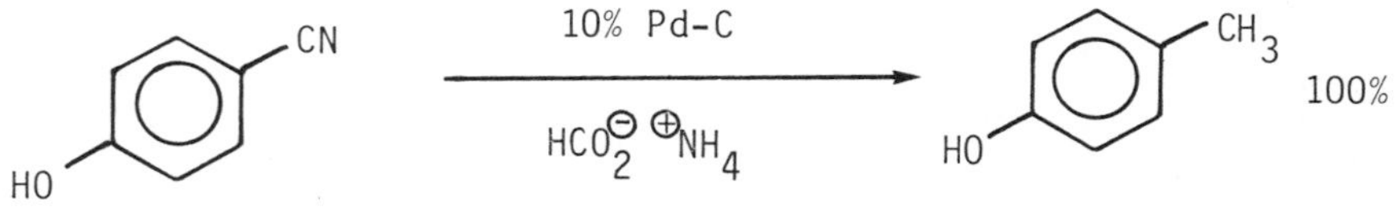

Synthesis, 1036 (1982)

Section 74 Alkyls, Methylenes and Aryls from Olefins

The following reaction types are included in this section:

A.   Hydrogenation of olefins (and aryls).

B.   Formation of aryls.

C.   Alkylations and arylations of olefins.

D.   Conjugate reductions of conjugated aldehydes, ketones, acids, esters, and nitriles.

E.   Conjugate alkylations.

F.   Cyclopropanations, including halocyclopropanations.

74A: Hydrogenation of Olefins (and aryls)

$$C_7H_{15}CH{=}CHCH_2OH \xrightarrow[\text{DMF}]{NH_2OH,\ EtOAc} C_7H_{15}CH_2CH_2CH_2OH \quad 96\%$$

Synth Comm, 12, 287 (1982)

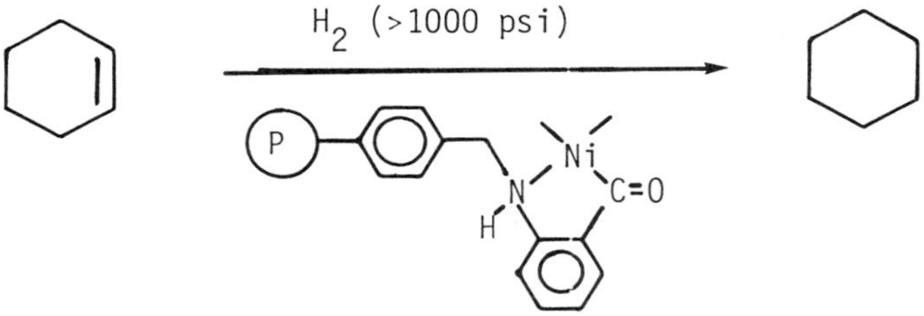

70%

JOC, 45, 1418 (1980)

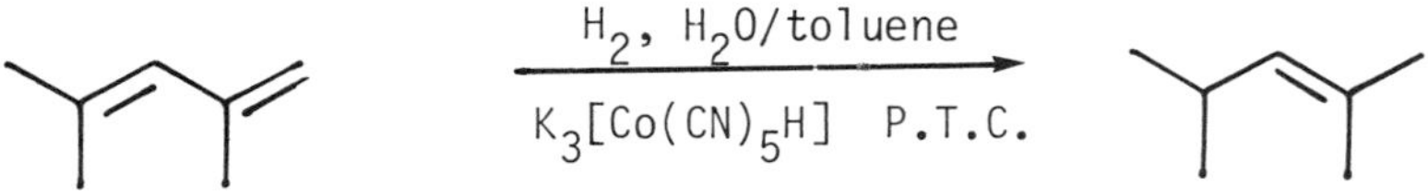

97%

JOC, 45, 3860 (1980)

$$\text{1,3-cyclohexadiene} \xrightarrow[\text{toluene}]{\text{Ni(acac)}_2 \cdot \text{Al}_2\text{Et}_3\text{Cl}_2 \cdot \text{PPh}_3} \text{cyclohexene} \quad 81\%$$

Bull Chem Soc Japan, **55**, 343 (1982)

$$\text{1,4-cyclohexadiene} \xrightarrow[\text{RONa, Ni(OAc)}_2]{\text{H}_2/\text{NaH}} \text{cyclohexene} \quad 98\%$$

JOC, **45**, 1937 (1980)

$$\underset{\text{OAc}}{\overset{\text{CF}_3}{\diagup}} \xrightarrow{[\text{Rh}(1,5\text{-cot}(R,R)\text{-di-PAMP})]^{\oplus}} \text{CH}_3\text{-}\overset{*}{\text{CH}}\underset{\text{OAc}}{\overset{\text{CF}_3}{\diagup}}$$

100%
77% ee (S)

JOC, **45**, 2362 (1980)

$$\text{tert-butylbenzene} \xrightarrow[\substack{\text{silica-bound} \\ \text{Rh-phosphine catalyst}}]{\text{H}_2 \text{ (80 atm)}} \text{tert-butylcyclohexane} \quad 96\%$$

Chem Lett, 603 (1982)

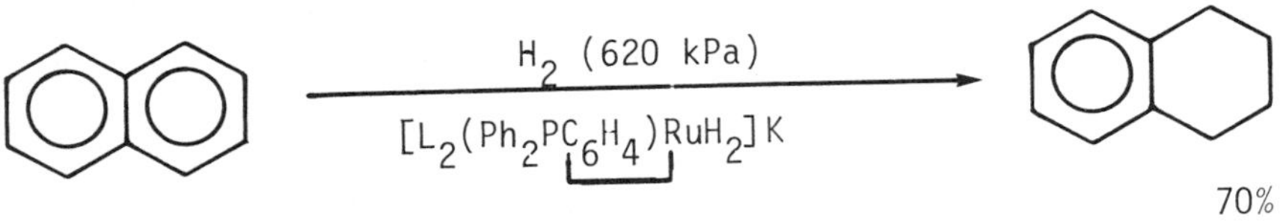
H_2 (620 kPa)
$[L_2(Ph_2PC_6H_4)RuH_2]K$
70%

JACS, 102, 5948 (1980)

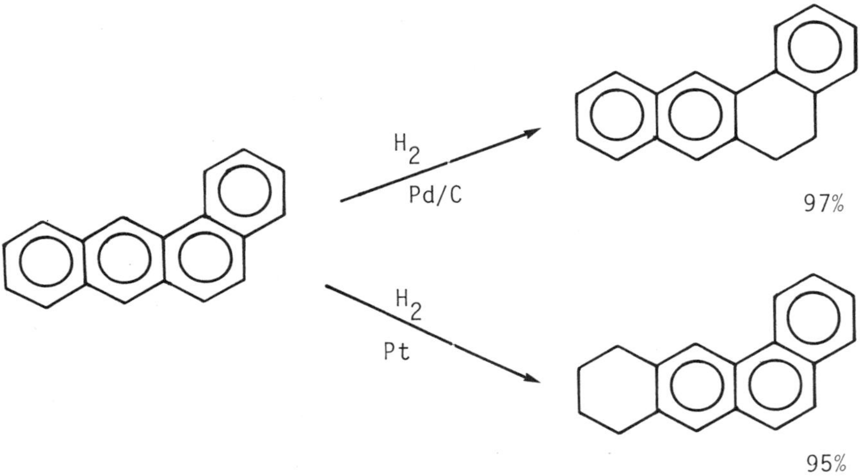
H_2
Pd/C
97%
H_2
Pt
95%

JOC, 45, 2797 (1980)

74B: Formation of Aryls

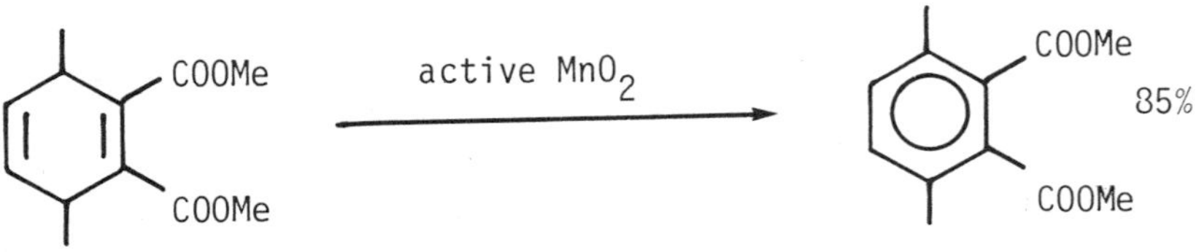
COOMe
COOMe
active MnO_2
COOMe
85%
COOMe

Synth Comm, 12, 637 (1982)

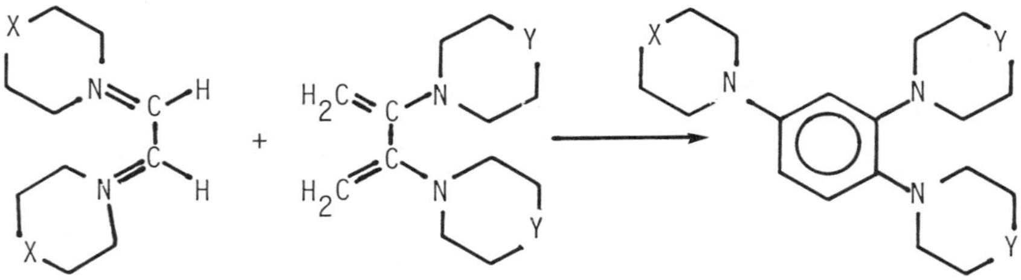

X = O, CH$_2$, NMe

Y = O, CH$_2$

46-70%

Tetr Lett, 21, 889 (1980)

74C: Alkylations and Arylations of Olefins

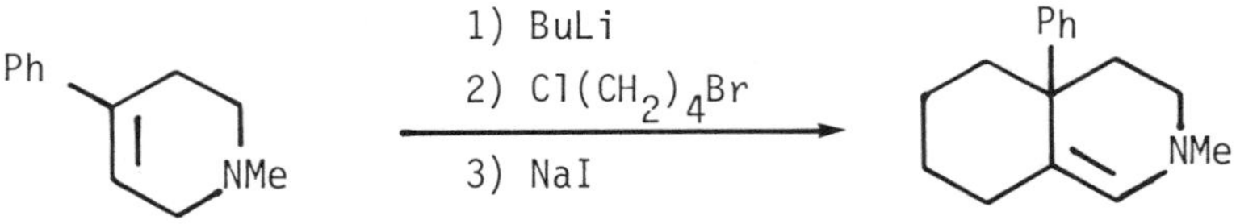

1) BuLi

2) Cl(CH$_2$)$_4$Br

3) NaI

61%

JACS, 102, 5955 (1980)

74D: Conjugate Reductions

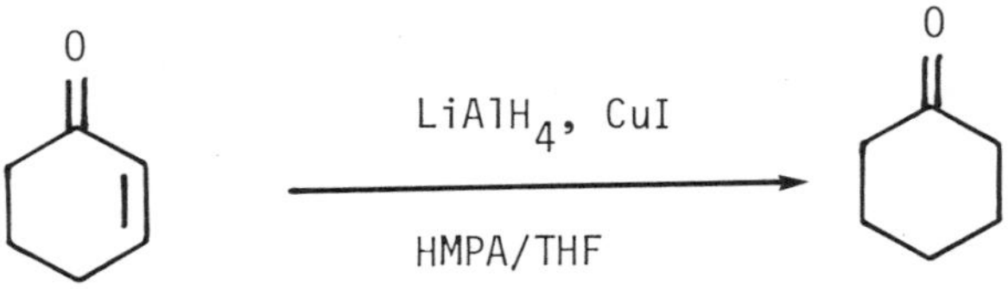

LiAlH$_4$, CuI

HMPA/THF

91%

JCS Chem Comm, 1013 (1980)

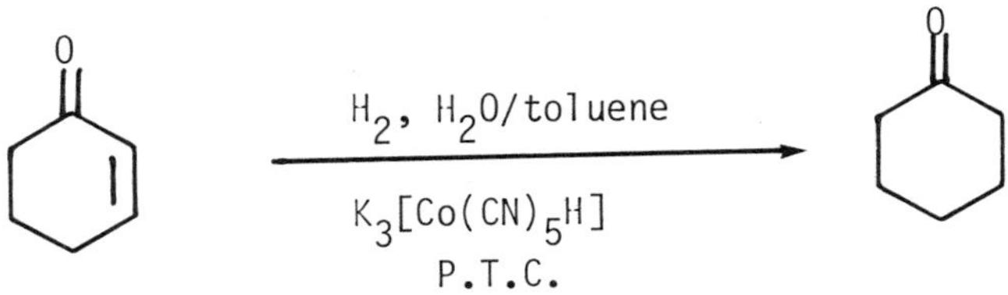
O
H_2, H_2O/toluene
$K_3[Co(CN)_5H]$
P.T.C.
O

75%

JOC, 45, 3860 (1980)

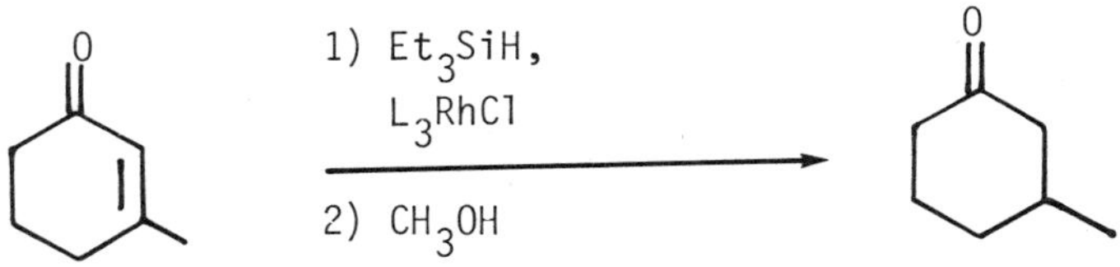
O
1) Et_3SiH,
L_3RhCl
2) CH_3OH
O

97%

Organometallics, 1, 1390 (1982)

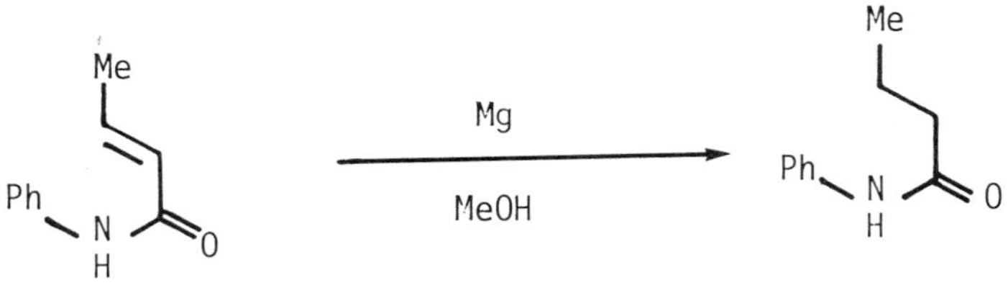
Me
Ph
N
H
O
Mg
MeOH
Me
Ph
N
H
O

77%

JCS Perkin I, 2912 (1981)

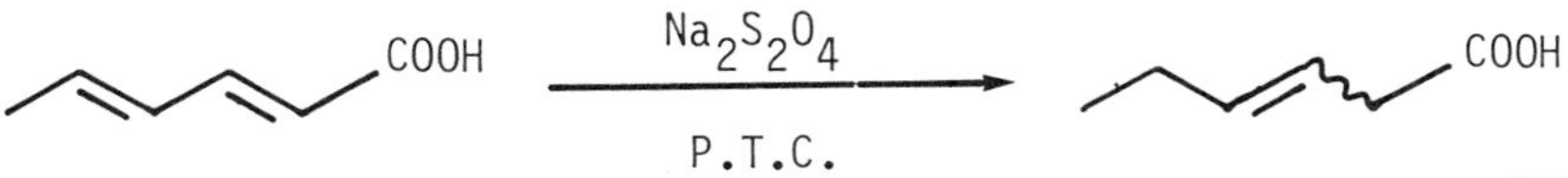
COOH
$Na_2S_2O_4$
P.T.C.
COOH

63%

Chem Lett, 715 (1982)

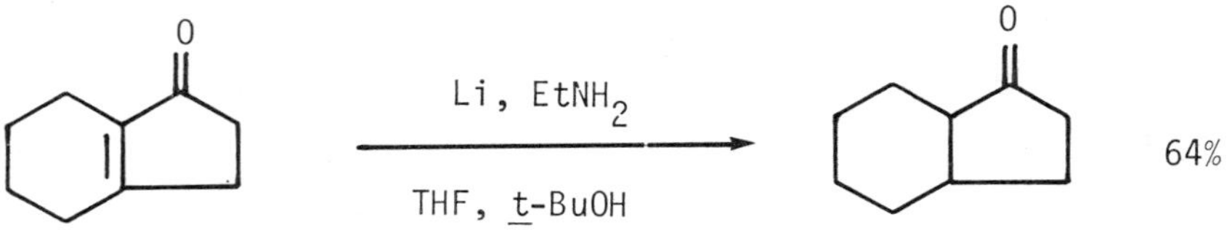

Li, EtNH₂

THF, t-BuOH

64%

Synthesis, 400 (1980)

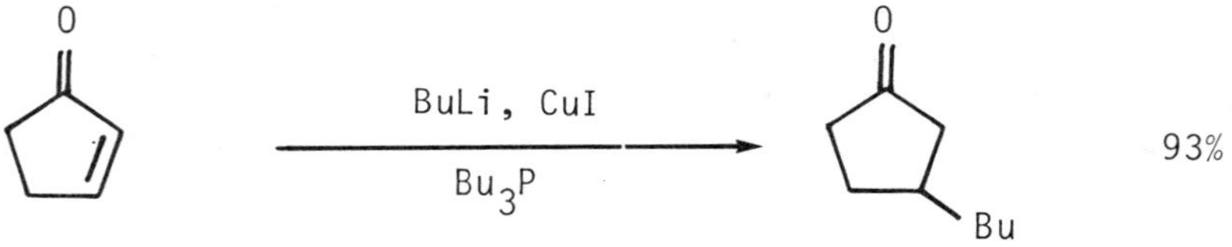

BuLi, CuI

Bu₃P

93%

Tetr Lett, 21, 1247 (1980)

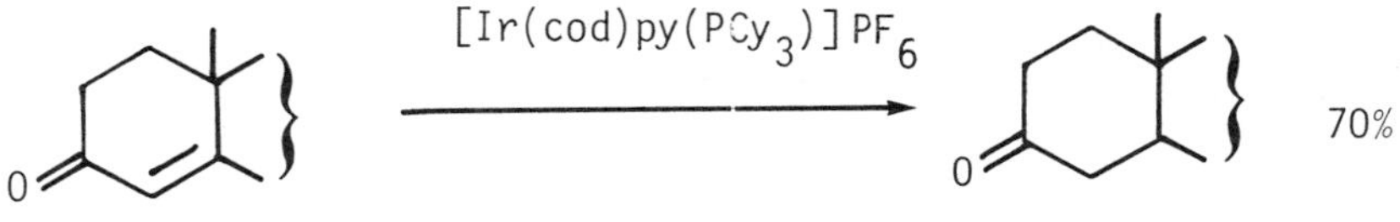

[Ir(cod)py(PCy₃)]PF₆

70%

Tetr Lett, 22, 303 (1981)

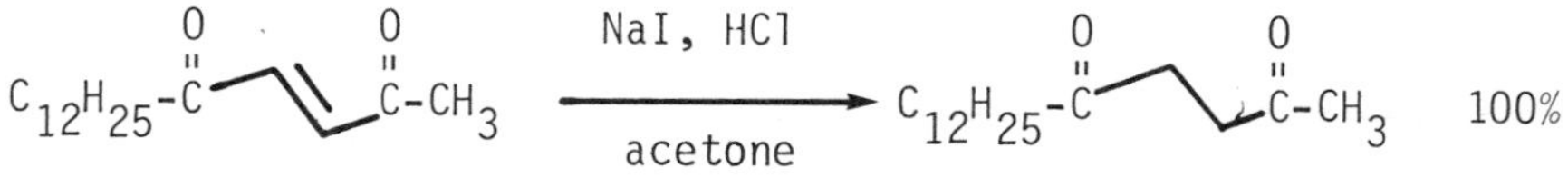

NaI, HCl

acetone

100%

Synthesis, 245 (1980)

Ph⌒COOH →[NH₂OH, CH₃-C-OEt / DMF] Ph⌒⌒COOH 67%

$$Ph\text{-}CH=CH\text{-}COOH \xrightarrow[\text{DMF}]{NH_2OH,\ CH_3\text{-}\overset{O}{\overset{\|}{C}}\text{-}OEt} Ph\text{-}CH_2CH_2\text{-}COOH \quad 67\%$$

Synth Comm, 12, 287 (1982)

$$Ph\text{-}CH=CH\text{-}\overset{O}{\overset{\|}{C}}\text{-}CH_3 \xrightarrow[\text{THF}]{HFe(CO)_4^{\ominus}} Ph\text{-}CH_2CH_2\text{-}\overset{O}{\overset{\|}{C}}\text{-}CH_3 \quad >98\%$$

Bull Chem Soc Japan, 55, 1329 (1982)

$$Ph\text{-}CH=CH\text{-}CHO \xrightarrow[\text{PdL}_4]{Bu_3SnH} Ph\text{-}CH_2CH_2\text{-}CHO \quad 99\%$$

Tetr Lett, 23, 477 (1982)

$$Ph\text{-}CH=CH\text{-}\overset{O}{\overset{\|}{C}}\text{-}Ph \xrightarrow[\text{RuCl}_2L_3,\ P.T.C.]{HCOONa} Ph\text{-}CH_2CH_2\text{-}\overset{O}{\overset{\|}{C}}\text{-}Ph$$

Tetr Lett, 22, 1709 (1981)

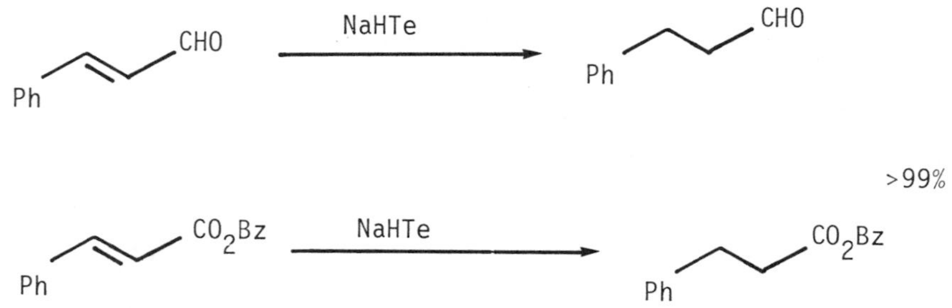

Chem Lett, 847 (1980)

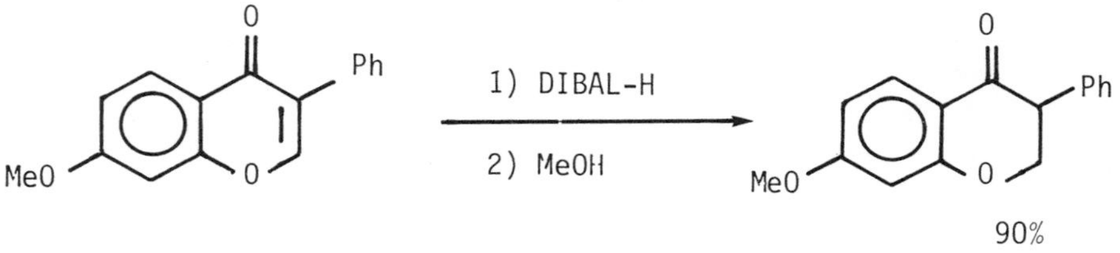

1) DIBAL-H

2) MeOH

90%

Synthesis, 574 (1981)

CuBr/Vitride

butanol/THF

92%

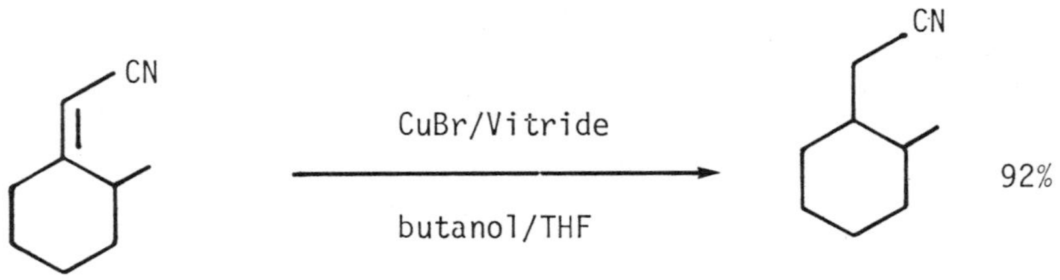

JOC, <u>45</u>, 167 (1980)

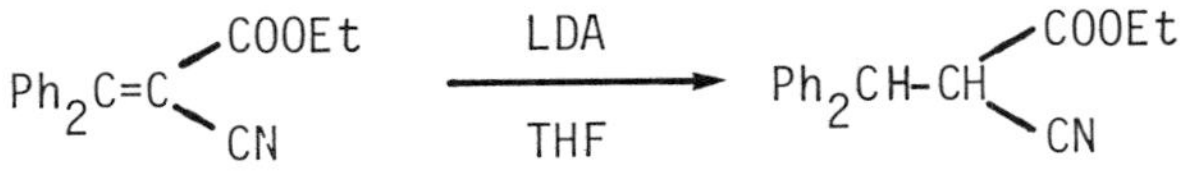

$$Ph_2C=C{<}^{COOEt}_{CN} \xrightarrow[\text{THF}]{\text{LDA}} Ph_2CH-CH{<}^{COOEt}_{CN} \qquad 81\%$$

JCS Perkin I, 1267 (1980)

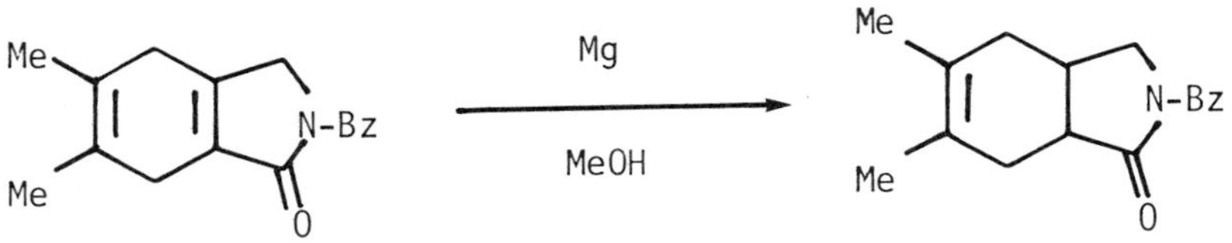

Tetr Lett, 21, 2915 (1980)

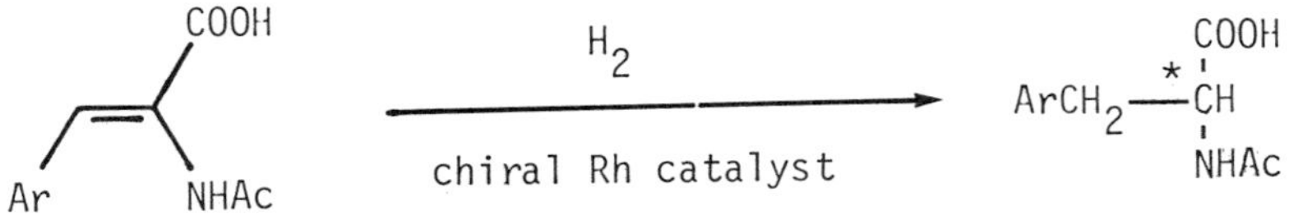

Ar = subst. Ph

>90%
>90% ee

Chem Ber, 113, 2323 (1980)
JOC, 45, 5187 (1980)
Chem Lett, 7 (1980)
JACS, 102, 988 (1980)
Synthesis, 76 (1981)
JOC, 46, 2954 and 2960 (1981)
JACS, 103, 2273 (1981)
J Chem Res (S), 117 (1982)

74E: Conjugate Alkylations

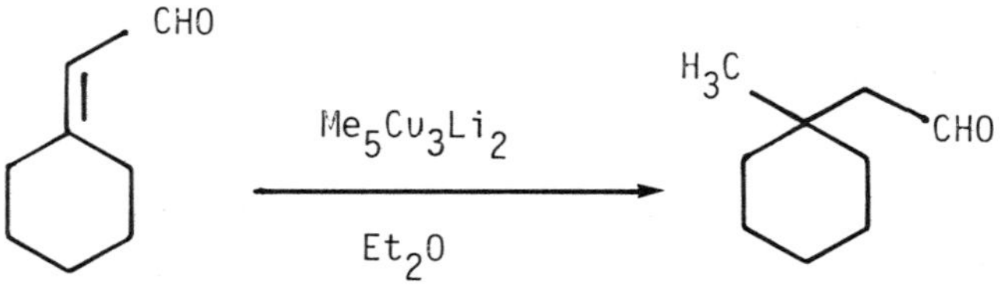
CHO
$Me_5Cu_3Li_2$
Et_2O
H_3C
CHO

90%

JCS Chem Comm, 643 (1981)
JOC, 47, 2572 (1982)

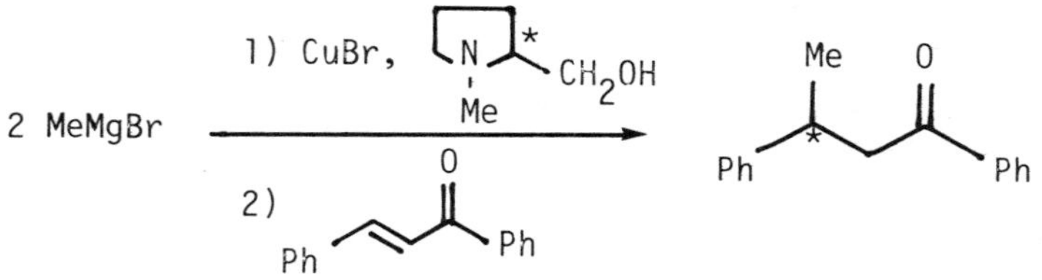
2 MeMgBr
1) CuBr,
N
Me
*
CH_2OH
2)
O
Ph
Ph
Me
O
Ph
*
Ph

88%
61% ee

Chem Lett, 45 (1980)

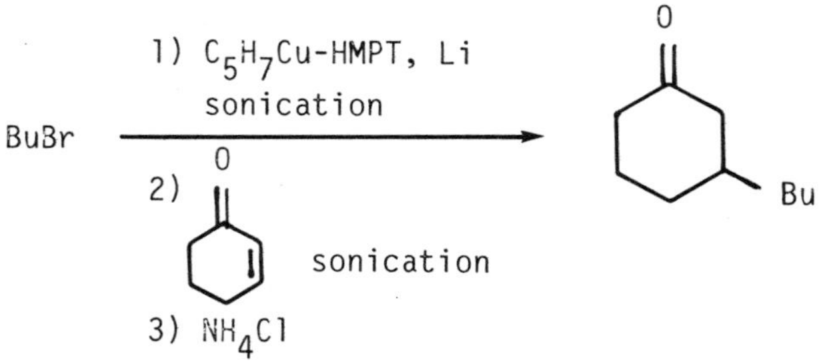
BuBr
1) C_5H_7Cu-HMPT, Li
sonication
2)
O
sonication
3) NH_4Cl
O
Bu

91%

JOC, 47, 3805 (1982)

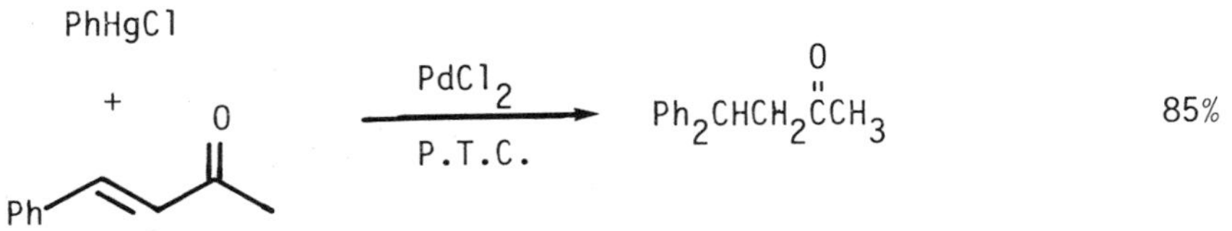

PhHgCl

+

$\xrightarrow[\text{P.T.C.}]{\text{PdCl}_2}$ $Ph_2CHCH_2\overset{\overset{\displaystyle O}{\|}}{C}CH_3$ 85%

Tetrahedron, <u>37</u>, 2941 (1981)

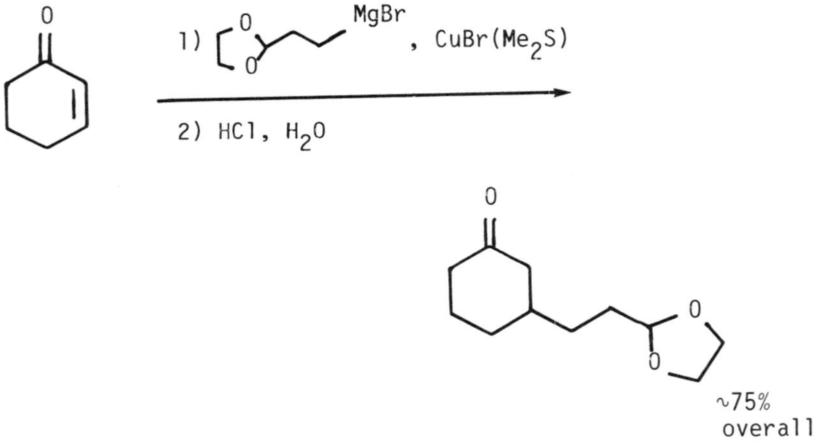

1) dioxolane MgBr , $CuBr(Me_2S)$

2) HCl, H_2O

~75%
overall

JCC, <u>47</u>, 5045 (1982)

$\xrightarrow[\text{2) Raney Ni}]{CH_2(COSEt)_2, \text{ DABCO}}$ 75%

CH_2CH_2OH

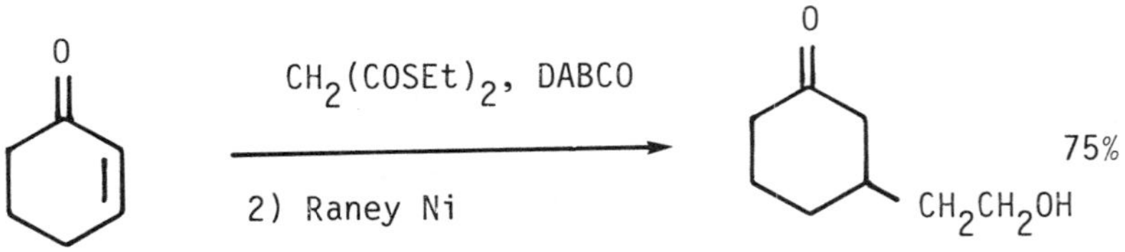

Can J Chem, <u>60</u>, 94 (1982)

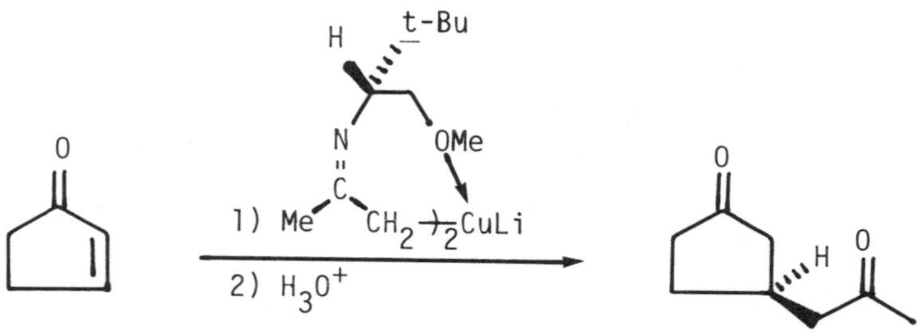

89%

75% ee

Tetr Lett, 23, 3711 (1982)

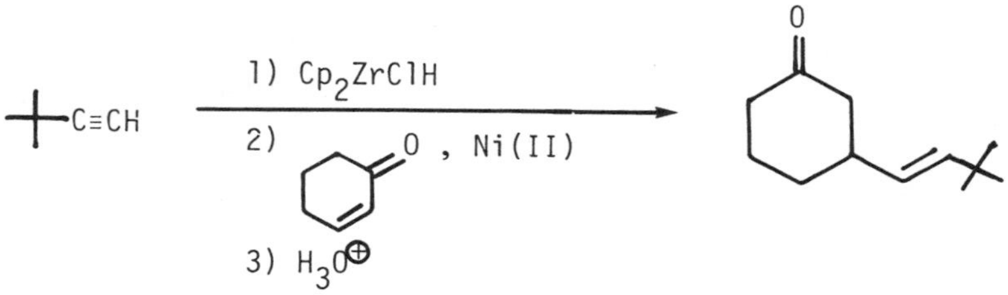

78%

JACS, 102, 1334 (1980)

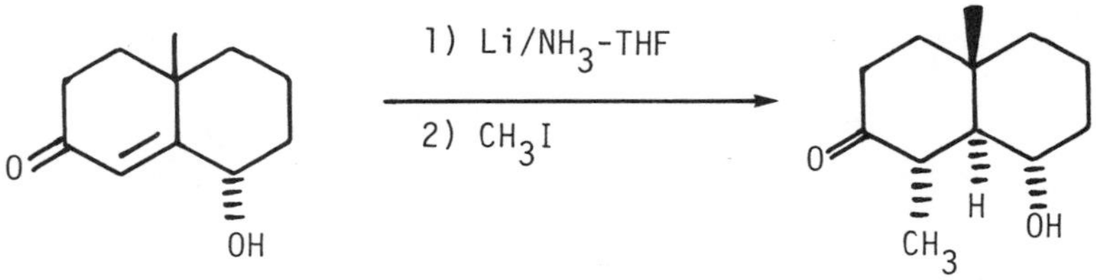

66%

JACS, 102, 1218 and 1219 (1980)

Tetr Lett, <u>21</u>, 3237 (1980)

Tetr Lett, <u>21</u>, 4823 (1980)

Tetr Lett, <u>23</u>, 5531 (1982)

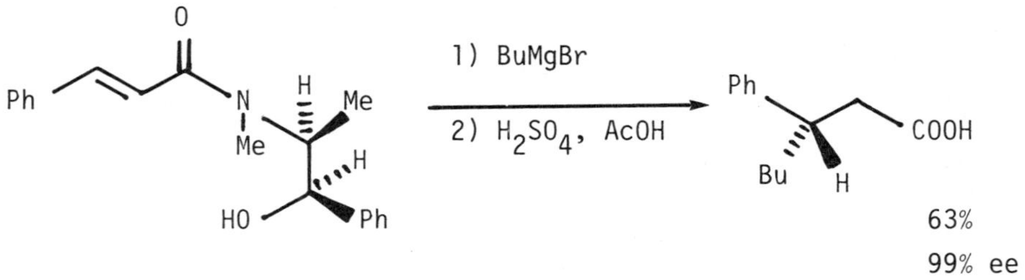
O
Ph
N
Me
H
Me
H
HO
Ph
1) BuMgBr
2) H_2SO_4, AcOH
Ph
COOH
Bu
H
63%
99% ee

Chem Lett, 913 (1981)

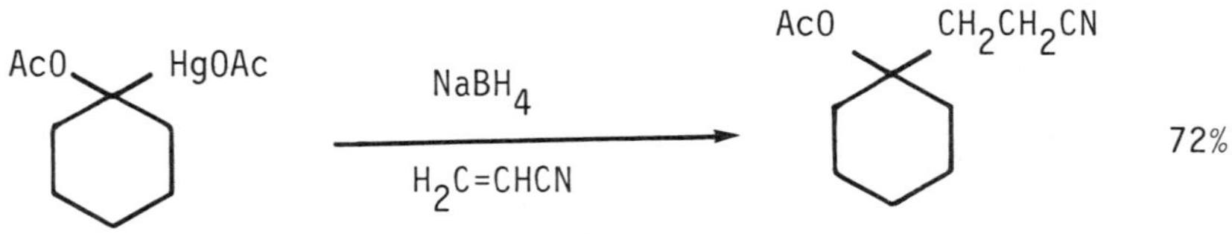
AcO
HgOAc
$NaBH_4$
$H_2C=CHCN$
AcO
CH_2CH_2CN
72%

Angew Chem Int Ed, 21, 130 (1982)

74F: Cyclopropanations

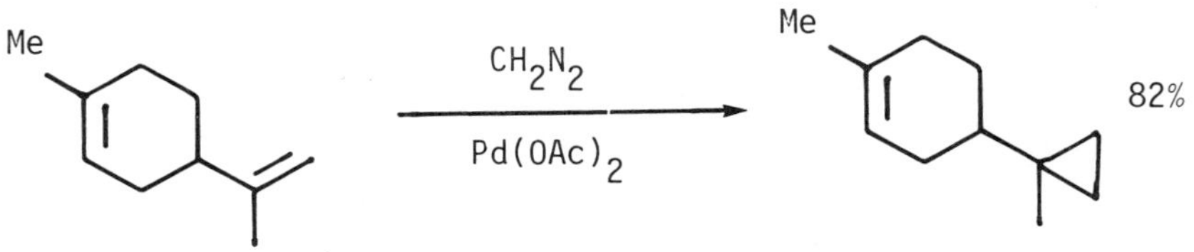
Me
CH_2N_2
$Pd(OAc)_2$
Me
82%

Synthesis, 714 (1981)

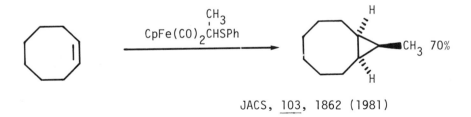

JACS, <u>103</u>, 1862 (1981)

CH$_3$CHI$_2$ / Zn-Cu, ether   63%

JOC, <u>47</u>, 1615 (1982)

Ph-CH=CH-C-Ph $\xrightarrow{[Me_3\overset{\oplus}{N}-\overset{\ominus}{CH}-CN]}$ Ph-CH-CH-C-Ph   61%

5:1 trans:cis

Synthesis, 301 (1982)

EtOOCCHN$_2$ / Rh, Cu catalysts   COOEt   88%

JOC, <u>45</u>, 695 and 1538 (1980)
Tetr Lett, <u>22</u>, 1783 (1981)

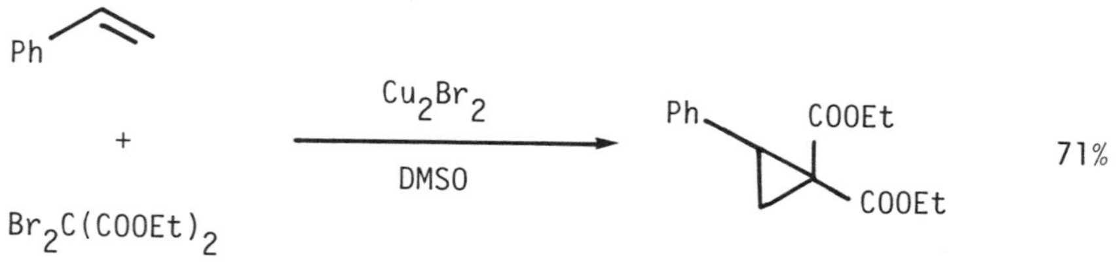

$$Ph \diagup\diagdown + Br_2C(COOEt)_2 \xrightarrow[\text{DMSO}]{Cu_2Br_2} \text{Ph} \triangle \begin{array}{l} COOEt \\ COOEt \end{array}$$

71%

Bull Chem Soc Japan, <u>54</u>, 2539 (1981)

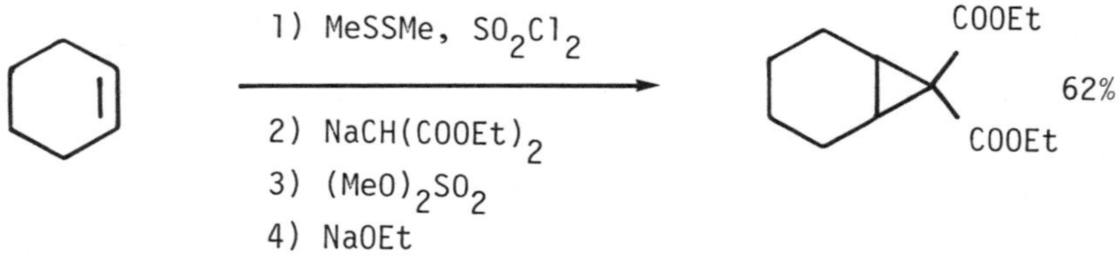

1) MeSSMe, SO$_2$Cl$_2$

2) NaCH(COOEt)$_2$
3) (MeO)$_2$SO$_2$
4) NaOEt

COOEt

62%

COOEt

Synthesis, 690 (1980)

hν, Ni(acac)$_2$

, THF

85%

Indian J Chem, <u>20B</u>, 911 (1981)

Section 75 <u>Alkyls and Methylenes from Miscellaneous Compounds</u>

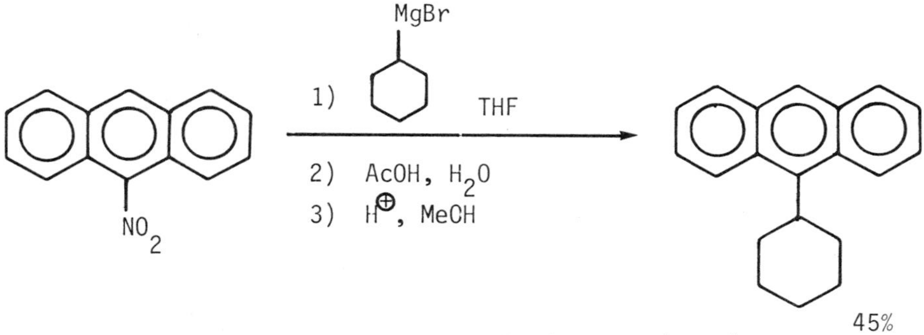

Synthesis, 836 (1982)

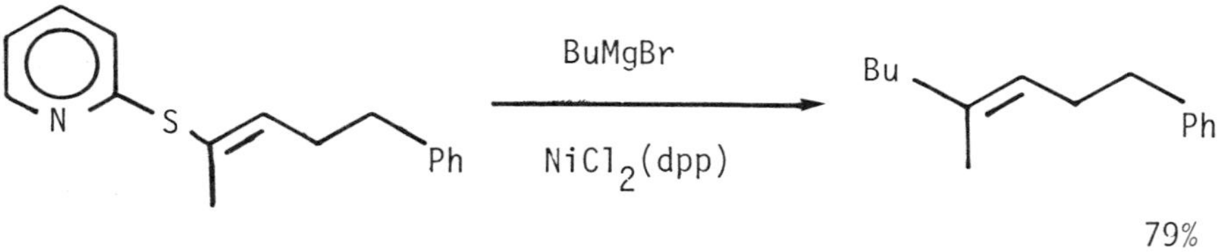

45%

Chem Lett, 1209 (1980)

79%

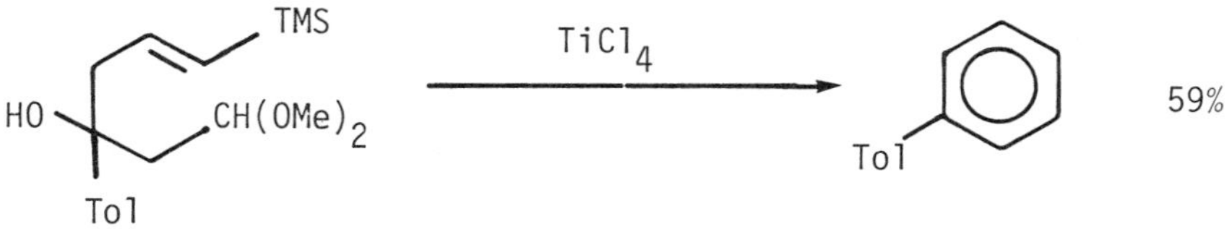

Tetr Lett, <u>22</u>, 3335 (1981)

59%

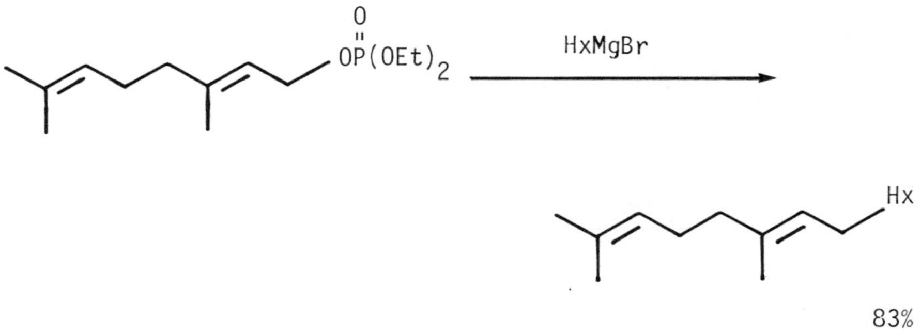

83%

JCS Chem Comm, 265 (1982)

CHAPTER 6
PREPARATION OF AMIDES

Section 76 <u>Amides from Acetylenes</u>

No additional examples

Section 77 <u>Amides from Carboxylic Acids, Acid Halides, and</u>
<u>Anhydrides</u>

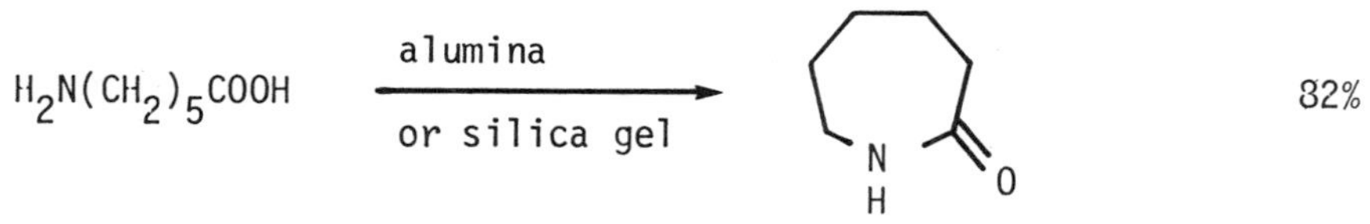

$H_2N(CH_2)_5COOH$ $\xrightarrow[\text{or silica gel}]{\text{alumina}}$

82%

Tetr Lett, <u>21</u>, 2443 (1980)

$H_2N(CH_2)_5COOH$ $\xrightarrow[Bu_2SnO]{\Delta}$

>95%

JACS, <u>102</u>, 7578 (1980)

$$\underset{\substack{\text{Pr}\\ \\ \text{HNBz}}}{\overset{}{\diagdown}}\underset{\text{COOH}}{\diagup}\quad\xrightarrow[\text{H}_2\text{O, P.T.C.}]{\text{MeSO}_2\text{Cl, CHCl}_3}\quad \underset{\text{Bz}}{\overset{\text{Pr}}{\diagdown}}$$

Chem Lett, 443 (1981)

$$\text{CH}_3(\text{CH}_2)_{16}\text{COOH}\quad\xrightarrow[\text{3) CH}_3\text{SO}_2\text{Cl}]{\substack{\text{1) CH}_3\text{SO}_2\text{Cl}\\ \text{2) NH}_3}}\quad\text{CH}_3(\text{CH}_2)_{16}\overset{\overset{\text{O}}{\|}}{\text{C}}\text{-NH}_2\qquad 78\%$$

Org Prep Proc Int, <u>14</u>, 396 (1982)

$$\underset{}{\overset{\text{COOH}}{\diagup}}\quad\xrightarrow[\text{2)}\ \ \ \text{Et}_3\text{N}]{\substack{\text{1) ClSO}_2\text{F, Et}_3\text{N}\\ \text{NH}_2}}\quad \underset{}{\overset{\text{O=C-NH}-\bigcirc}{}}\qquad 81\%$$

Synthesis, 661 (1980)

$$\text{Ph-COOH}\quad\xrightarrow[]{\substack{\text{1) ClSO}_2\text{NCO, Et}_3\text{N}\\ \text{2)}\ \bigcirc-\text{NH}_2}}\quad\text{Ph-}\overset{\overset{\text{O}}{\|}}{\text{C}}\text{-NH}-\bigcirc\qquad 82\%$$

Synthesis, 506 (1982)

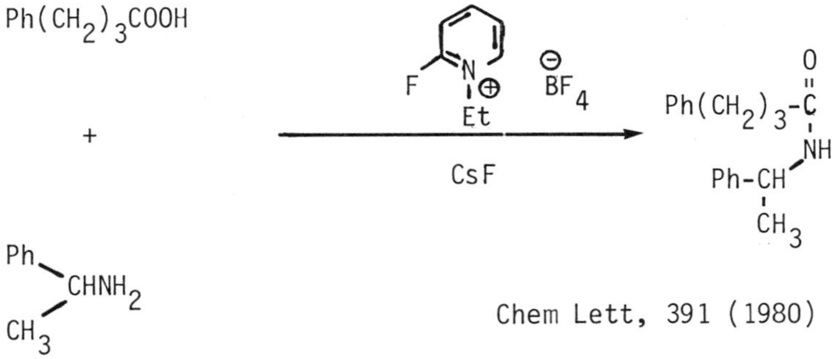

$Ph(CH_2)_3COOH$

+

Ph,
  CHNH$_2$
CH$_3$

94%

Chem Lett, 391 (1980)

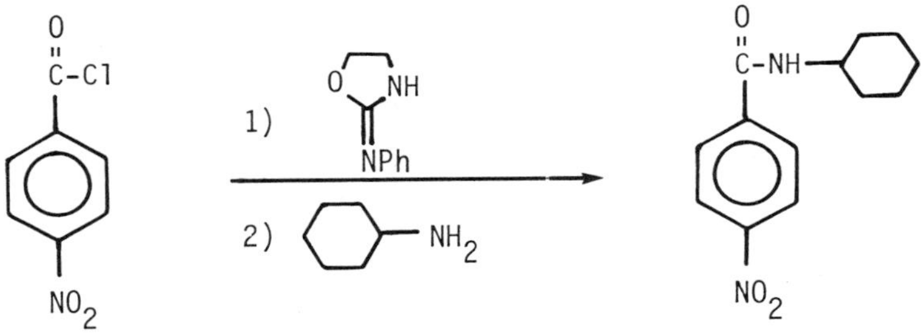

~75%

Bull Soc Chim France II, 167 (1982)

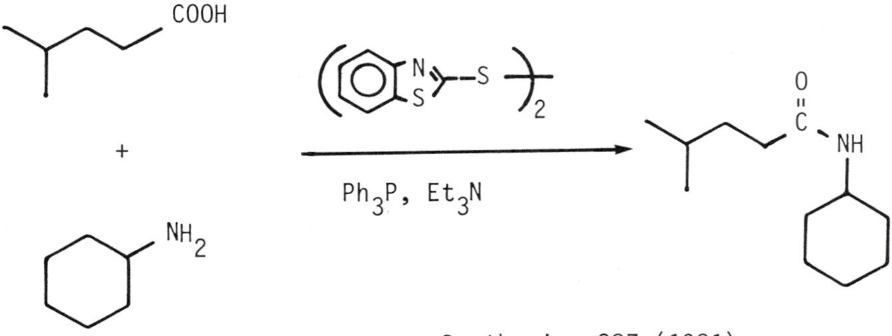

Synthesis, 287 (1981)

$C_5H_{11}-COOH$

1) [benzisothiazole-S-benzothiazole reagent structure]

2) $BzNH_2$

$\longrightarrow$ $C_5H_{11}-\overset{O}{\overset{\|}{C}}-NHBz$ 82%

Synthesis, 933 (1982)

[hexanoic acid structure] COOH

+

$Ph-\underset{NH_2}{\underset{|}{C}HCH_3}$

$\xrightarrow[\substack{PhP(\overset{O}{\overset{\|}{})-O-\text{(ring)}-NO_2)_2}]{KOH,\ Bu_4N^{\oplus}\ HSO_4^{\ominus}}}$

[hexanamide structure] $\overset{O}{\overset{\|}{C}}-NH-\underset{CH_3}{\overset{Ph}{\overset{|}{C}H}}$

89%

Chem Lett, 285 (1981)

$Ph_3P(OCH_2CF_3)_2$

+

CH_3COOH

$\xrightarrow[\text{2) } BuNH_2]{}$

$CH_3\overset{O}{\overset{\|}{C}}-NHBu$ 80%

JOC, 45, 5052 (1980)

Synthesis, 385 (1980)

Ph-O-P may be used as a carbonyl-activating group for the synthesis of amides and anhydrides from carboxylic acids. Yields are >90%

Synthesis, 288 (1982)

can be used to replace DCC as a coupling agent in peptide synthesis. No sparingly soluble secondary products are formed.

Angew Chem Int Ed, 19, 133 (1980)

Diphenyl phosphorazidate (DPPA) and diethyl phosphorocyanidate (DEPC) can serve as coupling reagents for solid-phase peptide synthesis in DMF.

Chem Pharm Bull, 28, 3064 (1980)

Use of SDPP in place of DCC for preparation of active esters
in peptide synthesis.

$$N-O-\overset{\overset{\textstyle O}{\|}}{P}(OPh)_2$$

SDPP

Tetr Lett, 21, 1467 (1980)

Use of NDPP as an activating reagent in the mixed anhydride
method of peptide synthesis.

$$NDPP = (PhO)_2\overset{\overset{\textstyle O}{\|}}{P}-O-N$$

JCS Chem Comm, 1029 (1980)

Use of polymeric N-hydroxysuccinimide (NHS) as a coupling
agent in peptide synthesis. Yields are ~90% with little racemization.

Acta Chem Scand B, 33, 311 (1979)

Use of phenyltetrazolinethione/isocyanide as an activating
group for peptide formation. Yields are 60-84%, without racemization.

Angew Chem Int Ed, 21, 143 (1982)

Use of thiazoline-2-thione as a carbonyl-activating reagent for peptide synthesis.

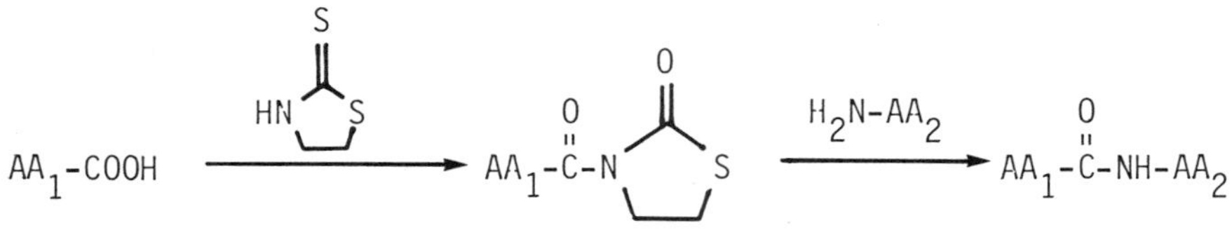

Chem Pharm Bull, <u>28</u>, 3140 (1980)

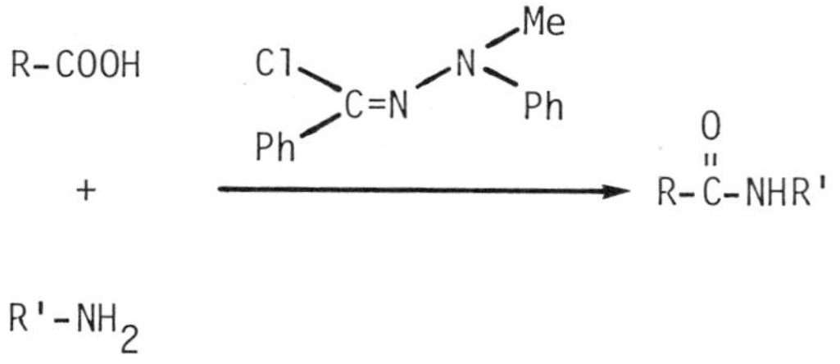

Allows peptide synthesis using unprotected amino acids.

JACS, <u>102</u>, 4537 (1980)

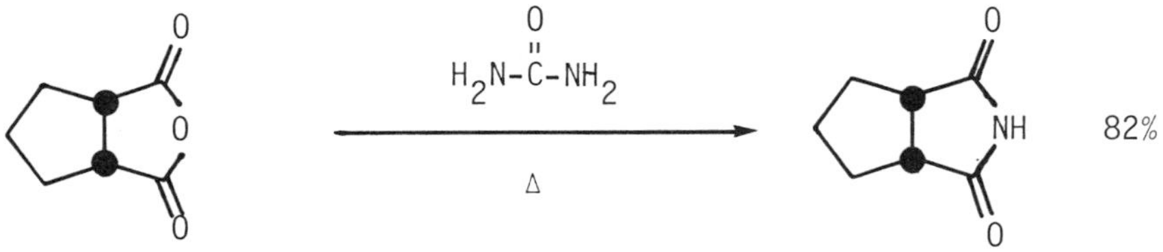

82%

Synth Comm, <u>11</u>, 447 (1981)

Related methods: Amides from Amines (Section 82)

Section 78 Amides from Alcohols

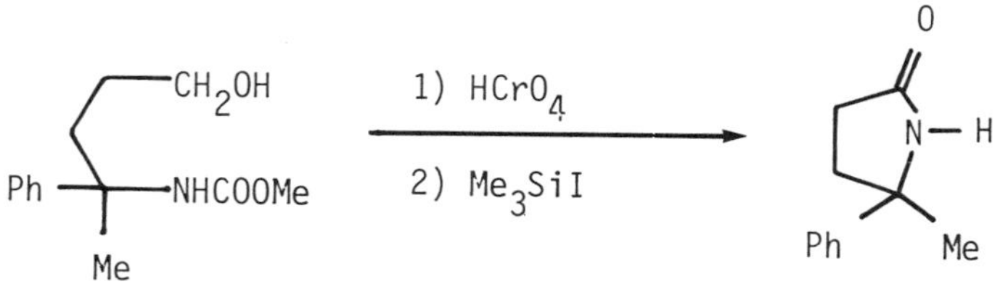

JOC, <u>46</u>, 1616 (1981)

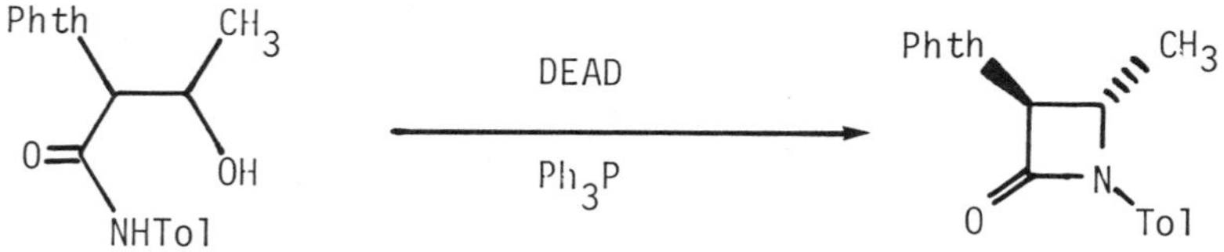

Phth = phthalimido

JOC, <u>46</u>, 1229 (1981)

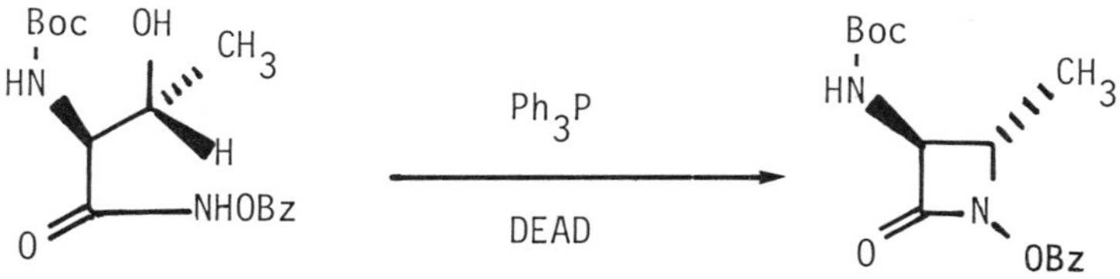

JACS, <u>102</u>, 7026 (1980)

Section 79 <u>Amides from Aldehydes</u>

Ph-C(=O)-⟨benzene⟩-CHO   →   1) Me$_3$SiCN 2) LDA 3) Ph$_2$PO$_2$NMe$_2$ 4) H$_3$O$^\oplus$ →   Ph-C(=O)-⟨benzene⟩-C(=O)-NMe$_2$ 51%

Tetr Lett, <u>23</u>, 3255 (1982)

PhO-CH$_2$-C(=O)-O-P(=O)(OCH$_2$CCl$_3$)$_2$ →   Ph-CH=NPh / Et$_3$N   →   [β-lactam: PhO and Ph substituents, N-Ph]   59%

Synthesis, 1053 (1982)

[imine: isopropyl-N-CH*-COOMe, C(=O)H-Bu]   →   1) TiCl$_4$ 2) Me$_2$C=C(OMe)(OTMS)   3) LDA   →   [β-lactam: Bu*, N-CH*(isopropyl)(COOMe)]   77%   90% ee

Tetr Lett, <u>21</u>, 2077 and 2081 (1980)

Synthesis, 545 (1981)

Section 80 <u>Amides from Alkyls, Methylenes and Aryls</u>

No additional examples

Section 81 <u>Amides from Amides</u>

Conjugate reductions of unsaturated amides are listed in Section 74 (Alkyls from Olefins).

$$\underset{\text{Ph-C-NH}_2}{\overset{\overset{\displaystyle O}{\|}}{\text{Ph-C-NH}_2}} \xrightarrow[\text{NaOH, P.T.C.}]{\text{2 BuBr}} \underset{\text{Ph-C-N}}{\overset{\overset{\displaystyle O}{\|}}{\text{Ph-C-N}}}\overset{\text{Bu}}{\underset{\text{Bu}}{}} \qquad 93\%$$

Synthesis, 1005 (1981)

81%

Chem Lett, 1143 (1981)

$$\underset{\underline{n}\text{-}C_8H_{17}\overset{\overset{\displaystyle O}{\|}}{C}\text{NHCH}_3}{} \xrightarrow[\text{2)}\ \triangleright\text{-NH}_2]{\text{1) N}_2O_4,\ \text{NaOAc}} \underset{\underline{n}\text{-}C_8H_{17}\overset{\overset{\displaystyle O}{\|}}{C}\text{NH}}{} \qquad 92\%$$

Tetr Lett, 23, 1127 (1982)

$C_{15}H_{31}-\overset{O}{\underset{}{C}}-N$ (thiazolidinone ring) $+$ piperidine (NH)

$\longrightarrow$

$C_{15}H_{31}-\overset{O}{\underset{}{C}}-N$ (piperidine ring)

99%

Tetr Lett, <u>21</u>, 841 (1980)

$Me_3C-\underset{\underset{Br}{|}}{CH}-\overset{O}{\underset{}{C}}-NHCMe_3$

$\xrightarrow[\text{18-crown-6}]{\text{KOH, benzene}}$

(aziridinone ring with tert-butyl groups)

~80-90%

Synthesis, 586 (1982)

(Boc-NH, HO, CH_3 substituted structure with C(=O)NHOBz)

$\xrightarrow[\text{DEAD}]{Ph_3P}$

(β-lactam ring with Boc-HN, CH_3, N-OBz)

79%

JACS, <u>102</u>, 7026 (1980)

$$CH_3(CH_2)_6-\overset{\overset{\displaystyle O}{\|}}{C}NHNH_2 \xrightarrow[\substack{\overset{\displaystyle \text{NH}}{\bigcirc}}]{Cu^{++}} CH_3(CH_2)_6-\overset{\overset{\displaystyle O}{\|}}{C}-N\overset{\displaystyle \bigcirc}{}$$

94%

Tetrahedron, <u>36</u>, 1311 (1980)

$$Me_2CH-\overset{\overset{\displaystyle O}{\|}}{C}NHNHPh \xrightarrow[\substack{Pd/C \\ EtOH/HOAc}]{H_2 \ (50 \ psi)} Me_2CH\overset{\overset{\displaystyle O}{\|}}{C}-NH_2$$

100%

Synth Comm, <u>10</u>, 253 (1980)

electrolysis

CH_3OH

71%

Chem Lett, 565 (1982)

$$H_3C-\overset{\overset{\displaystyle O}{\|}}{C}-N(CHO)_2$$

75%

Synthesis, 264 (1982)

O_2N-⬡$-\overset{\overset{O}{\|}}{C}-NH_2$

$+$

$Me_2N-\underset{\underset{Me}{|}}{C}(OMe)_2$

$\xrightarrow[\text{2) } H_2O]{\text{1) } 120^{\circ}}$

$\overset{\overset{O}{\|}}{C}-NH-\overset{\overset{O}{\|}}{C}-Me$ ⬡ NO_2 89%

Synthesis, 119 (1980)

$Bz-\underset{\underset{H}{|}}{N}-\overset{\overset{S}{\|}}{C}-Me$ $\xrightarrow[\text{2) } H_2O]{\text{1) } CSCl_2}$ $Bz-\underset{\underset{H}{|}}{N}-\overset{\overset{O}{\|}}{C}-Me$ 75%

Indian J Chem, 19B, 211 (1980)

Section 82 Amides from Amines

⬡$-NH_2$ →   ⬡$-HN-\overset{\overset{O}{\|}}{C}-CH_3$ 95%

Synth Comm, 12, 709 (1982)

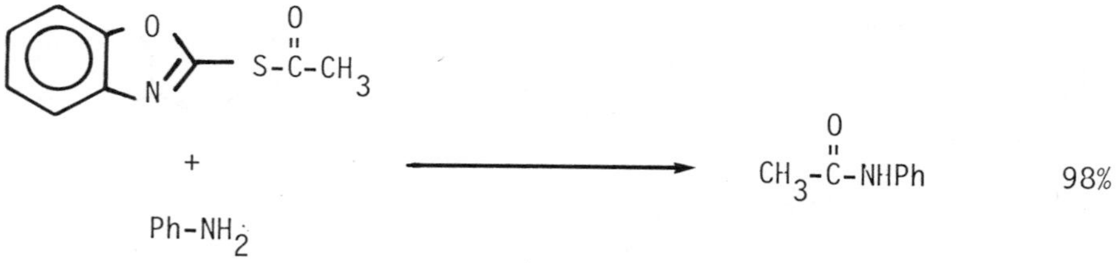

CH$_3$-$\overset{\overset{O}{\|}}{C}$-NHPh   98%

Synthesis, 991 (1981)

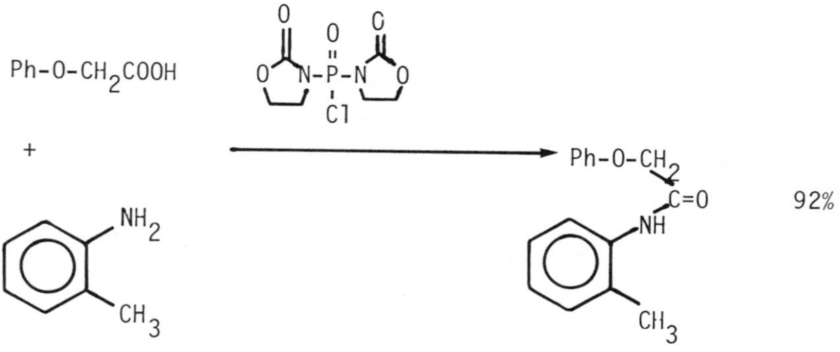

Synthesis, 547 (1980)

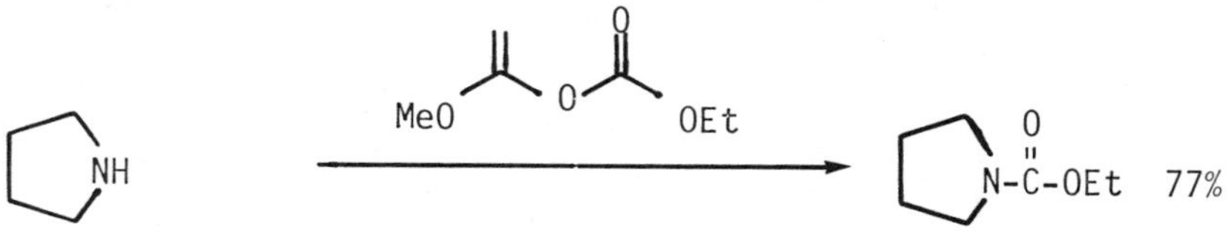

JOC, <u>45</u>, 4519 (1980)

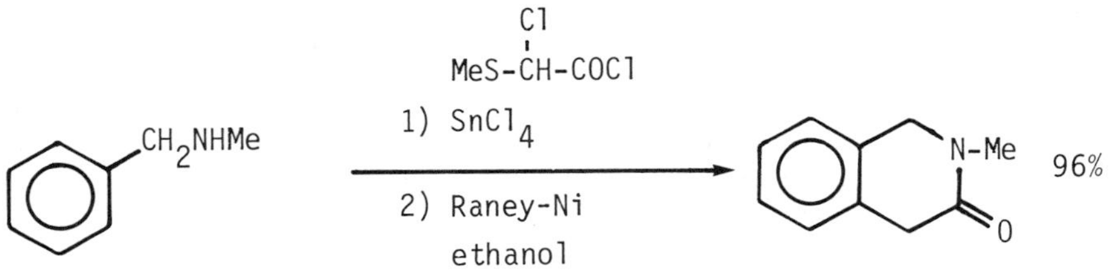

Synthesis, 534 (1981)

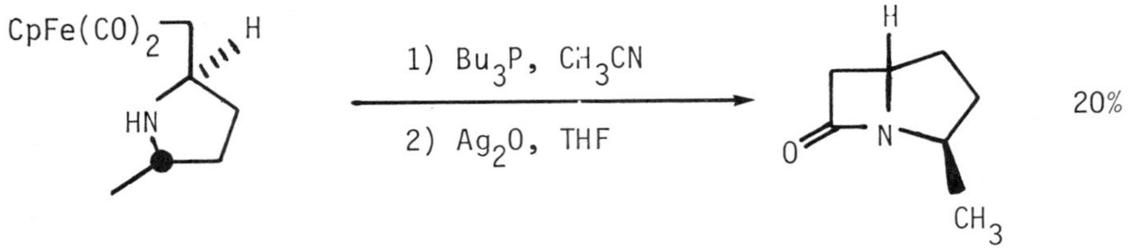

JOC, 45, 1984 (1980)

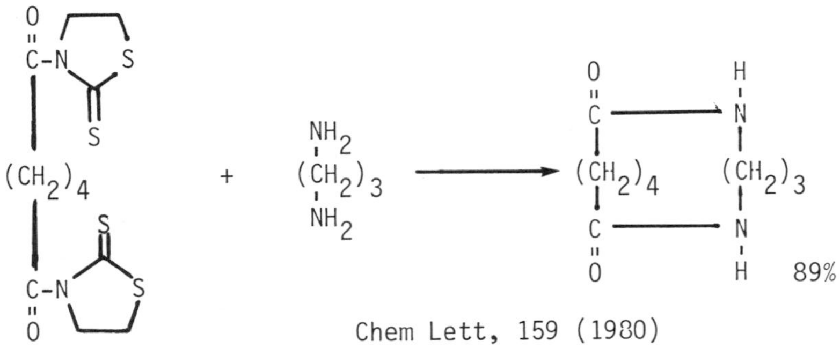

Chem Lett, 159 (1980)

Related methods: Amides from Carboxylic Acids (Section 77)
                  Protection of Amines (Section 105A)

Section 83 <u>Amides from Esters</u>

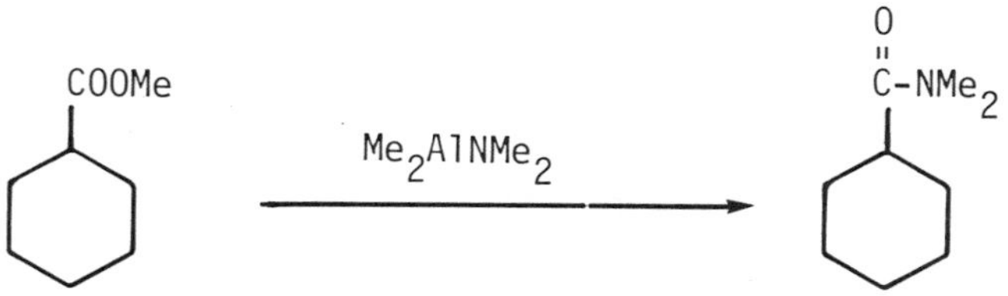

Org Syn, <u>59</u>, 49 (1980)

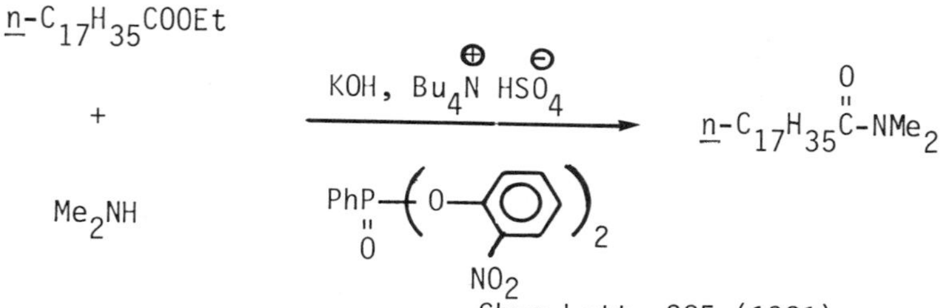

Chem Lett, 285 (1981)

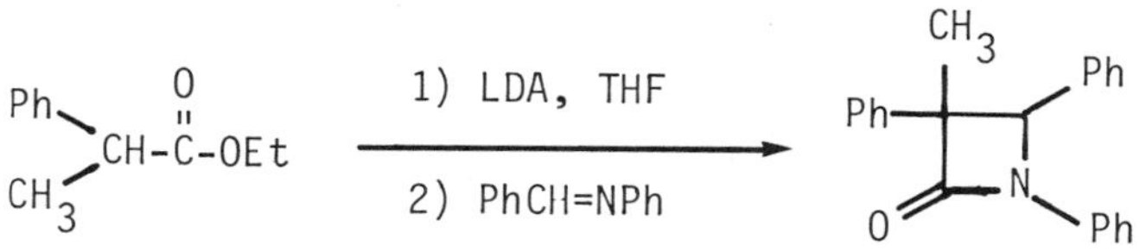

JOC, <u>45</u>, 3413 (1980)

Section 84 <u>Amides from Ethers and Epoxides</u>

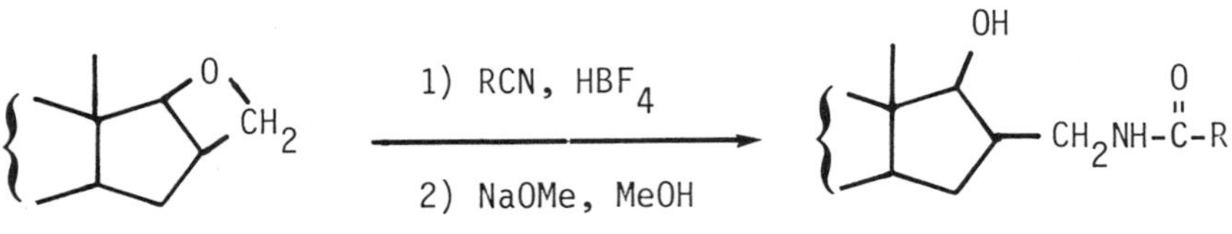

Tetr Lett, <u>22</u>, 341 (1981)

Section 85 <u>Amides from Halides</u>

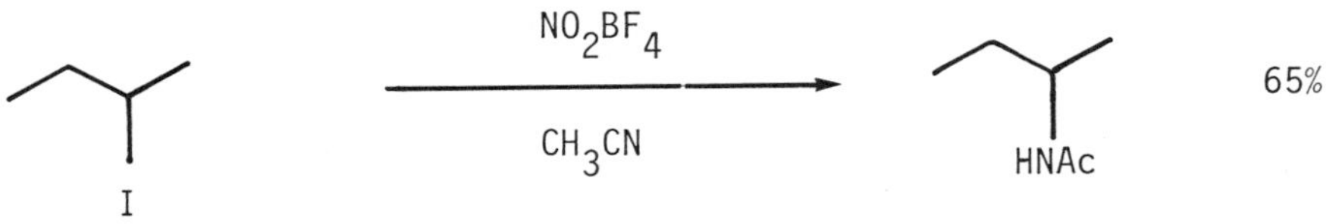

65%

JOC, <u>45</u>, 165 (1980)

94%

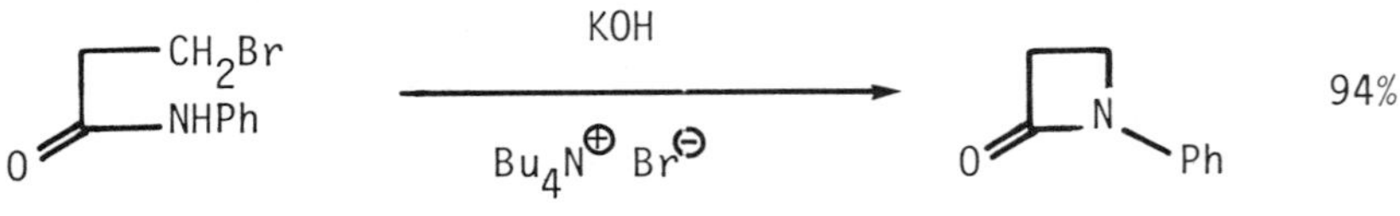

Chem Pharm Bull, <u>29</u>, 1063 (1981)

Section 86 <u>Amides from Hydrides</u>

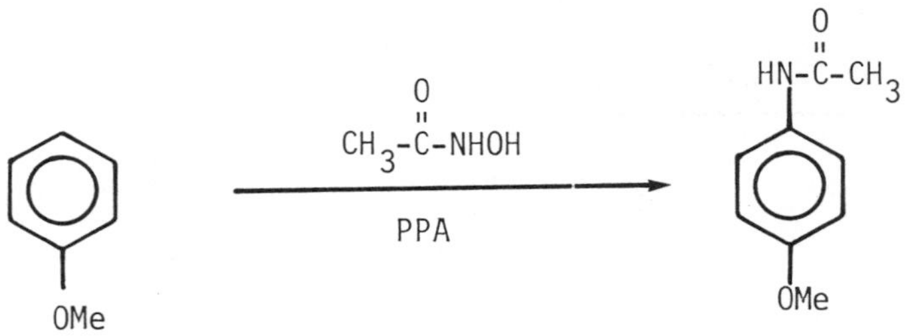

$$CH_3-\overset{\overset{\displaystyle O}{\|}}{C}-NHOH$$

PPA

59%

JOC, <u>46</u>, 4304 (1981)

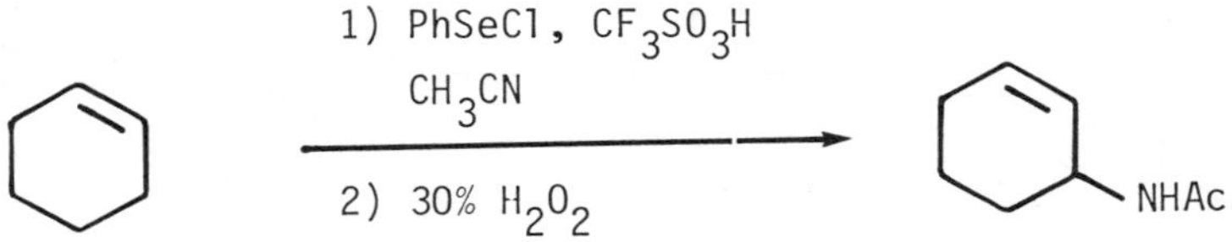

1) PhSeCl, CF_3SO_3H
   CH_3CN

2) 30% H_2O_2

JOC, <u>46</u>, 4727 (1981)

OTMS

+

Cl

$N-CO_2Me$
Me

$TiCl_4$

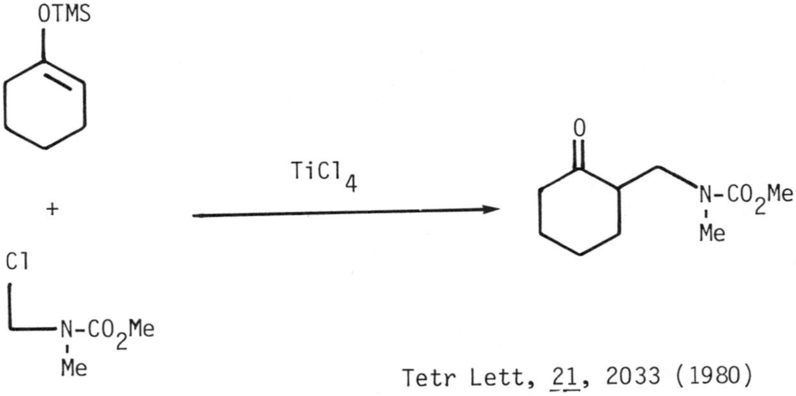

78%

Tetr Lett, <u>21</u>, 2033 (1980)

Section 87 Amides from Ketones
==

$$CH_3\text{-}\overset{\overset{\displaystyle O}{\|}}{C}\text{-}Ph \xrightarrow[\text{2) } \Delta]{\text{1) } H_2NOSO_3H,\ H_2O} CH_3\text{-}\overset{\overset{\displaystyle O}{\|}}{C}\text{-}NHPh \qquad \sim 60\%$$

JOC (USSR), 17, 2284 (1982)

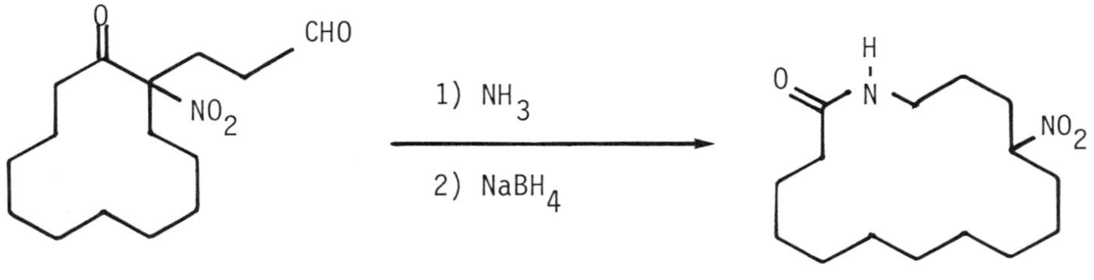

41%

Helv Chim Acta, 65, 2299 (1982)

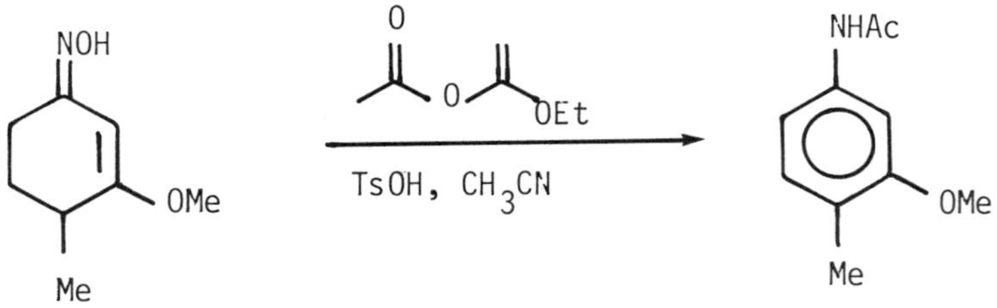

76%

Synthesis, 483 and 887 (1980)

Section 88 <u>Amides from Nitriles</u>

$$Bu_4\overset{\oplus}{N} \ H\overset{\ominus}{SO_4}, \ CH_2Cl_2$$

$$H_2O_2, \ H_2O, \ NaOH$$

85%

Synthesis, 243 (1980)

$$Ph-CH_2-C\equiv N \xrightarrow[\text{alumina}]{KF} Ph-CH_2-\overset{\overset{O}{\|}}{C}-NH_2$$

74%

Synth Comm, <u>12</u>, 177 (1982)

$$Cu(0), \ N_2, \quad \text{(HO—⟍◯⟍—OH)}$$

$$H_2O, \ 90°$$

89%

JOC, <u>47</u>, 4812 (1982)

Section 89 <u>Amides from Olefins</u>

$$Hx_3B \xrightarrow[\text{NaOH, P.T.C.}]{O_2N-\langle O \rangle-SO_3NHCOOEt} Hx-NH-\overset{O}{\underset{\parallel}{C}}-OEt$$

Synth Comm, <u>11</u>, 475 (1981)

Ph-CH=CH$_2$

+

$$\xrightarrow[\text{2) NaBH}_4]{\text{1) Hg(NO}_3)_2} \text{Me-}\overset{O}{\underset{\parallel}{C}}\text{-NH-CH-CH}_3 \quad 84\%$$
$$\overset{|}{\underset{Ph}{}}$$

CH$_3$-C-NH$_2$
  ∥
  O

JCS Chem Comm, 670 (1981)

$$\xrightarrow[\text{NaBH}_4]{\text{TsNH}_2, \text{Hg(NO}_3)_2}$$

NHTs

66%

JCS Chem Comm, 1178 (1981)

Section 90 Amides from Miscellaneous Compounds

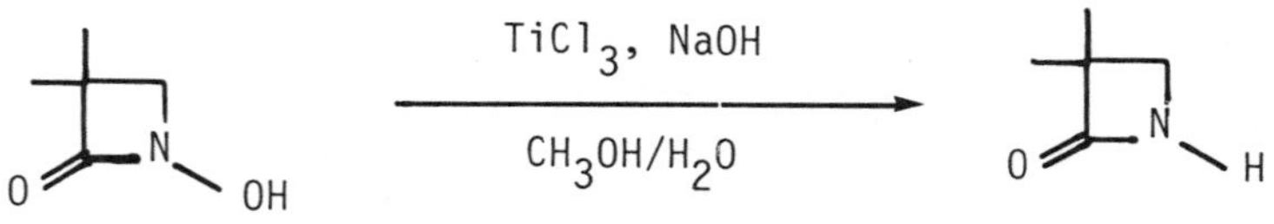

JOC, 45, 410 (1980)

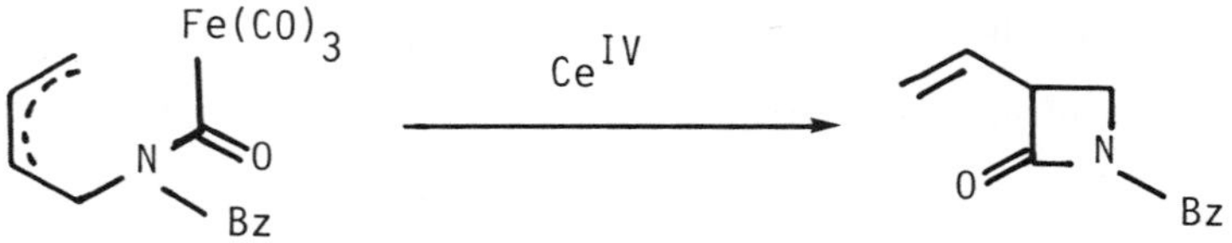

JCS Chem Comm, 297 (1980)

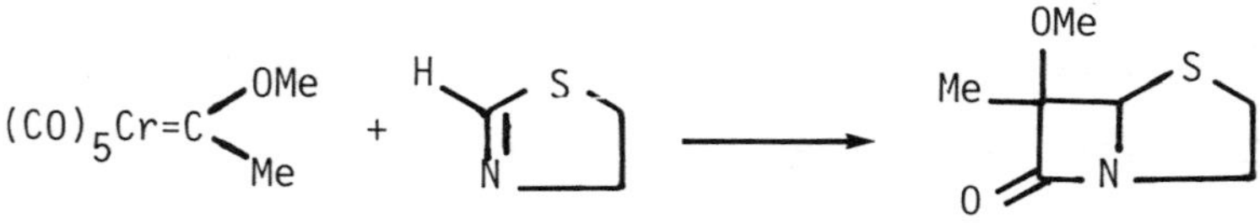

JACS, 104, 5538 (1982)

Review: "Prominent Aspects of Electroorganic Synthesis in β-
           lactam Chemistry."

Bull Soc Chim Belges, 91, 951 (1982)

Section 90A Protection of Amides

Synth Comm, 11, 787 (1981)

A study of N-acyl protecting groups for deoxynucleosides, including substituted phenylacetyl, phenoxyacetyl, and benzoyl protecting groups.

Tetrahedron, 37, 363 (1981)

The N-acyl protecting groups in nucleoside derivatives can be selectively removed by treatment with $ZnBr_2$ in the presence of alcohols to give O-protected nucleosides.

Tetr Lett, 22, 3761 (1981)

CHAPTER 7
PREPARATION OF AMINES

Section 91 <u>Amines from Acetylenes</u>

No additional examples

Section 92 <u>Amines from Carboxylic Acids and Acid Halides</u>

No additional examples

Section 93 <u>Amines from Alcohols</u>

$Ph_3P(OCH_2CF_3)_2$

$+$ $\xrightarrow{\text{2) BuNH}_2}$ $\underline{n}\text{-}C_8H_{17}NHBu$ 73%

$\underline{n}\text{-}C_8H_{17}OH$

JOC, <u>45</u>, 5052 (1980)

Review: "Catalytic Amination of Alcohols, Aldehydes, and Ketones"

Russ Chem Rev, <u>49</u>, 14 (1980)

N-OH

$\xrightarrow[\text{H}_2, \text{ Cu|MgO}]{\underline{i}\text{-Bu-CH}_2\text{OH}}$

HN-CH$_2$-$\underline{i}$-Bu

48%

Doklady Chem, **244**, 69 (1979)

Section 94 Amines from Aldehydes

CH=NOH

NO$_2$

$\xrightarrow[\text{TiCl}_4]{\text{NaBH}_4}$

CH$_2$NH$_2$

NH$_2$

82%

Synthesis, 695 (1980)

Ph-CH=NPh $\xrightarrow[\text{Rh catalyst}]{\text{i-PrOH}}$ Ph-CH$_2$NH-Ph

85%

Synthesis, 442 (1981)

N-OH

$\xrightarrow[\text{H}_2, \text{ Cu/MgO}]{\text{CH}_3\text{-CHO}}$

HN-CH$_2$CH$_3$

63%

Doklady Chem, **244**, 69 (1979)

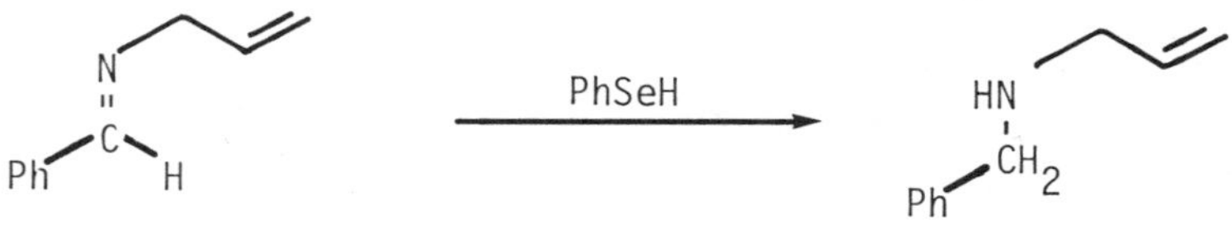

83%

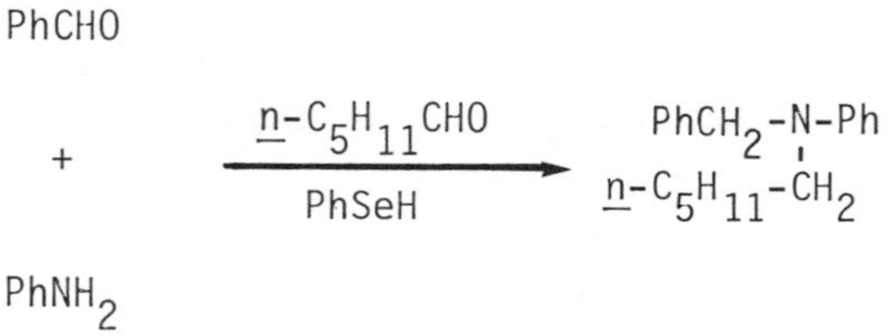

PhCHO

+

PhNH$_2$

Tetr Lett, <u>21</u>, 3385 (1980)

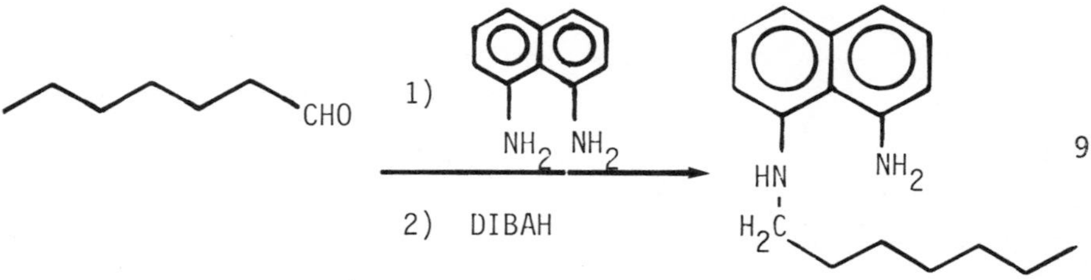

94%

JACS, <u>103</u>, 4186 (1981)

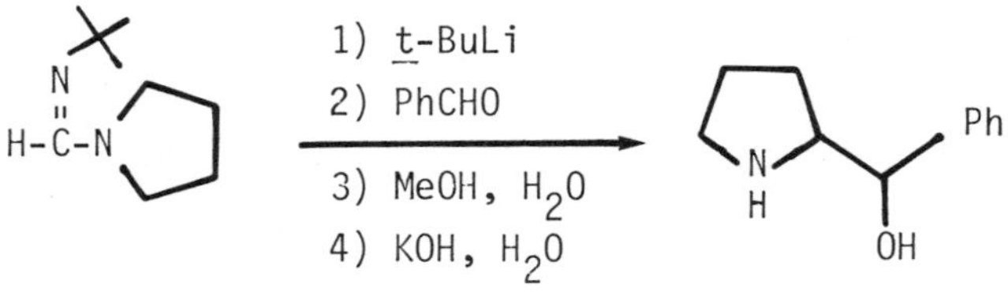

57%

JACS, <u>102</u>, 7125 (1980)

JOC, <u>46</u>, 3119 (1981)

Review: "Catalytic Amination of Alcohols, Aldehydes, and Ketones"

Russ Chem Rev, <u>49</u>, 14 (1980)

Related methods: Amines from Ketones (Section 102)

Section 95 <u>Amines from Alkyls, Methylenes, and Aryls</u>

No examples.

Section 96 <u>Amines from Amides</u>

$$\underline{n}\text{-}C_9H_{19}\text{-}\overset{\overset{\text{O}}{\|}}{C}\text{-NHBz} \quad \xrightarrow[\text{TiCl}_4]{\text{NaBH}_4} \quad \underline{n}\text{-}C_9H_{19}CH_2NHBz \qquad 93\%$$

Synthesis, 695 (1980)

$$\xrightarrow[\text{H}^{\oplus}, \text{ DMSO}]{\text{NaBH}_4} \qquad 73\%$$

JOC, <u>46</u>, 2579 (1981)

$$\xrightarrow[\text{2) ether, HCl}]{\text{1) BH}_3 \cdot \text{SMe}_2} \qquad 77\%$$

Synthesis, 439 (1981)

$$CH_3\text{-}\overset{\overset{\text{S}}{\|}}{C}\text{-NHPh} \quad \xrightarrow{\text{NaBH}_4} \quad CH_3\text{-}CH_2NHPh \qquad 65\%$$

JOC, <u>46</u>, 3730 (1981)

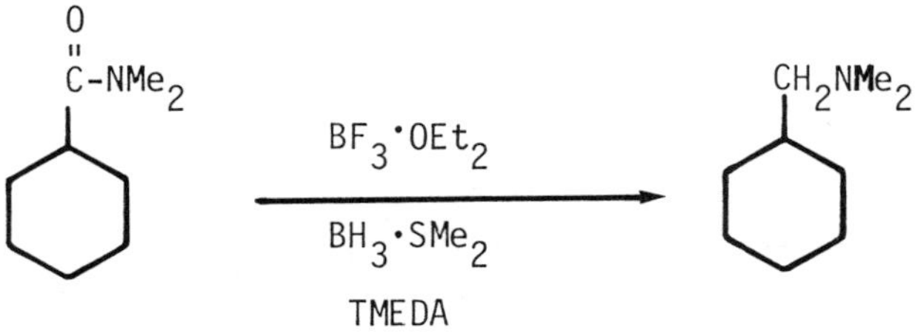

Synthesis, 996 (1981)

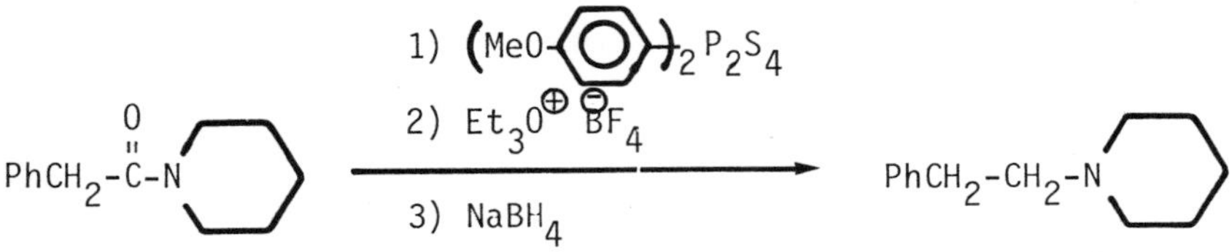

~70%

Tetr Lett, <u>21</u>, 4061 (1980)

Related methods: Protection of Amines (Section 105A)

Section 97 <u>Amines from Amines</u>

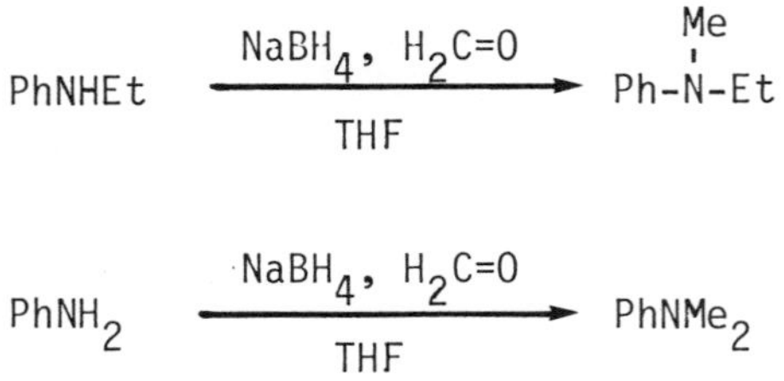

PhNHEt $\xrightarrow[\text{THF}]{\text{NaBH}_4, \text{ H}_2\text{C=O}}$ Ph-N-Et (with Me on N)   87%

PhNH$_2$ $\xrightarrow[\text{THF}]{\text{NaBH}_4, \text{ H}_2\text{C=O}}$ PhNMe$_2$ 98%

Synthesis, 743 (1980)

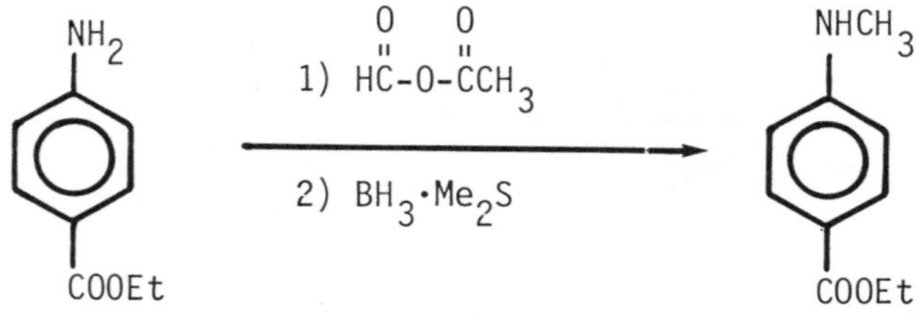

$$\text{HC-O-CCH}_3$$

93%

Tetr Lett, <u>23</u>, 3315 (1982)

$$\text{Bz}_2\text{NH} \xrightarrow{\text{Bu}_2\text{CuLi}} \text{Bz}_2\text{NBu}$$

62%

JOC, <u>45</u>, 2739 (1980)

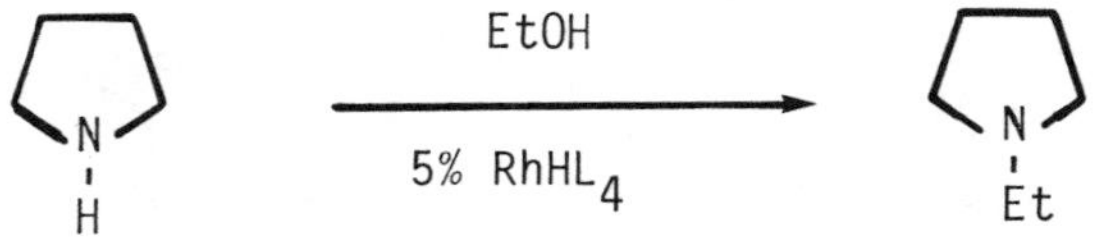

74%

JCS Chem Comm, 611 (1981)

$$\text{Ph-NH}_2 \xrightarrow[\text{RuCl}_2\text{L}_3]{\text{BuOH}} \text{PhNBu}_2$$

79%

Tetr Lett, <u>22</u>, 2667 (1981)

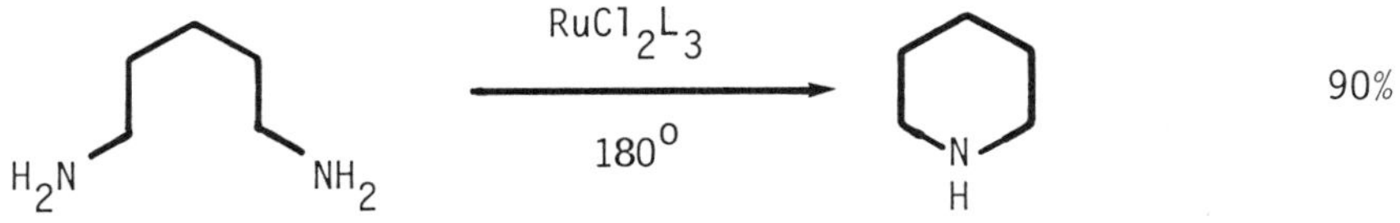

JOC, <u>46</u>, 1759 (1981)

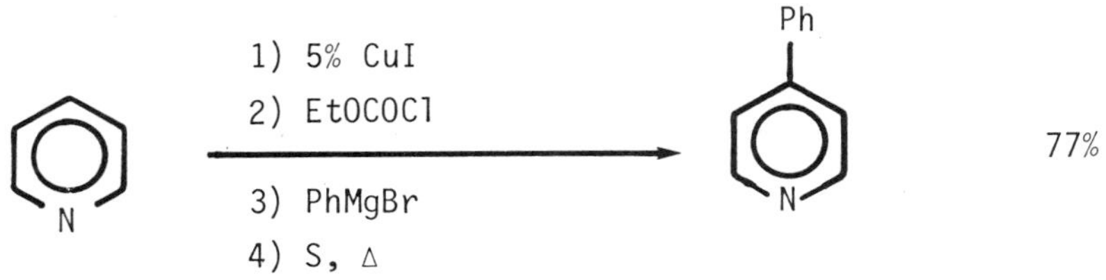

1) 5% CuI
2) EtOCOCl
3) PhMgBr
4) S, Δ

77%

JOC, <u>47</u>, 4315 (1982)

2) H₂O

44%

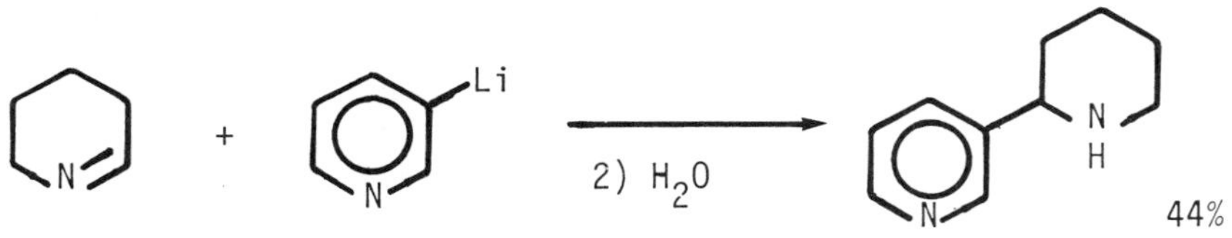

JOC, <u>45</u>, 1515 (1980)

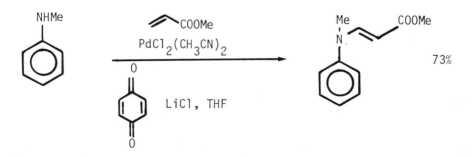

N=CHPh

| 1) (MeO)$_2$SO$_2$ |
| 2) H$_2$O, H$_2$SO$_4$ |
| 3) $^-$OH |

NHMe

NHMe   →   NHMe   99%

N=CHPh

Synthesis, 303 (1980)

1) HCl, ether

2) ⟨N-H⟩

3) H$_3$O$^{\oplus}$

4) NaOH

CH$_2$CH$_2$NH$_2$ 72%

Synthesis, 541 (1981)

NHMe

⌄COOMe

PdCl$_2$(CH$_3$CN)$_2$

LiCl, THF

Me COOMe
N

73%

JOC, <u>46</u>, 2561 (1981)

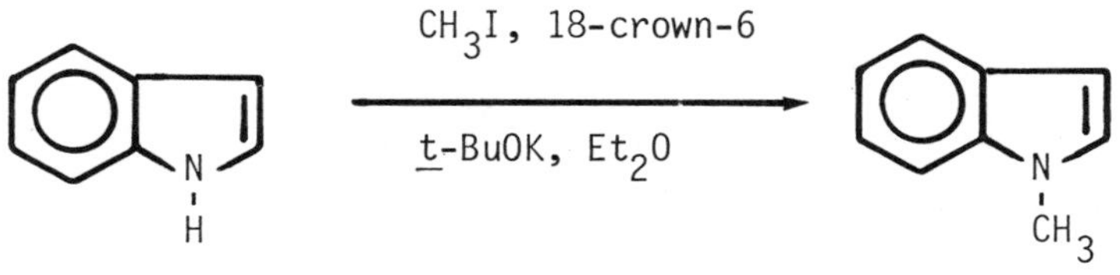
CH_3I, 18-crown-6
t-BuOK, Et_2O
N
H
N
CH_3

87%

JOC, 45, 3172 (1980)

Section 98 Amines from Esters

No additional examples.

Section 99 Amines from Ethers

No additional examples.

Section 100 Amines from Halides

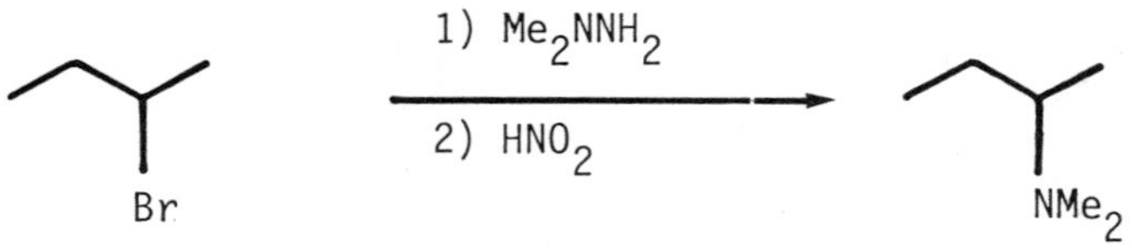
1) Me_2NNH_2
2) HNO_2
Br
NMe_2

60%

Synth Comm, 12, 801 (1982)

Ph-N=PPh$_3$

+

BuI

$\xrightarrow{\quad\text{2)}\quad H_2O, \ ^\ominus OH\quad}$

Ph-NH
|
Bu

72%

Synthesis, 295 (1980)

Me-NH-CHO

KOH

90%

Synthesis, 39 (1980)

$\underline{n}$-C$_8$H$_{17}$-N$_3$ $\xrightarrow[\text{Toluene, P.T.C.}]{\text{NaBH}_4, \ H_2O}$ $\underline{n}$-C$_8$H$_{17}$-NH$_2$ 92%

(from the halide)

JOC, <u>47</u>, 4327 (1982)

Li

$\xrightarrow{\quad CH_3ONH_2-CH_3Li\quad}$

NH$_2$

67%

(isolated as the benzamide)

JOC, <u>47</u>, 2822 (1982)

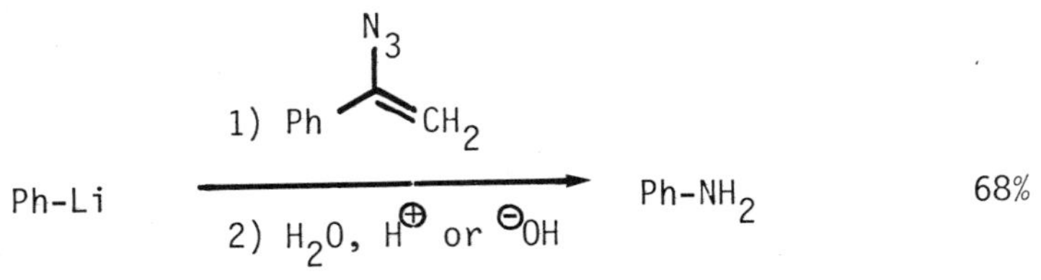

1) Ph $\overset{N_3}{\underset{}{\diagdown}}$ CH$_2$

Ph-Li → Ph-NH$_2$ 68%

2) H$_2$O, H$^\oplus$ or $^\ominus$OH

Many aromatic and heterocyclic examples.

Tetr Lett, <u>23</u>, 699 (1982)

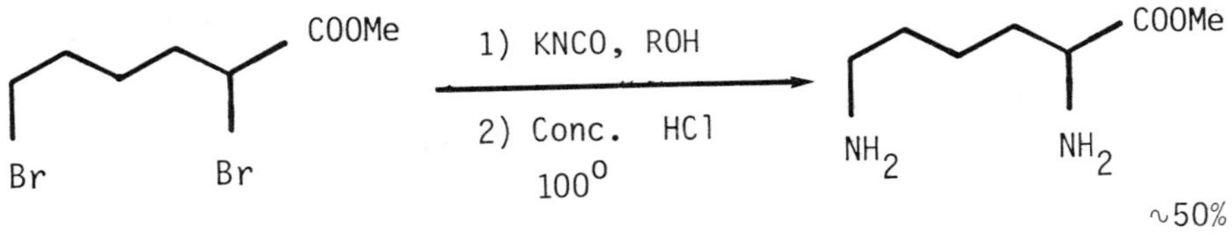

1) KNCO, ROH

2) Conc. HCl
   100°

~50%

Chem Ber, <u>114</u>, 173 (1981)

CH$_3$CN

61%

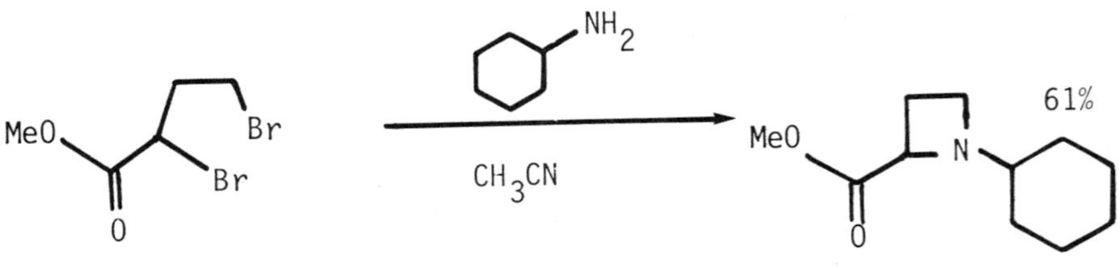

JOC, <u>46</u>, 2991 (1981)

n-C$_8$H$_{17}$Br $\xrightarrow[\text{18-crown-6, toluene}]{}$ ... N-C$_8$H$_{17}$ 94%

Bull Chem Soc Japan, **55**, 1671 (1982)

Hx-CH=CF$_2$ $\xrightarrow[\text{Et}_2\text{O}]{\text{LDA}}$ Hx-C≡C-N($\underline{i}$-Pr)$_2$ 70%

Chem Lett, 935 (1980)

Section 101 Amines from Hydrides

OTMS ... $\overset{\oplus}{\text{Me}_2\text{N}}$=CH$_2$ $\xrightarrow{\hspace{2cm}}$ ... NMe$_2$

79%

Tetr Lett, **21**, 805 (1980)

Me$_3$N

+ $\xrightarrow{\text{electrolysis}}$ Me$_2$NCH$_2$CH(COOEt)$_2$ 53%

H$_2$C(COOEt)$_2$

Tetrahedron, **37**, 2297 (1981)

Section 102 Amines from Ketones

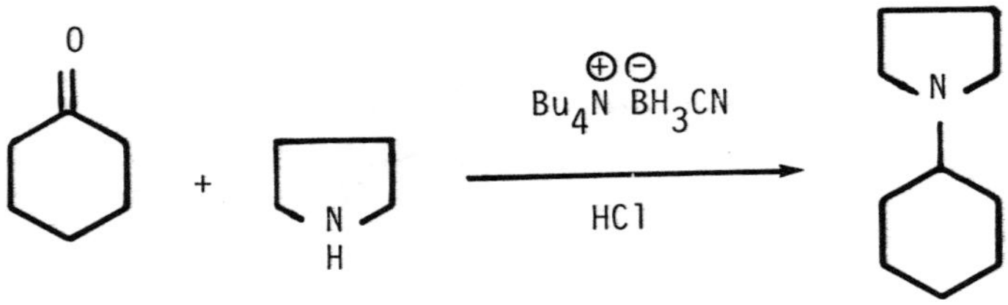

JOC, 46, 3571 (1981)

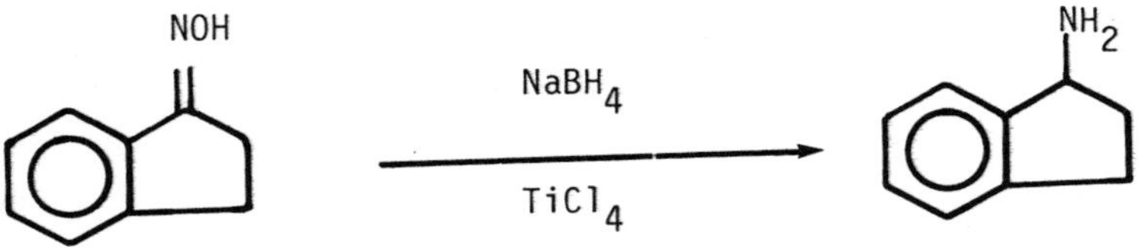

Synthesis, 695 (1980)

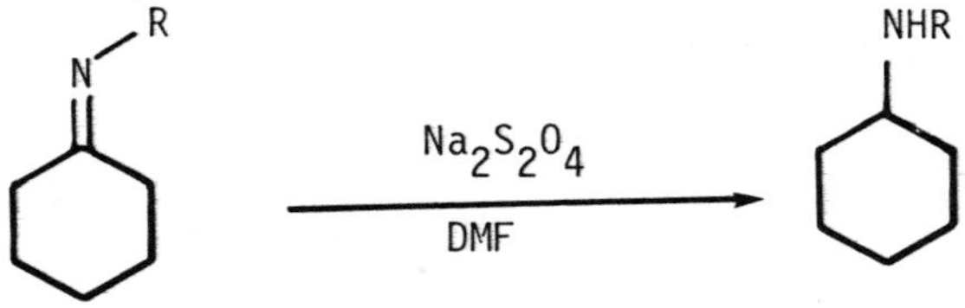

R = Bz, c-Hx

Aust J Chem, 32, 201 (1979)

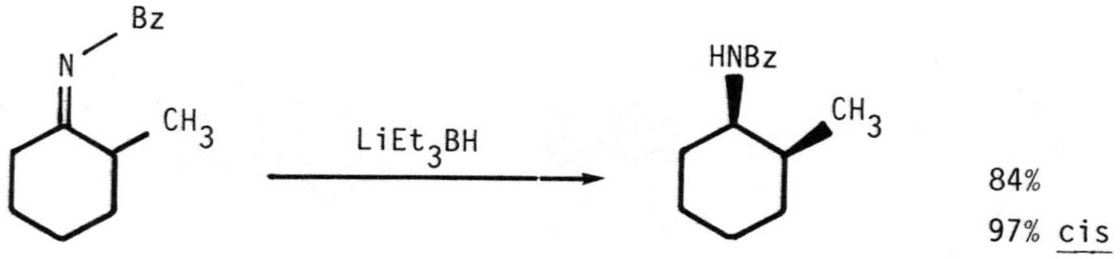

Tetr Lett, 22, 3447 (1981)

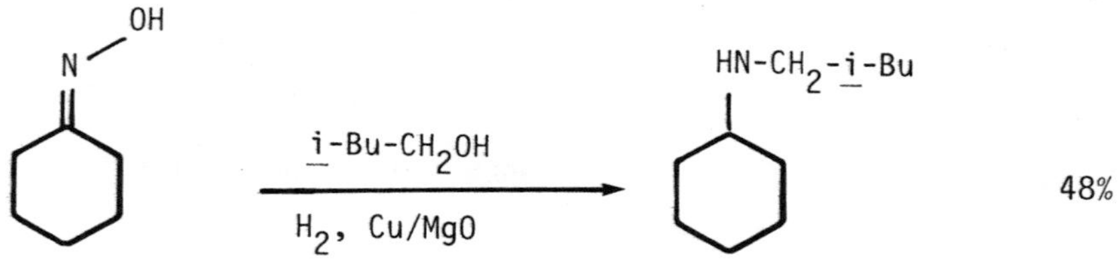

84%
97% cis

48%

Doklady Chem, 244, 69 (1979)

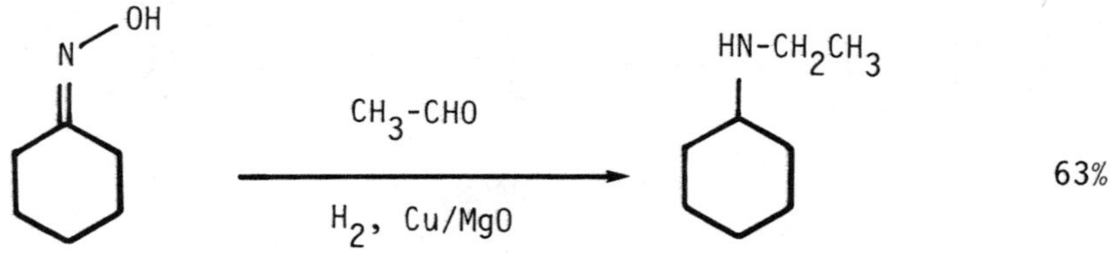

63%

Doklady Chem, 244, 69 (1979)

1) t-BuLi
2) (cyclohexanone) = O
3) MeOH, H₂O
4) KOH, H₂O

45%

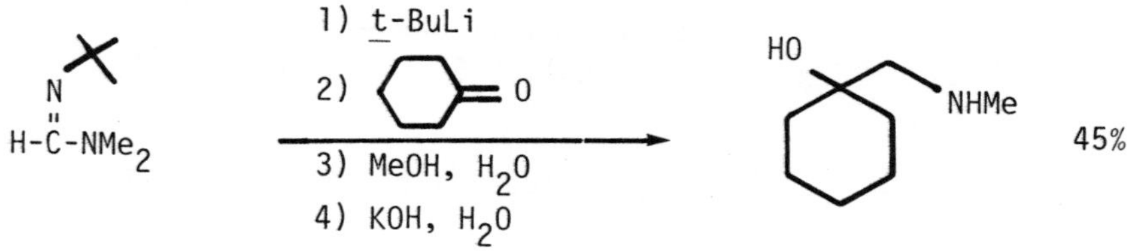

JACS, 102, 7125 (1980)

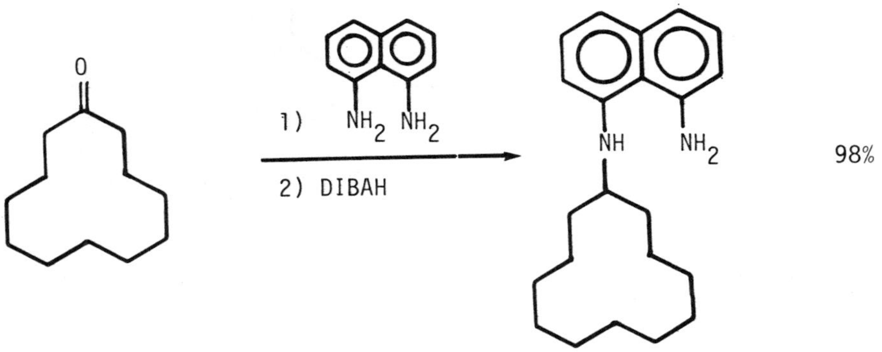

JACS, 103, 4186 (1981)

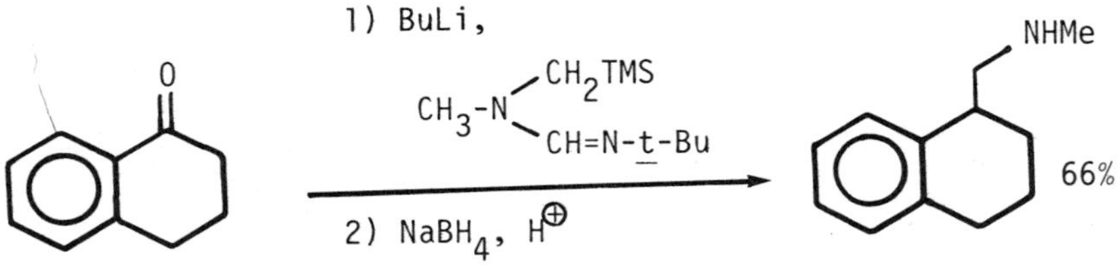

JACS, 104, 877 (1982)

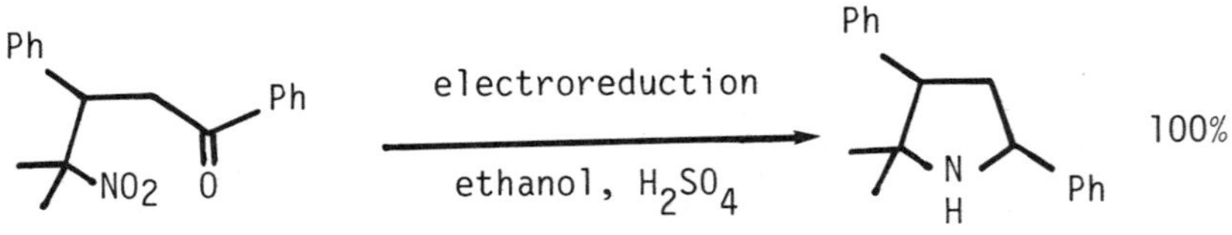

Tetr Lett, 22, 3961 (1981)

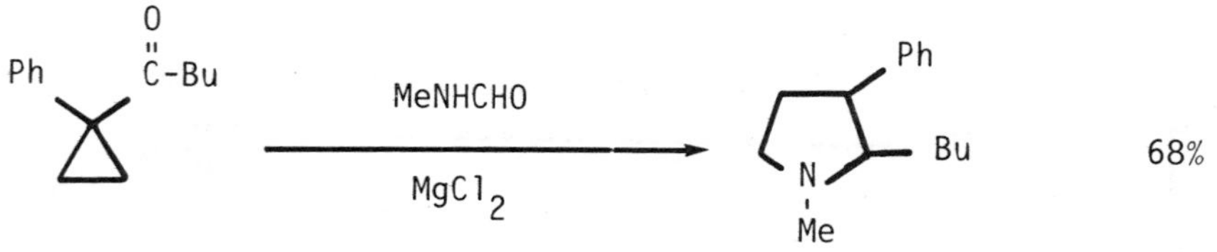

JCS Perkin I, 700 (1981)

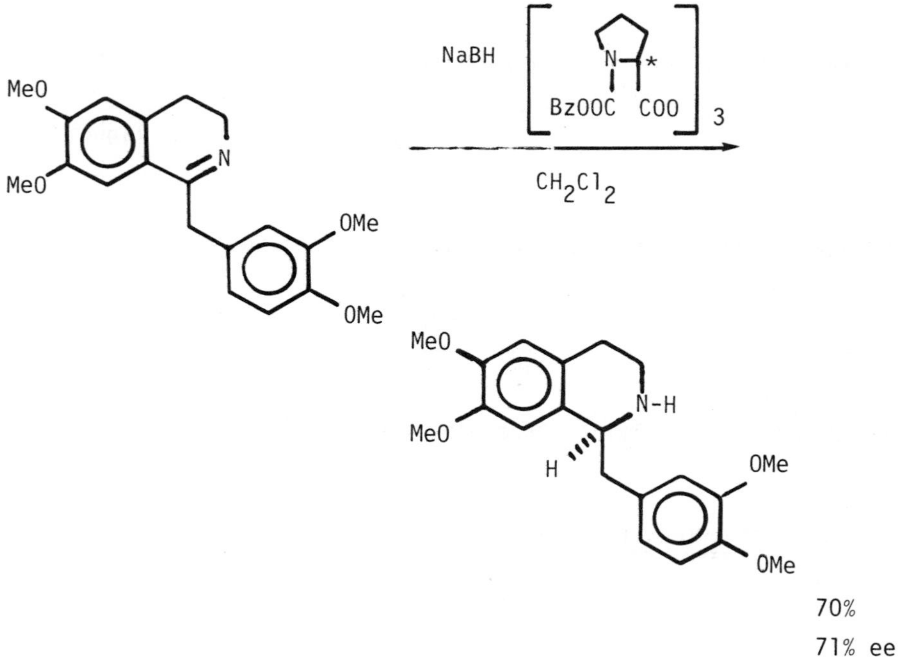

70%

71% ee

Tetr Lett, 22, 3869 (1981)

(S)-pyradoxamine analog

Zn^{++}

68%

96% ee

Chem Lett, 1765 and 1769 (1982)

Review: "Catalytic Amination of Alcohols, Aldehydes, and Ketones"

Russ Chem Rev, 49, 14 (1980)

Review: "The Friedländer Synthesis of Quinolines"

Org React, 28, 37 (1982)

Related methods: Amines from Aldehydes (Section 94)

Section 103 Amines from Nitriles

1) Me$_2$S·BH$_3$
2) HCl, H$_2$O

3) NaOH

72%

Synthesis, 605 (1981)

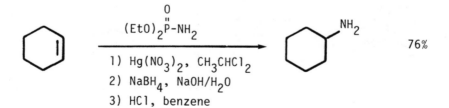

Synthesis, 270 (1981)

Section 104 Amines from Olefins

JCS Chem Comm, 62 (1982)
JOC, 46, 4296 (1981)

Synthesis, 918 (1982)

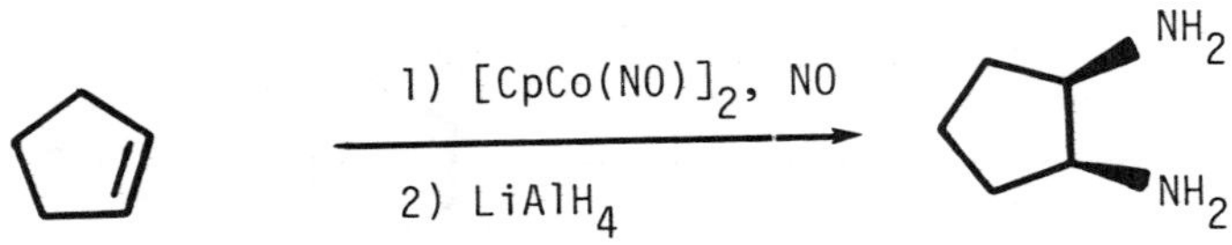
1) [CpCo(NO)]$_2$, NO
2) LiAlH$_4$
NH$_2$
NH$_2$

70%

JACS, 102, 5676 (1980)

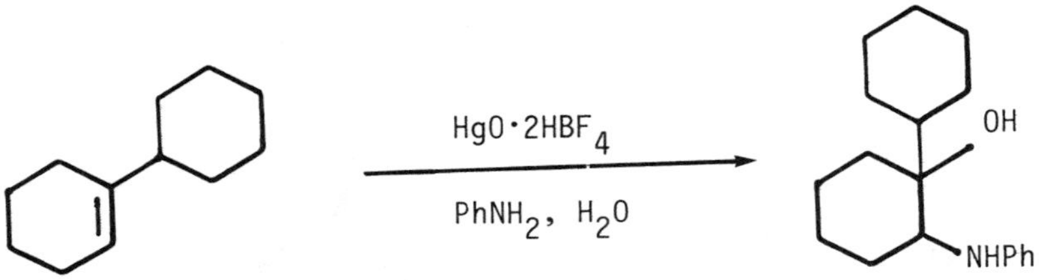
HgO·2HBF$_4$
PhNH$_2$, H$_2$O
OH
NHPh

80%

Synthesis, 376 (1981)

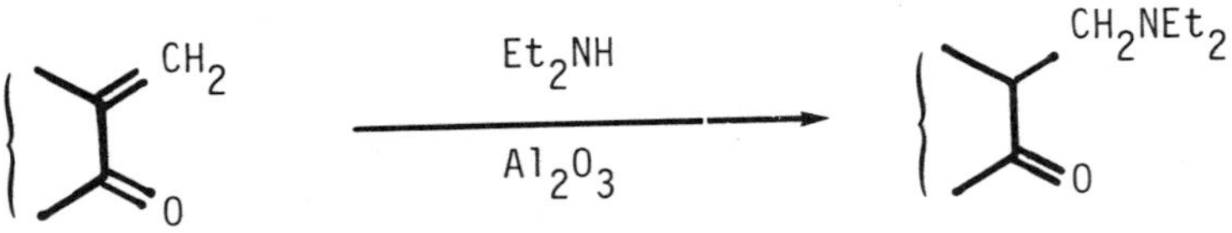
CH$_2$
O
Et$_2$NH
Al$_2$O$_3$
CH$_2$NEt$_2$
O

100%

Tetr Lett, 21, 809 (1980)

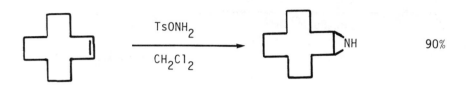

Tetrahedron, <u>37</u>, 2895 (1981)

JCS Chem Comm, 560 (1980)

Tetrahedron, <u>36</u>, 73 (1980)

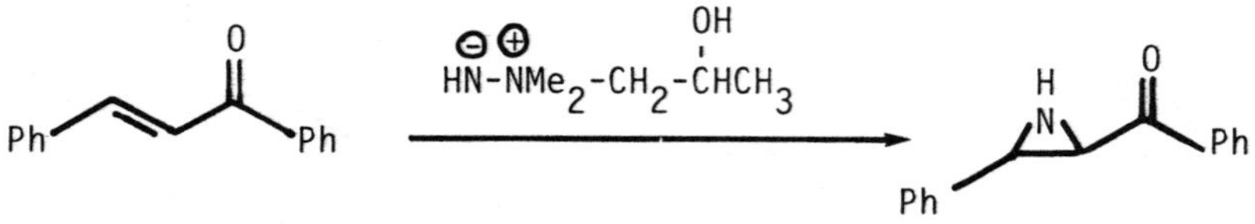

74%

Synthesis, 650 (1980)

Section 105 Amines from Miscellaneous Compounds

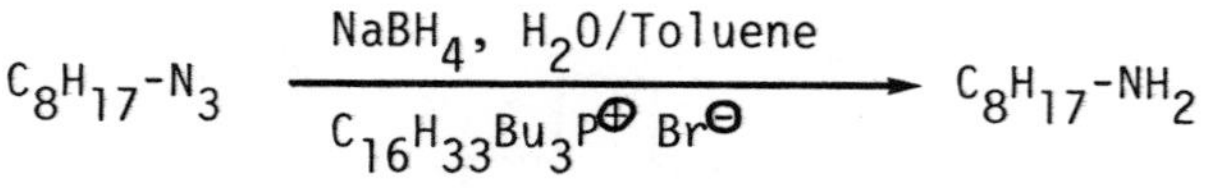

92%

JOC, 47, 4327 (1982)

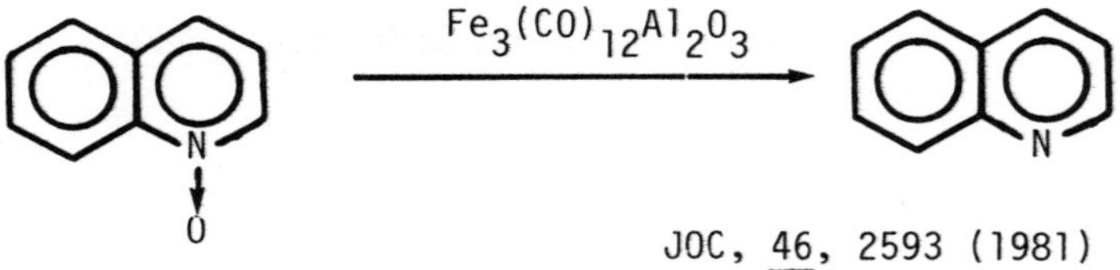

79%

JOC, 46, 2593 (1981)

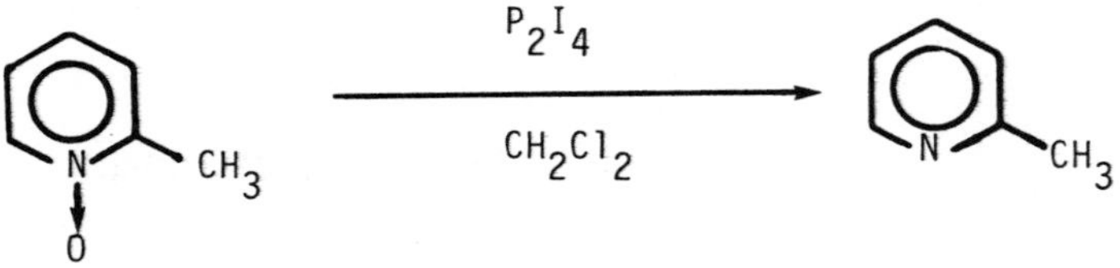

95%

Chem Lett, 459 (1980)

Synthesis, 741 (1980)

$Ph-NO_2 \xrightarrow[\text{EtOH}]{\text{NaBH}_4\text{-SnCl}_2} Ph-NH_2$

Chem Pharm Bull, <u>29</u>, 1443 (1981)

Chem Pharm Bull, <u>29</u>, 1159 (1981)

Tetr Lett, <u>23</u>, 147 (1982)

Synth Comm, <u>11</u>, 925 (1981)

JOC, <u>45</u>, 4992 (1980)

Synth Comm, <u>12</u>, 293 (1982)

$$Ph-NO_2 \xrightarrow{\ Al_2Te_3,\ H_2O\ } Ph-NH_2 \qquad 90\%$$

Angew Chem Int Ed, <u>19</u>, 1008 and 1010 (1980)

COPh
$Fe_3(CO)_{12}$
————————→
Al_2O_3, hexane
COPh   68%
NH_2
NO_2

JCS Chem Comm, 821 (1980)

NO_2
$TiCl_3$
————————→
H_2O
N-Me
O

NH_2 86%
N-Me
O

Chem Pharm Bull, 28, 2515 (1980)

COPh
CO, $Ru_3(CO)_{12}$
————————→
NaOH, H_2O/benzene
P.T.C.
NO_2

COPh   100%
NH_2

Tetr Lett, 21, 2603 (1980)

Review: "Dihydropyridine Equivalents as Intermediates for the
          Synthesis of Alkaloids"

Bull Soc Chim Belges, 91, 985 (1982)

Section 105A Protection of Amines

Related methods: Amides from Amines (Section 82); Amines from
                   Amides (Section 96)

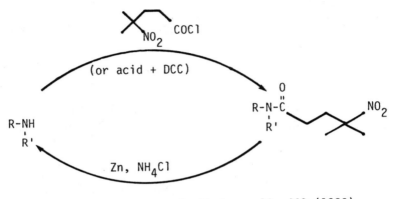

Synth Comm, <u>10</u>, 469 (1980)

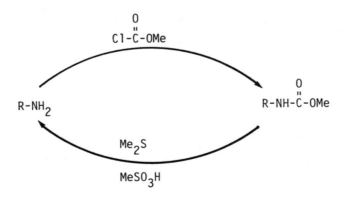

Chem Lett, 705 (1980)

$$\underset{\substack{|\\ H}}{\overset{\overset{O}{\|}}{R-N-C}}-OBz \xrightarrow[Me_2NAc, \ NH_3(1)]{H_2, \ Pd, \ Et_3N} R-NH_2 \qquad \sim 80\%$$

R may contain a wide variety of other functional and protecting groups.

Org Syn, <u>59</u>, 159 (1980)

$$\underset{\substack{|\\ H}}{\overset{\overset{O}{\|}}{R-N-C}}-OBz \xrightarrow[10\% \ Pd/C]{\overset{\oplus}{NH_4}, \ HCOO^{\ominus}} RNH_2 \qquad >90\%$$

Synthesis, 929 (1980)

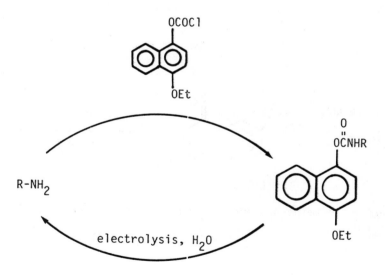

Tetr Lett, <u>22</u>, 3719 (1981)

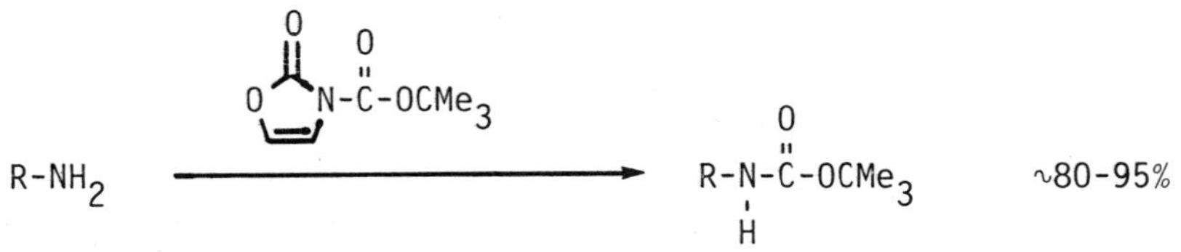

R-NH$_2$ $\longrightarrow$ R-N-C-OCMe$_3$ ~80-95%

Tetr Lett, 21, 3065 (1980)

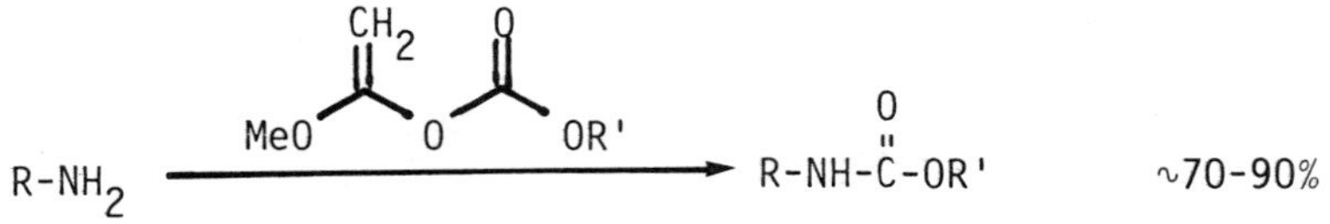

R-NH$_2$ $\longrightarrow$ R-NH-C-OR'   ~70-90%

R' = t-Bu, Bz, -CH$_2$CH$_2$TMS

JOC, 47, 2697 (1982)

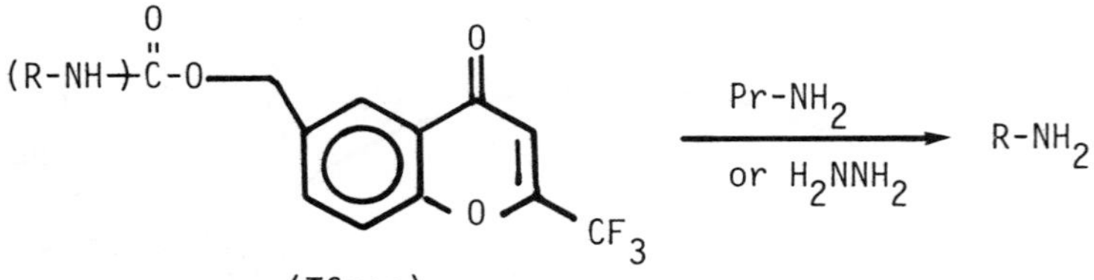

(TCroc)

$\xrightarrow[\text{or H}_2\text{NNH}_2]{\text{Pr-NH}_2}$ R-NH$_2$

Used as an amine-protecting group in peptide synthesis.

JOC, 46, 4971 (1981)

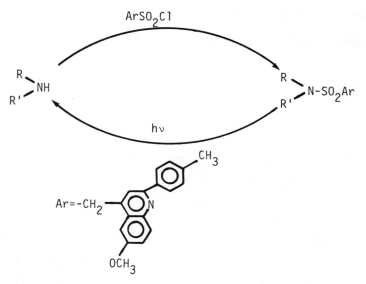

Tetr Lett, 23, 3843 (1982)

$$R-\underset{\underset{R'}{|}}{N}-Ts \xrightarrow[\text{diglyme}]{\text{K, crown ether}} R-\underset{\underset{R'}{|}}{N}H$$

Chem Pharm Bull, 30, 3178 (1982)

Ac$_2$O / pyridine   ~100%

JOC, 45, 547 (1980)

Use of various multisubstituted benzenesulfonyl protecting groups for the guanidino function of arginine.   Removed by TFA-thioanisole.

Chem Pharm Bull, 29, 2825 (1981)

Used for tryptophan in peptide synthesis.

Chem Pharm Bull, 30, 2825 (1982)

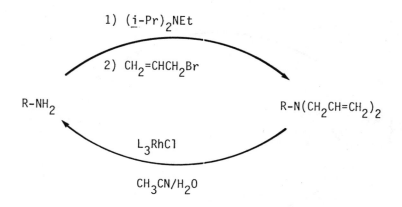

1) $(\underline{i}\text{-Pr})_2$NEt

2) CH_2=CHCH$_2$Br

R-NH$_2$ R-N(CH$_2$CH=CH$_2$)$_2$

L$_3$RhCl

CH$_3$CN/H$_2$O

Tetr Lett, <u>22</u>, 1483 (1981)

$$\overset{\oplus}{H_3}N\text{-Asn-}\overset{\overset{O}{\|}}{C}\text{-OTMS} \quad \xrightarrow[\text{3) MeOH}]{\substack{\text{1) Et}_3\text{N} \\ \text{2) Ph}_3\text{CCl}}} \quad Ph_3C\text{-}\overset{H}{\underset{|}{N}}\text{-Asn-COOH} \qquad 75\%$$

JOC, <u>47</u>, 1324 (1982)

The DPPM group is used to protect histidine in peptide synthesis. It is stable to acid, but cleaved by Zn/HOAc or electrolytic reduction.

N⬡—CPh$_2$— = DPPM

JCS Perkin I, 522 (1981)

$\overset{\oplus}{R}$-NMe$_3$ $\xrightarrow{\text{L-Selectride}}$ R-NMe$_2$ ~80-90%

I$^-$

Org Prep Proc Int, <u>12</u>, 345 (1980)

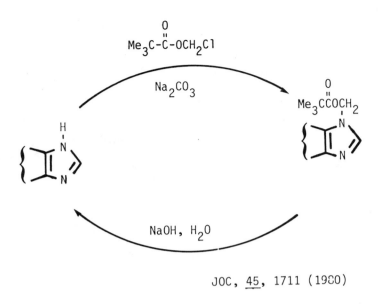

JOC, <u>45</u>, 1711 (1980)

Use of the ε-phenylacetyl protecting group for lysine in peptide synthesis. Deprotected by a penicillin aminohydrolase enzyme.

Coll Czech Chem Comm, <u>46</u>, 1983 (1981)

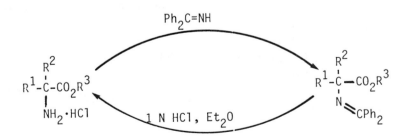

JOC, <u>47</u>, 2663 (1982)

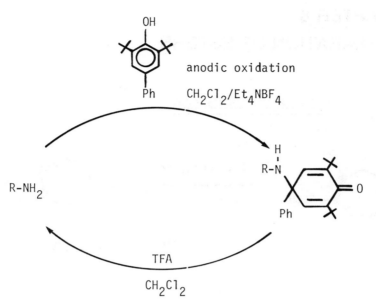

anodic oxidation

CH_2Cl_2/Et_4NBF_4

TFA

CH_2Cl_2

Used for amino acid esters.

Angew Chem Int Ed, <u>19</u>, 712 (1980)

CHAPTER 8
PREPARATION OF ESTERS

Section 106 Esters from Acetylenes

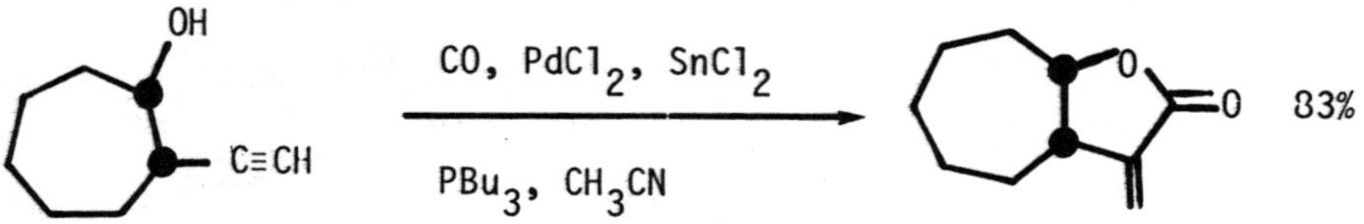
OH
C≡CH
CO, PdCl2, SnCl2
PBu3, CH3CN
O
O
83%

JACS, 103, 7520 (1981)

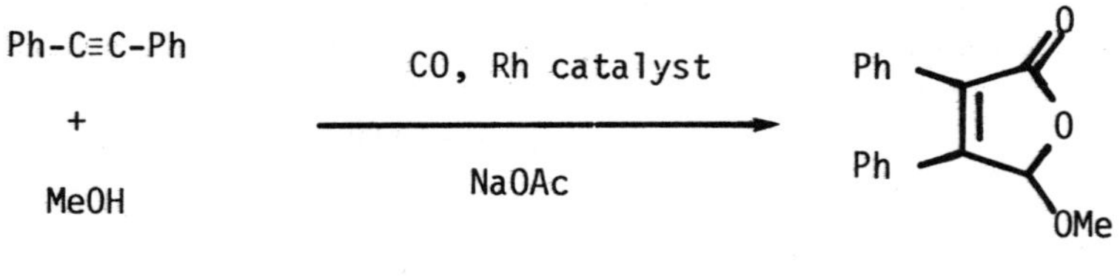
Ph-C≡C-Ph
+
MeOH
CO, Rh catalyst
NaOAc
Ph
Ph
O
O
OMe

Chem Lett, 993 (1981)

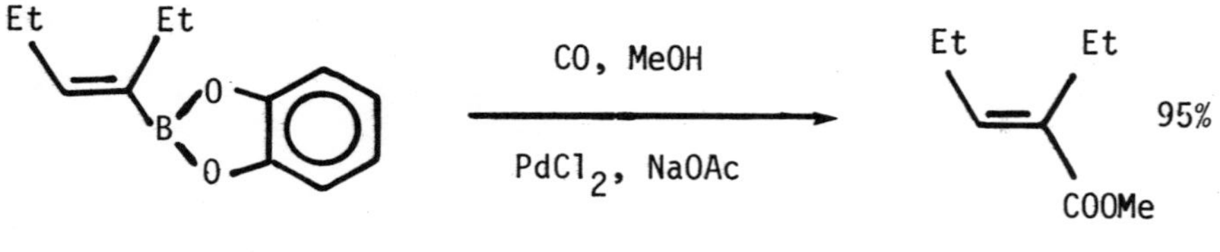
Et
Et
O
B
O
CO, MeOH
PdCl2, NaOAc
Et
Et
COOMe
95%

Chem Lett, 879 (1981)

Section 107 <u>Esters from Carboxylic Acids, Acid Halides, and</u>

<u>Anhydrides</u>

The following types of reactions are found in this section:

1.   Esters from the reaction of alcohols with carboxylic acids,
     acid halides, and anhydrides.

2.   Lactones from hydroxy acids.

3.   Esters from carboxylic acids and halides, sulfoxides, and
     miscellaneous compounds.

R-COOH

+ $\xrightarrow[\text{polymer-bound}]{\text{DCC}}$ $R\text{-}\overset{\displaystyle O}{\overset{\|}{C}}\text{-}OR'$

R'-OH

NMe$_2$

Bull Chem Soc Japan, <u>54</u>, 631 (1981)

$\overset{*}{\text{NPS-NH-CH-COOH}}$ $\xrightarrow[\text{BzOH}]{\text{DCC}}$ $\overset{*}{\text{NPS-NH-CH}}\text{-}\overset{\displaystyle O}{\overset{\|}{C}}\text{-}OBz$ 92%

CH$_2$Ph   CH$_2$Ph

No racemization.

JCS Chem Comm, 1132 (1982)

Z-Pro-COOH $\xrightarrow[\text{DMAP, CH}_2\text{Cl}_2]{\underline{t}\text{-BuOH, carbodiimide}}$ Z-Pro-COO-$\underline{t}$-Bu   88%

No racemization.

JOC, <u>47</u>, 1962 (1982)

Ph-COOH $\xrightarrow[\text{2) BzOH, Et}_3\text{N}]{\text{1) SO}_2\text{ClF, Et}_3\text{N}}$ Ph-$\overset{\overset{\text{O}}{\|}}{\text{C}}$-OBz   80%

Synthesis, 790 (1981)

Synth Comm, <u>12</u>, 727 (1982)

$C_{11}H_{23}COOH$ $\xrightarrow[\text{2) EtOH}]{\begin{array}{c}\text{1) ClSO}_2\text{NCO,}\\\text{Et}_3\text{N}\end{array}}$ $C_{11}H_{23}COOEt$ 80%

Synthesis, 506 (1982)

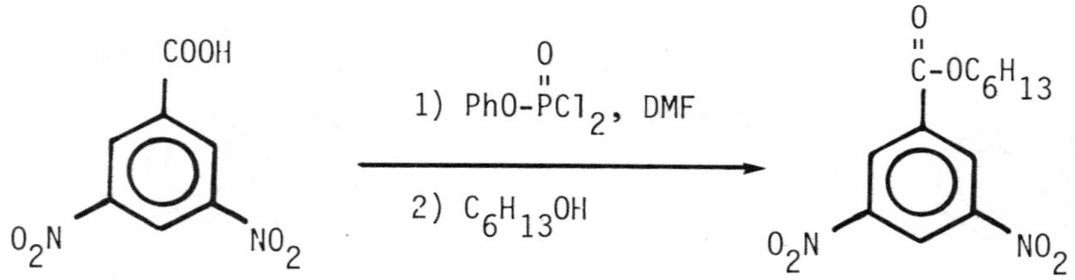

Synth Comm, 12, 681 (1982)

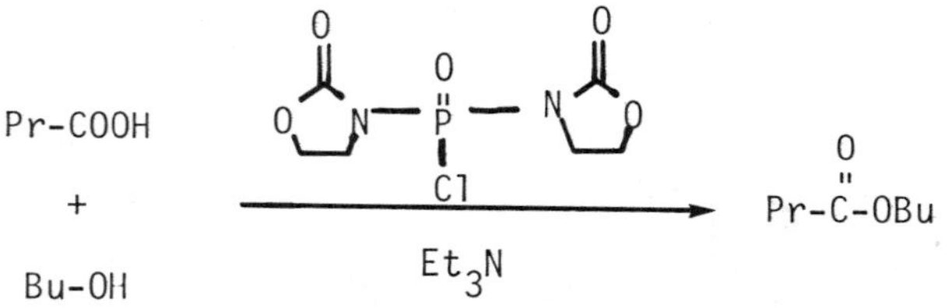

Synthesis, 547 (1980)

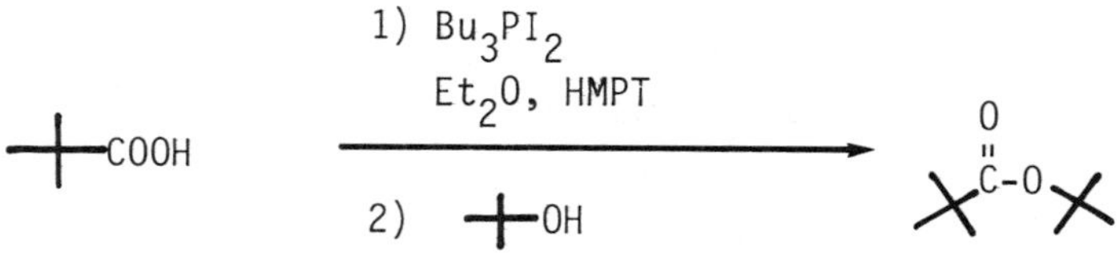

Aust J Chem, 35, 517 (1982)

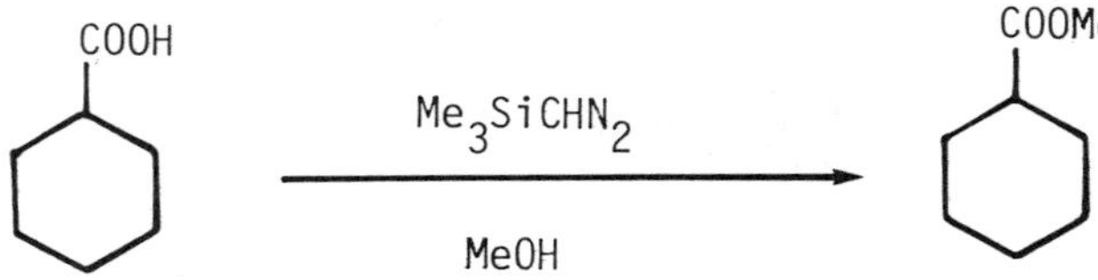

Chem Pharm Bull, 29, 1475 (1981)

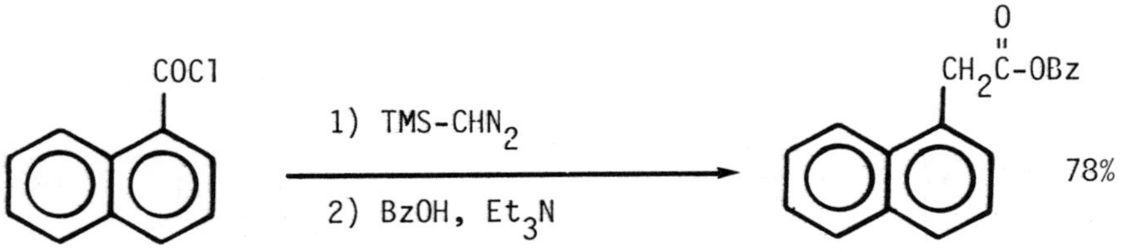

78%

Chem Pharm Bull, 29, 3249 (1981)
Tetr Lett, 21, 4461 (1980)

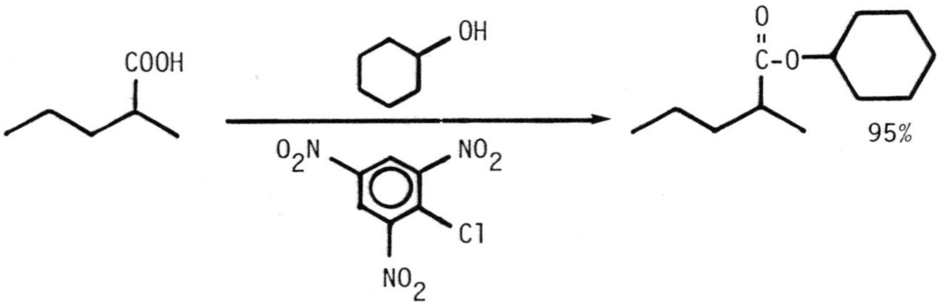

95%

Bull Chem Soc Japan, 54, 1470 (1981)

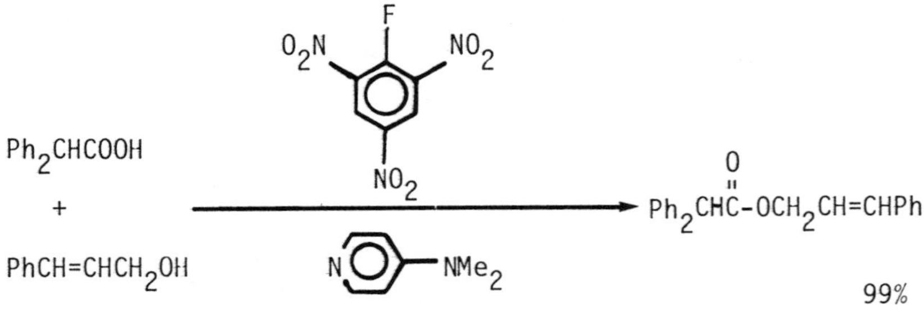

99%

Synth Comm, 11, 121 (1981)

$\underline{n}\text{-}C_5H_{11}COOH$

+

$\underline{t}\text{-}BuOH$

Cl
   \
    C=NMe$_2$ ⊕
   /
Me$_2$N

pyridine

→

$\underline{n}\text{-}C_5H_{11}\text{-}\overset{\displaystyle O}{\overset{\|}{C}}\text{-}O\text{-}\underline{t}\text{-}Bu$

77%

Chem Lett, 1891 (1982)

$Ph(CH_2)_3COOH$

+

$Ph(CH_2)_3OH$

F—[pyridine ring] ⊕
      |
      Et

BF$_4$ ⊖

CsF

→

$Ph(CH_2)_3\text{-}\overset{\displaystyle O}{\overset{\|}{C}}$
$Ph(CH_2)_3\text{-}O$

94%

Chem Lett, 391 (1980)

Ph—CH=CH—COOH

[benzisothiazole-pyridone reagent with NO$_2$]
O$_2$N

BzOH, Et$_3$N

→

Ph—CH=CH—$\overset{\displaystyle O}{\overset{\|}{C}}$-OBz

92%

Chem Lett, 1161 (1980)

$Me_3C-COOH$

1) [8-quinolinesulfonyl tetrazole reagent]

2) BzOH

$\longrightarrow$ $Me_3C-\overset{O}{\overset{\|}{C}}-OBz$ 79%

Chem Pharm Bull, 30, 2633 (1982)

$Ph-CH_2CH_2CH_2COOH$

1) $\left(\overset{\frown}{\underset{N}{N}}N-\right)_2 C=O$

2) t-BuOH, DBU

$\longrightarrow$ $Ph-CH_2CH_2CH_2\overset{O}{\overset{\|}{C}}-O-\underline{t}-Bu$ 75%

Synthesis, 833 (1982)

$\underline{n}-C_{17}H_{35}\overset{O}{\overset{\|}{C}}-O-\overset{O}{\overset{\|}{C}}-CF_3$ $\xrightarrow{CH_3OH}$ $\underline{n}-C_{17}H_{35}\overset{O}{\overset{\|}{C}}-OMe$ 91%

Can J Chem, 59, 2617 (1981)

$HO(CH_2)_{14}COOH$ $\xrightarrow[Bu_2SnO]{\Delta}$ $(CH_2)_{14} \overset{C=O}{\underset{O}{\Big|}}$ 63%

JACS, 102, 7578 (1980)

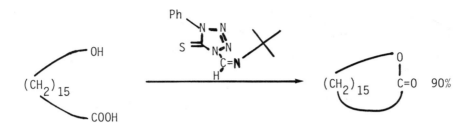

Angew Chem Int Ed, 20, 771 (1981)

Tetr Lett, 21, 1893 (1980)   86%

Org Prep Proc Int, 12, 225 (1980)

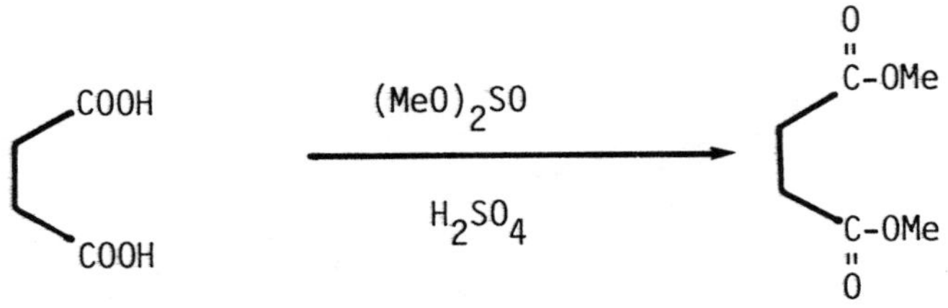

Indian J Chem, 21B, 259 (1982)

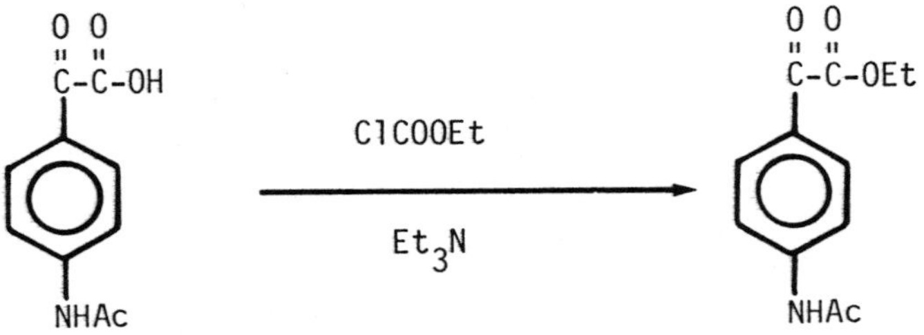

Tetr Lett, 21, 4997 (1980)

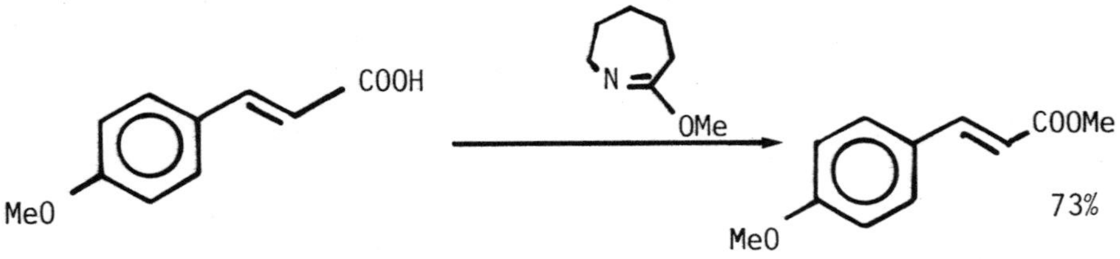

Synth Comm, 12, 453 (1982)

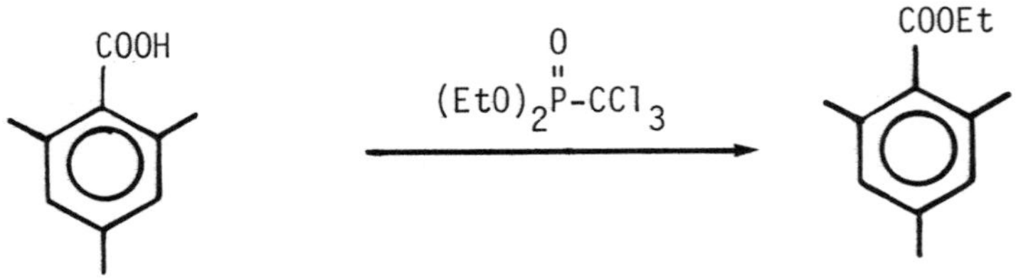

Tetrahedron, 38, 1457 (1982)

Review: "Recent Developments in Methods for the Esterification and
          Protection of the Carboxyl Group"

                              Tetrahedron, <u>36</u>, 2409 (1980)

Further examples of the reaction RCOOH + ROH → RCOOR are included
in Section 108 (Esters from Alcohols and Phenols) and Section 10A
(Protection of Carboxylic Acids).

Section 108 <u>Esters from Alcohols and Phenols</u>

EtOAc

neutral alumina

71%

Synthesis, 789 (1981)
Tetr Lett, <u>22</u>, 5003 and 5007 (1981)

$$\underline{n}\text{-}C_8H_{17}OH \xrightarrow[\text{DBU, } CH_2Cl_2]{Ph\text{-}\overset{O}{\overset{\|}{C}}\text{-}\overset{O}{\overset{\|}{P}}(OEt)_2} \underline{n}\text{-}C_8H_{17}\text{-}O\text{-}\overset{O}{\overset{\|}{C}}\text{-}Ph$$ 85%

                    Tetr Lett, <u>22</u>, 3617 (1981)

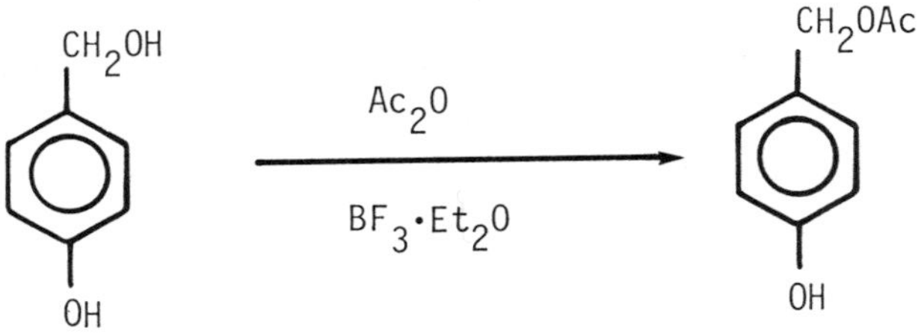

68%

Chem Pharm Bull, 29, 3202 (1981)

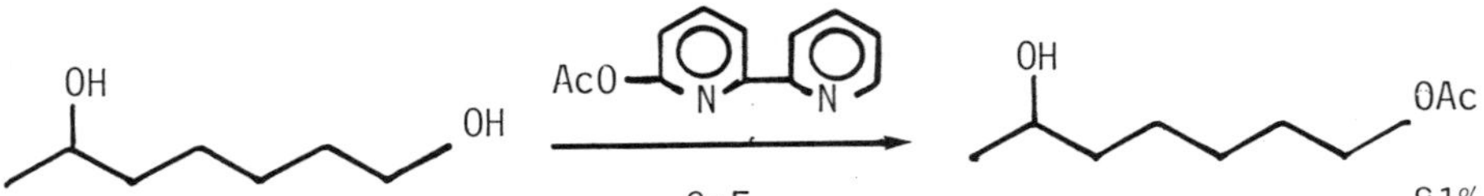

81%

Chem Lett, 563 (1981)

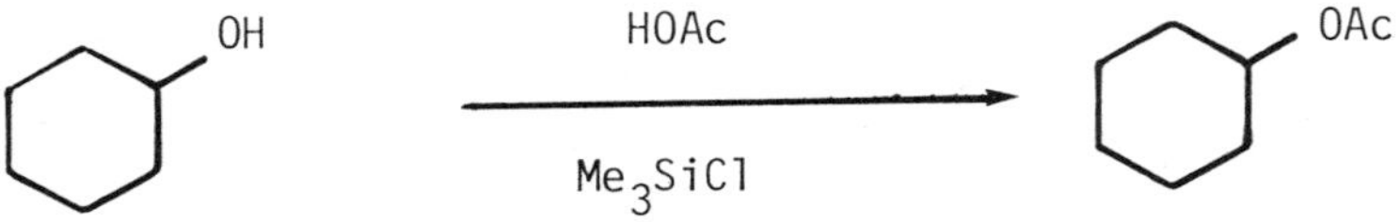

81%

Bull Chem Soc Japan, 54, 1267 (1981)

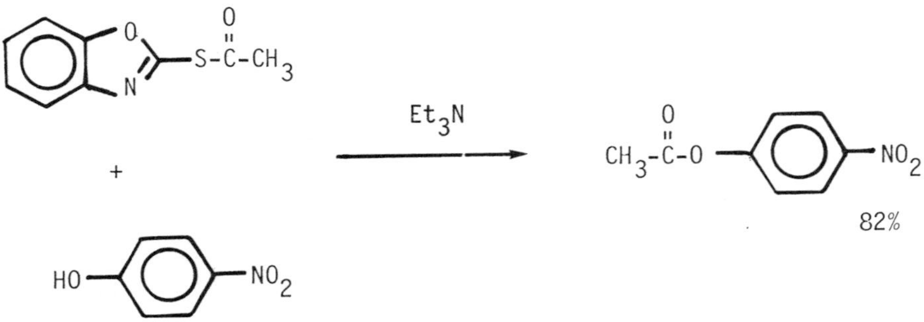

82%

Synthesis, 991 (1981)

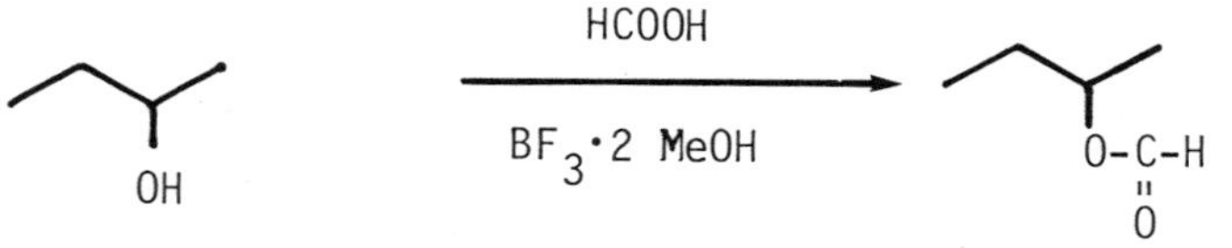

Org Prep Proc Int, 14, 177 (1982)

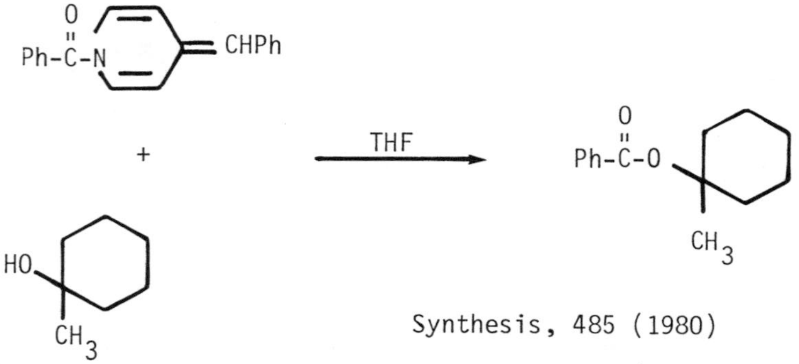

Synthesis, 485 (1980)

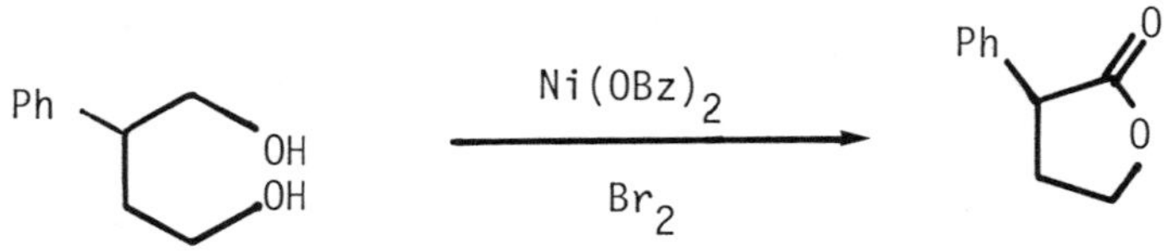

Synth Comm, 10, 881 (1980)

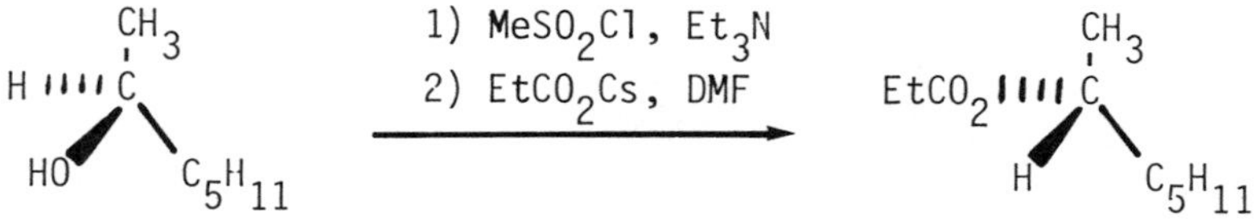

JOC, 46, 4321 (1981)

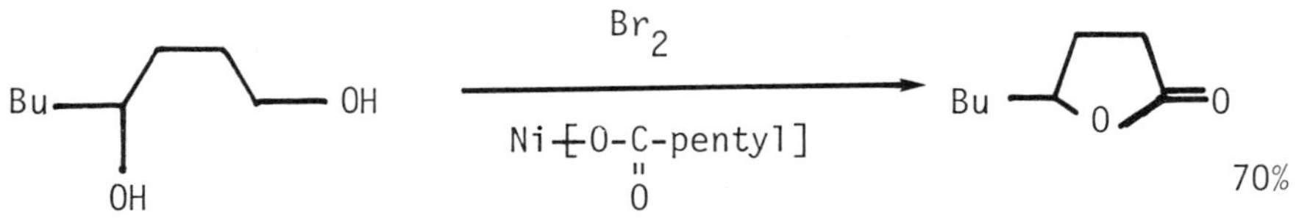

JOC, 46, 4806 (1981)

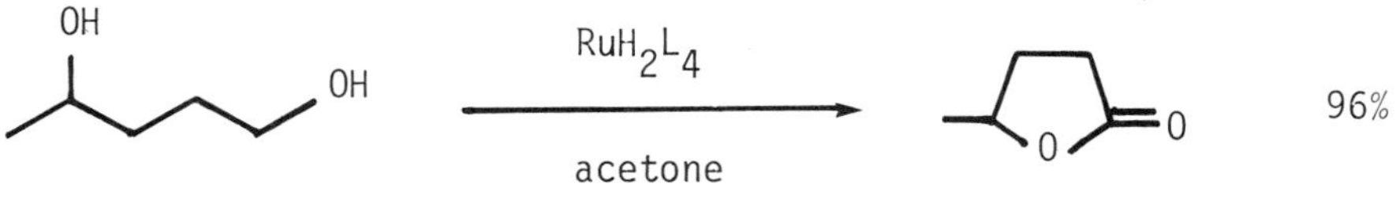

Tetr Lett, 22, 5327 (1981)

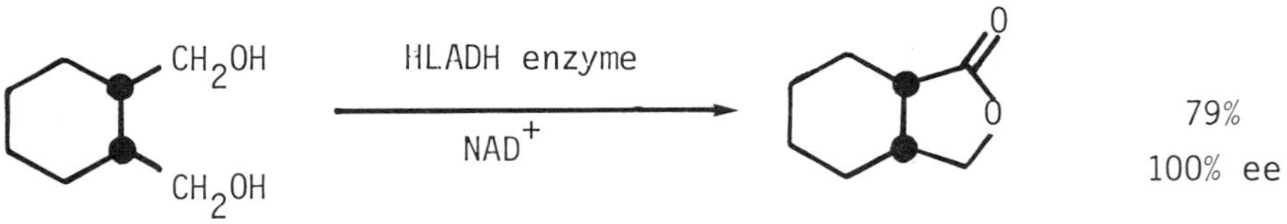

JCS Chem Comm, 515 (1980)
JACS, 104, 4659 (1982)

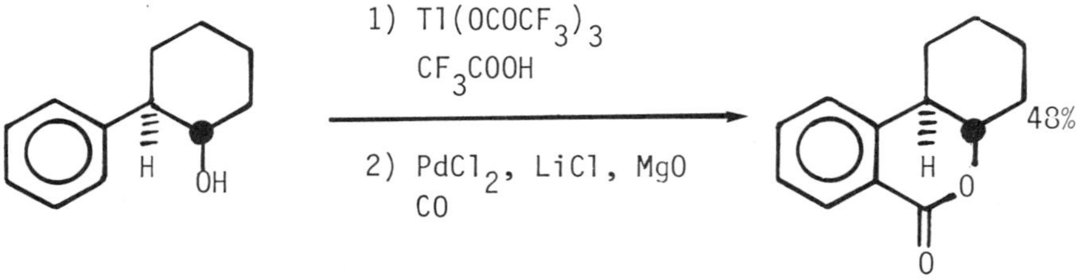

JACS, 104, 1900 (1982)

Further examples of the reaction ROH →R'COOR are included in
Section 107 (Esters from Carboxylic Acids and Acid Halides) and
Section 45A (Protection of Alcohols and Phenols).

Section 109 Esters from Aldehydes

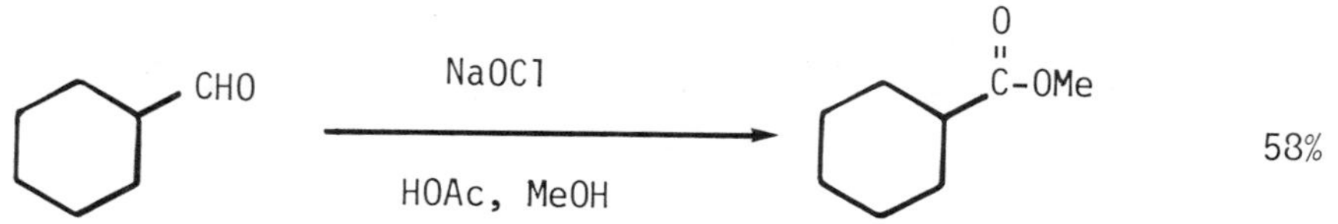

Tetr Lett, 23, 4647 (1982)

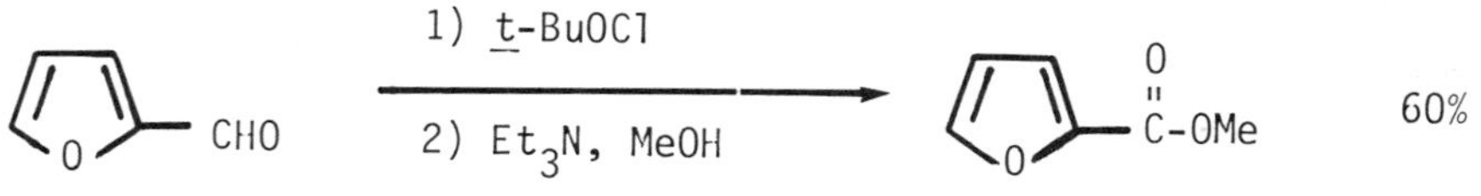

JOC, 47, 1360 (1982)

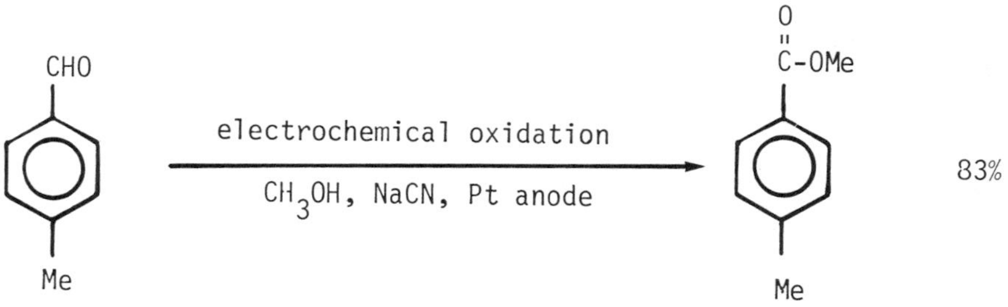

Bull Chem Soc Japan, 55, 335 (1982)

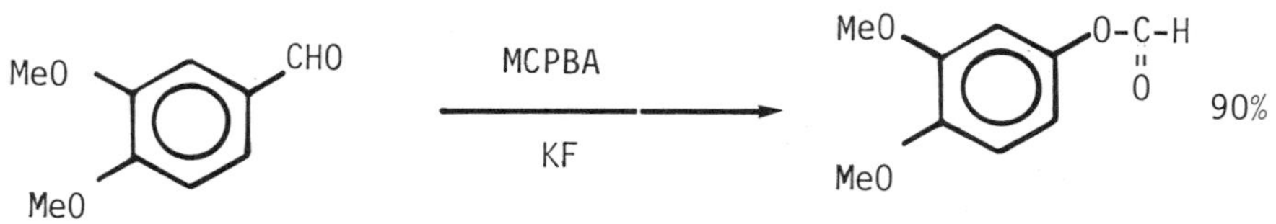

Tetr Lett, 22, 3895 (1981)

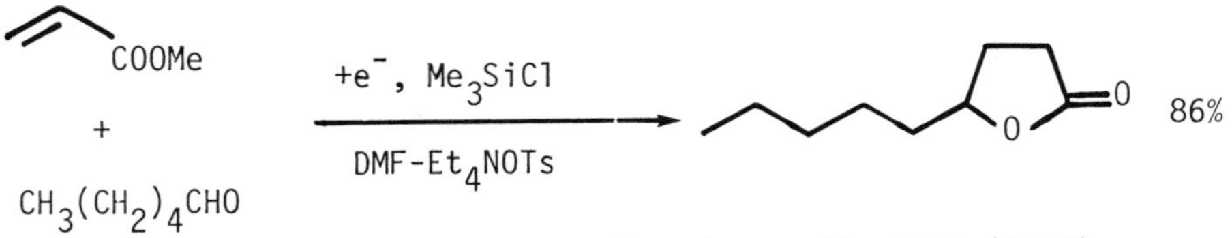

Tetr Lett, _21_, 5029 (1980)

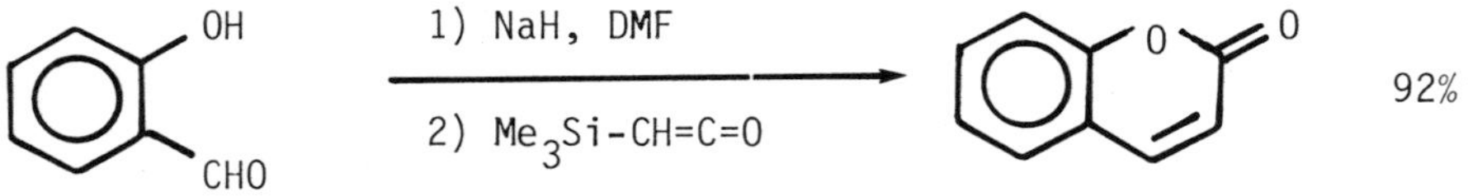

Synthesis, 672 (1982)

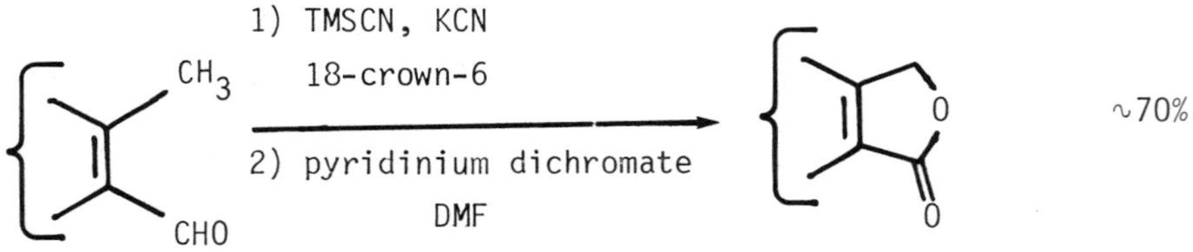

Tetr Lett, _21_, 731 (1980)

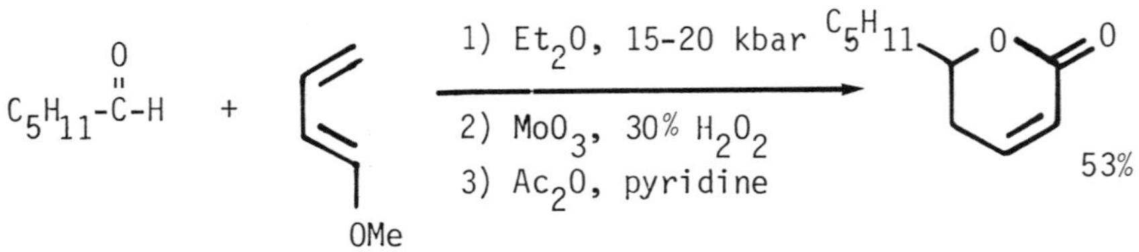

JOC, _46_, 2230 (1981)

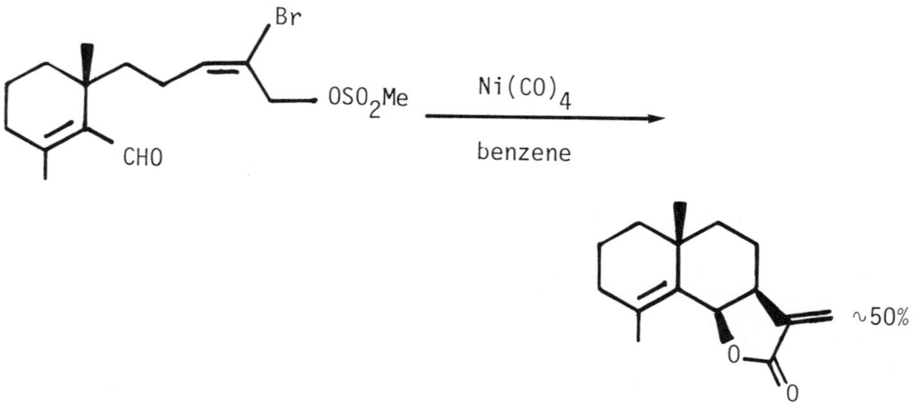

JACS, 103, 3945 (1981)

Related methods: Esters from Ketones (Section 117)

Section 110 Esters from Alkyls, Methylenes and Aryls

No examples of the reaction RR → RCOOR' or R'COOR (R,R' = alkyl,
aryl, etc.) occur in the literature. For the reaction RH → RCOOR'
or R'COOR see Section 116 (Esters from Hydrides).

Section 111 Esters from Amides

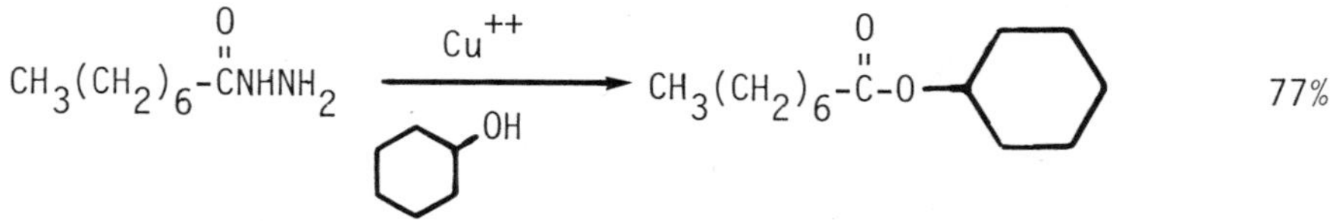

Tetrahedron, 36, 1311 (1980)

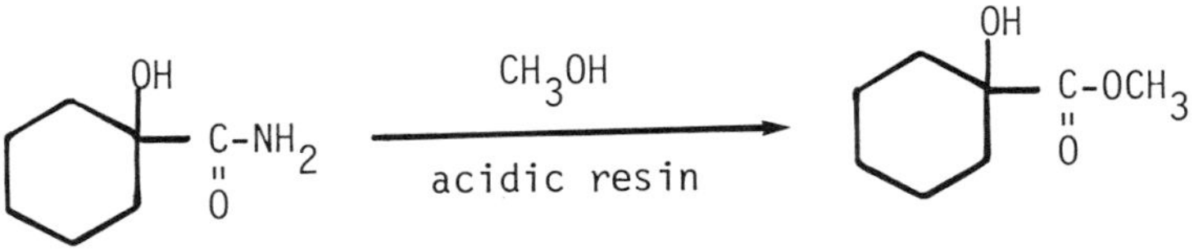

63%

JOC, 46, 5351 (1981)

Section 112 Esters from Amines

No additional examples.

Section 113 Esters from Esters

Conjugate reductions and conjugate alkylations of unsaturated esters are found in Section 74 (Alkyls from Olefins).

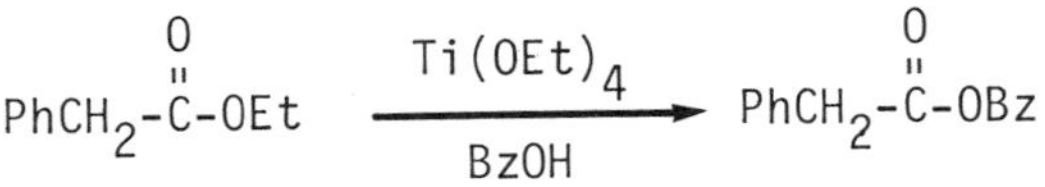

85%

Synthesis, 138 (1982)

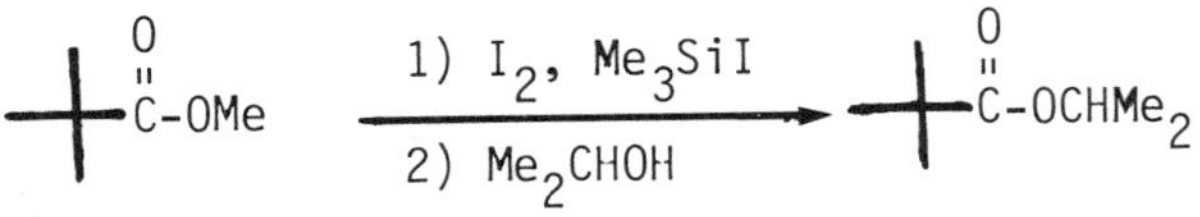

71%

Synthesis, 142 (1981)

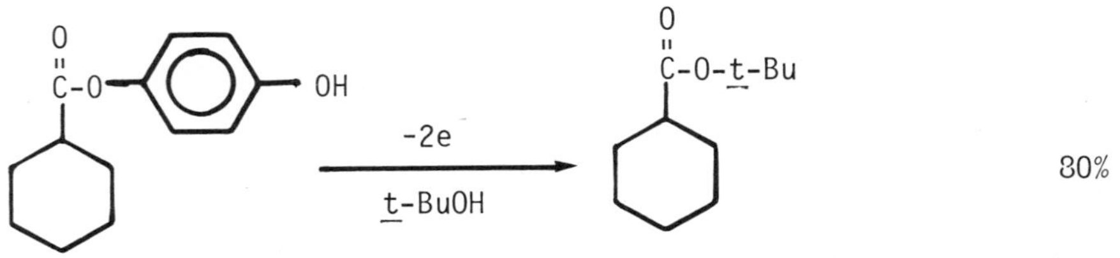

Tetr Lett, 22, 3715 (1981)

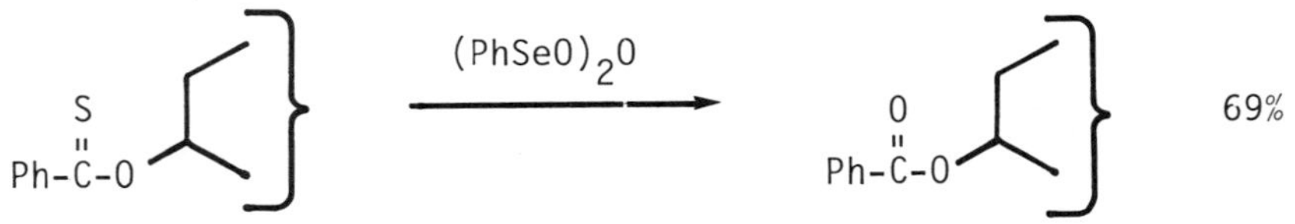

JCS Perkin I, 1650 (1980)

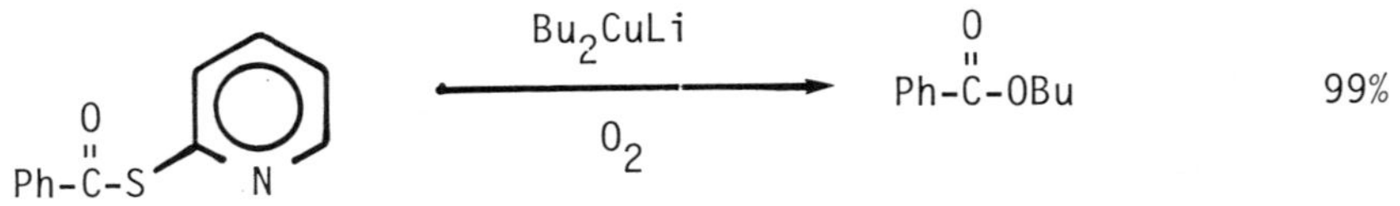

JCS Chem Comm, 1231 (1981)

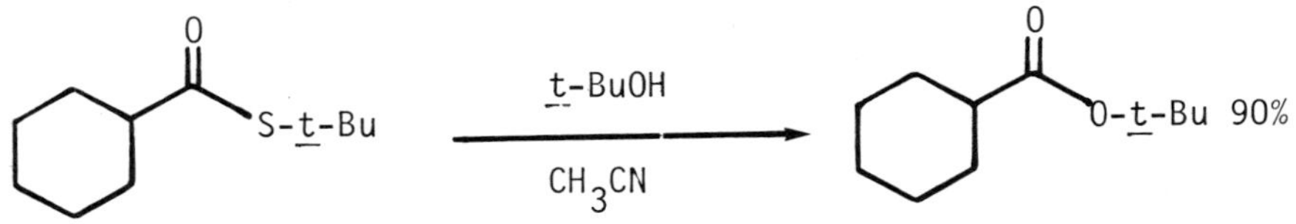

Org Syn, 61, 48 (1983)

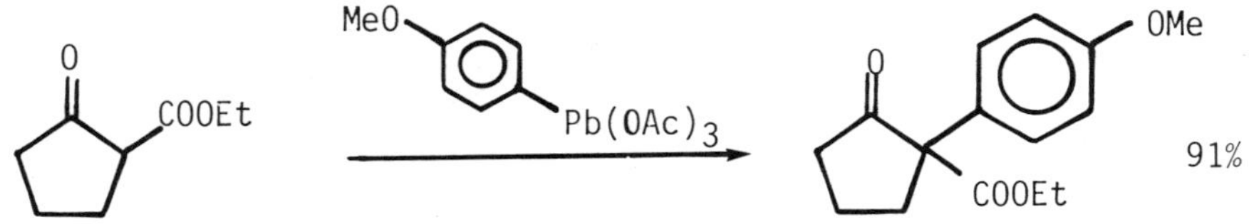

Aust J Chem, 33, 113 (1980)

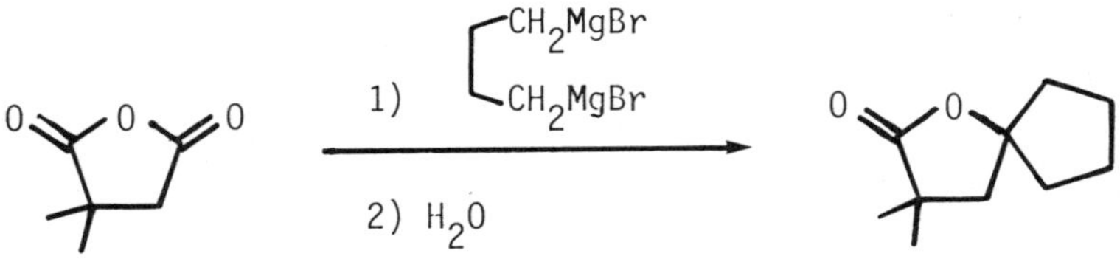

Tetr Lett, 21, 4167 (1980)

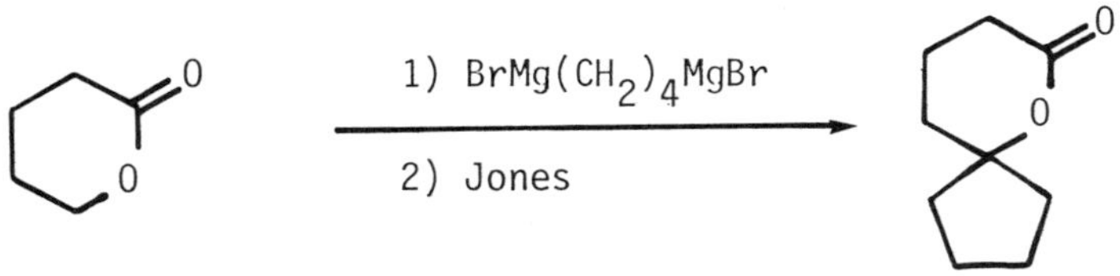

JOC, 45, 1828 (1980)

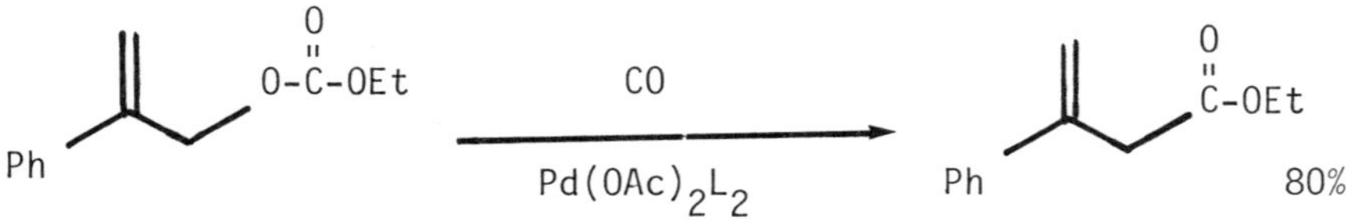

Tetr Lett, 23, 5189 (1982)

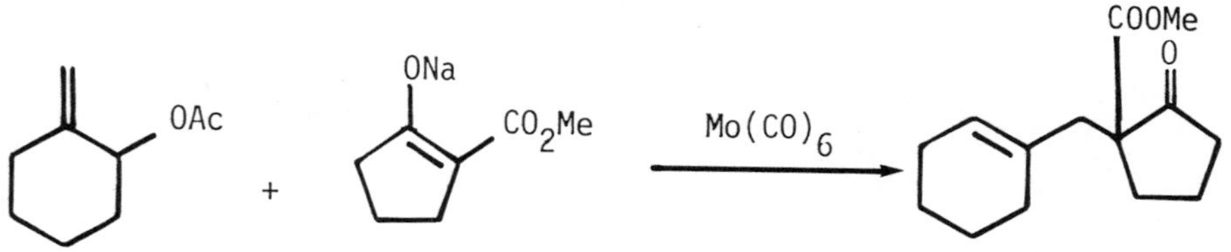

82%

JACS, 104, 5543 (1982)

91%

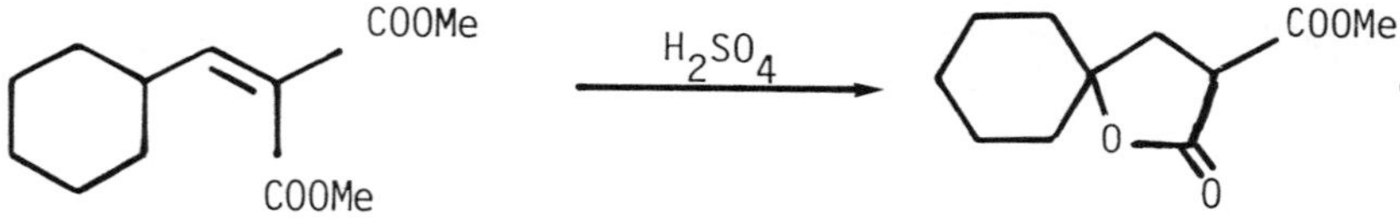

Synth Comm, 11, 35 (1981)

66%

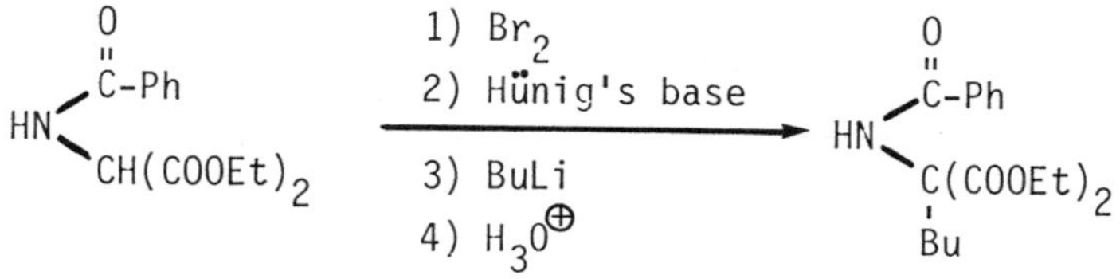

Angew Chem Int Ed, 21, 203 (1982)

Section 114 Esters from Ethers and Epoxides

$$C_8H_{17}\text{-}CH_2SPh \xrightarrow[\text{2) MeOH/}H_2O]{\text{1) }SO_2Cl_2\text{, pyridine}} C_8H_{17}\text{-}\overset{O}{\overset{\|}{C}}\text{-OMe}$$ 76%

JCS Chem Comm, 857 (1982)

$$H_3PO_5, \quad O = \bigcirc = O$$

$$\xrightarrow{}$$

$$CH_3CN$$

44%

JOC, <u>45</u>, 1320 (1980)

$$\xrightarrow{RuO_2/NaIO_4}$$

$$CCl_4/H_2O$$

59%

Synth Comm, <u>10</u>, 205 (1980)

$$\xrightarrow{\underline{t}\text{-BuCOCl}}$$

NaI

84%

Tetr Lett, <u>23</u>, 681 (1982)

$$\xrightarrow{PhCH_2PdCl(PPh_3)_2}$$

$$Bu_3SnCl, \quad Ph\text{-}\overset{O}{\underset{\|}{C}}\text{-}Cl$$

81%

JOC, <u>47</u>, 1215 (1982)

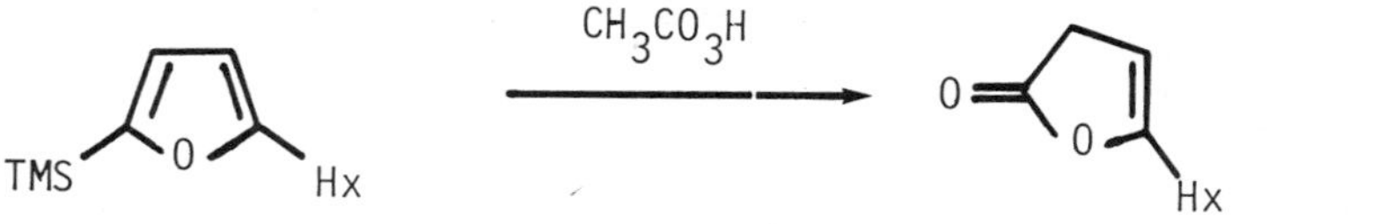

CH$_3$CO$_3$H

84%

Tetr Lett, <u>22</u>, 5191 (1981)

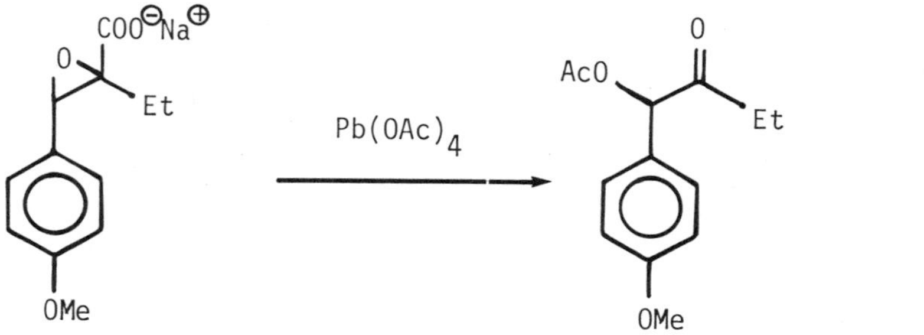

Pb(OAc)$_4$

100%

Chem Lett, 879 (1982)

Section 115 Esters from Halides

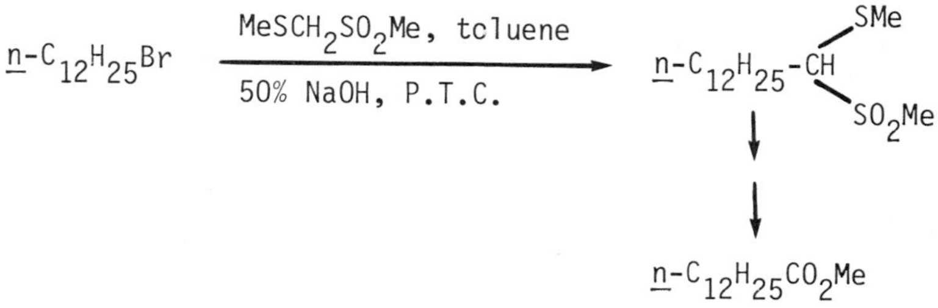

$\underline{n}$-C$_{12}$H$_{25}$Br

MeSCH$_2$SO$_2$Me, toluene

50% NaOH, P.T.C.

$\underline{n}$-C$_{12}$H$_{25}$-CH$\begin{smallmatrix}SMe\\SO_2Me\end{smallmatrix}$

$\underline{n}$-C$_{12}$H$_{25}$CO$_2$Me   83%

Tetr Lett, <u>22</u>, 4499 (1981)

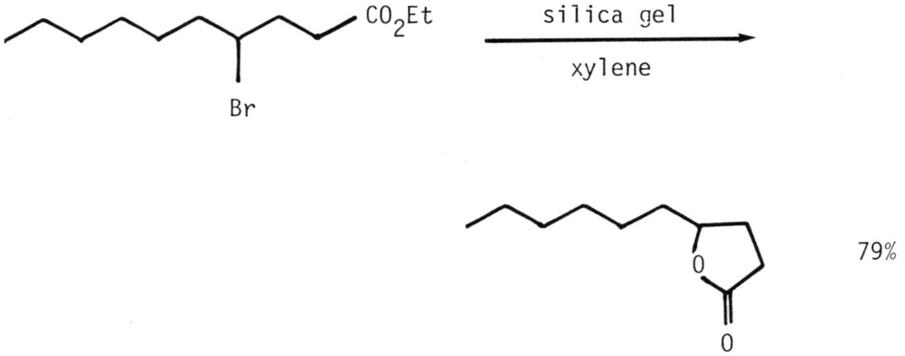

Chem Lett, 1909 (1982)

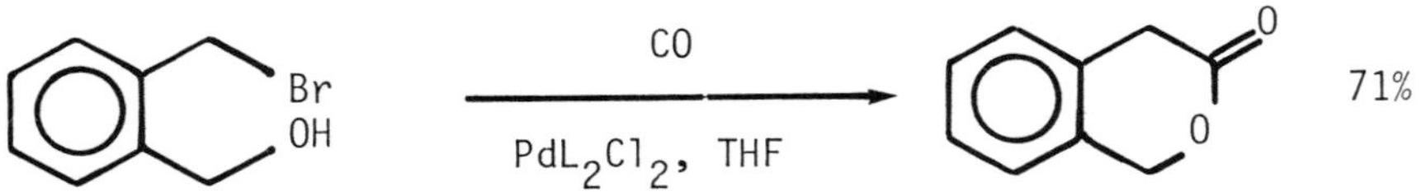

JACS, 102, 4193 (1980)

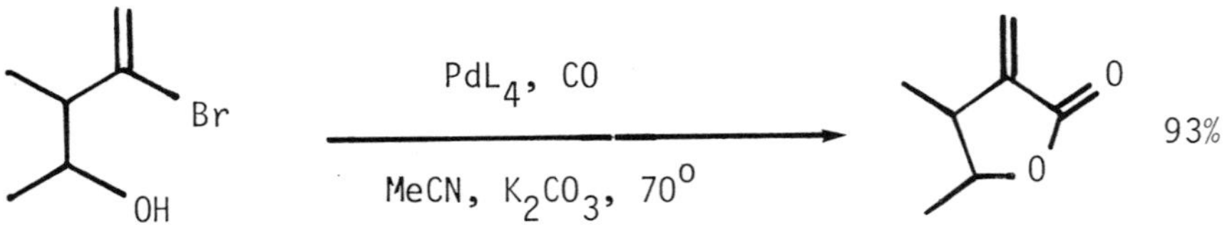

JOC, 47, 3630 (1982)

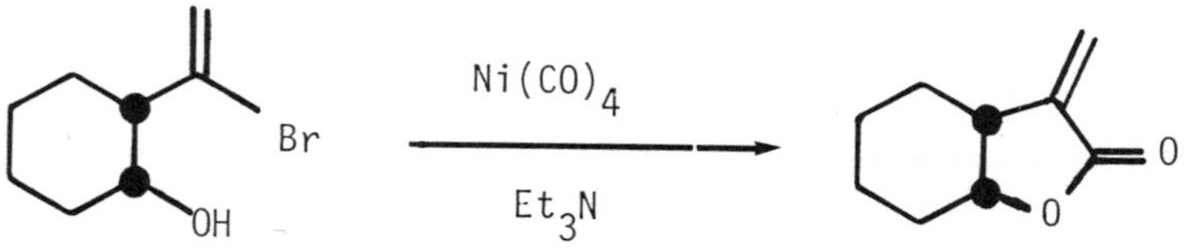

JOC, _46_, 1723 (1981)

Related methods: Carboxylic Acids from Halides (Section 25)

Section 116 Esters from Hydrides

This section contains examples of the reaction RH → RCOOR' or
R'COOR (R = alkyl, aryl, etc.).

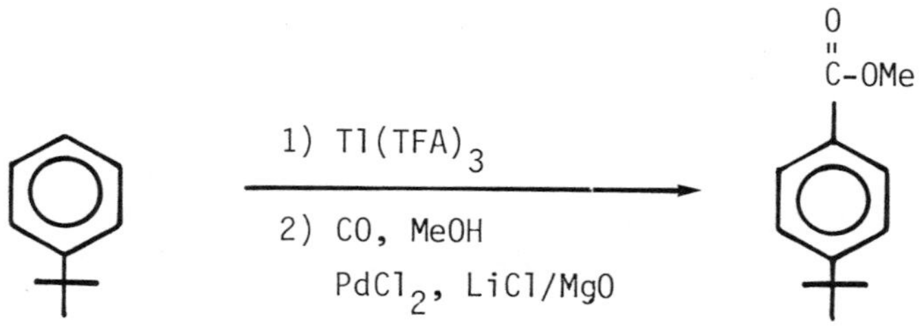

80%

JACS, _104_, 1900 (1982)
JOC, _45_, 363 (1980)

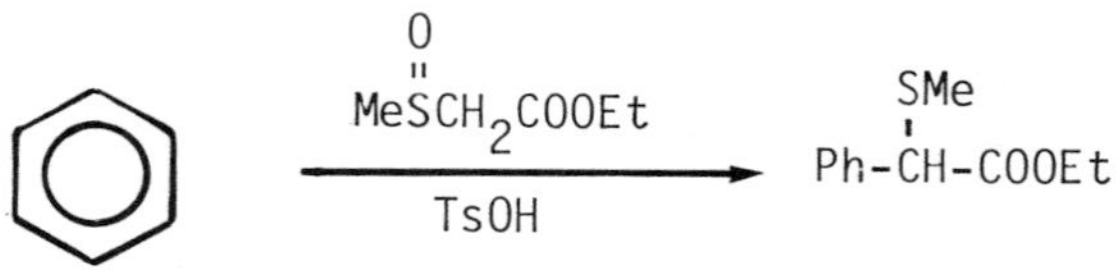

88%

Tetr Lett, _22_, 81 (1981)

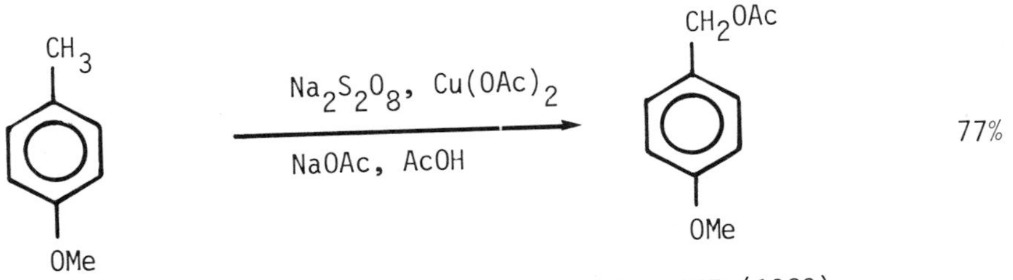

$$Na_2S_2O_8, \; Cu(OAc)_2$$
$$NaOAc, \; AcOH$$

77%

Synthesis, 477 (1980)

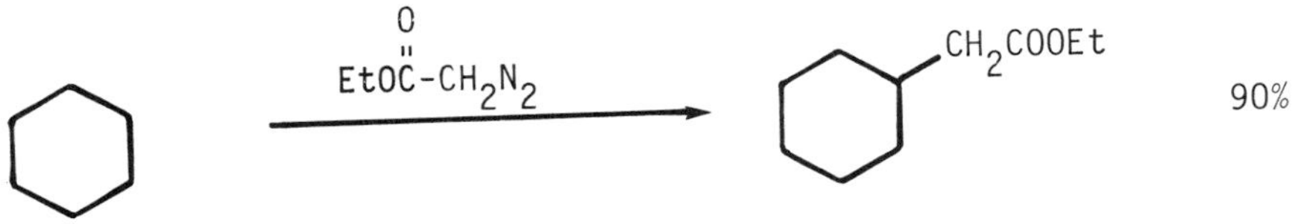

$$EtO\overset{O}{\overset{\|}{C}}-CH_2N_2$$

90%

JCS Chem Comm, 688 (1981)

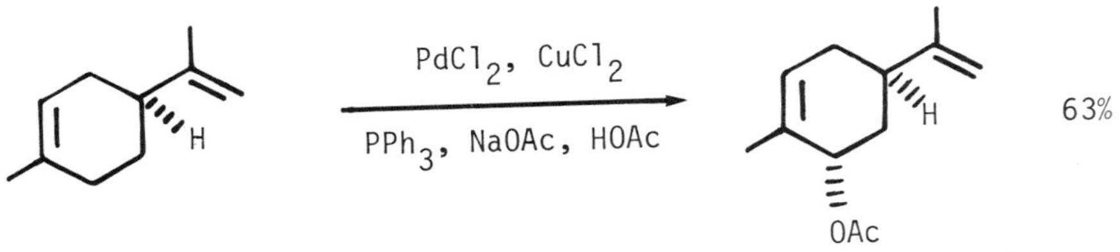

$$PdCl_2, \; CuCl_2$$
$$PPh_3, \; NaOAc, \; HOAc$$

63%

Angew Chem Int Ed, 21, 366 (1982)

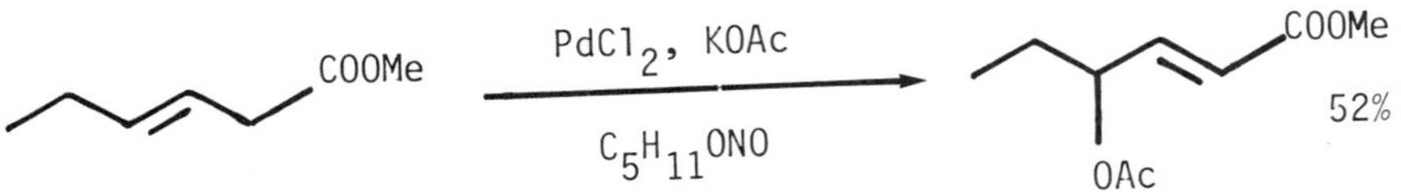

$$PdCl_2, \; KOAc$$
$$C_5H_{11}ONO$$

52%

Tetr Lett, 22, 131 (1981)

$$C_5H_{11}-C\equiv CH \xrightarrow[\text{NaOAc, MeOH}]{\substack{CO \\ PdCl_2,\ CuCl_2}} C_5H_{11}-C\equiv C-CO_2Me$$

74%

Tetr Lett, <u>21</u>, 849 (1980)

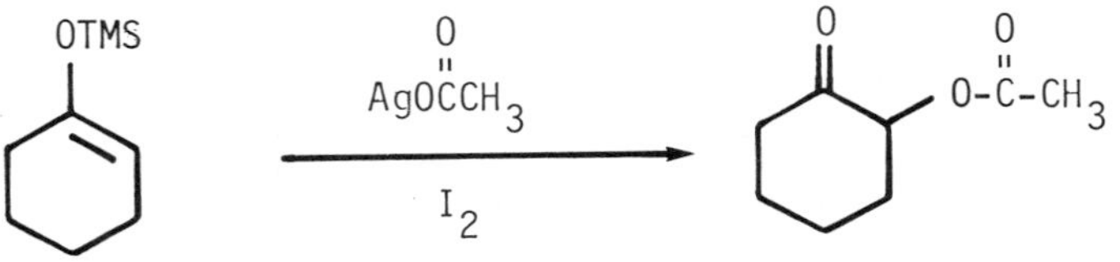

90%

JOC, <u>46</u>, 2717 (1981)

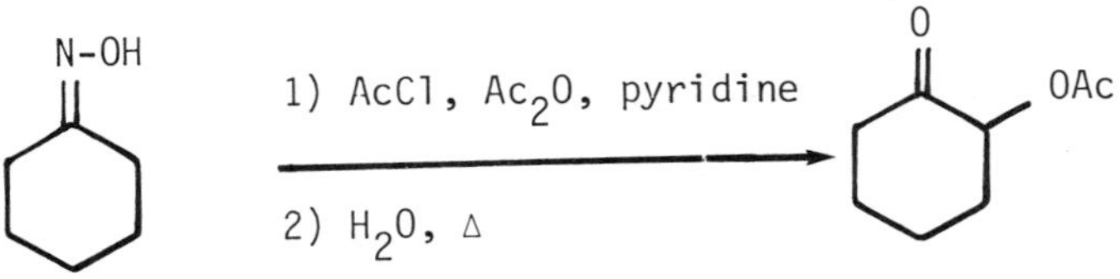

60%

Synthesis, 223 (1981)

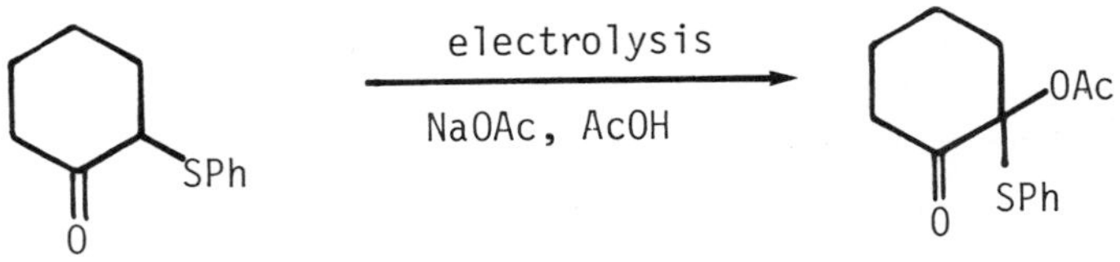

62%

Tetr Lett, <u>21</u>, 2557 (1980)

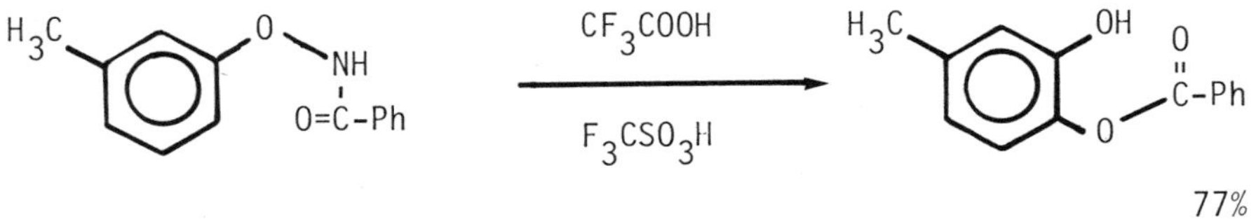
CF3COOH
F3CSO3H
77%

Synthesis, 461 (1980)

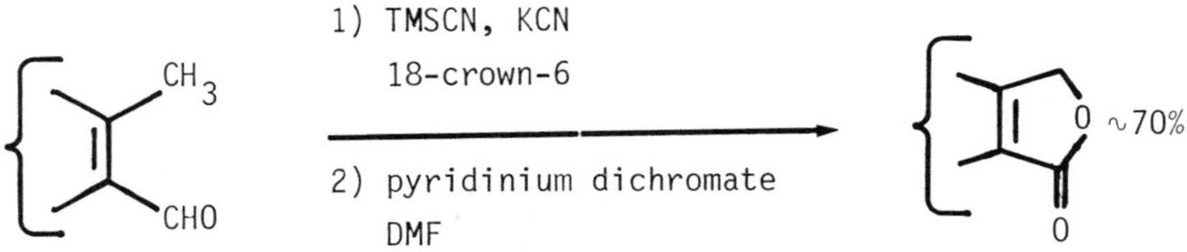
1) TMSCN, KCN
18-crown-6
2) pyridinium dichromate
DMF
~70%

Tetr Lett, 21, 731 (1980)

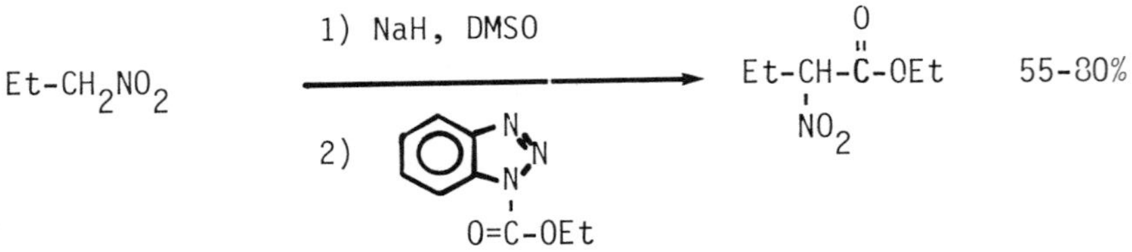
Et-CH2NO2
1) NaH, DMSO
2)
O=C-OEt
Et-CH-C-OEt
NO2
55-80%

Angew Chem Int Ed, 21, 139 (1982)

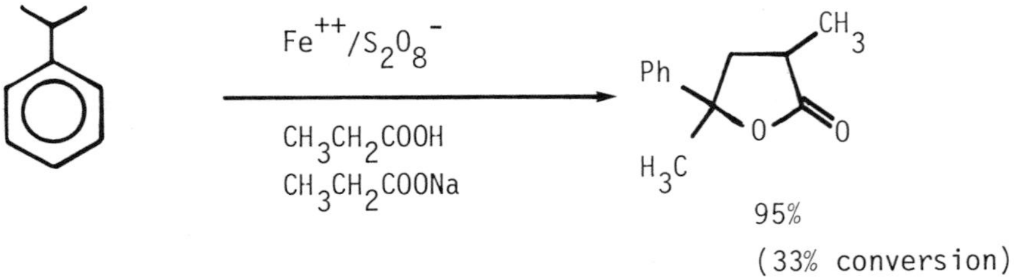
Fe++/S2O8-
CH3CH2COOH
CH3CH2COONa
95%
(33% conversion)

Tetrahedron, 36, 3559 (1980)

Also via: Carboxylic acids, Section 26; Alcohols, Section 41

Section 117 Esters from Ketones

$$\underset{\substack{\text{O}\\ \|}}{\text{Ph-C-CH}_3} \quad \xrightarrow[\text{2) Pb(OAc)}_4,\ \text{MeOH}]{\text{1) BF}_3\cdot\text{OEt}_2} \quad \underset{\substack{\text{O}\\ \|}}{\text{Ph-CH}_2\text{-C-OMe}} \qquad 86\%$$

Synthesis, 126 (1981)

68%

Synthesis, 456 (1982)

49%

JOC, 46, 1914 (1981)

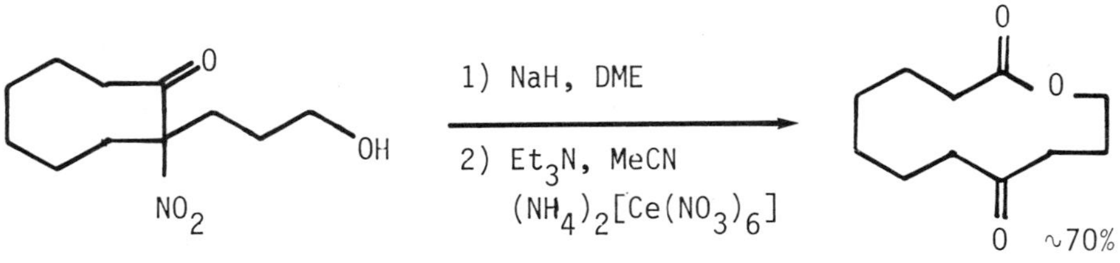
O
OH
NO_2
1) NaH, DME
2) Et_3N, MeCN
$(NH_4)_2[Ce(NO_3)_6]$
O
O
O
~70%

Tetr Lett, 23, 3521 (1982)

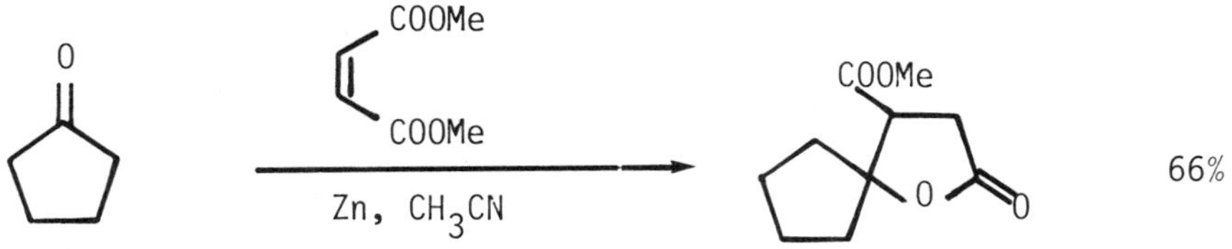
O
COOMe
COOMe
Zn, CH_3CN
COOMe
O
O
66%

Chem Lett, 1217 (1981)

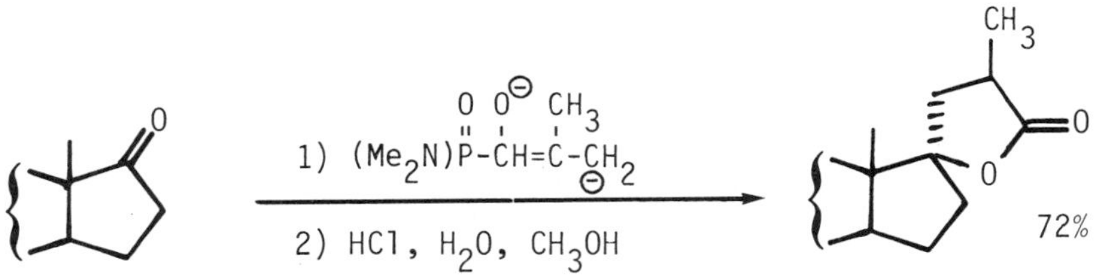
O
1) $(Me_2N)\overset{O}{P}-CH=\overset{O^{\ominus}}{C}-\overset{CH_3}{C}H_2^{\ominus}$
2) HCl, H_2O, CH_3OH
CH_3
O
O
72%

Synthesis, 289 (1980)

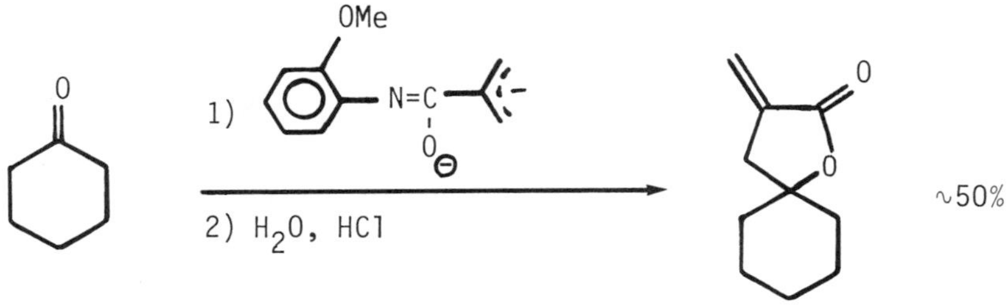
O
OMe
1)
N=C
O
2) H_2O, HCl
O
O
~50%

Chem Lett, 1567 (1980)

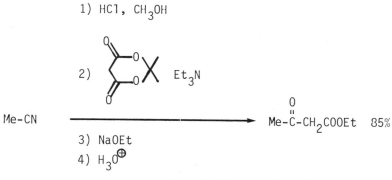

JACS, <u>104</u>, 6879 (1982)

Also via Carboxylic Acids, Section 27

Section 118 <u>Esters from Nitriles</u>

1) HCl, CH$_3$OH

2) [image of di-tert-butyl malonate structure] Et$_3$N

Me-CN $\xrightarrow{\hspace{3cm}}$ Me-$\overset{\overset{\text{O}}{\|}}{\text{C}}$-CH$_2$COOEt   85%

3) NaOEt
4) H$_3$O$^{\oplus}$

Synthesis, 130 (1981)

Section 119 <u>Esters from Olefins</u>

$N_2CH-COOEt$

$Rh_2(OAc)_4$

COOEt 88%

Tetr Lett, <u>22</u>, 1783 (1981)

JOC, <u>45</u>, 695 and 1538 (1980)

Ph

+

$Br_2C(COOEt)_2$

Cu_2Br_2

DMSO

Ph   COOEt

COOEt   71%

Bull Chem Soc Japan, <u>54</u>, 2539 (1981)

COO⁻

COO⁻

$Pb(OAc)_4$

CH_3CN

98%

Tetr Lett, <u>21</u>, 1819 (1980)

$CF_2=CH_2$

$$C_2F_5\overset{O}{\overset{\|}{C}}-OCl \xrightarrow[\text{NaF}]{\text{ClF}} C_2F_5\overset{O}{\overset{\|}{C}}-OCF_2CH_2Cl \quad 80\%$$

JOC, <u>45</u>, 1214 (1980)

Also via Alcohols, Section 44

Section 120 <u>Esters from Miscellaneous Compounds</u>

No additional examples.

Section 120A <u>Protection of Esters</u>

$$\underset{\text{COCl}}{\bigcirc}\text{I} \quad \xrightarrow[\substack{\text{2) AgNO}_3,\ \text{MeOH} \\ \text{collidine}}]{\text{1) EtSH, AlCl}_3} \quad \underset{\text{C(OMe)}_3}{\bigcirc}\text{I} \qquad 76\%$$

JOC, <u>45</u>, 740 (1980)

CHAPTER 9
PREPARATION OF ETHERS
AND EPOXIDES

Section 121 <u>Ethers and Epoxides from Acetylenes</u>

No examples.

Section 122 <u>Epoxides from Acid Halides</u>

1) ClCH=CHCH$_2$TMS
   AlCl$_3$

2) NaBH$_4$, MeOH
3) NaOH

74%

Tetr Lett, <u>21</u>, 4369 (1980)

Section 123 <u>Ethers and Epoxides from Alcohols and Phenols</u>

1) NaOH, H$_2$O

2) BuCl

70%

JOC, <u>45</u>, 1095 (1980)

$$C_8H_{17}OH \xrightarrow{\text{MeI, KF-alumina}} C_8H_{17}OMe \qquad 90\%$$

Bull Chem Soc Japan, **55**, 2504 (1982)

$$Ph-OH \xrightarrow[\substack{\underline{t}\text{-BuCl} \\ Ni(acac)_2 \\ NaHCO_3}]{} Ph-O-\underline{t}-Bu \qquad 32\%$$

Synthesis, 186 (1982)

$$R-OH \xrightarrow[\substack{\overset{\overset{NH}{\|}}{BzOCCCl_3} \\ TfOH, hexane/CH_2Cl_2}]{} R-O-Bz \qquad 82\text{-}98\%$$

JCS Chem Comm, 1240 (1981)

$$Ar-OH \xrightarrow{\text{BzOTs}} Ar-OBz$$

Synth Comm, **11**, 853 (1981)

80%

JOC, **47**, 4374 (1982)

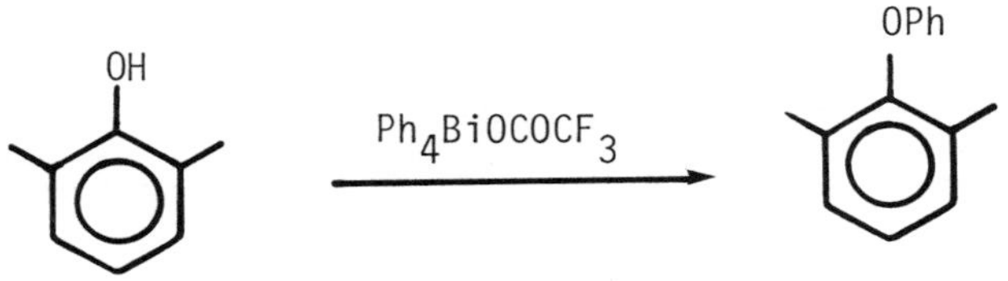

JCS Chem Comm, 503 (1981)

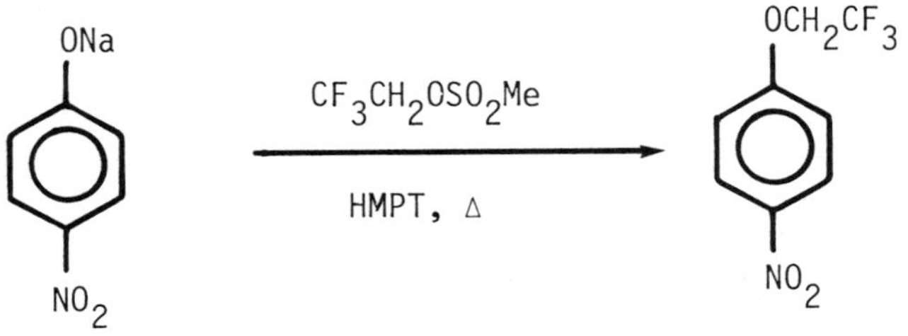

Synthesis, 727 (1980)

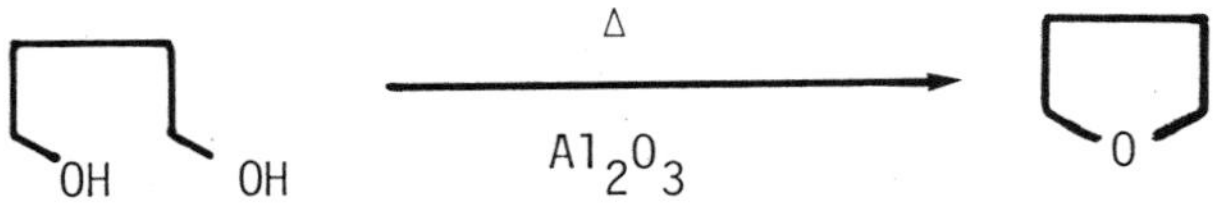

Bull Chem Soc Japan, 53, 3031 (1980)

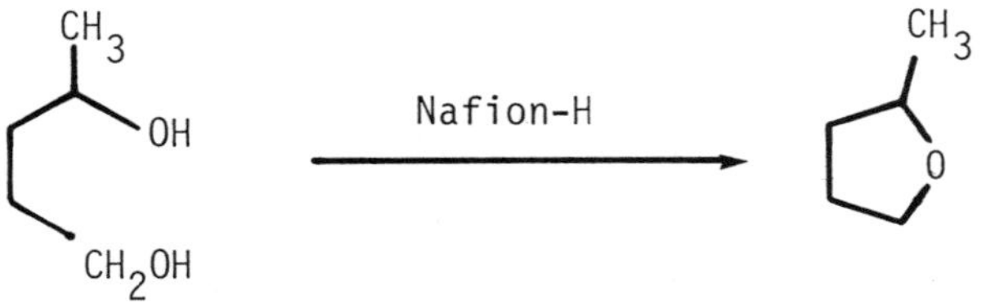

Synthesis, 474 (1981)

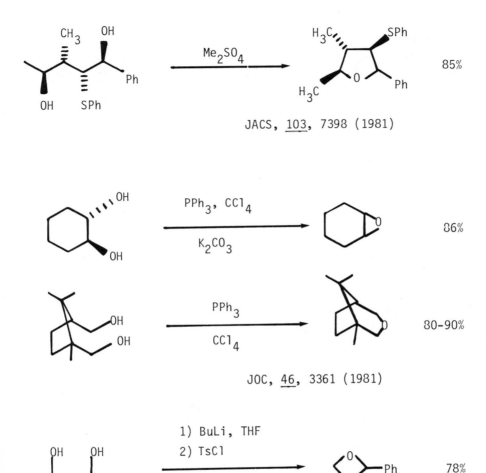

JACS, <u>103</u>, 7398 (1981)

85%

86%

80-90%

JOC, <u>46</u>, 3361 (1981)

1) BuLi, THF
2) TsCl
3) BuLi

78%

Synthesis, 550 (1981)

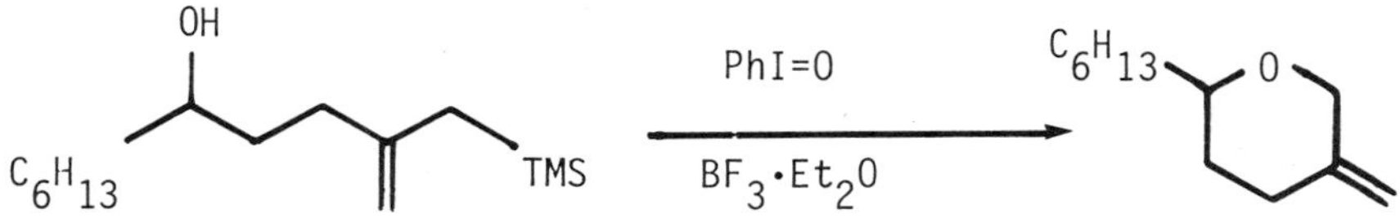

65%

JCS Chem Comm, 1108 (1982)

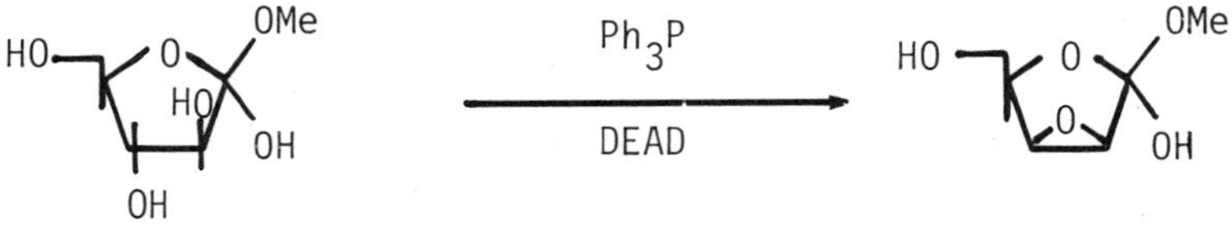

JCS Chem Comm, 784 (1980)

Related methods: Protection of Alcohols and Phenols (Section 45A)

Section 124 Epoxides from Aldehydes

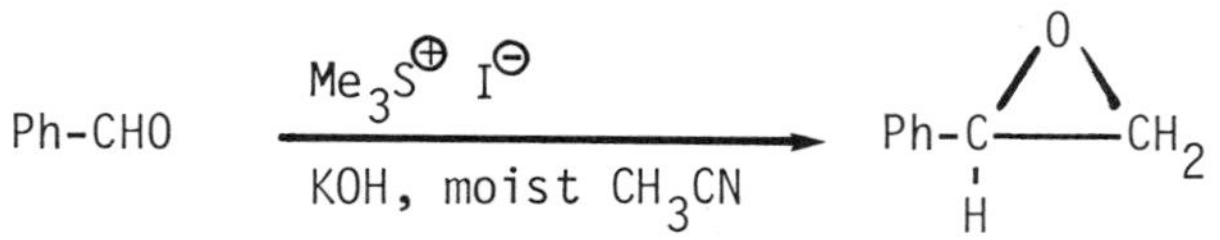

96%

Tetr Lett, 23, 5283 (1982)

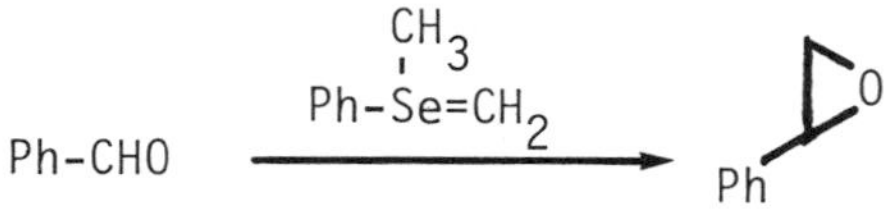

94%

Angew Chem Int Ed, 20, 671 (1981)

$$\underline{n}\text{-}C_7H_{15}\text{-CHO} \xrightarrow[\text{KN(SiMe}_3)_2]{\overset{\oplus}{Ph_3AsCH_2CHMe_2}}$$

72%

JACS, 103, 1283 (1981)

Related methods:　Ethers and Epoxides from Ketones (Section 132)

Section 125 Ethers and Epoxides from Alkyls, Methylenes and Aryls

No examples of the preparation of ethers and epoxides by replace-
ment of alkyl, methylene, and aryl groups occur in the literature.
For the conversion of RH → ROR' (R, R' = alkyl) see Section 131
(Ethers from Hydrides)

Section 126 Ethers and Epoxides from Amides

No additional examples.

Section 127 Ethers and Epoxides from Amines

No additional examples.

Section 128 Ethers from Esters

$$\text{Ph-OAc} \xrightarrow[\text{K}_2\text{CO}_3,\ 18\text{-crown-6}]{\text{CH}_3\text{CH}_2\text{CH}_2\text{I}} \text{Ph-O-CH}_2\text{CH}_2\text{CH}_3$$

79%

JCS Chem Comm, 815 (1982)

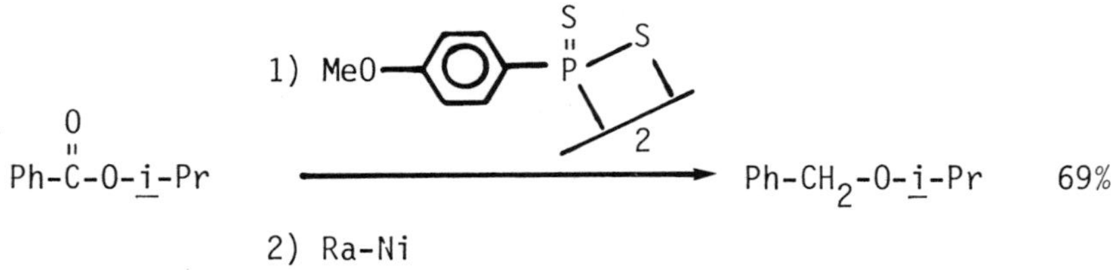

$$\underset{\substack{\text{O}\\ \|}}{\text{Ph-C-O-\underline{i}-Pr}} \xrightarrow[\text{2) Ra-Ni}]{\text{1) MeO-} \bigcirc \text{-P(S)(S)}} \text{Ph-CH}_2\text{-O-\underline{i}-Pr} \quad 69\%$$

JOC, 46, 832 (1981)

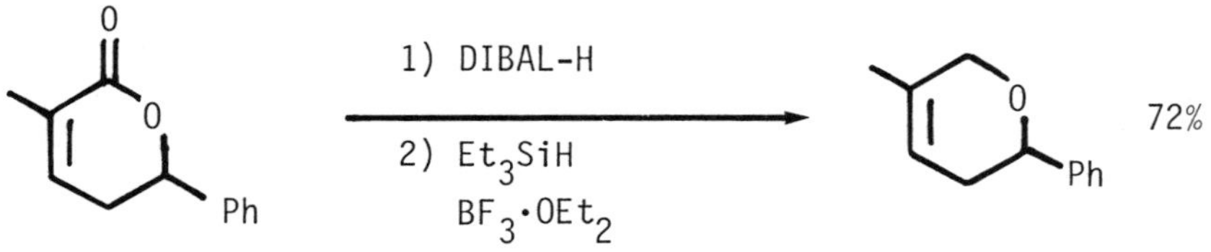

JOC, 46, 2417 (1981)

Section 129 Ethers from Epoxides

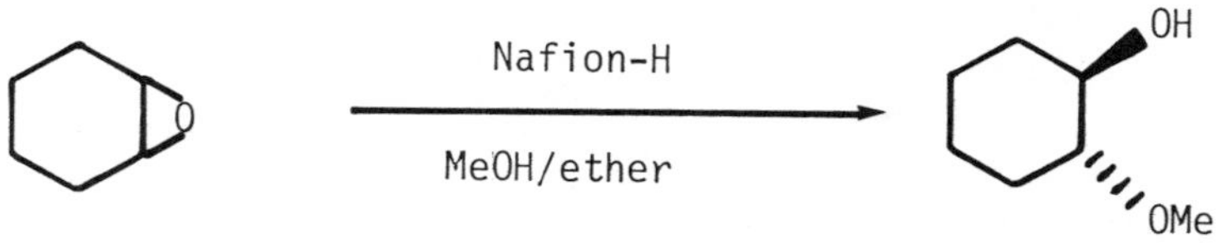

Nafion-H

MeOH/ether

74%

Synthesis, 280 (1981)

Section 130 Ethers from Halides

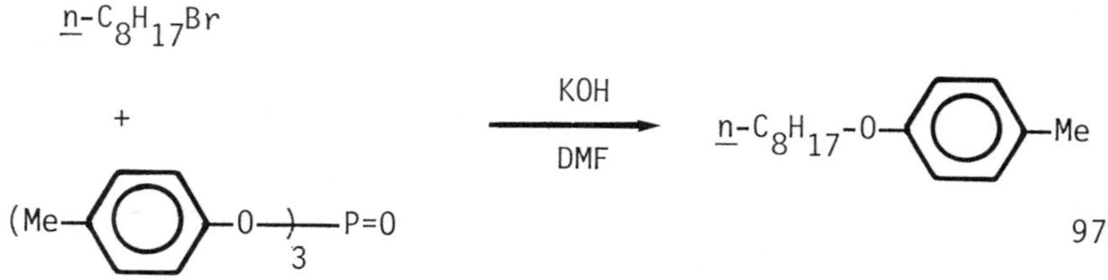

$\underline{n}$-C$_8$H$_{17}$Br

+

$(Me-\bigcirc-O-)_3-P=O$

KOH

DMF

$\underline{n}$-C$_8$H$_{17}$-O-$\bigcirc$-Me

97%

Also can be used to form diaryl ethers.

Synthesis, 828 (1982)

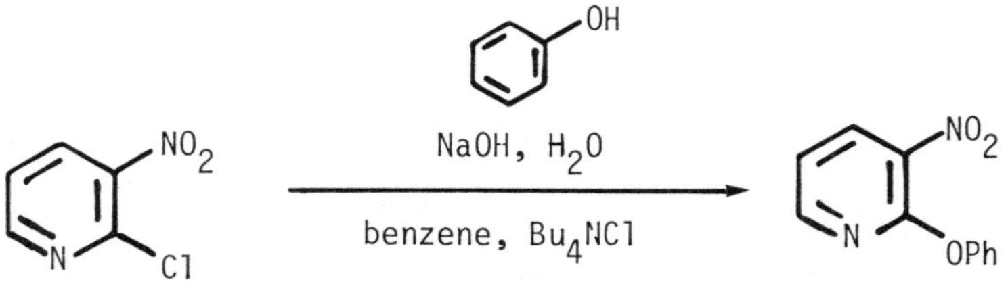

NaOH, H$_2$O

benzene, Bu$_4$NCl

84%

Synthesis, 921 (1980)

$$Ph-OH \xrightarrow[\text{NaHCO}_3]{\underset{\text{Ni(acac)}_2}{\underline{\text{t-BuCl}}}} Ph-O-\underline{t}-Bu \qquad 32\%$$

Synthesis, 186 (1982)

Related methods: Ethers from Alcohols (Section 123)

Section 131 <u>Ethers from Hydrides</u>

$$Ph-CH_2COOMe \xrightarrow[\text{2) NaOMe, MeOH}]{\text{1) PhI(OAc)}_2} \underset{\overset{|}{\text{OMe}}}{Ph-CHCOOMe} \qquad 70\%$$

Tetr Lett, <u>22</u>, 2747 (1981)

Synthesis, 315 (1980)

88%

86%
(92% erythro)

Tetr Lett, <u>21</u>, 2527 (1980)
JACS, <u>102</u>, 3248 (1980)

Section 132 <u>Ethers and Epoxides from Ketones</u>

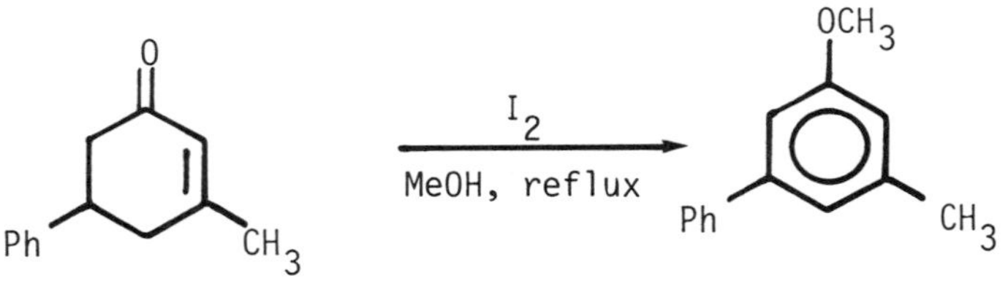

86%

Chem and Ind, 888 (1980)

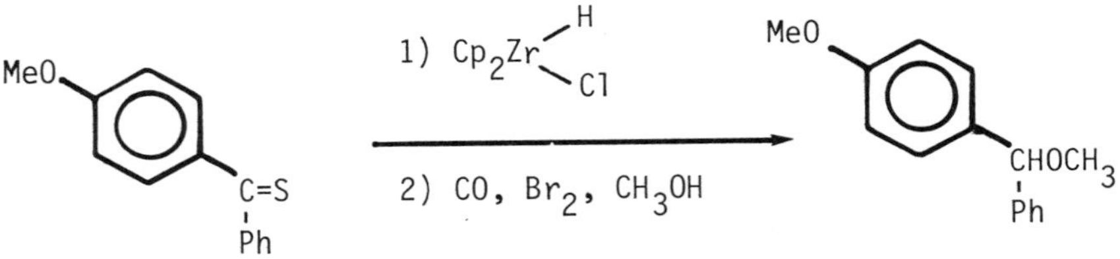

JOC, <u>46</u>, 289 (1981)

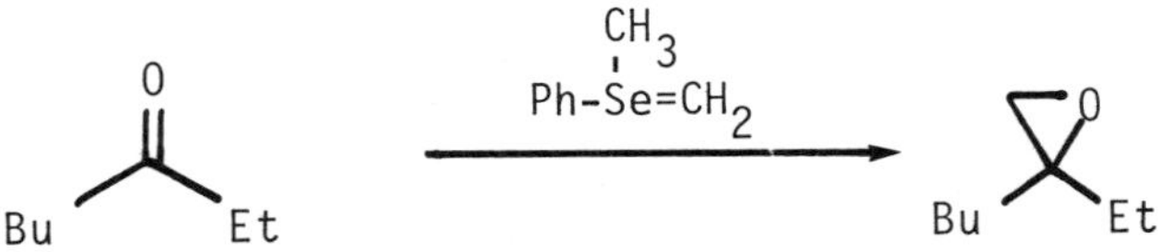

Angew Chem Int Ed, <u>20</u>, 671 (1981)

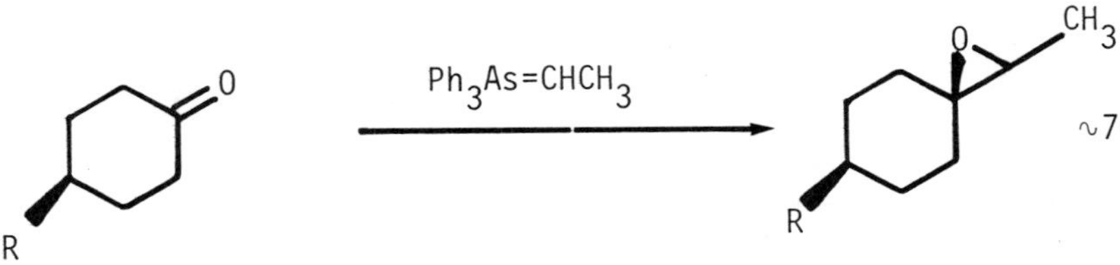

~70%

JACS, <u>103</u>, 1283 (1981)

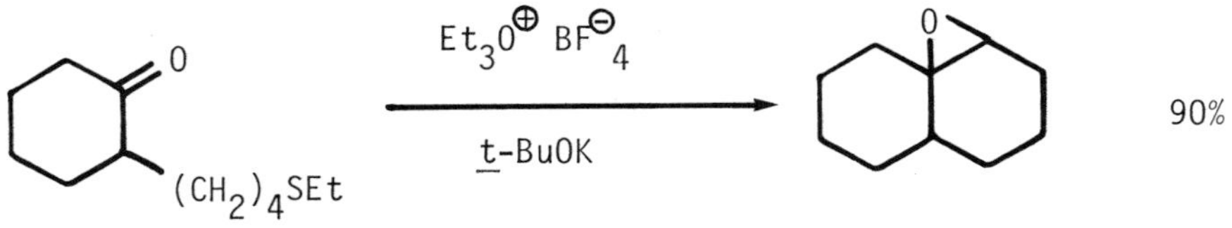

Tetr Lett, 21, 4807 and 4811 (1980)

Related methods: Epoxides from Aldehydes (Section 124)

Section 133 Ethers and Epoxides from Nitriles

No additional examples.

Section 134 Ethers and Epoxides from Olefins

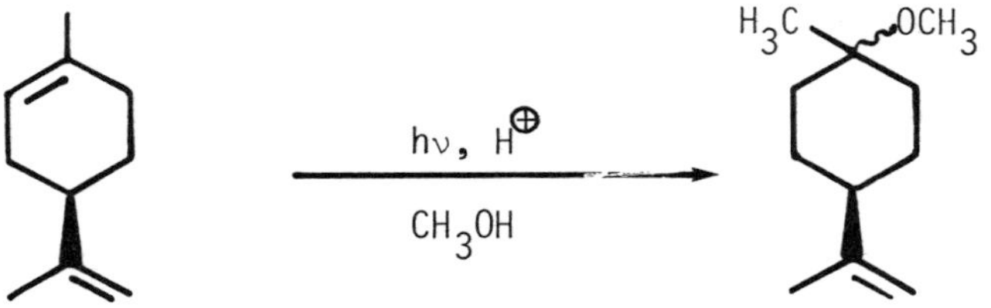

Org Syn, 61, 112 (1983)

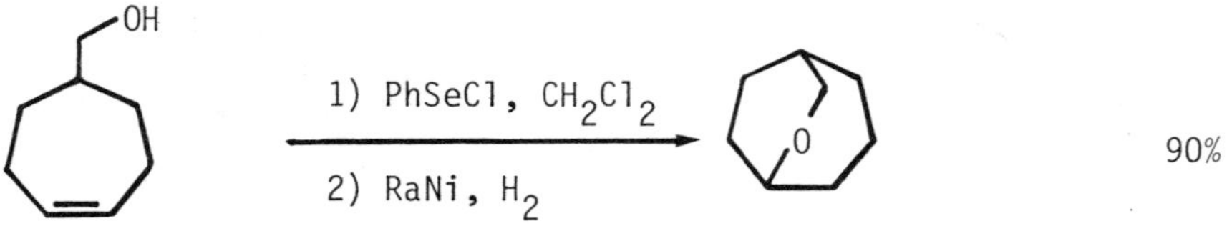

JACS, 102, 3784 (1980)

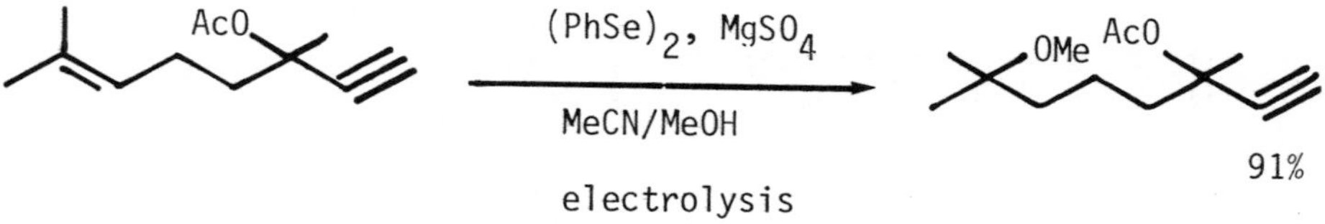

JACS, <u>103</u>, 4606 (1981)

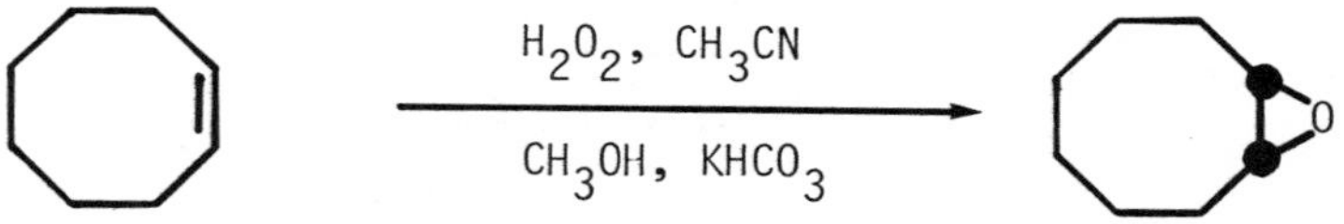

61%

Org Syn, <u>60</u>, 63 (1981)

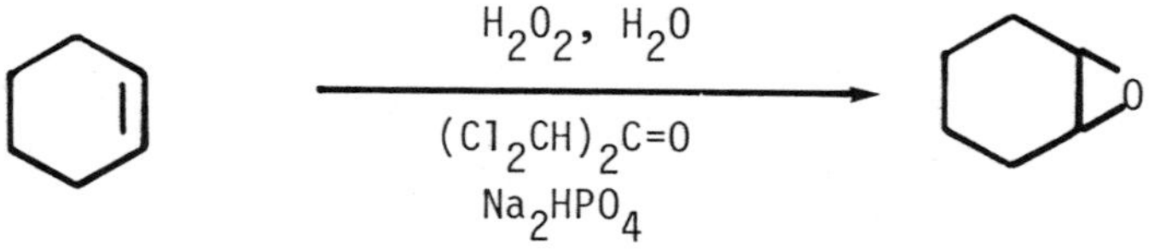

80%

Tetr Lett, <u>22</u>, 2089 (1981)

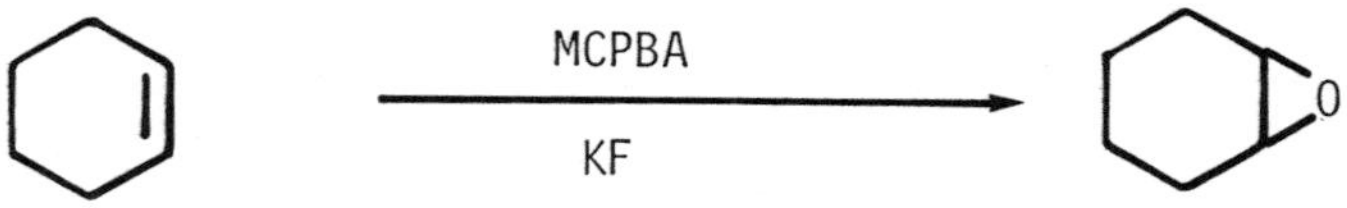

>95%

Tetr Lett, <u>22</u>, 3895 (1981)

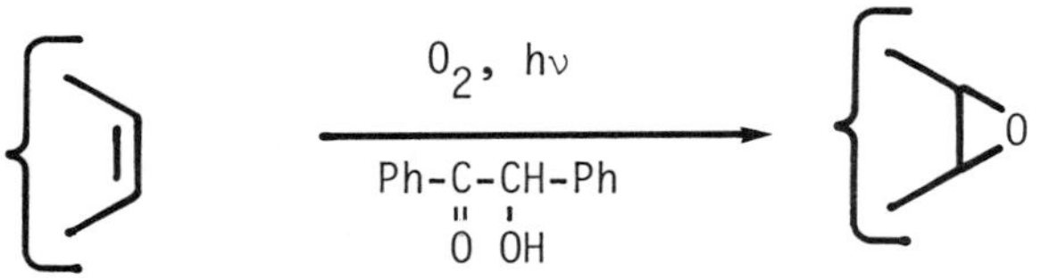

JACS, 103, 2049 (1981)

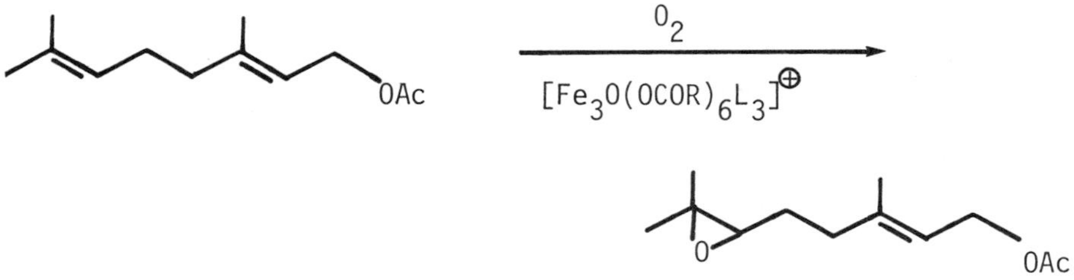

~70%

JACS, 104, 6450 (1982)

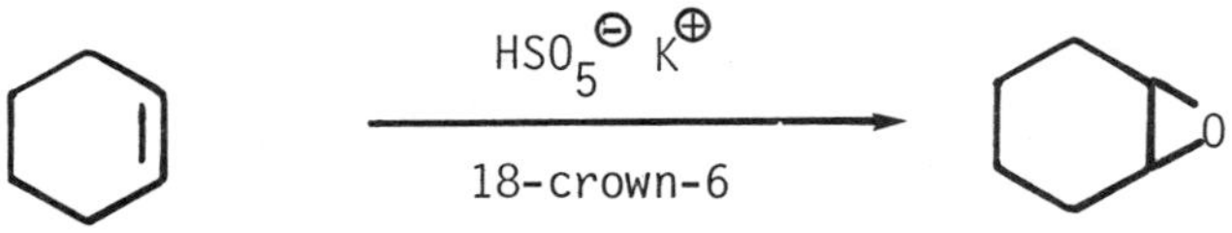

98%

JOC, 45, 4758 (1980)

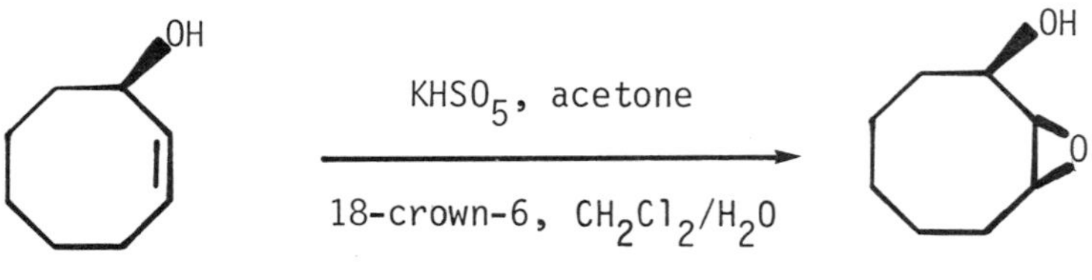

91%

JOC, 47, 2670 (1982)

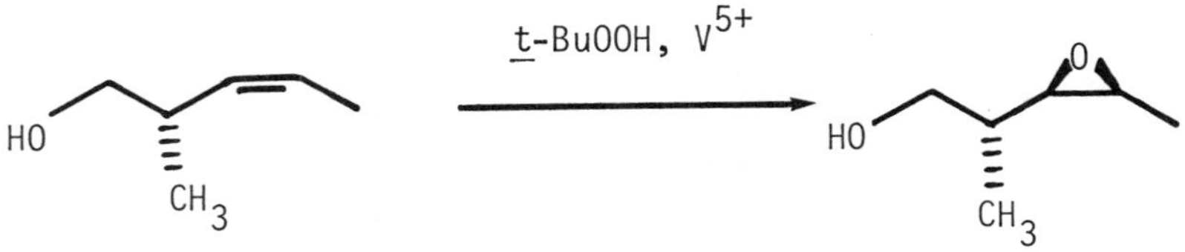
t-BuOOH, V5+
HO
CH3
O
HO
CH3

JACS, 103, 7690 (1981)

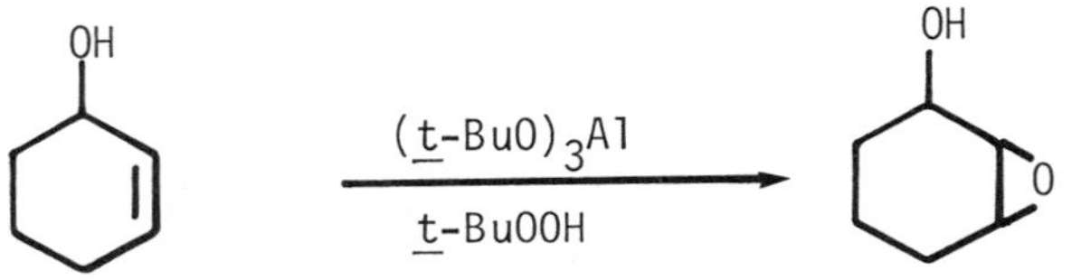
OH
(t-BuO)3Al
t-BuOOH
OH
O

72%

Tetr Lett, 21, 1657 (1980)

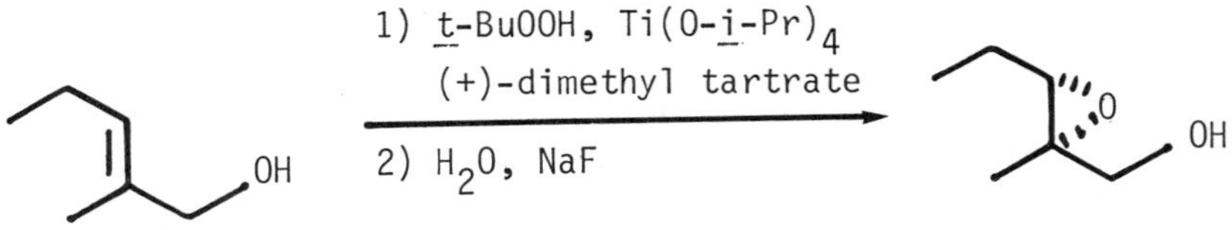
1) t-BuOOH, Ti(O-i-Pr)4
(+)-dimethyl tartrate
2) H2O, NaF
OH
O
OH

79%

95% ee

JACS, 102, 5974 (1980)

JACS, 103, 464 (1981)

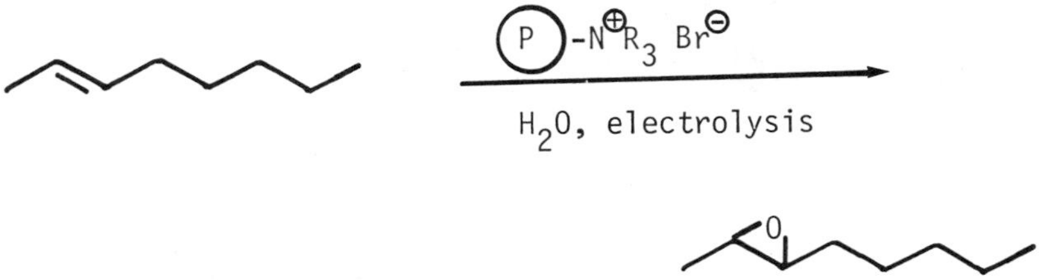
P -N⊕R3 Br⊖
H2O, electrolysis
O

73%

JOC, 47, 3575 (1982)

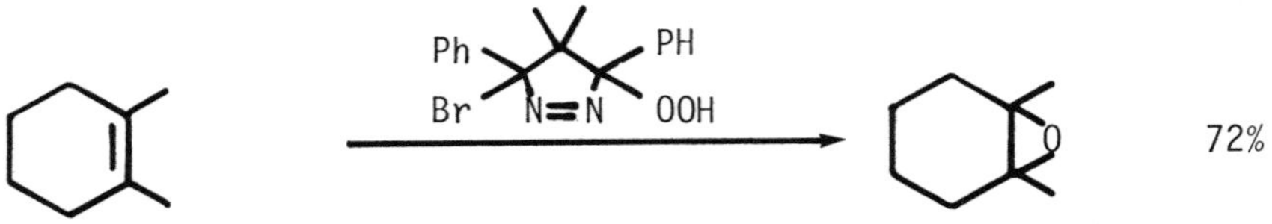
Ph
PH
Br
N=N
OOH
O
72%

JOC, 47, 1141 (1982)

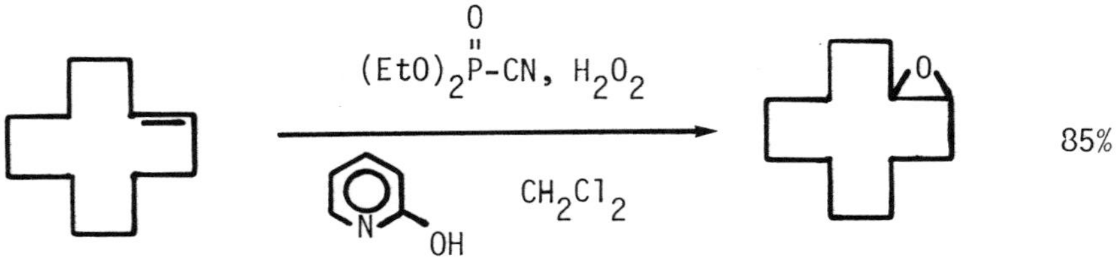
O
$(EtO)_2P$-CN, H_2O_2
N
OH
CH_2Cl_2
O
85%

Chem Pharm Bull, 29, 1774 (1981)

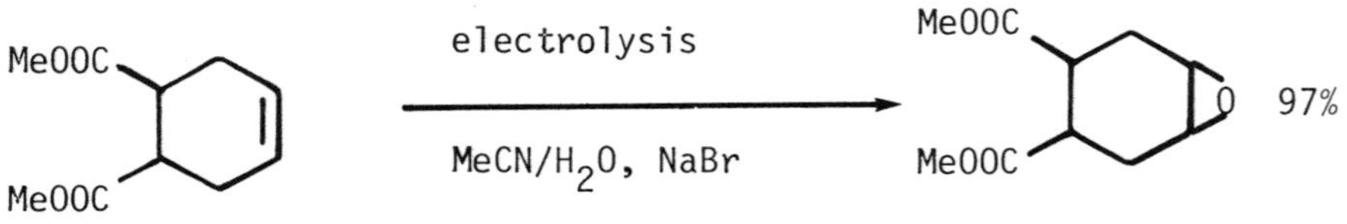
MeOOC
MeOOC
electrolysis
MeCN/H_2O, NaBr
MeOOC
MeOOC
O
97%

JOC, 46, 3312 (1981)

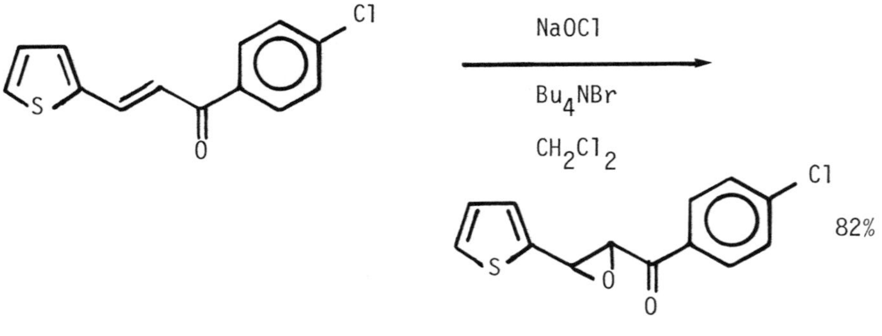
Cl
S
O
NaOCl
Bu_4NBr
CH_2Cl_2
Cl
S
O
O
82%

Gazz Chim Ital, 110, 267 (1980)

Section 135 <u>Ethers and Epoxides from Miscellaneous Compounds</u>

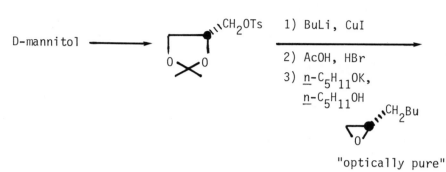

"optically pure"

Angew Chem Int Ed, <u>19</u>, 198 (1980)

CHAPTER 10
PREPARATION OF HALIDES AND SULFONATES

Section 136 Halides from Acetylenes

$$Bu-C\equiv CH \xrightarrow{\begin{array}{c}\text{Amberlyst A-26}\\\text{(perbromide form)}\end{array}} \begin{array}{c}Bu\\Br\end{array}{>}C=CHBr \qquad 94\%$$

Synthesis, 143 (1980)

$$Ph-C\equiv C-Me \xrightarrow{SO_2Cl_2} \qquad 85\%$$

(structure: Ph and Cl on one carbon, Cl and Me on the other, C=C)

Bull Chem Soc Japan, 54, 2843 (1981)

$$HC\equiv C-(CH_2)_8COOMe \xrightarrow{\begin{array}{l}1)\ \text{catecholborane}\\2)\ H_2O\\3)\ ICl,\ NaOAc\end{array}} \begin{array}{c}H\\I\end{array}{>}C=CH(CH_2)_8COOMe \quad 70\%$$

Synth Comm, 11, 247 (1981)

Section 137 Halides from Carboxylic Acids

(bicyclic COOH) $\xrightarrow[\text{Freon 113}]{t\text{-BuOI, }h\nu}$ (bicyclic I)   58%

JOC, 45, 4226 (1980)

Section 138 Halides and Sulfonates from Alcohols

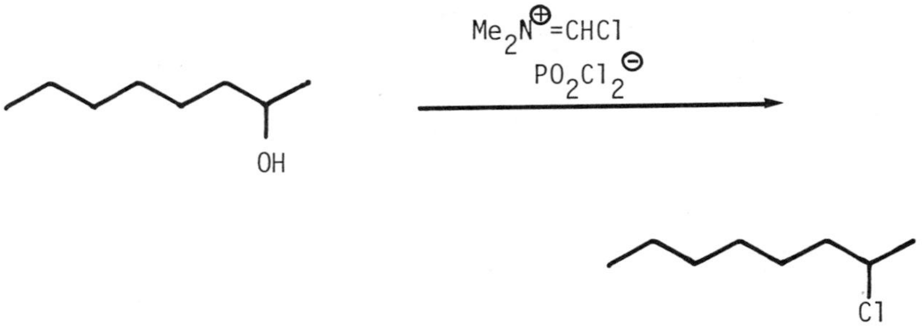

$Me_2N^{\oplus}$=CHCl
$PO_2Cl_2^{\ominus}$
OH
Cl

90%

Synthesis, 746 (1980)

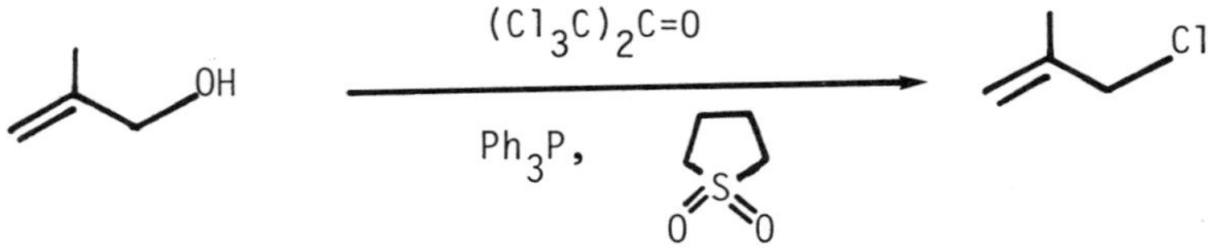

OH
$(Cl_3C)_2C=O$
Ph_3P,
S
O
O
Cl

>90%

JOC, 46, 824 (1981)

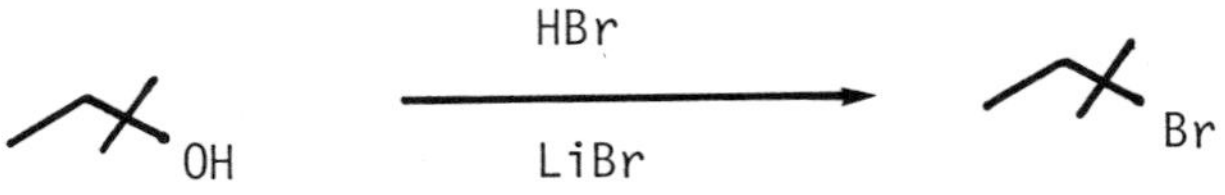

OH
HBr
LiBr
Br

91%

Bull Chem Soc Japan, 53, 1181 (1980)

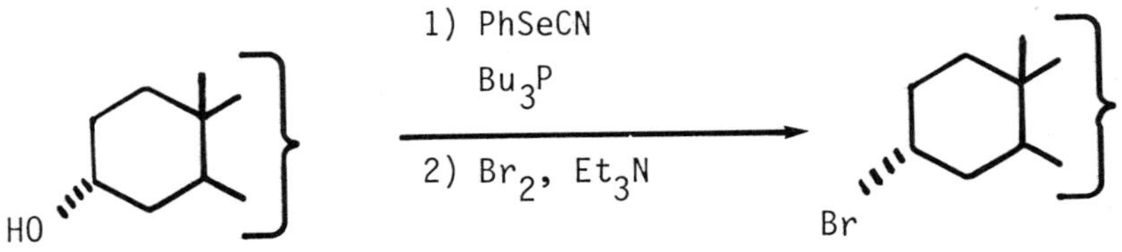

HO
1) PhSeCN
Bu_3P
2) Br_2, Et_3N
Br

JCS Chem Comm, 656 (1980)

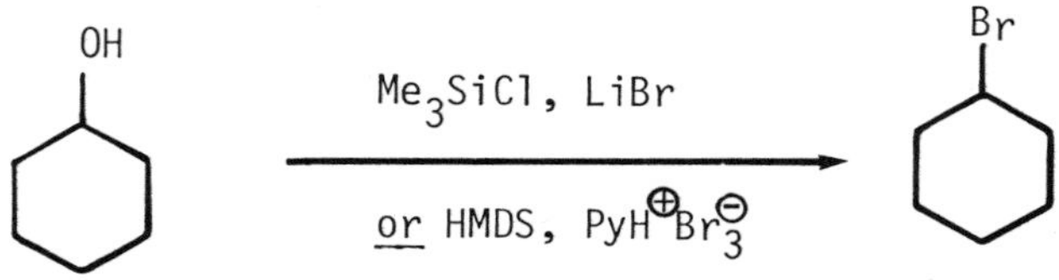

98%

JOC, 45, 1638 (1980)

82%

Aust J Chem, 35, 517 (1982)

$H\text{-}C{\equiv}C\text{-}CH_2OH \xrightarrow[\text{HMPT}]{Bu_3PI_2} H\text{-}C{\equiv}C\text{-}CH_2I$ 60%

Aust J Chem, 35, 517 (1982)

~85%

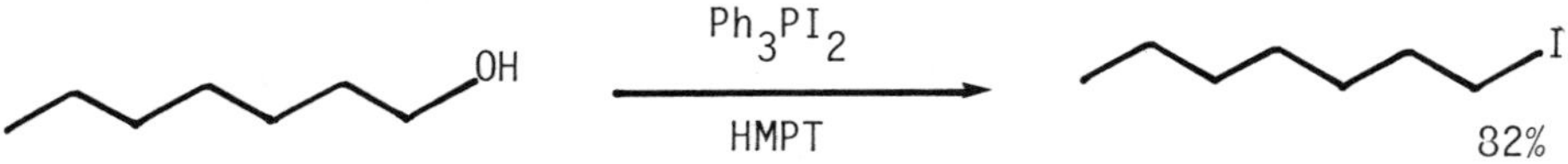

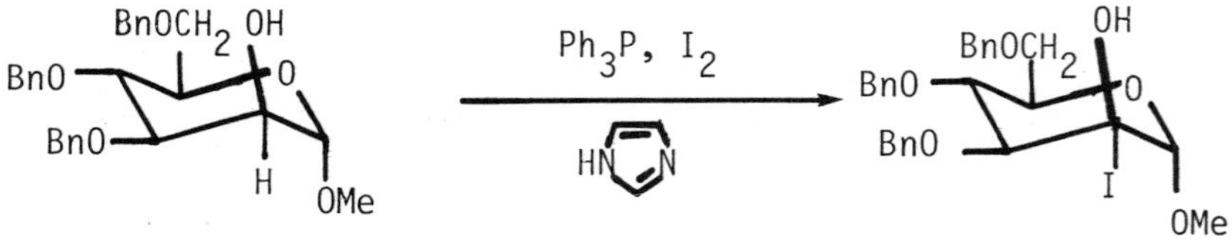

JCS Perkin I, 2866 (1980)

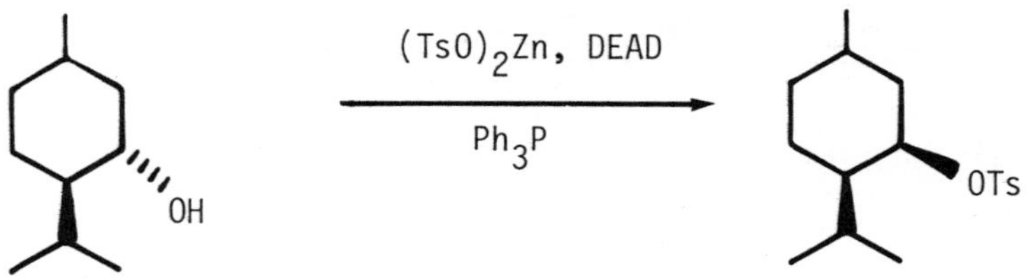

(TsO)$_2$Zn, DEAD
———————————→
Ph$_3$P

94%

Tetr Lett, 23, 4461 (1982)

Section 139 Halides from Aldehydes

No additional examples.

Section 140 Halides from Alkyls

For the conversion RH → RHal see Section 146 (Halides from Hydrides)

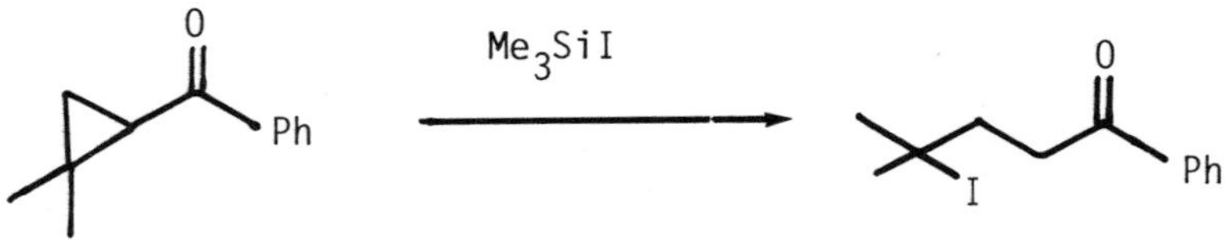

Me$_3$SiI
———————————→

98%

JOC, 46, 2412 (1981)

Section 141 <u>Halides and Sulfonates from Amides</u>

No additional examples.

Section 142 <u>Halides from Amines</u>

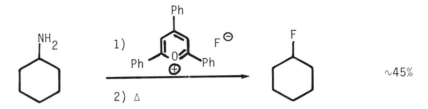

JCS Perkin I, 1890 and 2901 (1980)

Bull Chem Soc Japan, <u>53</u>, 1065 (1980)

JOC, <u>46</u>, 5239 (1981)

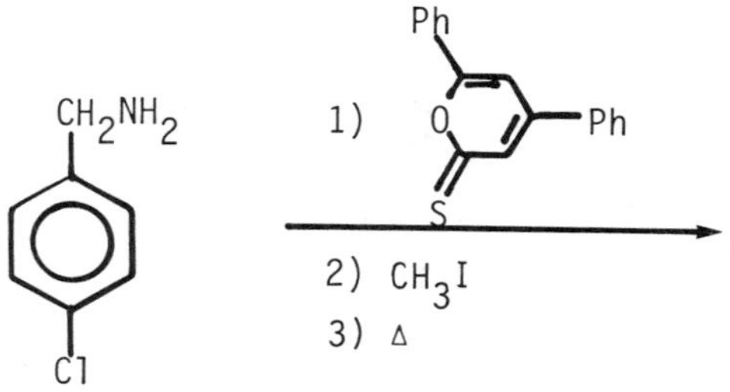

Synthesis, 853 (1980)

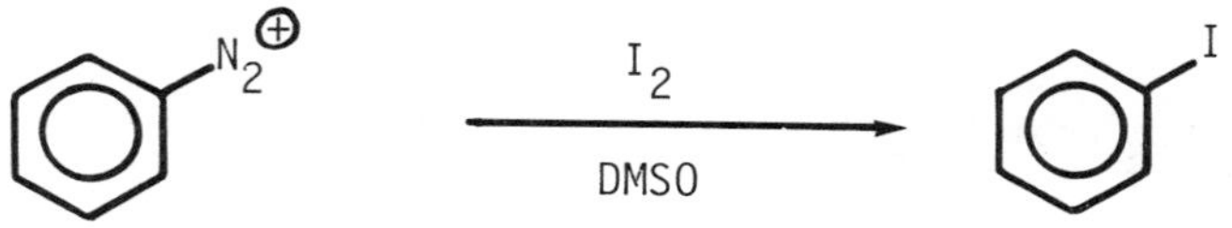

Synth Comm, <u>11</u>, 639 (1981)

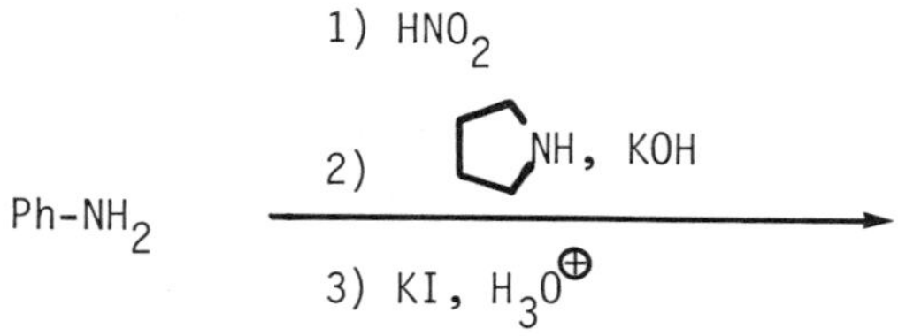

1) HNO$_2$

Ph-NH$_2$ 2) [pyrrolidine] NH, KOH   ────────→   Ph-I   78%

3) KI, H$_3$O$^{\oplus}$

Synthesis, 572 (1980)

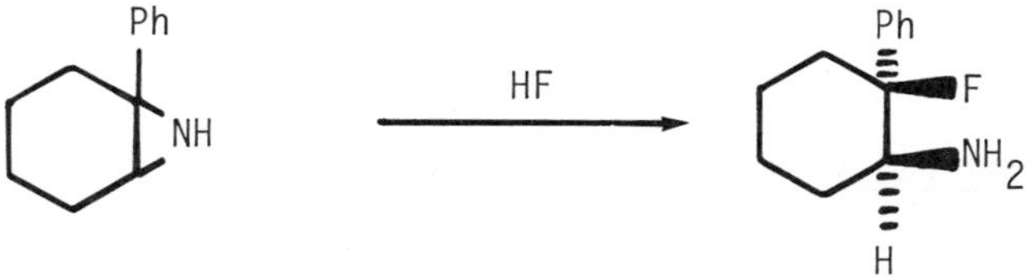

JOC, <u>45</u>, 5328 and 5333 (1980)

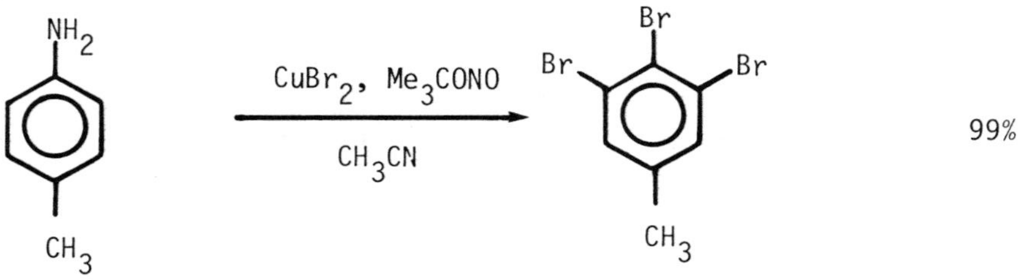

JOC, **45**, 2570 (1980)

Section 143 Halides from Esters

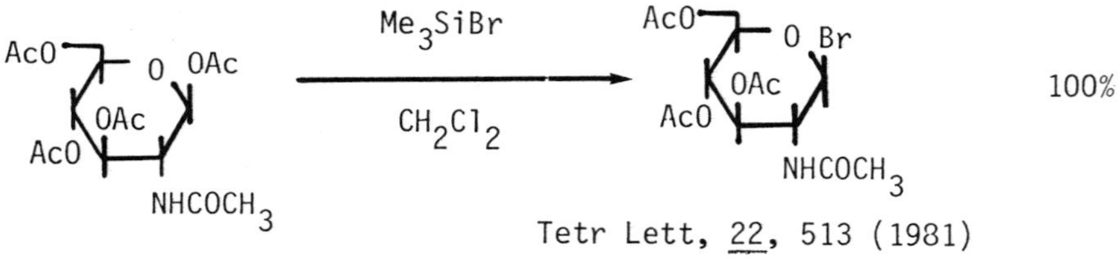

Tetr Lett, **22**, 513 (1981)

Synth Comm, **11**, 763 (1981)

Section 144 Halides from Ethers and Epoxides

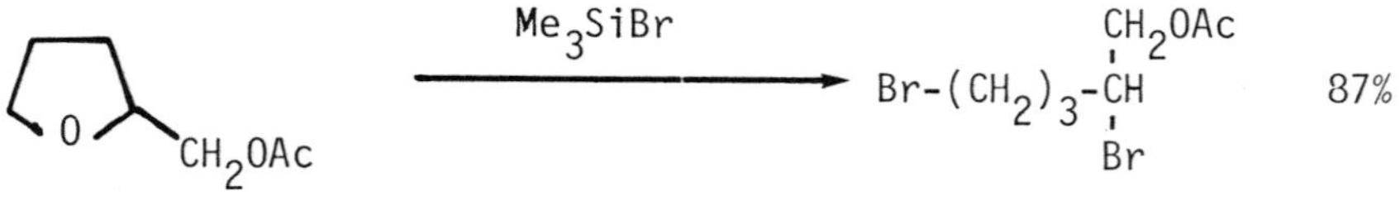

Synthesis, 383 (1981)

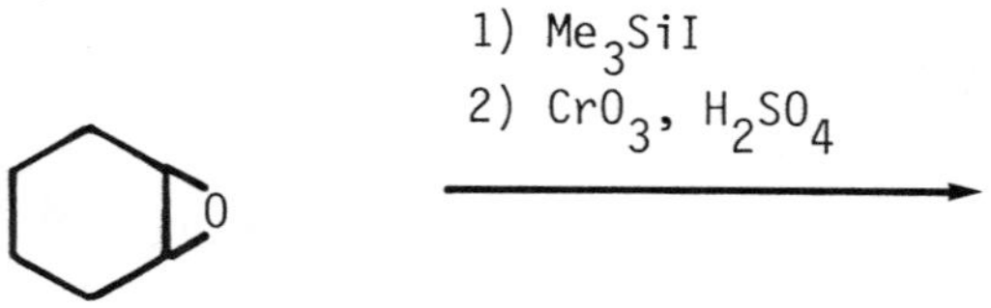
1) Me3SiI
2) CrO3, H2SO4
O
I
O

91%

Tetr Lett, 22, 1429 (1981)

Section 145 Halides from Halides and Sulfonates

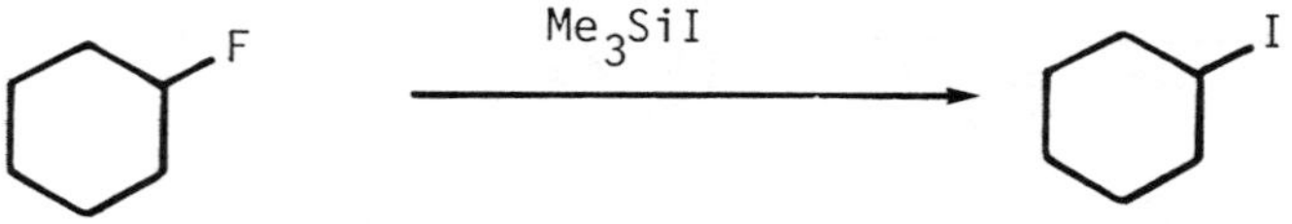
F
Me3SiI
I

72%

JOC, 46, 3727 (1981)

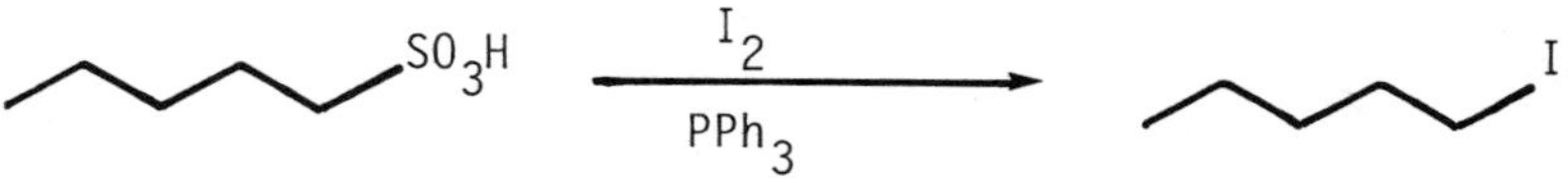
SO3H
I2
PPh3
I

81%

Synthesis, 371 (1981)

$$\text{R-X} \xrightarrow[\text{3) } 100^{\circ}\text{, DMF}]{\text{1) } LiCH_2TePh \quad \text{2) } SO_2Cl_2\text{, } Br_2\text{, or } I_2} \text{R-}CH_2X$$

59-95%

R = 1° alkyl

Chem Lett, 1081 (1982)

Section 146 Halides from Hydrides

α-Halogenations of aldehydes, ketones, esters, and acids are found in Sections 338 (Haloaldehydes), 369(Haloketones), 359(Halo-esters), and 319 (Haloacids).

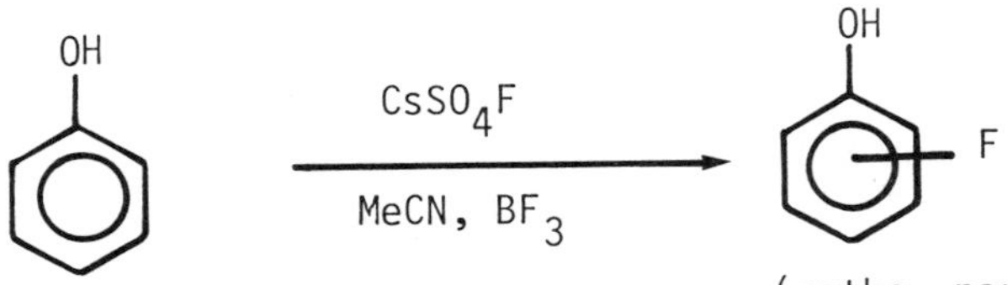

70%

(ortho, para)

JCS Chem Comm, 148 (1981)

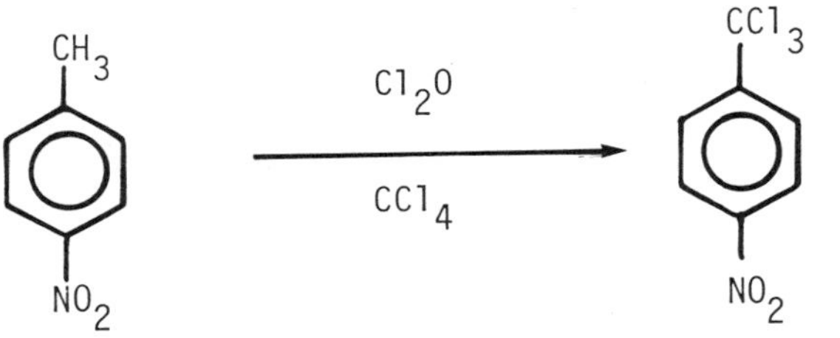

99%

JACS, 104, 4680 (1982)

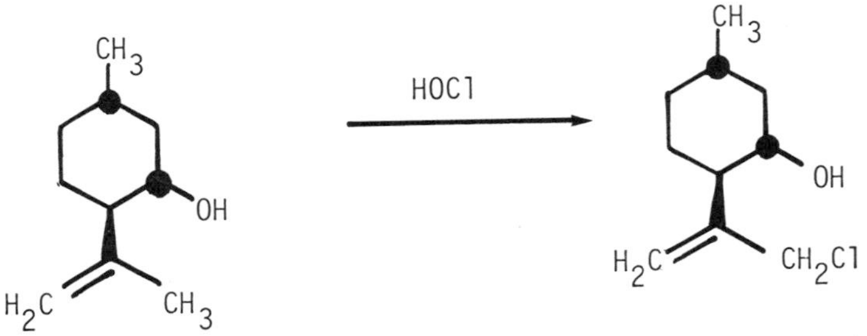

67%

Tetr Lett, 21, 445 (1980)

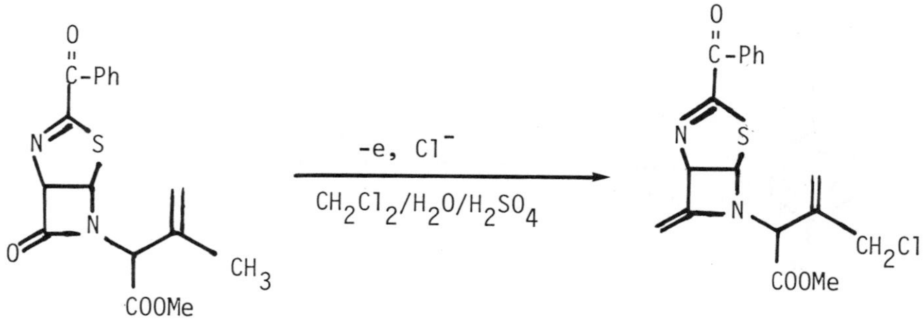

Tetr Lett, 22, 3193 (1981)

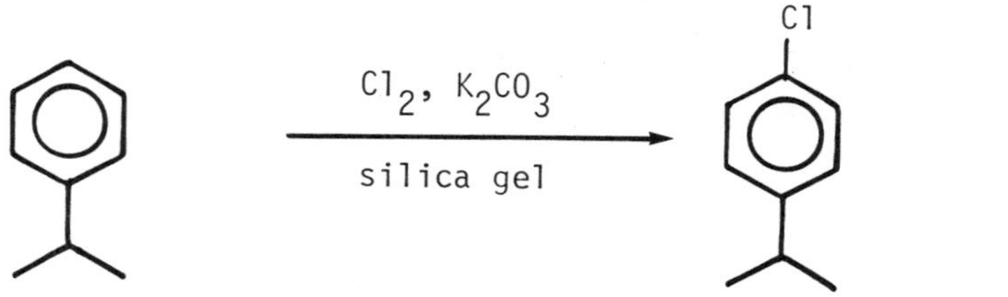

80%

para:ortho = 3.53

Chem Lett, 1423 (1980)

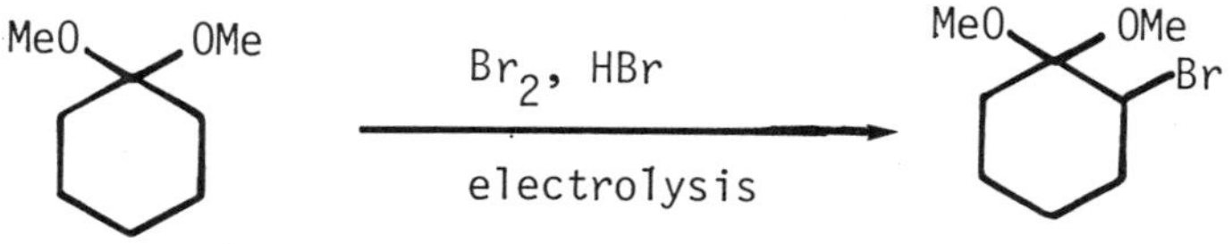

73%

Synth Comm, 10, 821 (1980)

68%

$Ph-CH_2-CH_3$ $\xrightarrow[\text{La(OAc)}_3]{\text{Br}_2}$ $Ph-CH-CH_3$ with Br substituent   90%

Bull Soc Chim France II, 327 (1982)

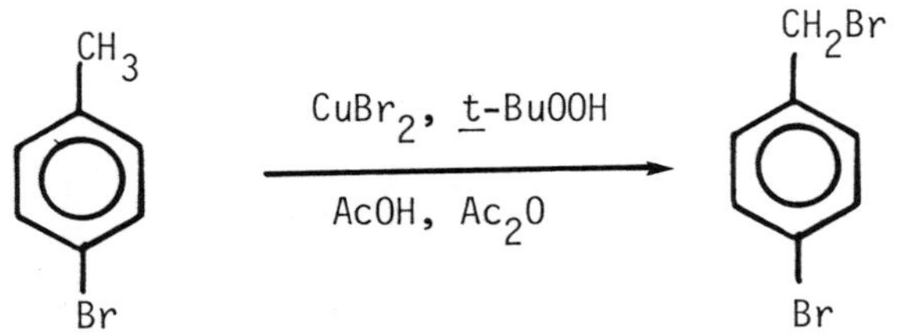

95%

Synth Comm, 11, 669 (1981)

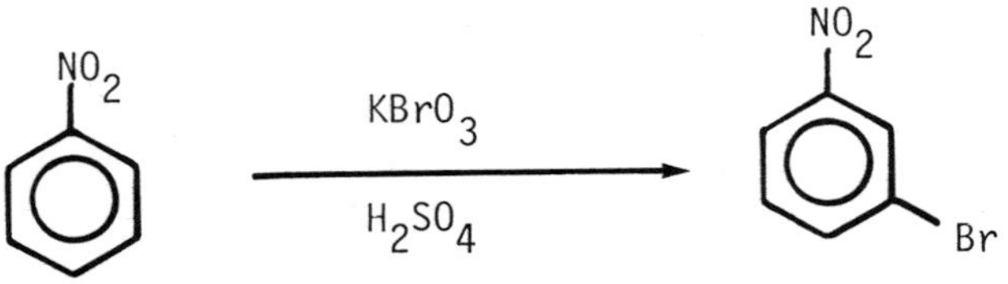

88%

JOC, 46, 2169 (1981)

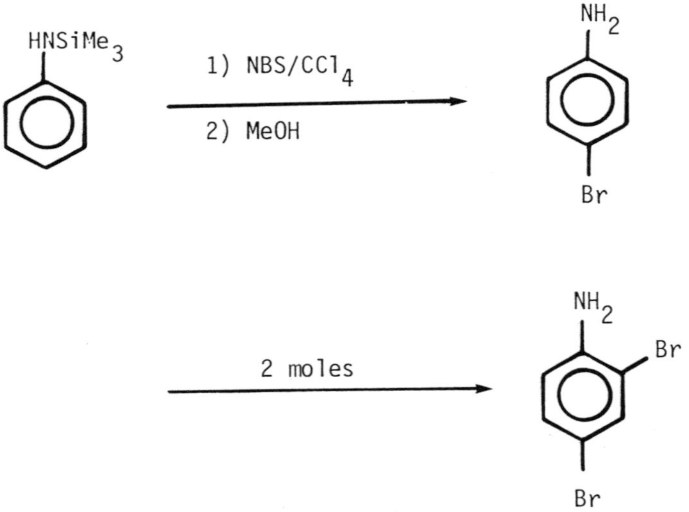

92%

93%

Synthesis, 263 (1982)

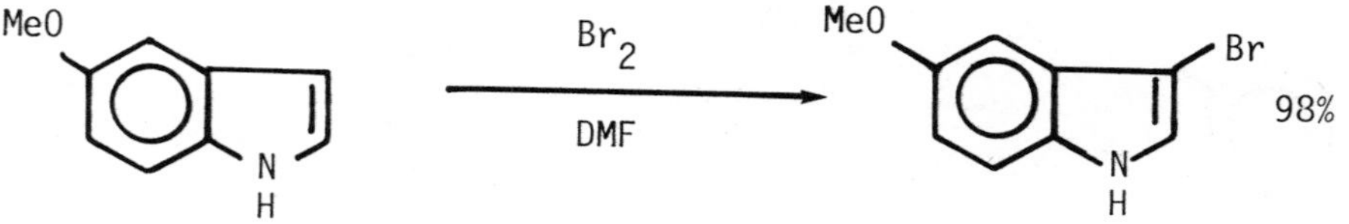

Synthesis, 1096 (1982)

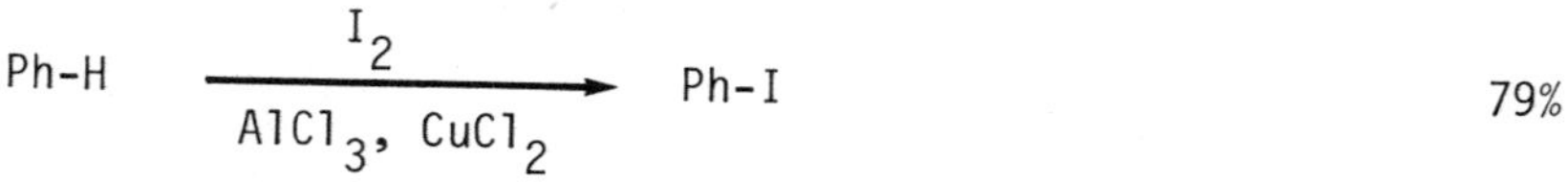

Chem Lett, 1481 (1982)

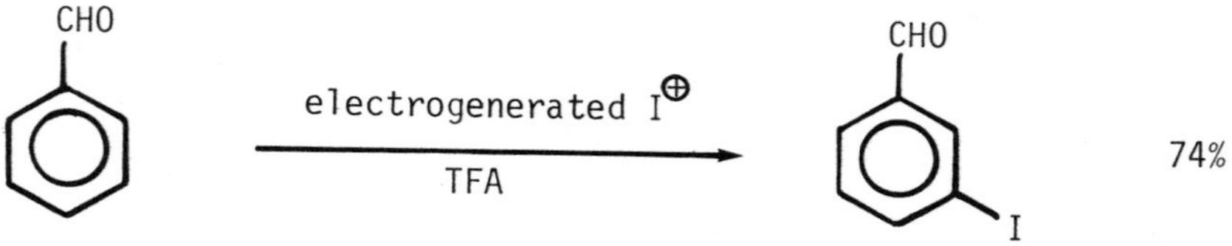

Acta Chem Scand B, 34, 47 (1980)

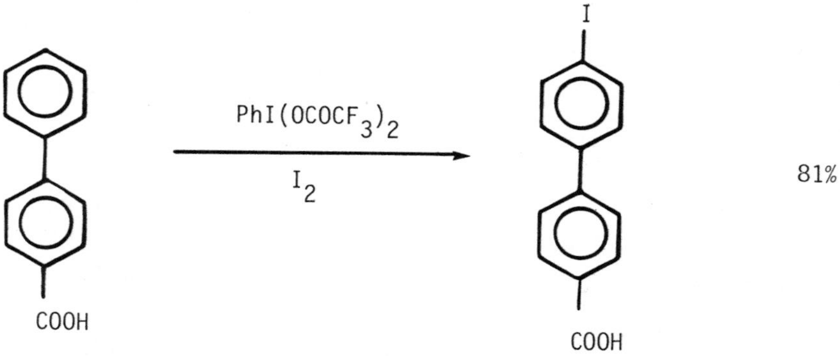

JOC (USSR), 17, 2320 (1982)

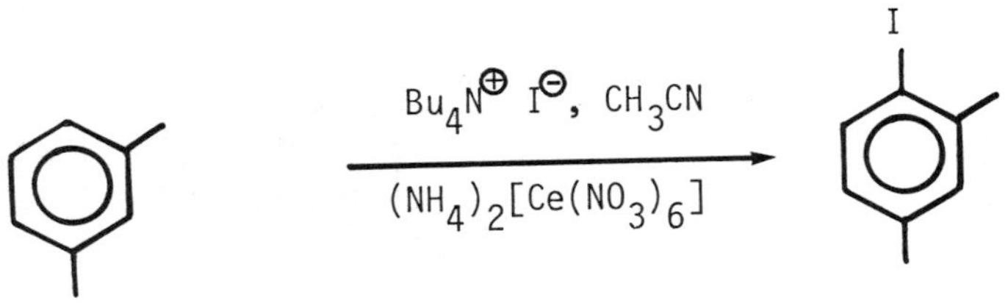

71%

Bull Chem Soc Japan, 54, 2847 (1981)

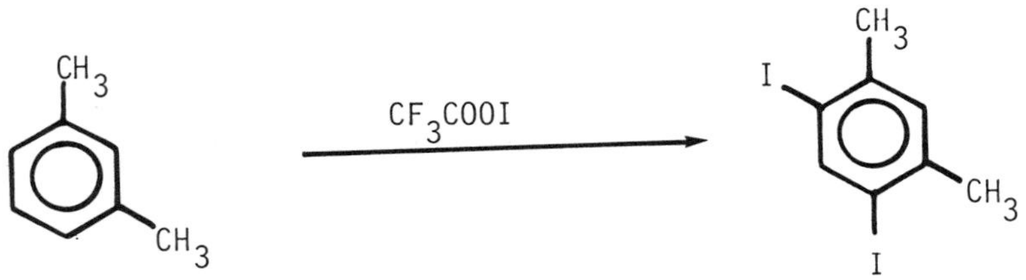

77%

Synthesis, 486 (1980)

Section 147 Halides from Ketones

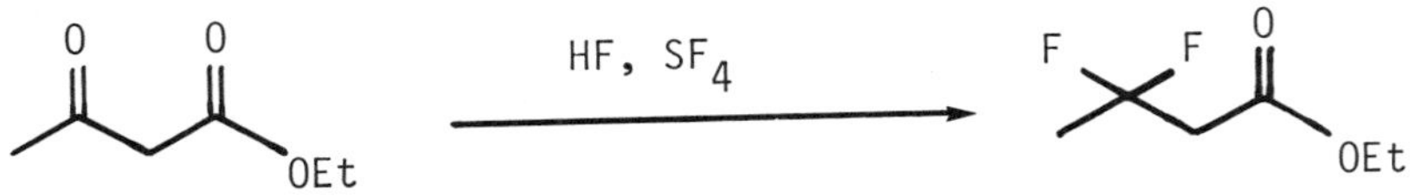

85%

JCC (USSR), 17, 1260 (1981)

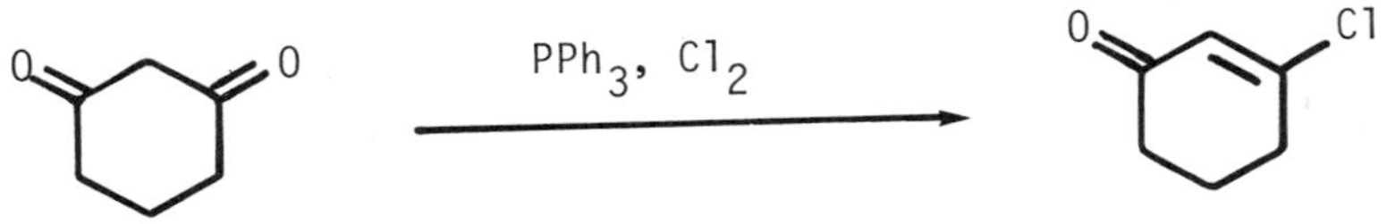

91%

Can J Chem, 60, 210 (1982)

Section 148 Halides and Sulfonates from Nitriles

No examples.

Section 149 Halides from Olefins

For halocyclopropanations see Section 74 (Alkyls from Olefins).

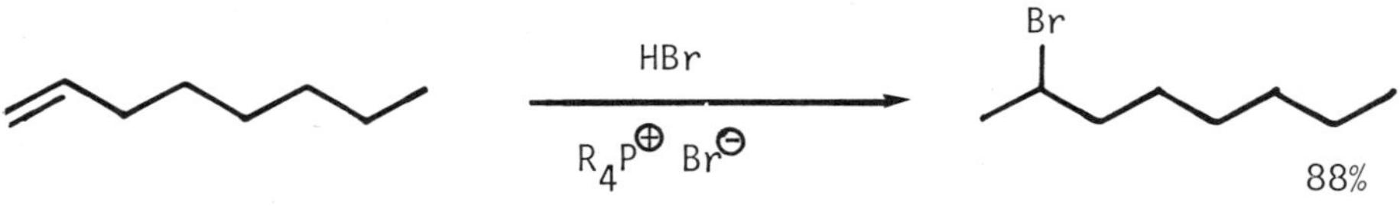

JOC, 45, 3527 (1980)

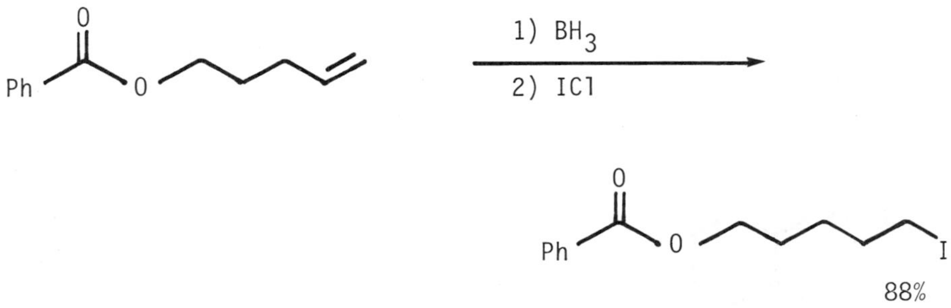

JOC, 45, 3578 (1980)

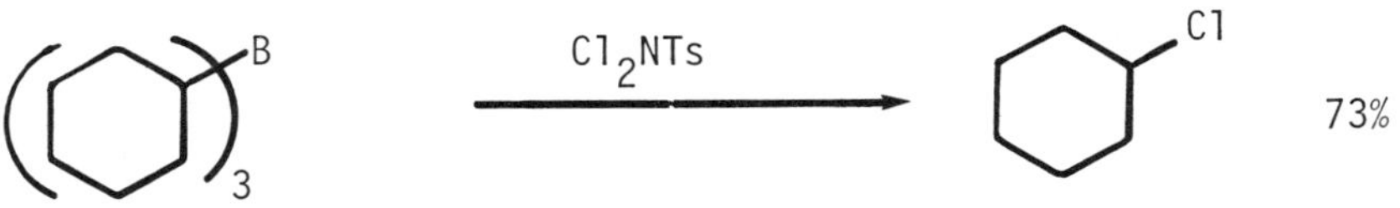

J Chem Res (S), 376 (1981)

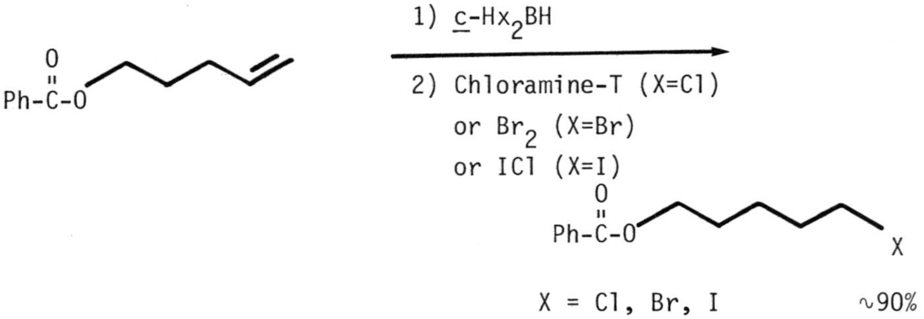
O
Ph-C-O
1) c-Hx2BH
2) Chloramine-T (X=Cl)
or Br2 (X=Br)
or ICl (X=I)
O
Ph-C-O
X
X = Cl, Br, I
~90%

JOC, 46, 2582 and 3113 (1981)
Synth Comm, 11, 521 (1981)

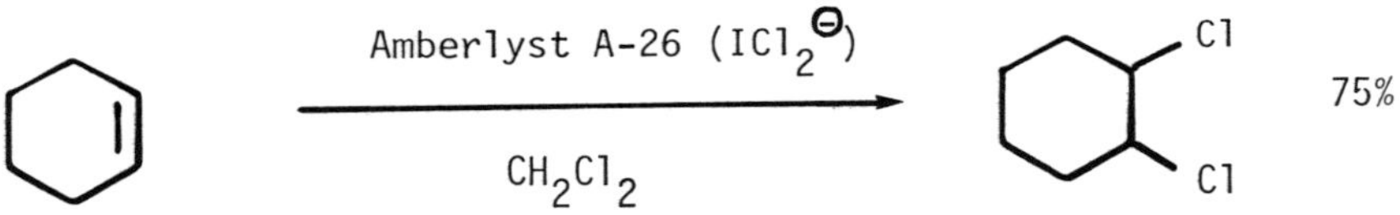
Amberlyst A-26 (ICl2⊖)
CH2Cl2
Cl
Cl
75%

JCS Chem Comm, 1278 (1980)

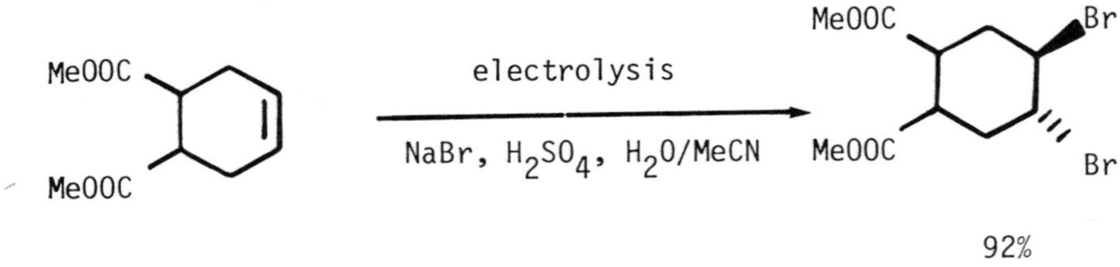
MeOOC
MeOOC
electrolysis
NaBr, H2SO4, H2O/MeCN
MeOOC
MeOOC
Br
Br
92%

JOC, 46, 3312 (1981)

45%

Tetr Lett, <u>21</u>, 4543 (1980)

$$Bu-CH=CH_2 \xrightarrow[\text{AIBN}]{\text{CH}_2\text{ClI}} Bu-\underset{\underset{I}{|}}{C}H-CH_2-\underset{\underset{Cl}{|}}{C}H_2 \qquad 81\%$$

Bull Chem Soc Japan, <u>53</u>, 770 (1980)

Section 150 Halides from Miscellaneous Compounds

1) CH₃MgX

2) ⁻OH, NaOCl
   H₂O/EtOH

60%

Synthesis, 616 (1980)

$$Ph-B(OH)_2 \xrightarrow[\text{chloramine-T}]{\text{NaI}} Ph-I \qquad 87\%$$

Org Prep Proc Int, <u>14</u>, 359 (1982)

Review: "Methods of Fluorination in Organic Chemistry"

Angew Chem Int Ed, 20, 647 (1981)

Review: "Radioiodination Techniques for Small Organic Molecules"

Chem Rev, 82, 575 (1982)

CHAPTER 11
PREPARATION OF HYDRIDES

This chapter lists hydrogenolysis and related reactions by which functional groups are replaced by hydrogen, e.g. $RCH_2X \rightarrow RCH_2\text{-}H$ or R-H

Section 151 Hydrides from Acetylenes

No examples of the reaction $RC{\equiv}CR \rightarrow RH$ occurs in the literature.

Section 152 Hydrides from Carboxylic Acids

This section lists examples of decarboxylation (R-COOH $\rightarrow$ R-H) and related reactions.

Synthesis, 141 (1982)

Section 153 Hydrides from Alcohols and Phenols

This section lists examples of the hydrogenolysis of alcohols and phenols, ROH → RH

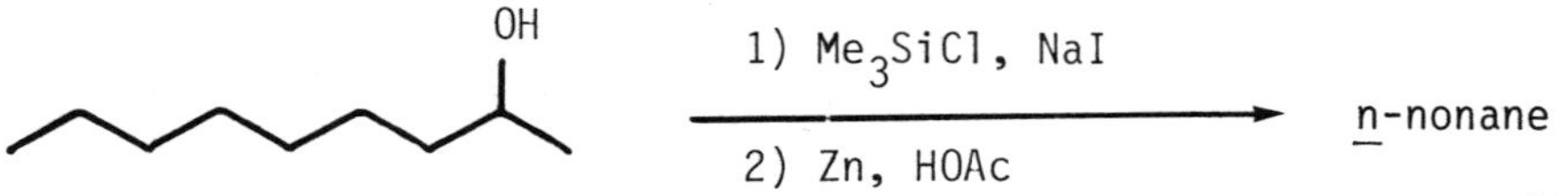

1) Me₃SiCl, NaI

2) Zn, HOAc

n-nonane

61%

Synthesis, 32 (1981)

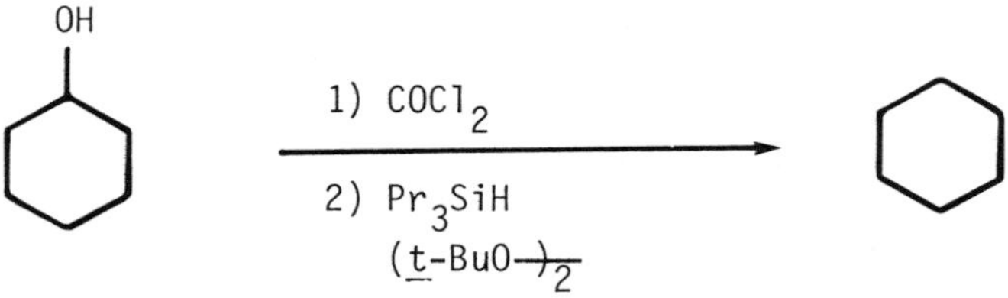

1) COCl₂

2) Pr₃SiH
(t-BuO)₂

91%

JCS Perkin I, 1207 (1980)

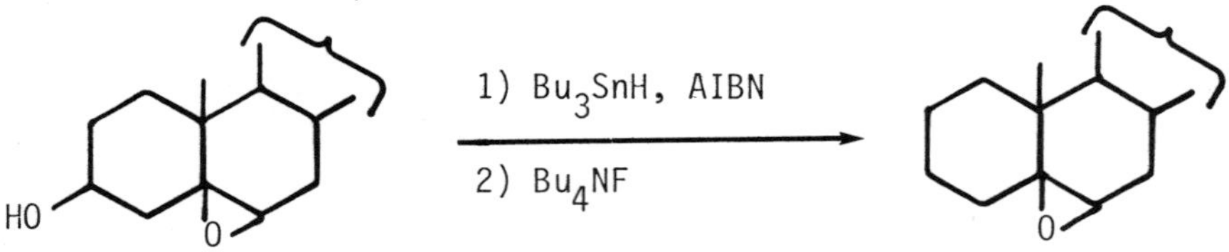

1) Bu₃SnH, AIBN

2) Bu₄NF

85%

JACS, 103, 932 (1981)

$$\underset{\text{Ph-CH-Ph}}{\overset{\text{OH}}{|}} \xrightarrow{\text{NaBH}_4\text{-PdCl}_2} \text{Ph-CH}_2\text{-Ph} \qquad\qquad 91\%$$

Chem Lett, 1029 (1981)

$$\text{Ph}_3\text{C-OH} \xrightarrow[\text{toluene}]{\text{Fe(CO)}_5} \text{Ph}_3\text{C-H} \qquad\qquad 90\%$$

Tetr Lett, $\underline{21}$, 801 (1980)

$$\text{R-OH} \xrightarrow[\text{2) Bu}_3\text{SnH}]{\substack{\text{1) N,N'-thiocarbonyl} \\ \text{diimidazole}}} \text{R-H} \qquad \sim 30\text{-}80\%$$

(deoxy sugar)

R = protected sugar

JOC, $\underline{46}$, 4843 (1981)

Also via Halides and Sulfonates, Section 160

Section 154 <u>Hydrides from Aldehydes</u>

No additional examples.

For the conversion RCHO → RMe etc. see Section 64 (Alkyls from Aldehydes)

Section 155 Hydrides from Alkyls

This section lists examples of the conversion R-R' → R-H or

Ar-R → Ar-H

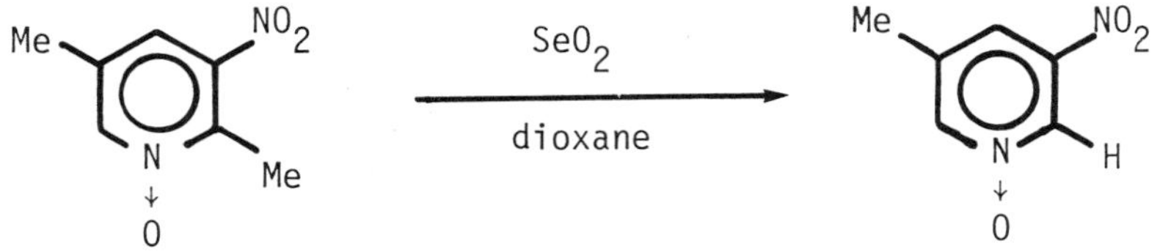

$$\xrightarrow[\text{dioxane}]{\text{SeO}_2}$$

70%

Tetr Lett, $\underline{21}$, 2433 (1980)

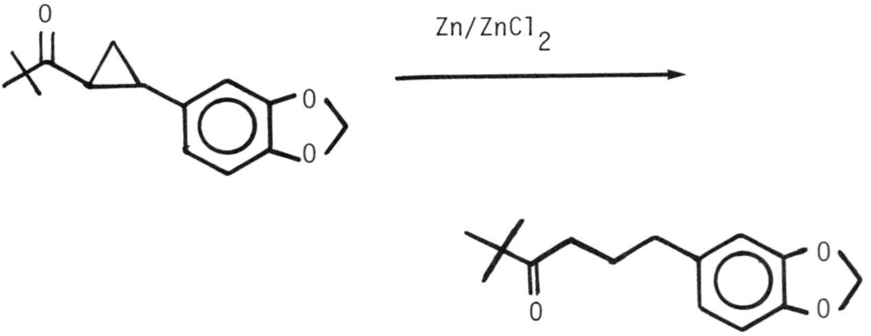

$$\xrightarrow{\text{Zn/ZnCl}_2}$$

88%

Tetr Lett, $\underline{22}$, 695 (1981)

Section 156 Hydrides from Amides

No additional examples.

Section 157 <u>Hydrides from Amines</u>

This section lists examples of the conversion $RNH_2 \rightarrow RH$

Indian J Chem, <u>20B</u>, 767 (1981)

Synthesis, 68 (1980)

Section 158 <u>Hydrides from Esters</u>

This section lists examples of the reactions $RCOOR' \rightarrow RH$ and $RCOOR' \rightarrow R'H$

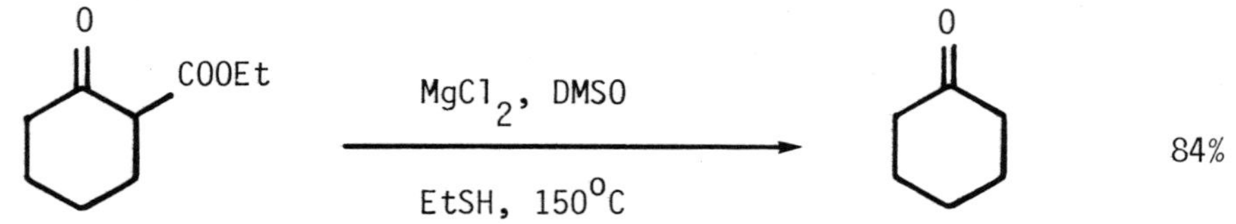

Synthesis, 119 (1981)

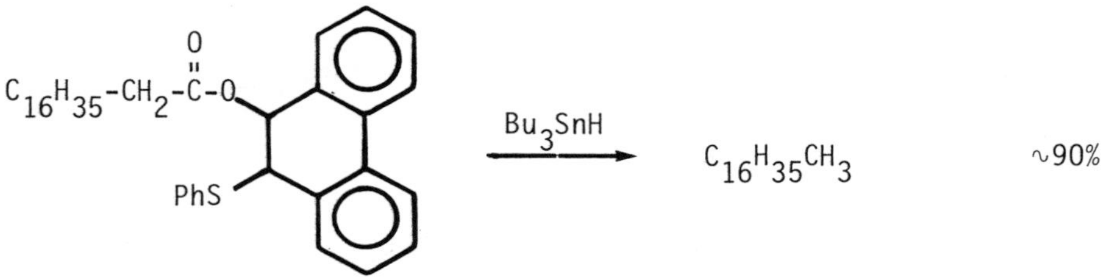

JCS Chem Comm, 732 (1980)

Section 159 Hydrides from Ethers

This section lists examples of the reaction R-O-R' → R-H.

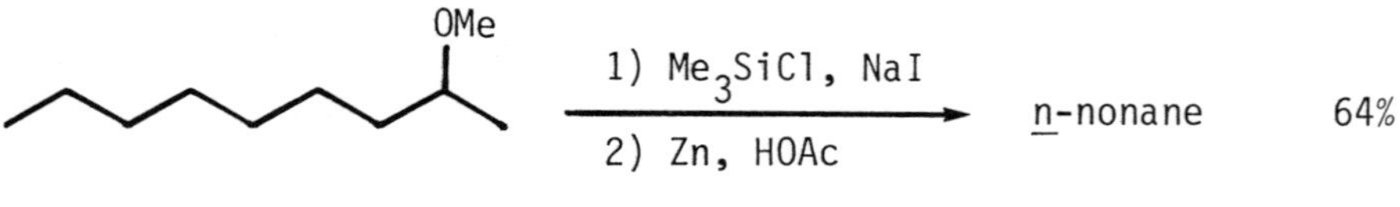

Synthesis, 32 (1981)

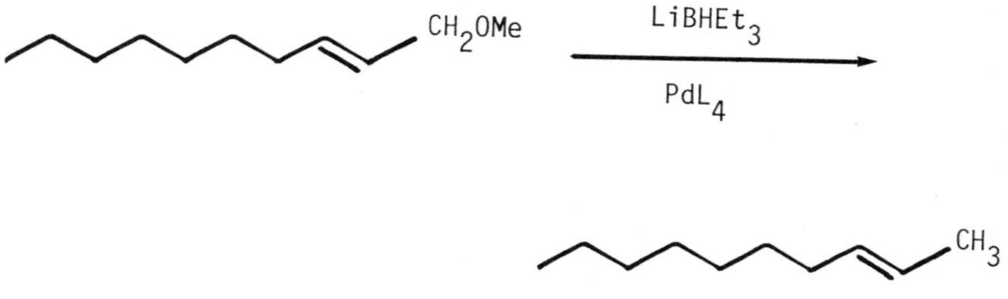

JOC, <u>47</u>, 4380 (1982)

Section 160 <u>Hydrides from Halides and Sulfonates</u>

This section lists the reductions of halides and sulfonates,
RX → RH

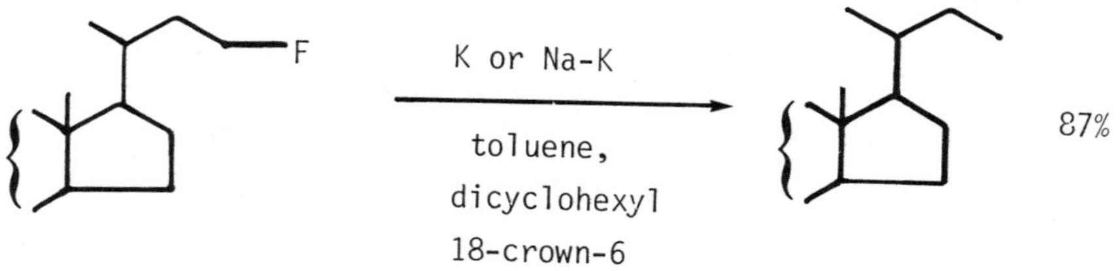

Tetr Lett, <u>22</u>, 2583 (1981)

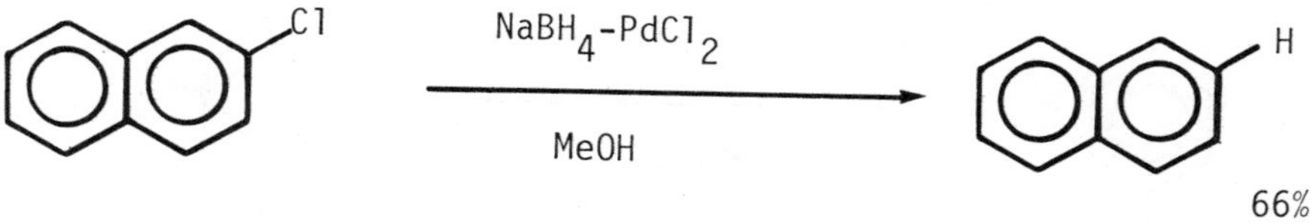

Chem Lett, 1029 (1981)

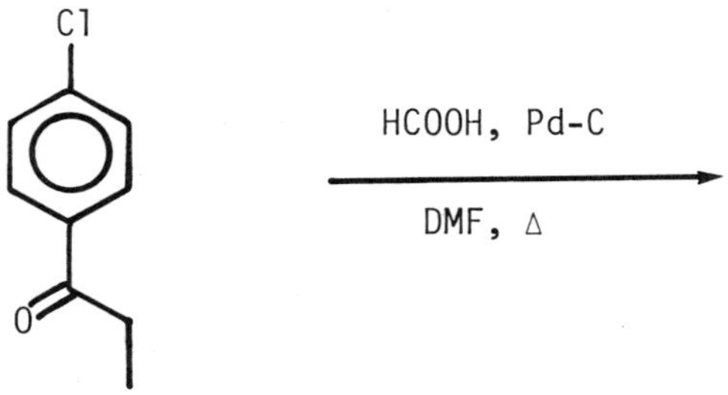
Cl
HCOOH, Pd-C
DMF, Δ
H
O
O

80%

Synthesis, 876 (1982)

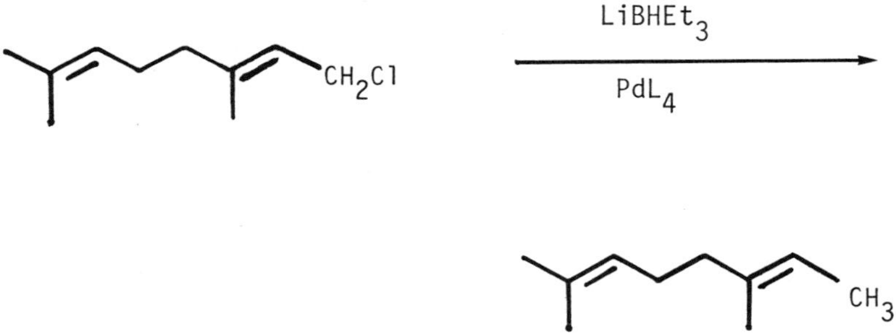
CH2Cl
LiBHEt3
PdL4
CH3

93%

JOC, 47, 4380 (1982)

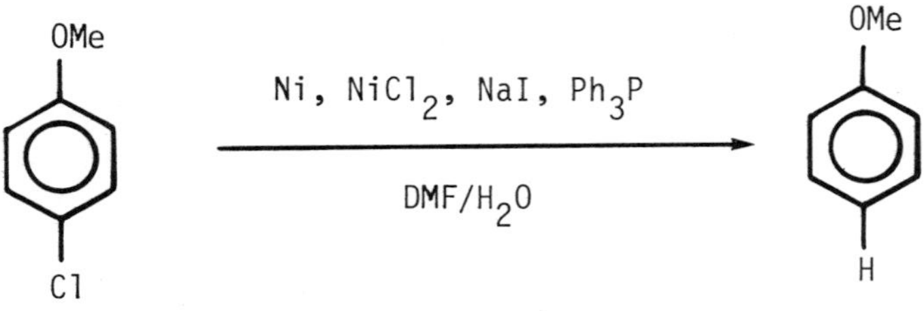
OMe
Ni, NiCl2, NaI, Ph3P
DMF/H2O
Cl
OMe
H

96%

JOC, 47, 2622 (1982)

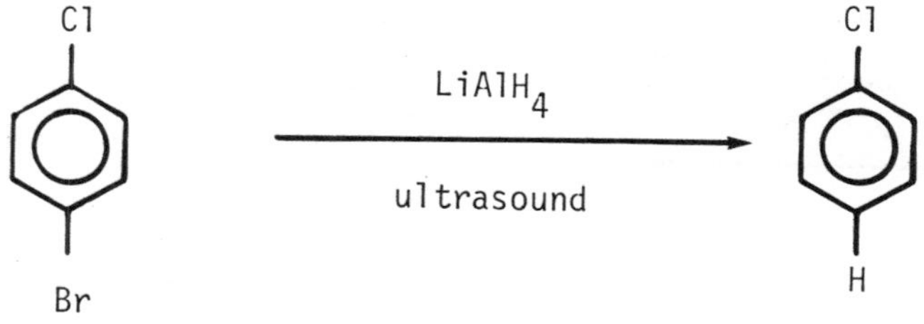
Cl
LiAlH4
ultrasound
Br
Cl
H

98%

Tetr Lett, 23, 1643 (1982)

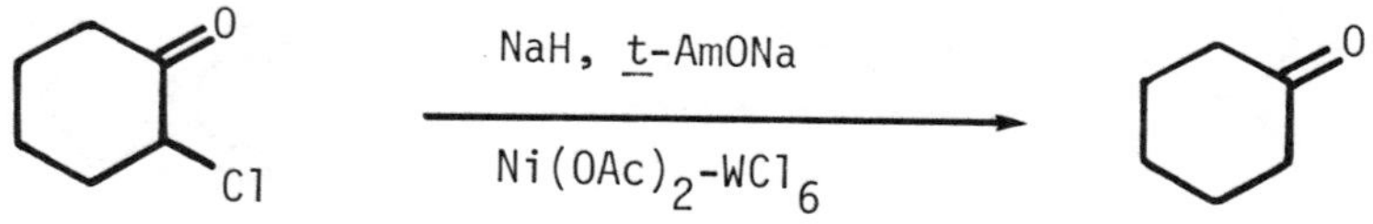
O
Cl
NaH, t-AmONa
Ni(OAc)2-WCl6
O

80%

JOC, 46, 1270 (1981)

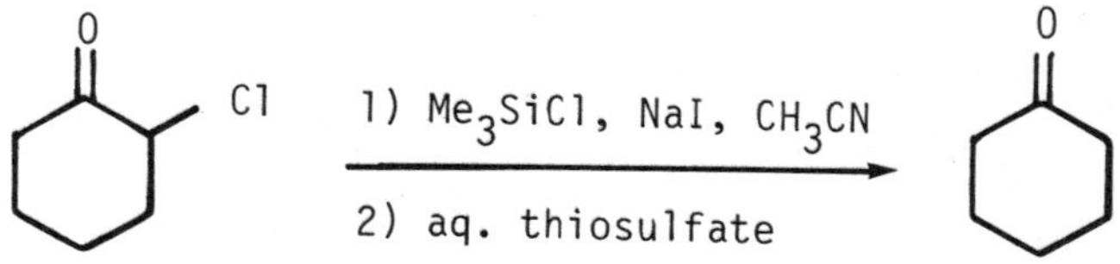
O
Cl
1) Me3SiCl, NaI, CH3CN
2) aq. thiosulfate
O

78%

JOC, 45, 3531 (1980)

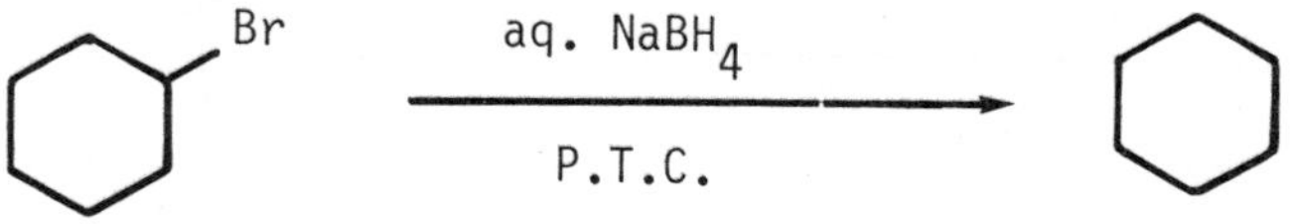
Br
aq. NaBH4
P.T.C.

83%

JOC, 46, 3909 (1981)

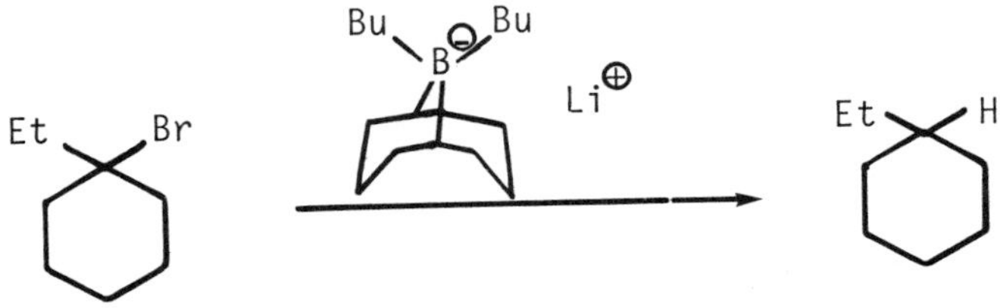

Tetrahedron, 37, 2261 (1981)

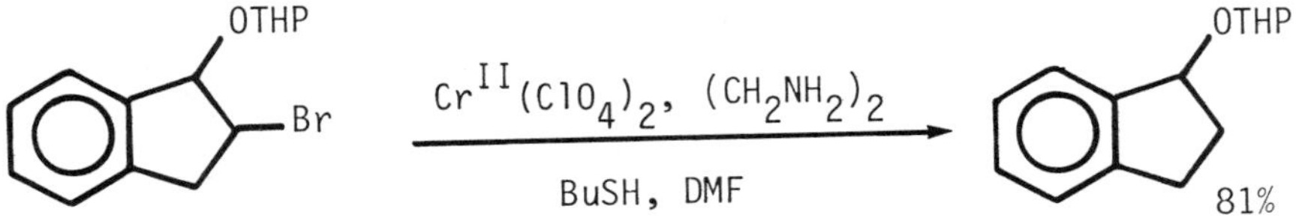

Angew Chem Int Ed, 19, 46 (1980)

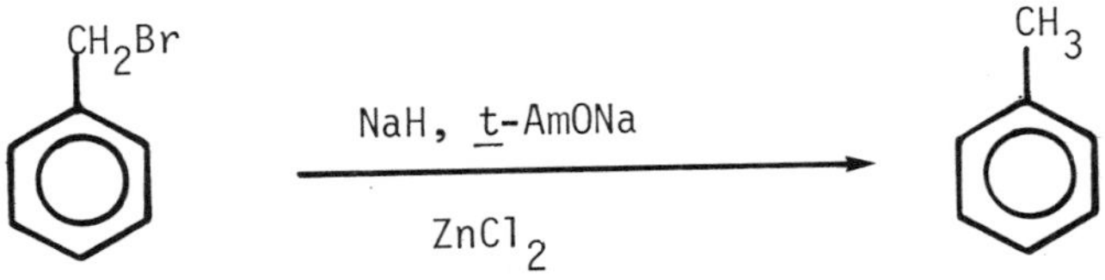

JOC, 46, 1270 (1981)

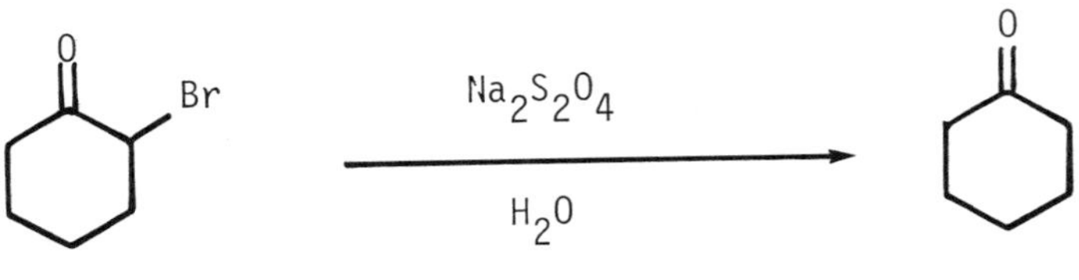

Synth Comm, 12, 261 (1982)

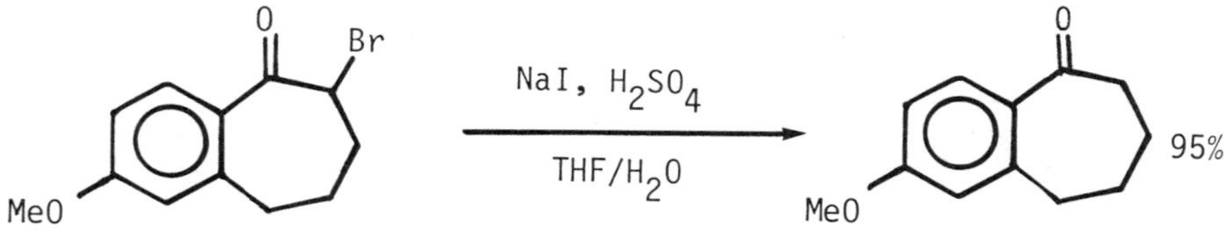

Tetr Lett, 21, 3195 (1980)

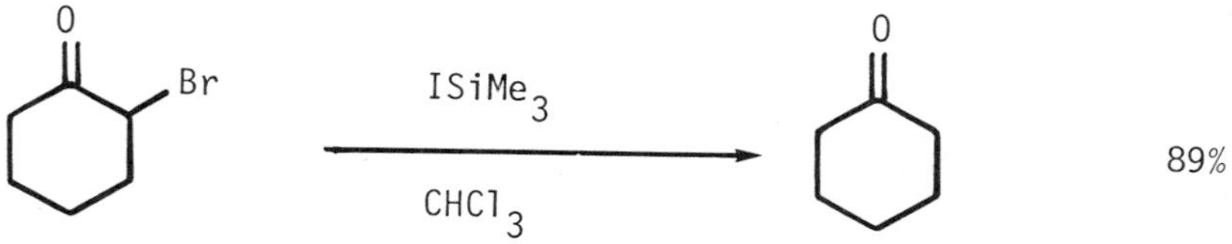

Synth Comm, 11, 101 (1981)

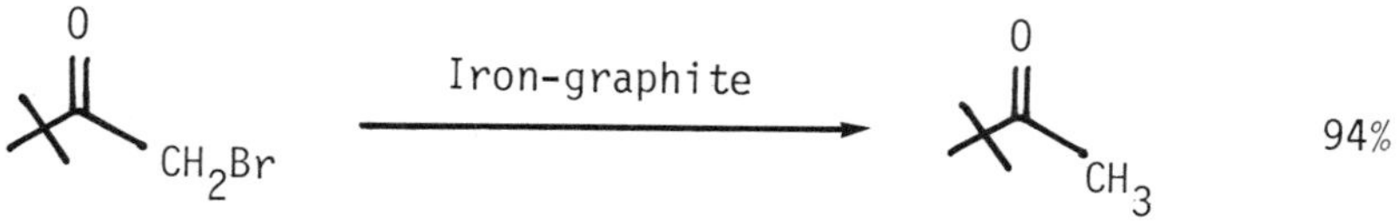

JOC, 47, 876 (1982)

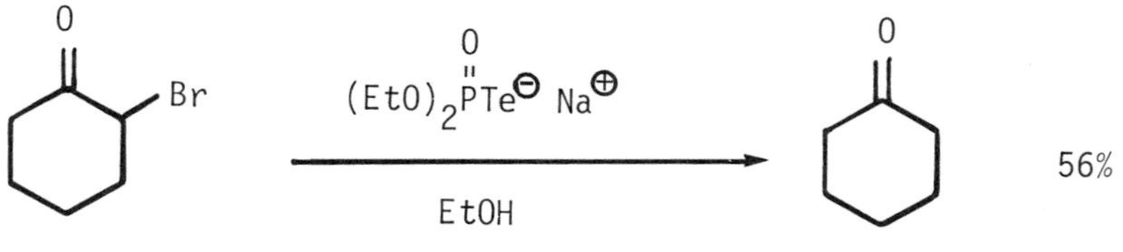

JOC, 47, 1124 (1982)

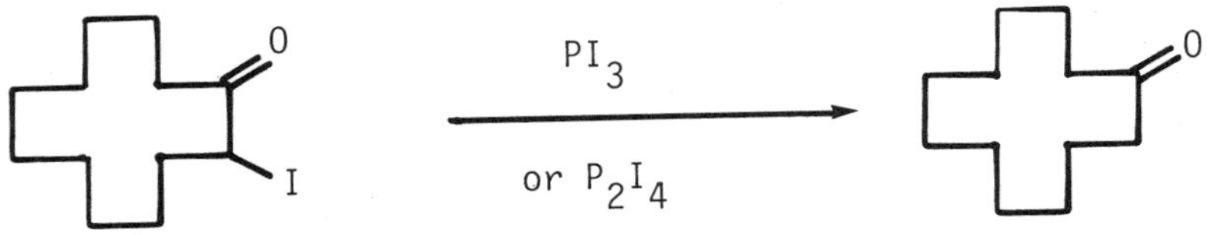

Tetr Lett, <u>22</u>, 1431 (1981)

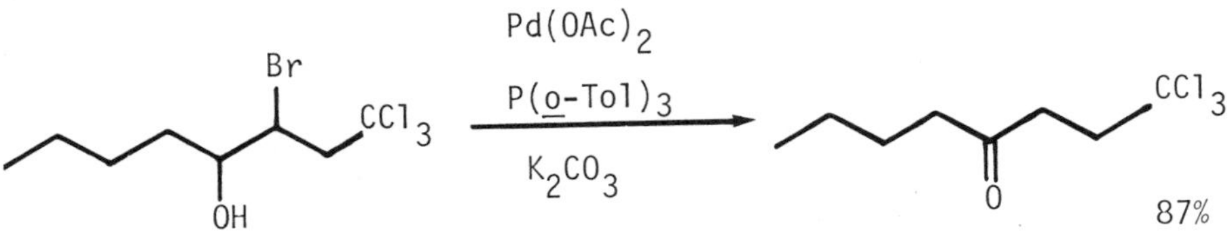

92%

Tetr Lett, <u>23</u>, 3085 (1982)

87%

Review: "The Hydrogenolysis of Organic Halides"

Synthesis, 425 (1980)

A study of the reduction of alkyl halides using $LiAlH_4$ in THF.

JOC, <u>47</u>, 276 (1982)

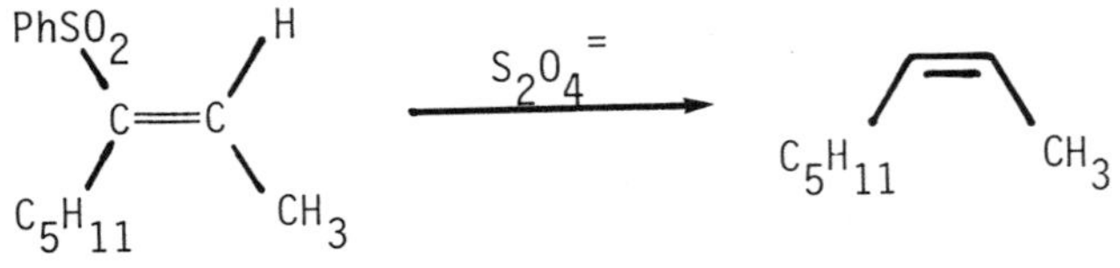

88%

Tetr Lett, <u>23</u>, 3265 (1982)

Section 161 Hydrides from Hydrides

No additional examples.

Section 162 Hydrides from Ketones

No additional examples.

For the conversion $R_2CO \rightarrow R_2CH_2$ or R_2CHR' see Section 72 (Alkyls and Methylenes from Ketones)

Section 163 Hydrides from Nitriles

$$CH_3(CH_2)_7-\overset{\overset{CH_3}{|}}{\underset{\underset{CH_3}{|}}{C}}-NC \xrightarrow[\text{AIBN}]{Bu_3SnH} CH_3(CH_2)_7-\overset{\overset{CH_3}{|}}{\underset{\underset{CH_3}{|}}{CH}} \qquad 91\%$$

JCS Perkin I, 2657 (1980)

$$\xrightarrow[\text{hexane}]{K/Al_2O_3} \qquad 88\%$$

JOC, 45, 3227 (1980)

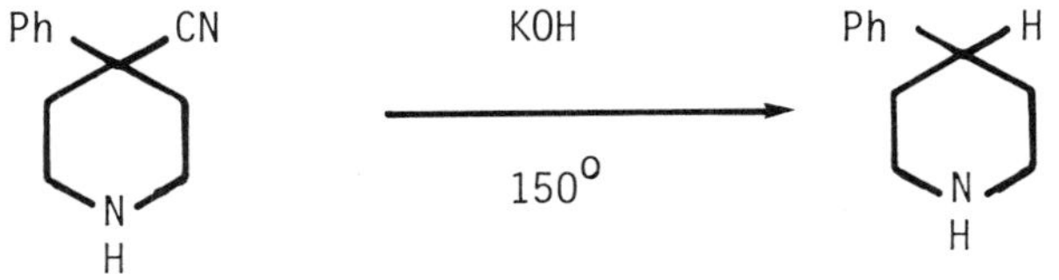

92%

Synth Comm, <u>10</u>, 939 (1980)

Section 164 <u>Hydrides from Olefins</u>

No additional examples.

Section 165 <u>Hydrides from Miscellaneous Compounds</u>

$$CH_3(CH_2)_{10}-CH_2SH \xrightarrow[\text{FeCl}_2]{\text{NaEt}_3\text{BH}} CH_3(CH_2)_{10}-CH_3$$ 78%

Angew Chem Int Ed, <u>19</u>, 315 (1980)

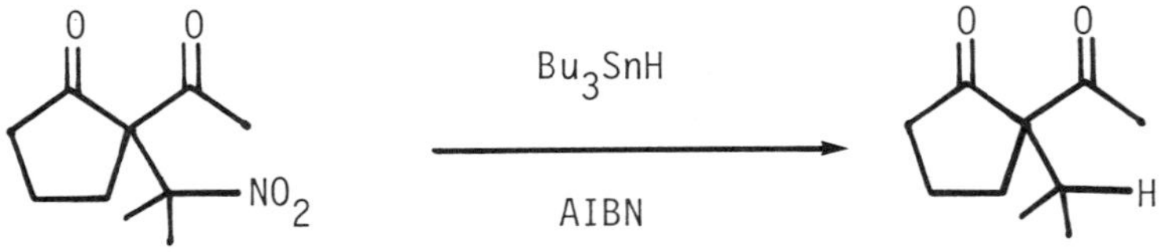

60%

Tetr Lett, <u>22</u>, 1705 (1981)

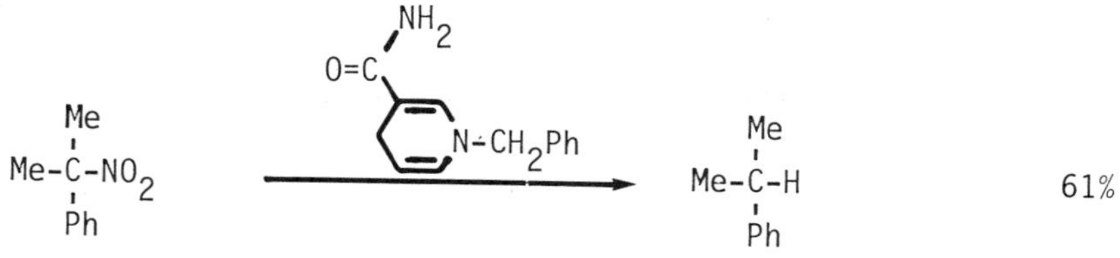

Me
Me-C-NO₂
Ph

⟶

Me
Me-C-H
Ph

61%

JACS, 102, 2851 (1980)

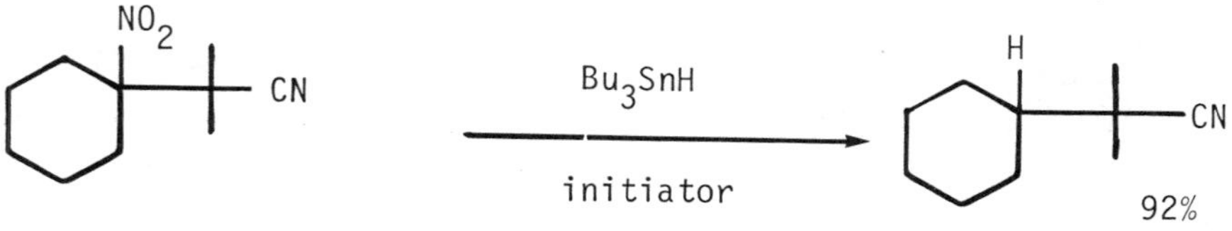

Bu₃SnH

initiator

92%

JACS, 103, 1557 (1981)

CHAPTER 12
PREPARATION OF KETONES

Section 166 <u>Ketones from Acetylenes</u>

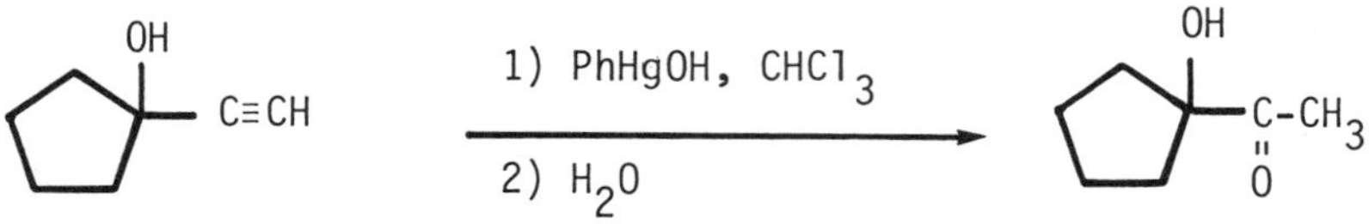

1) PhHgOH, CHCl$_3$

2) H$_2$O

JOC, <u>47</u>, 3331 (1982)

Nafion-H

H$^+$

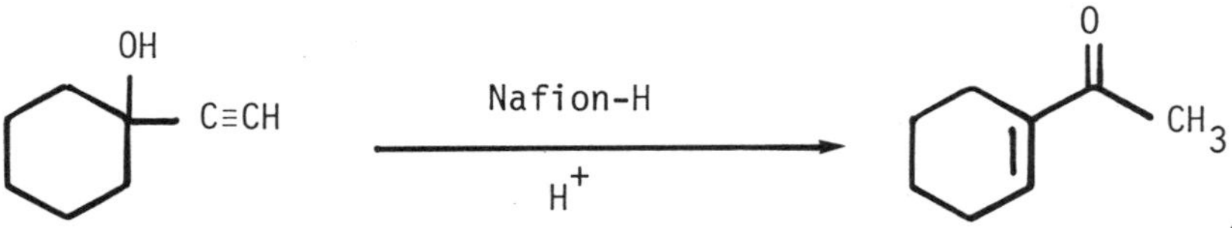

84%

Synthesis, 473 (1981)

Ph-C≡C-Ph $\xrightarrow[\text{CHCl}_3/\text{H}_2\text{O}]{\text{2 PhI(OCOCF}_3\text{)}_2}$ Ph-C-C-Ph

82%

Doklady Chem, <u>245</u>, 140 (1979)

$$Ph-C{\equiv}C-C_5H_{11} \xrightarrow[\text{RuCl}_2L_3]{\text{PhIO}}$$

Ph–CO–CO–C_5H_{11}

72%

Helv Chim Acta, 64, 2531 (1981)

Ph-I
+
H-C≡C-Ph

$$\xrightarrow[\text{CO}]{\text{PdCl}_2\text{(dppf)}}$$

$$Ph-\overset{O}{\overset{\|}{C}}-C{\equiv}C-Ph$$

~85%

JCS Chem Comm, 333 (1981)

Bu$_3$B
+
C_5H_{11}-C≡CH

$$\xrightarrow[\text{3) H}_2\text{O}_2]{\text{2) MVK, TiCl}_4}$$

Bu–CO–CH(C_5H_{11})–CH$_2$–CH$_2$–CO–CH$_3$

58%

Chem Lett, 221 (1980)

Ph$_2$CO
+
BrCH$_2$C≡CH

$$\xrightarrow[\text{2) H}^{\oplus}]{\text{1) Zn}}$$

Ph$_2$C=C(H)–CO–CH$_3$

78%

Synth Comm, 10, 637 (1980)

$$\text{CH}_2\text{-BH} \xrightarrow[\begin{array}{l}\text{2) NaOCH}_3 \\ \text{3) H}_2\text{O}_2, \text{ NaOH}\end{array}]{\text{1) C}_6\text{H}_{13}\text{-C}\equiv\text{C-Cl}} \text{C}_6\text{H}_{13}\text{-CH}_2 \overset{\overset{\text{O}}{\|}}{\text{C}} \text{CH}_2$$

87%

Synthesis, 193 (1982)

Section 167 Ketones from Carboxylic Acids and Acid Halides

COOLi

$$\xrightarrow[\text{MeO}\underset{\text{Cl}}{-}\text{C=NPh}_2]{\text{Ph} \diagup \text{MgBr}}$$

86%

Tetr Lett, 23, 5059 (1982)

$$\text{Ph} \diagup \overset{\overset{\text{O}}{\|}}{\text{C-Cl}} \xrightarrow{\text{Et}_3\text{Al}} \text{Ph} \diagup \overset{\overset{\text{O}}{\|}}{\diagup}$$

88%

J Gen Chem (USSR), 51, 1359 and 1434 (1982)

$$\begin{array}{c}\text{Hx}_3\text{Al} \\ + \\ \overset{\text{O}}{\underset{\|}{\text{Ph-C-Cl}}}\end{array} \xrightarrow[\text{CH}_2\text{Cl}_2]{\text{AlCl}_3} \overset{\overset{\text{O}}{\|}}{\text{Hx-C-Ph}}$$

66%

J Gen Chem (USSR), 50, 2195 (1980)

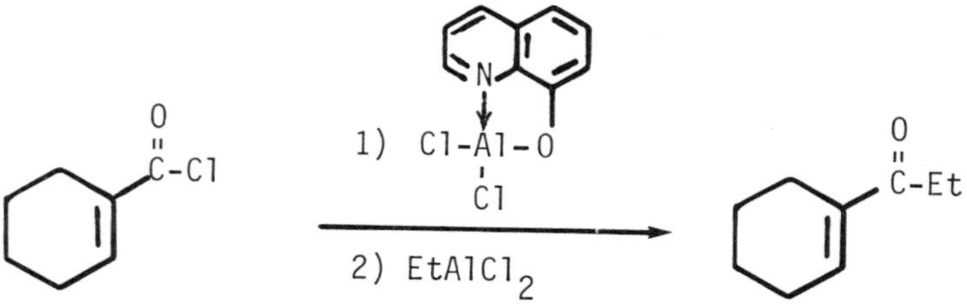

J Gen Chem (USSR), 52, 1170 (1982)

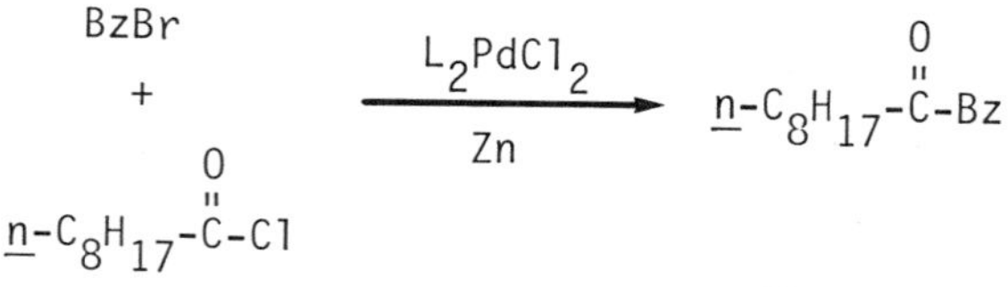

Chem Lett, 1135 (1981)

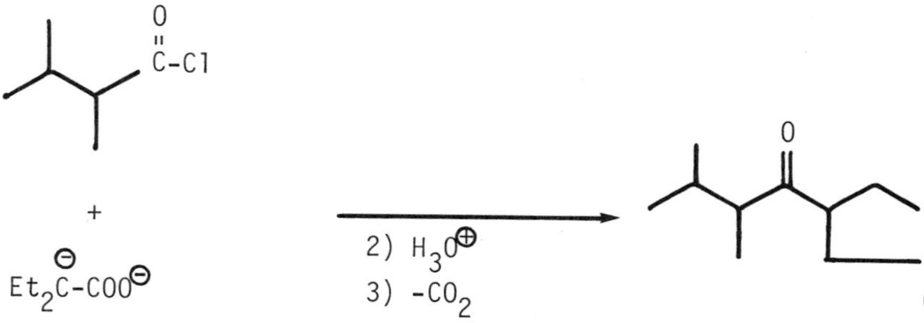

J Chem Research(S), 44 (1980)

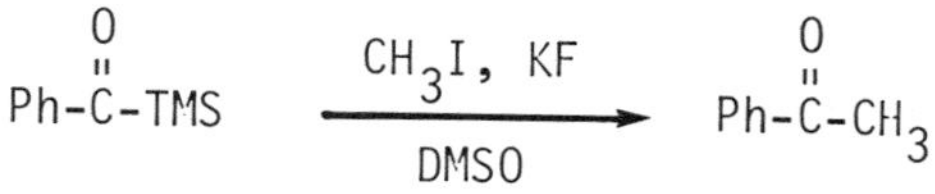

Tetr Lett, 22, 1881 (1981)

$$CH_2=CH(CH_2)_8-\overset{O}{\overset{\|}{C}}-Cl \quad \xrightarrow[\text{2) AcOH, H}_2O]{\text{1)}} \quad CH_2=CH(CH_2)_8-\overset{O}{\overset{\|}{C}}-CH_3 \quad 85\%$$

Indian J Chem, <u>20B</u>, 504 (1981)

Synth Comm, <u>10</u>, 221 (1980)

$$2 \text{ Ph}-\overset{O}{\overset{\|}{C}}-Cl \quad \xrightarrow{\text{SmI}_2} \quad \text{Ph}-\overset{O}{\overset{\|}{C}}-\overset{O}{\overset{\|}{C}}-\text{Ph} \quad 78\%$$

Tetr Lett, <u>22</u>, 3959 (1981)

94-96%

JOC, <u>47</u>, 3006 (1982)

Ph$_3$P=C=C=NPh

+

72%

Angew Chem Int Ed, <u>19</u>, 822 (1980)

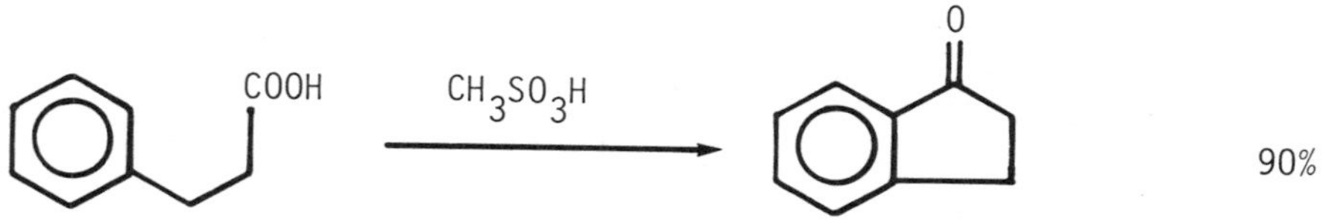

JOC, <u>46</u>, 2974 (1981)

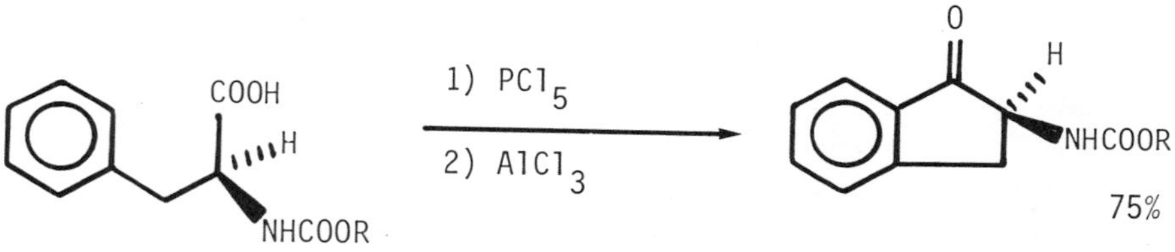

JOC, <u>46</u>, 2431 (1981)

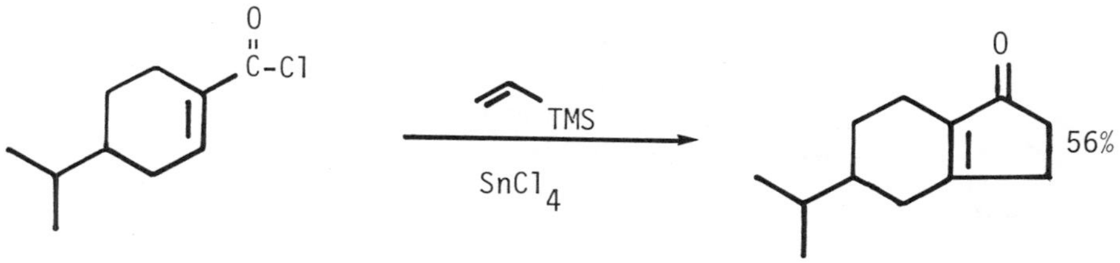

JOC, <u>45</u>, 1046 (1980)

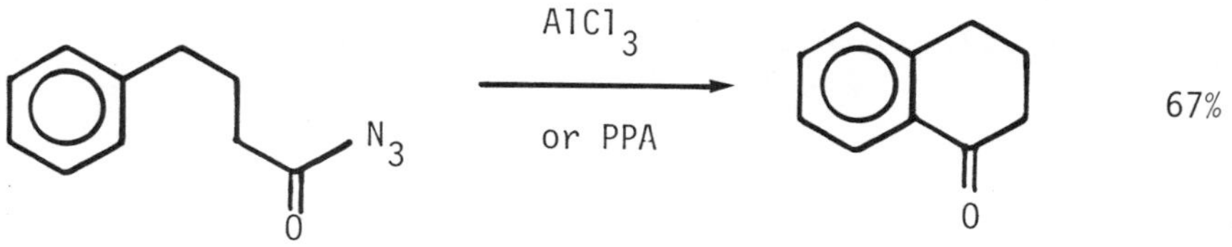

Chem Lett, 1501 (1980)

Section 168 Ketones from Alcohols

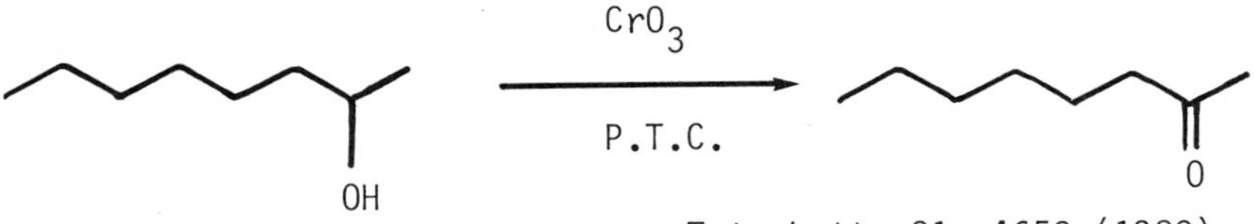

CrO_3
P.T.C.
OH
O

99%

Tetr Lett, 21, 4653 (1980)

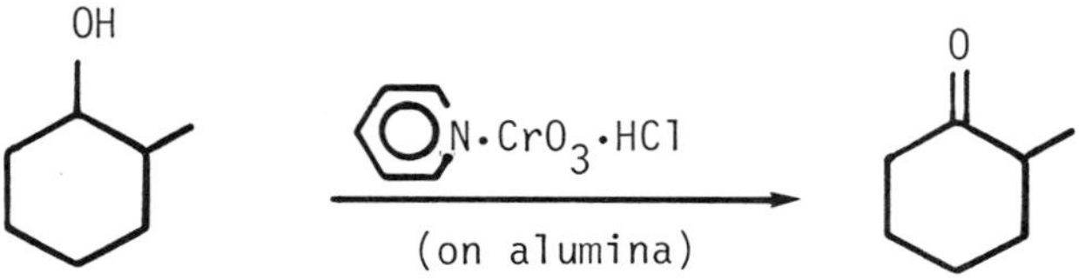

OH
$N \cdot CrO_3 \cdot HCl$
(on alumina)
O

83%

Synthesis, 223 (1980)

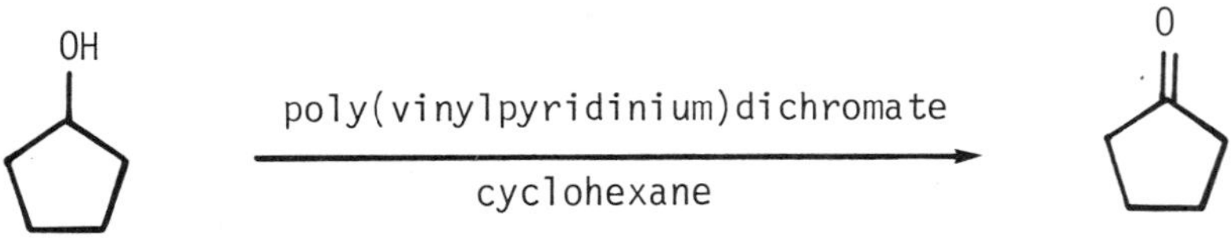

OH
poly(vinylpyridinium)dichromate
cyclohexane
O

69%

JOC, 46, 1728 (1981)

$$\text{Ph-}\underset{}{\overset{\text{OH}}{\text{CH}}}\text{-CH}_3 \xrightarrow[\text{polyethylene glycol}]{K_2Cr_2O_7} \text{Ph-}\overset{\text{O}}{\overset{\|}{\text{C}}}\text{-CH}_3 \qquad 82\%$$

Synthesis, 646 (1980)

Synthesis, 691 (1980)

OH
|
Ph-CH-CH$_3$ $\xrightarrow[\text{CH}_2\text{Cl}_2]{(\text{Bu}_4\text{N})_2\text{Cr}_2\text{O}_7}$ Ph-C-CH$_3$ 80%

Synth Comm, <u>10</u>, 75 (1980)

OH
|
Ph-CH-Ph $\xrightarrow[\text{HMPT}]{[\text{BzNEt}_3]_2 \ \text{Cr}_2\text{O}_7}$ Ph$\overset{\overset{\text{O}}{\|}}{\underset{}{\text{C}}}$Ph   99%

Synthesis, 1091 (1982)

Tetr Lett, <u>21</u>, 1583 (1980)

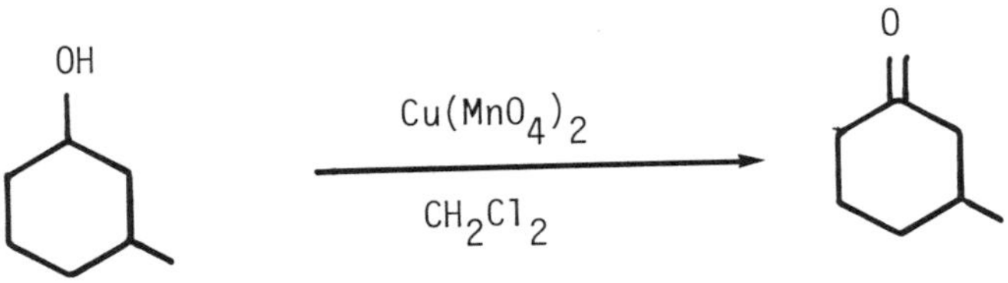

$$Cu(MnO_4)_2 \quad / \quad CH_2Cl_2$$

95%

JOC, _47_, 2790 (1982)

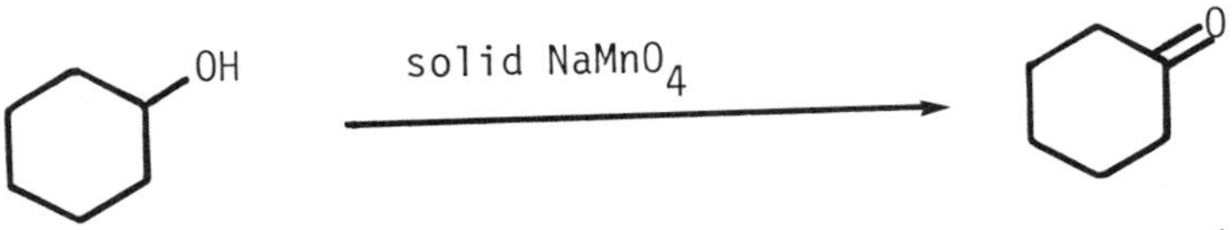

solid NaMnO$_4$

Tetr Lett, _22_, 1655 (1981)

NaOCl, HOAc

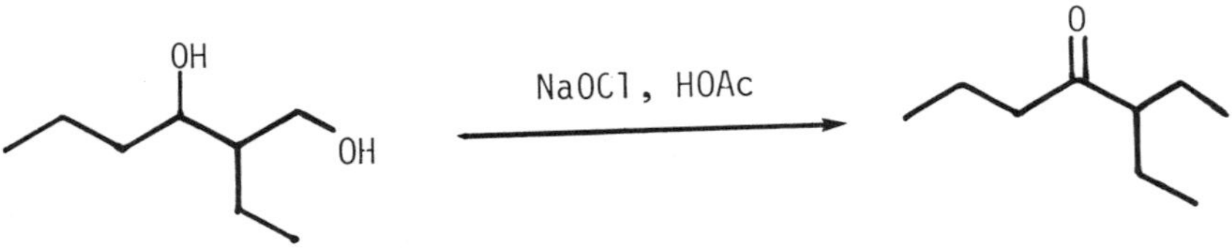

85%

JOC, _45_, 2030 (1980)
Tetr Lett, _23_, 4647 (1982)

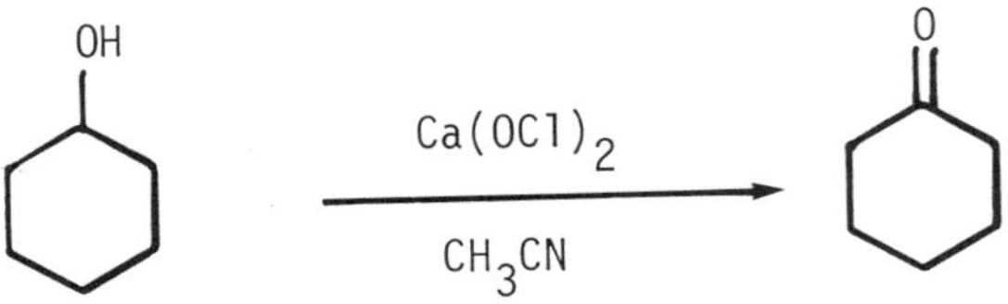

$$Ca(OCl)_2 \quad / \quad CH_3CN$$

93%

Tetr Lett, _23_, 35 (1982)

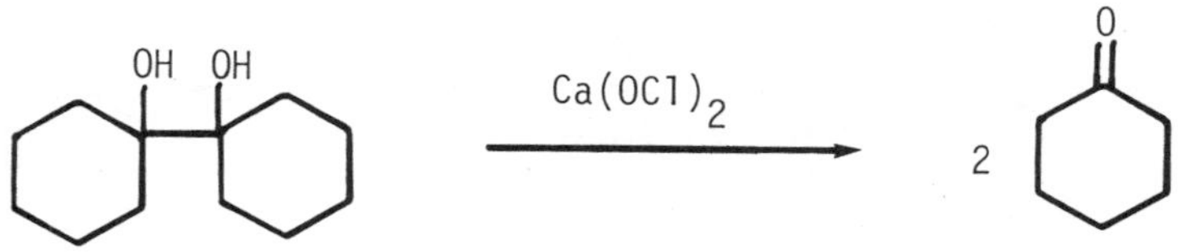

Tetr Lett, 23, 3135 (1982)

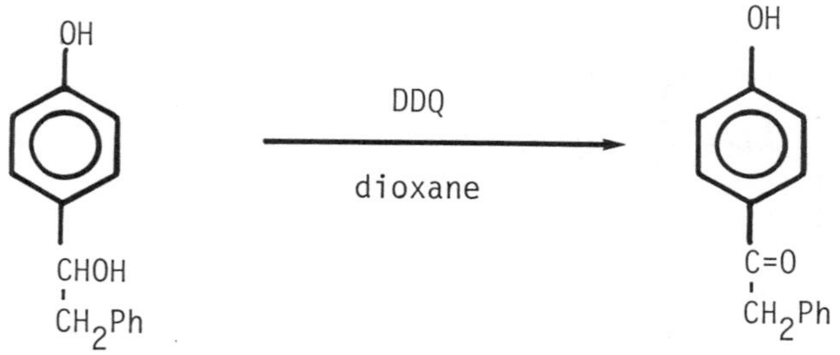

JOC, 45, 1596 (1980)

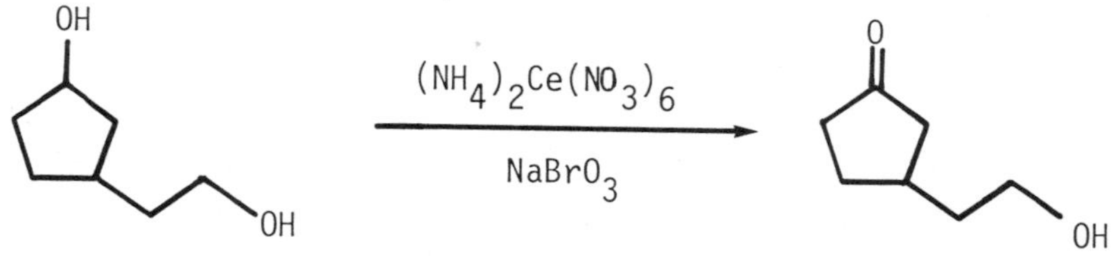

Tetr Lett, 23, 539 (1982)

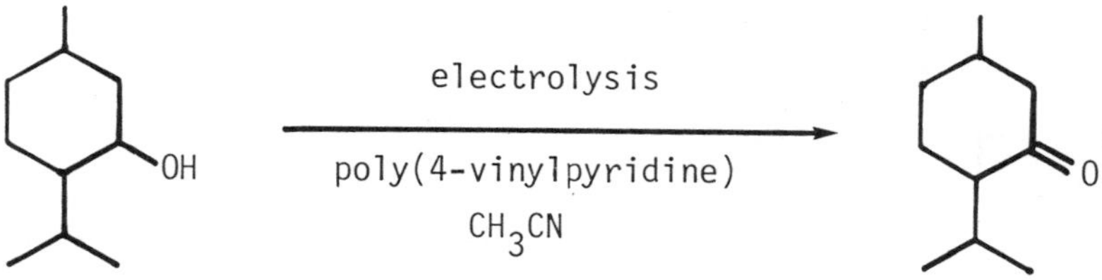

JOC, 45, 5269 (1980)

Tetr Lett, <u>21</u>, 1867 (1980)

Tetrahedron, <u>38</u>, 3299 (1982)

Synth Comm, <u>10</u>, 881 (1980)

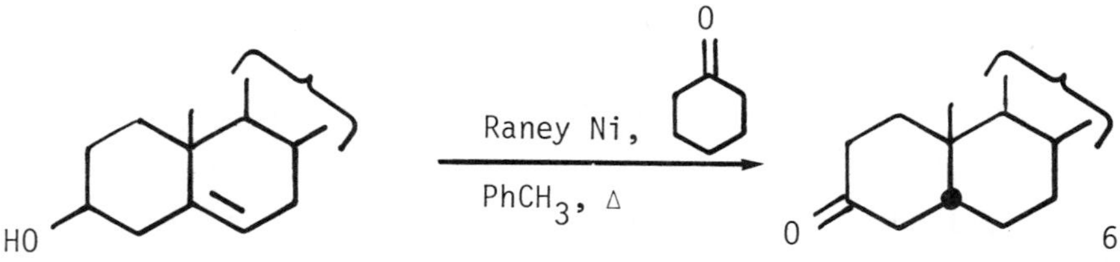

Tetr Lett, 21, 1071 (1980)

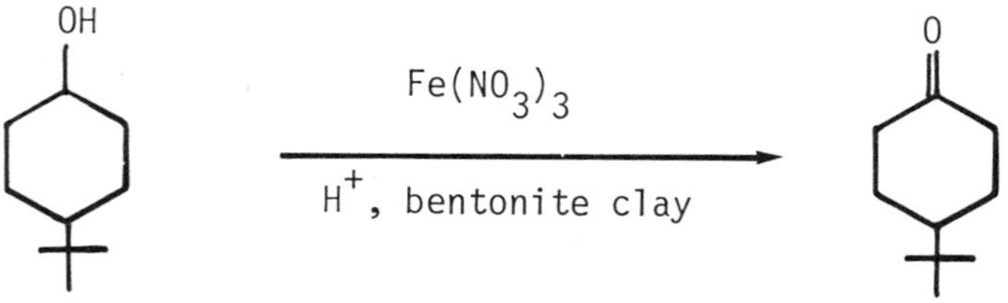

Synthesis, 849 (1980)

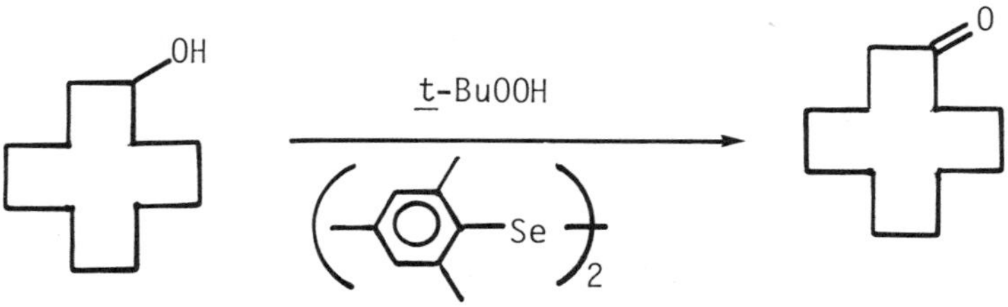

JOC, 47, 837 (1982)

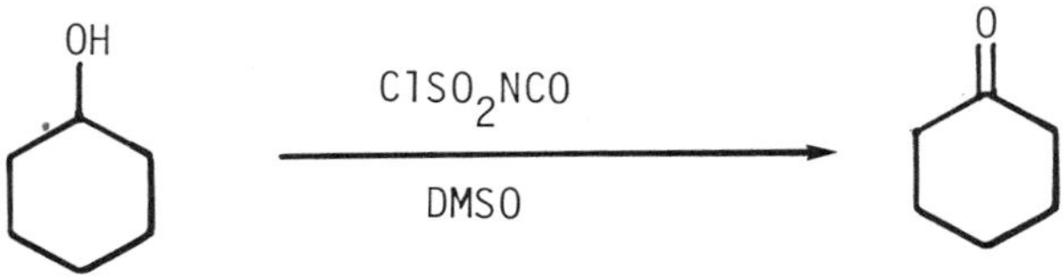

Synthesis, 141 (1980)

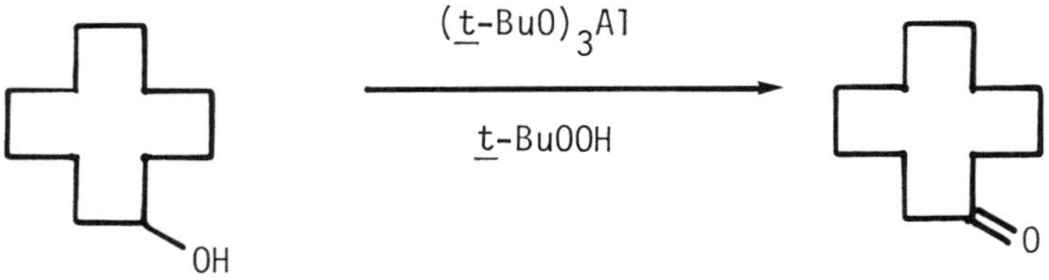

$$\xrightarrow[\underline{t}\text{-BuOOH}]{(\underline{t}\text{-BuO})_3\text{Al}}$$

85%

Tetr Lett, <u>21</u>, 1657 (1980)

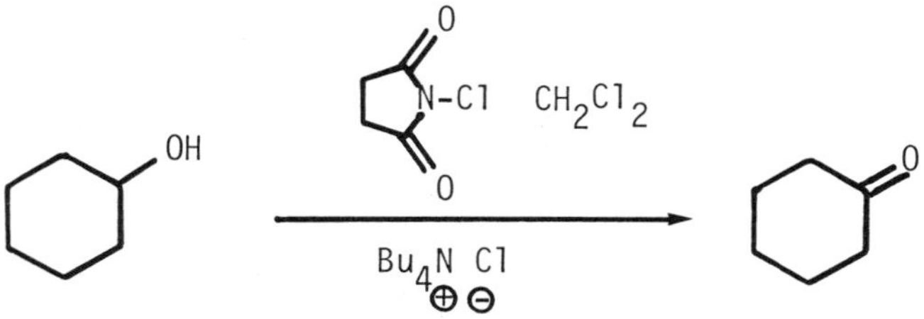

$$\xrightarrow[\underset{\oplus\ominus}{Bu_4N\ Cl}]{\text{N-Cl}\quad CH_2Cl_2}$$

90%

Synthesis, 394 (1981)

$$\xrightarrow{\underset{O}{\overset{2}{\text{N-I}}}} \quad 2\ CH_3\overset{O}{\overset{\|}{C}}CH_3$$

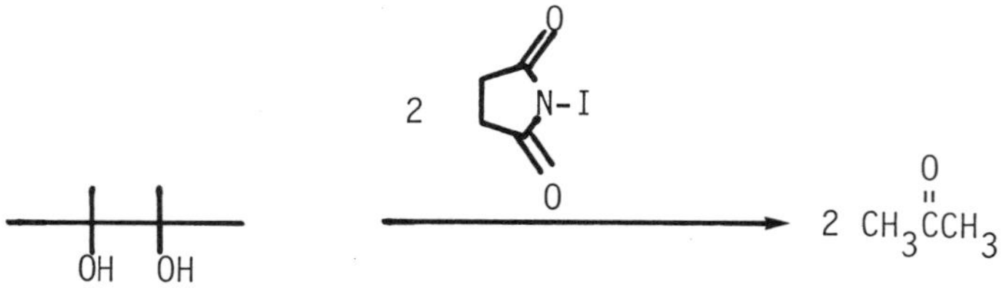

93%

JOC, <u>46</u>, 1927 (1981)

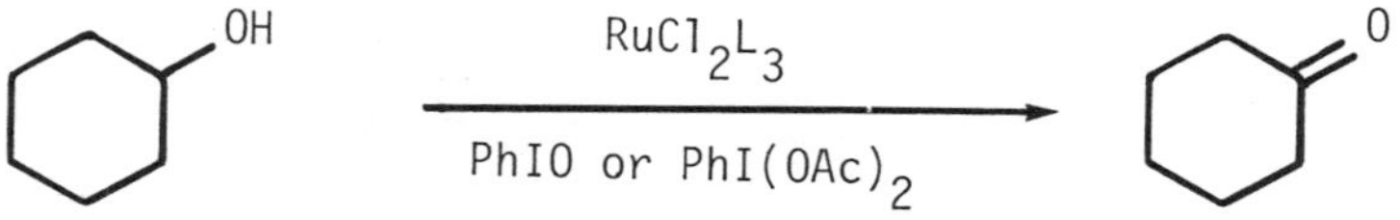

OH
RuCl2L3
PhIO or PhI(OAc)2
O

78%

Tetr Lett, 22, 2361 (1981)

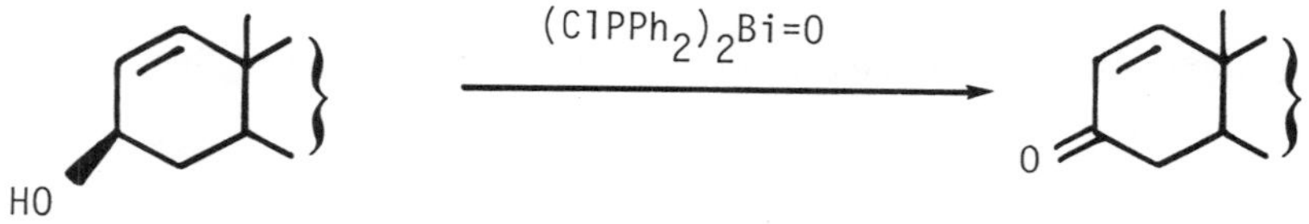

HO
(ClPPh2)2Bi=O
O

85%

Tetrahedron, 37, Suppl #1, 73 (1981)

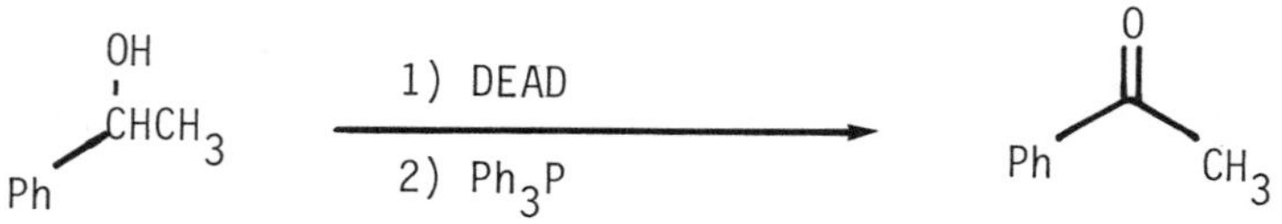

OH
CHCH3
Ph
1) DEAD
2) Ph3P
O
Ph
CH3

85%

Tetr Lett, 22, 2295 (1981)

OH
t-BuOK
t-BuOH
O

90%

Can J Chem, 58, 2730 (1980)

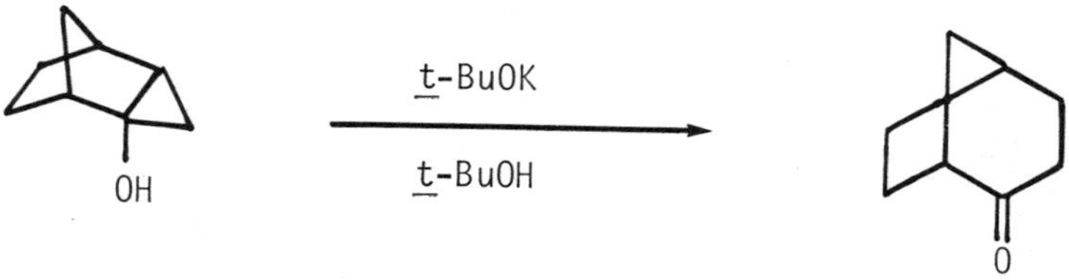

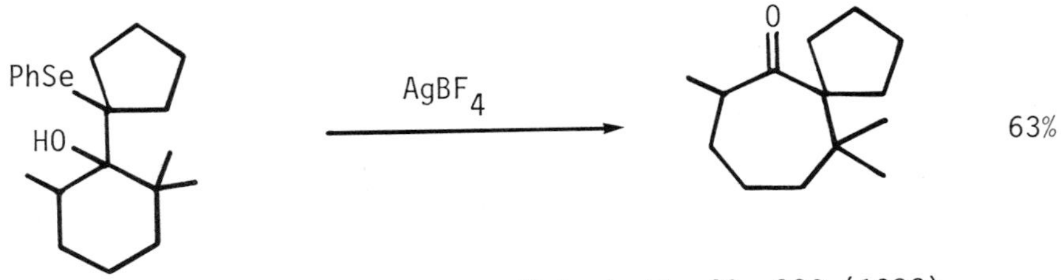

Tetr Lett, 23, 983 (1982)

Related Methods: Aldehydes from Alcohols and Phenols (Section 48)

Section 169 Ketones from Aldehydes

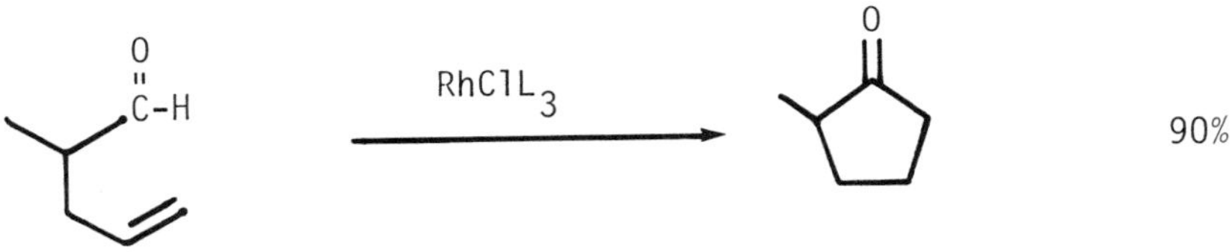

JACS, 102, 190 (1980)

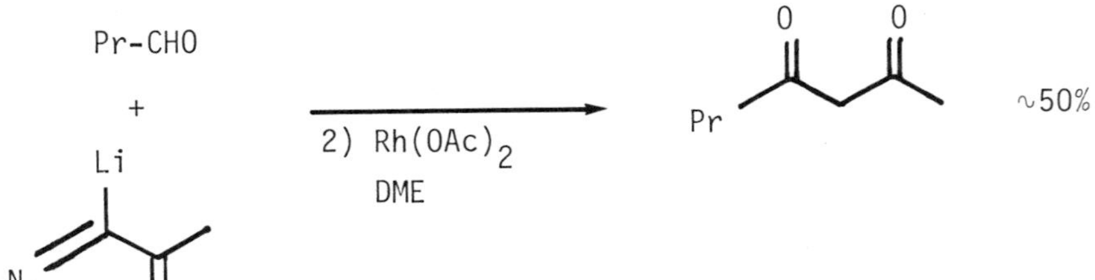

JCS Perkin I, 2566 (1981)

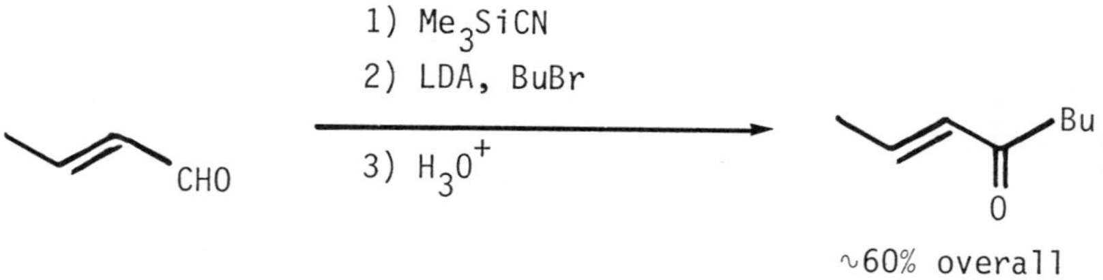
CHO
1) Me3SiCN
2) LDA, BuBr
3) H3O+
Bu
O
~60% overall

Chem Ber, 113, 3783 (1980)

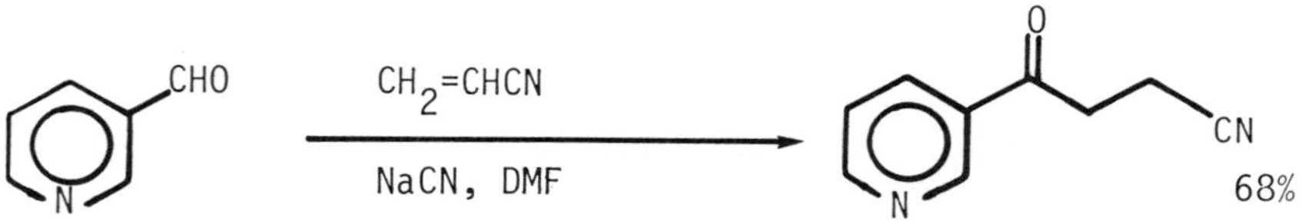
CHO
N
CH2=CHCN
NaCN, DMF
O
CN
N
68%

Org Syn, 59, 53 (1980)

Section 170 Ketones from Methylenes and Aryls

This section lists examples of the reaction $R\text{-}CH_2\text{-}R' \rightarrow R\text{-}\overset{O}{\overset{\|}{C}}\text{-}R'$

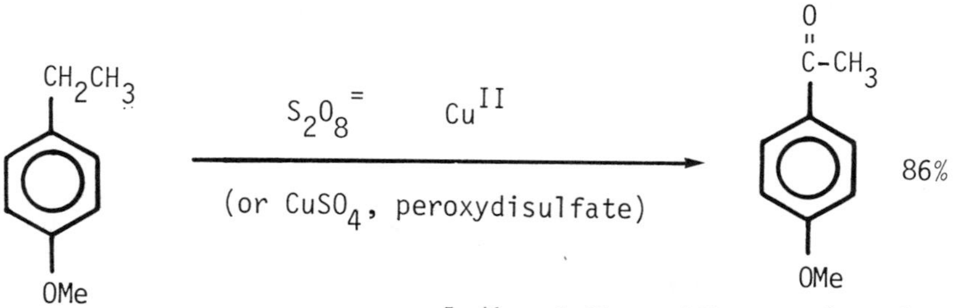
CH2CH3
OMe
S2O8=
CuII
(or CuSO4, peroxydisulfate)
O
C-CH3
OMe
86%

Indian J Chem, 20B, 153 (1981)

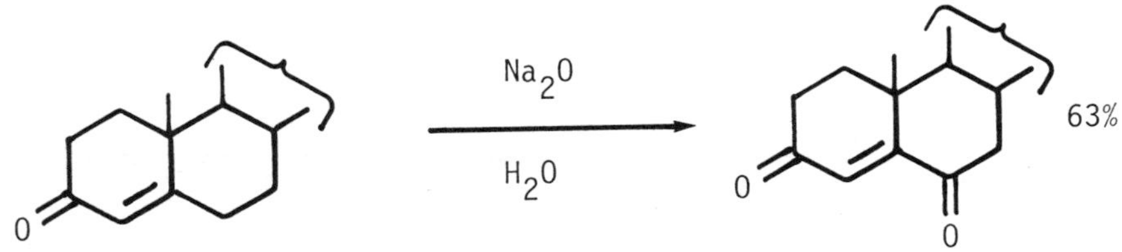

Tetr Lett, 22, 5127 (1981)

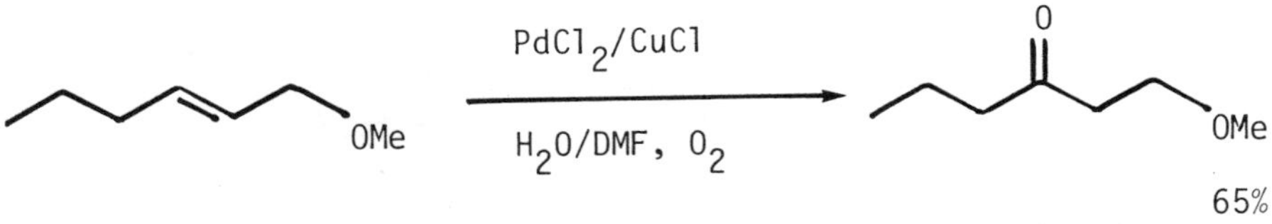

Tetr Lett, 23, 2679 (1982)

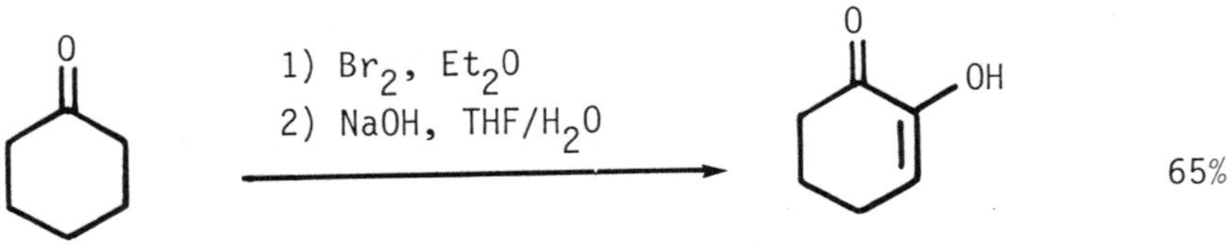

Chem Lett, 779 (1980)

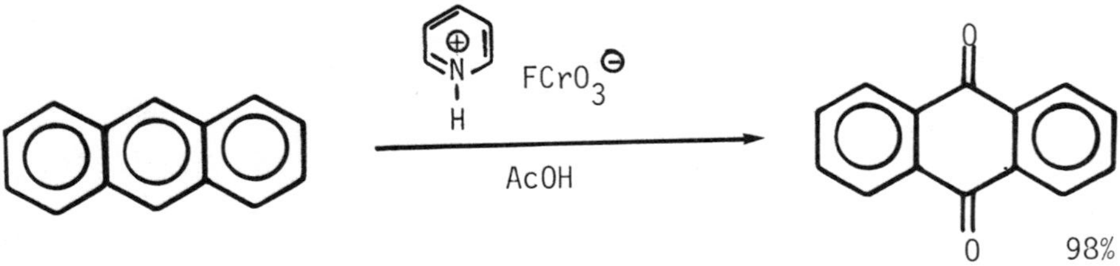

Synthesis, 588 (1982)

Section 171 <u>Ketones from Amides</u>

No additional examples.

Section 172 <u>Ketones from Amines</u>

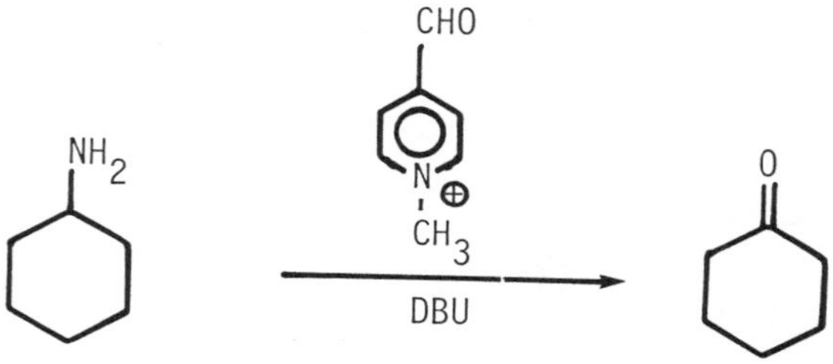

JACS, <u>104</u>, 4446 (1982)

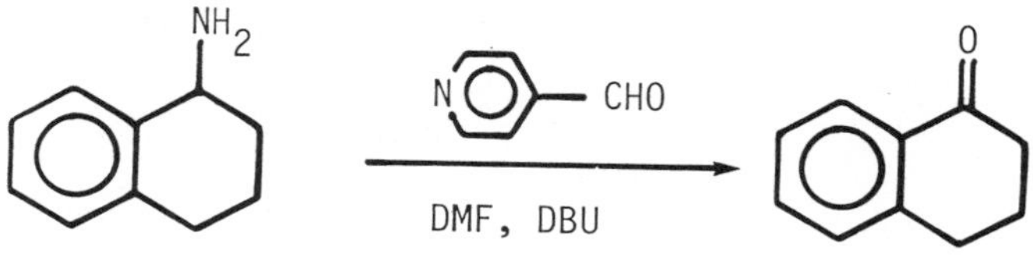

Synthesis, 756 (1982)

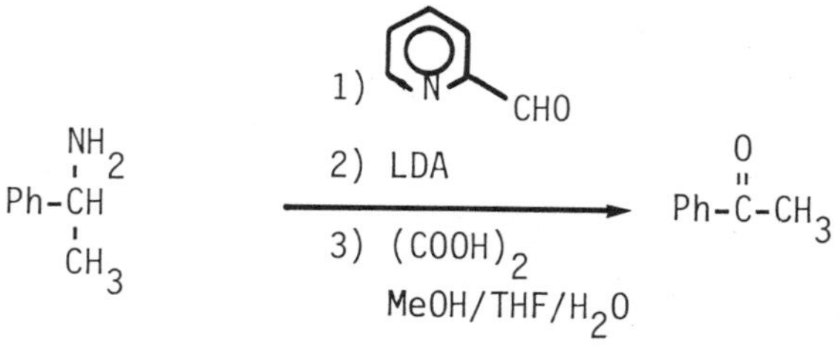

JOC, <u>46</u>, 1937 (1981)

90%

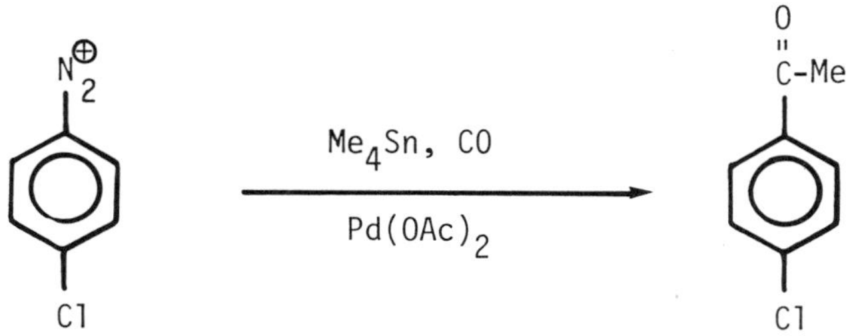

Chem Lett, 35 (1982)

Section 173 Ketones from Esters

~60%

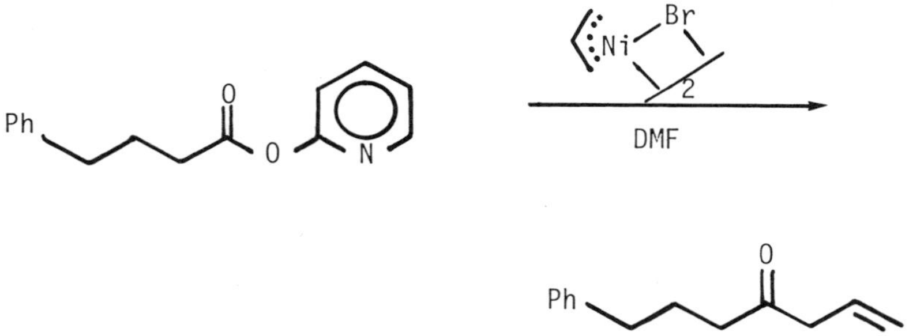

Chem Lett, 1483 (1979)

$$Bu\text{-}CCl_2Li + Et\text{-}COOMe \xrightarrow[Et_2O]{THF} BuCCl_2\text{-}\overset{O}{\overset{\|}{C}}\text{-}Et$$

71%

Comptes Rendus (C), 291, 105 (1980)

Ph-C(=O)-S-(2-pyridyl) →[Bu$_2$CuLi] Ph-C(=O)-Bu   99%

JCS Chem Comm, 1231 (1981)

+ CH$_3$COO-(tetrahydropyranyl) →[Me$_3$SiOTf]   96%

Tetr Lett, $\underline{23}$, 2601 (1982)

→[Pd(OAc)$_2$ / PPh$_3$]   100%

Tetr Lett, $\underline{21}$, 3199 (1980)

2 (3-Cl-C$_6$H$_4$)-C(=O)-S-(2-pyridyl) →[1) Ni(COD)$_2$, DMF 2) H$_3$O$^\oplus$]   90%

Chem Lett, 51 (1980)

Section 174 <u>Ketones from Ethers and Epoxides</u>

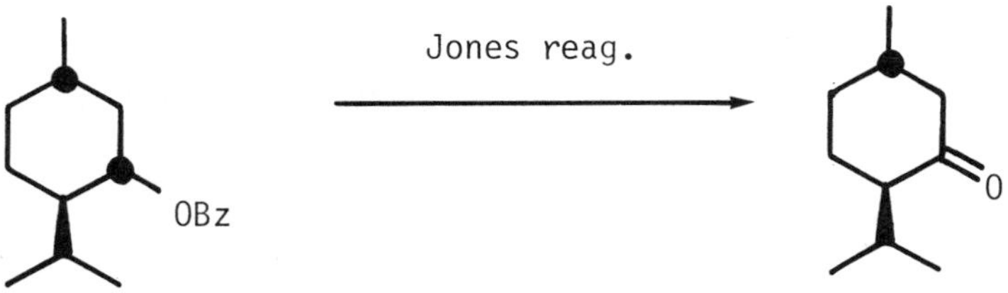

JOC, <u>46</u>, 1492 (1981)

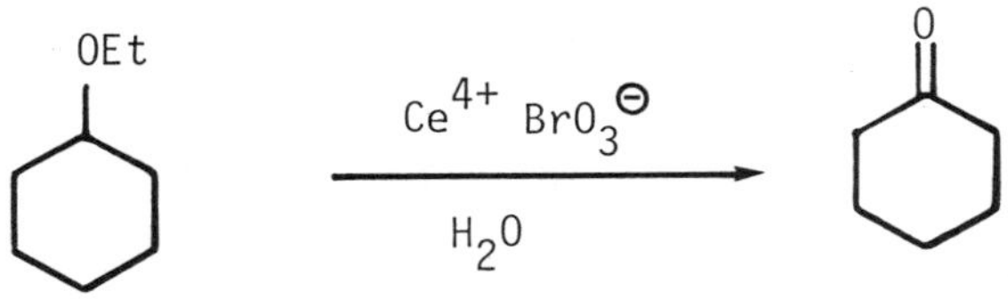

57%

Synthesis, 897 (1980)

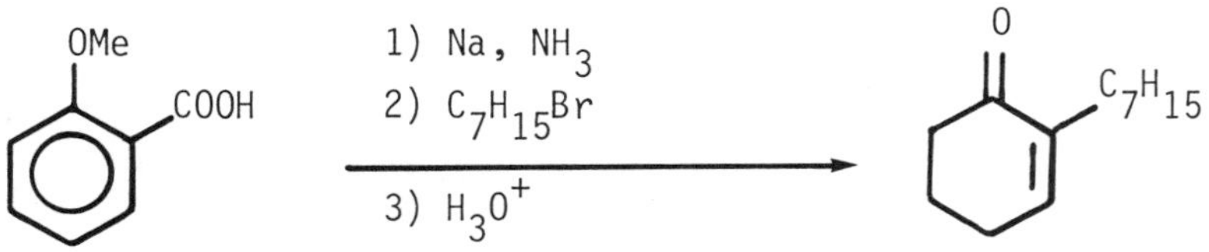

92%

Org Syn, <u>61</u>, 59 (1983)

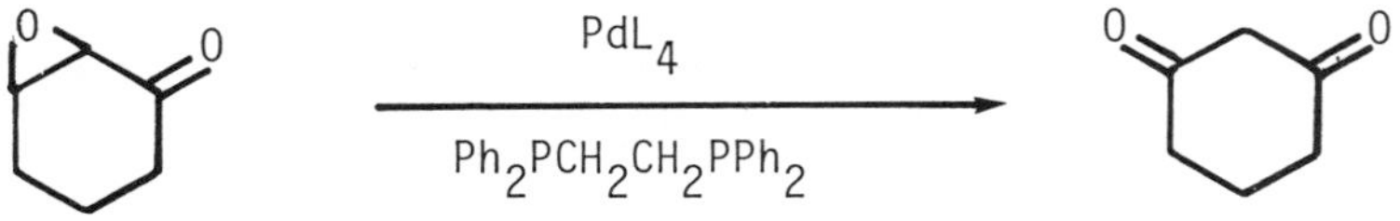

~50%

JACS, <u>102</u>, 2095 (1980)

62%

Tetr Lett, <u>21</u>, 4283 (1980)

Section 175 <u>Ketones from Halides and Sulfonates</u>

Bull Chem Soc Japan, <u>54</u>, 2221 (1981)

Tetr Lett, <u>21</u>, 3339 (1980)

Bull Chem Soc Japan, <u>53</u>, 3027 (1980)

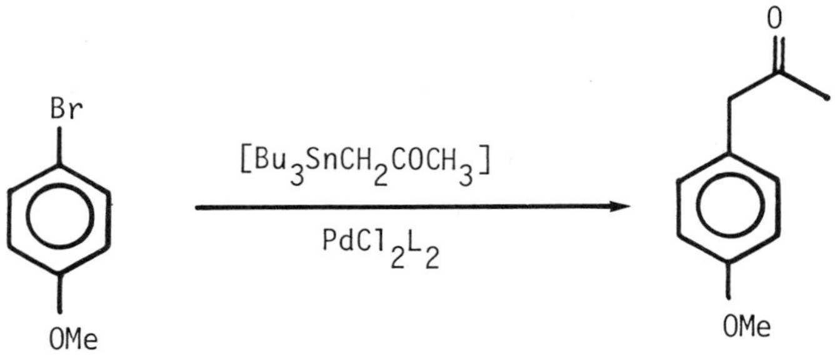

Chem Lett, 939 (1982)

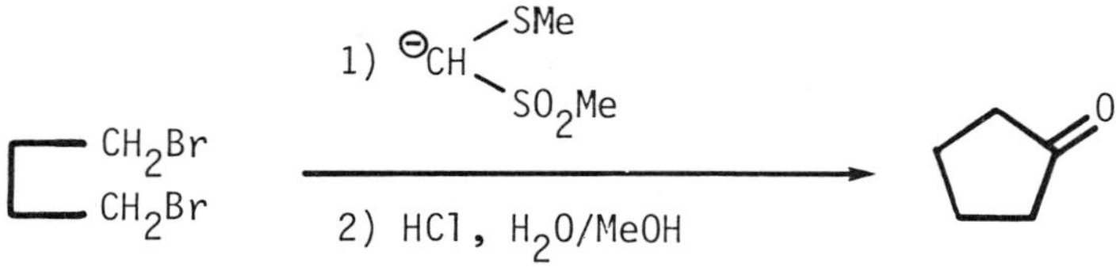

Chem Lett, 813 (1982)

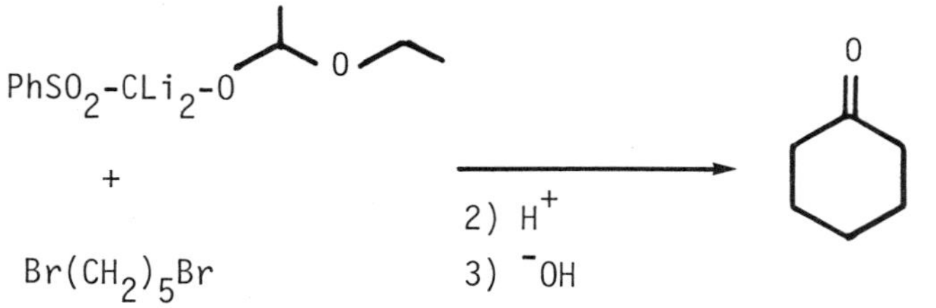

Bull Chem Soc Japan, 53, 3619 (1980)

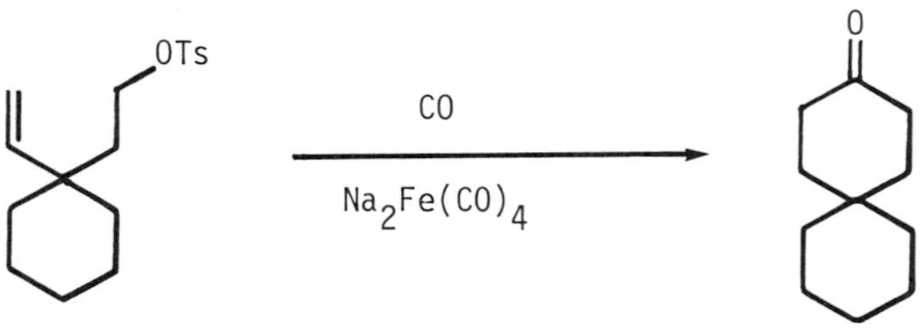

Tetr Lett, 21, 4687 (1980)

Chem Ber, <u>113</u>, 3706 (1980)

25%

JCS Chem Comm, 1121 (1981)

Related methods: Ketones from Ketones (Section 177), Aldehydes
from Halides (Section 55)

Section 176 <u>Ketones from Hydrides</u>

This section lists examples of the replacement of hydrogen by
ketonic groups, RH → RCOR'.   For the oxidation of methylenes
R_2CH_2 → R_2CO see Section 170 (Ketones from Alkyls and Methylenes)

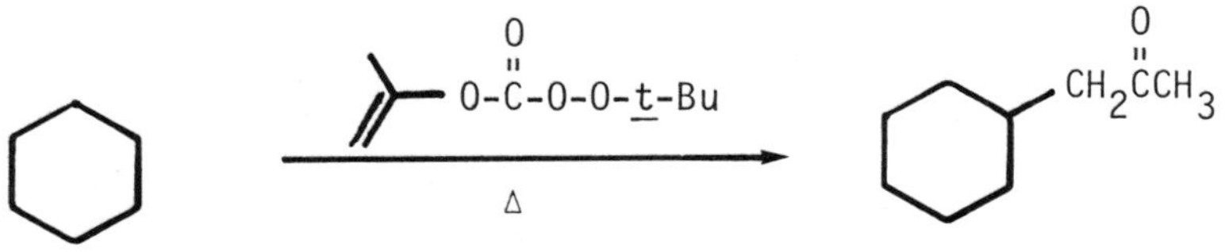

55%

Synthesis, 761 (1982)

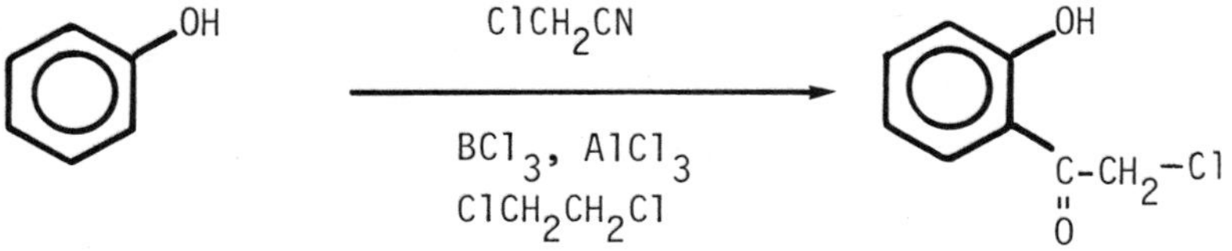

85%

JOC, 46, 189 (1981)

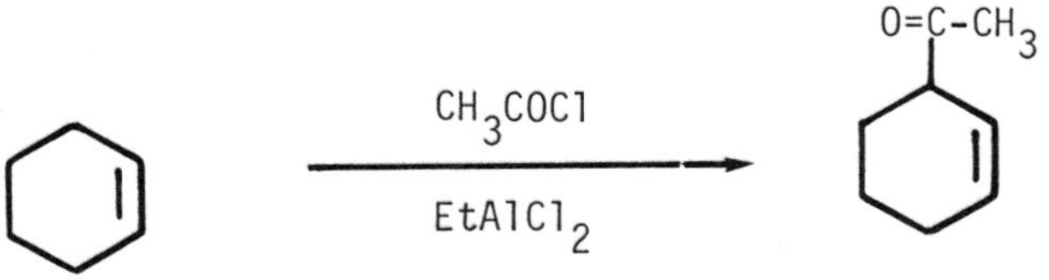

73%

JOC, 47, 5393 (1982)

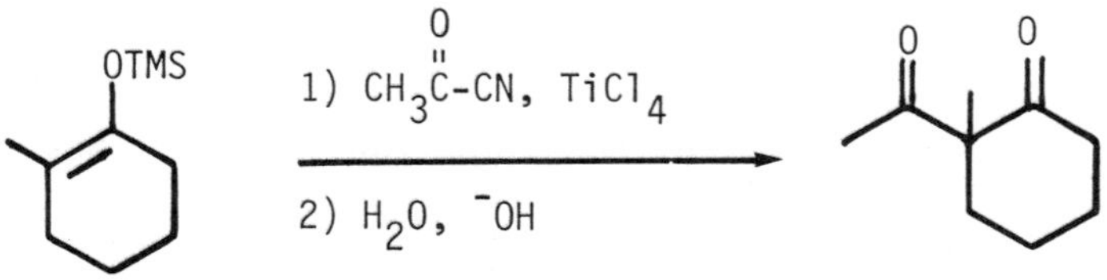

~90%

Tetr Lett, 22, 1171 (1981)

Tetr Lett, <u>22</u>, 4901 (1981)

Section 177 <u>Ketones from Ketones</u>

This section contains alkylations of ketones and protected ketones, ketone transpositions and annelations, ring expansions and ring openings, and dimerizations. Conjugate reductions and Michael alkylations of enones are listed in Section 74 (Alkyls from Olefins).

For the preparation of enamines from ketones see Section 356 (Amine-Olefin).

Chem Ber, <u>113</u>, 3734 and 3741 (1980)

OTMS

PhBr, Bu₃SnF
$$\xrightarrow{\text{PdCl}_2\text{L}_2}$$

Ph
60%

JACS, 104, 6831 (1982)

OTMS

PhCH₂Br
$$\xrightarrow{\text{BzMe}_3\overset{+}{\text{N}}\ \overset{-}{\text{F}}}$$

CH₂Ph
73%

JACS, 104, 1025 (1982)

Review: "Lewis Acid Induced α-Alkylation of Carbonyl Compounds"

Angew Chem Int Ed, 21, 96 (1982)

Ph

H
C
N
OMe

1) LDA, THF
$$\xrightarrow{\hspace{2cm}}$$
2) EtI

Et
82%
94% ee

JACS, 103, 3081 and 3088 (1981)

$$\underset{\substack{| \\ \text{BuCH=C-CH}_3}}{\overset{\text{OBBu}_2}{}} \quad \xrightarrow[\text{BzBr}]{\text{Me}_2\text{NCH}_2\text{CH}_2\text{OLi}} \quad \underset{\substack{| \\ \text{Bz}}}{\overset{\text{O}}{\text{Bu-CH-C-CH}_3}} \qquad 70\%$$

Synth Comm, <u>10</u>, 139 (1980)

$$\xrightarrow{\text{CH}_3\text{COBF}_4} \qquad 55\%$$

JOC, <u>46</u>, 3771 (1981)

$$\xrightarrow[\text{SbCl}_3]{\overset{\text{O}}{\text{CH}_3\text{C-Cl}}} \qquad 65\%$$

JOC, <u>47</u>, 5099 (1982)

$$\xrightarrow[\substack{\text{KF, crown ether,} \\ \text{CH}_2\text{Cl}_2}]{} \qquad 93\%$$

Tetr Lett, <u>23</u>, 3073 (1982)

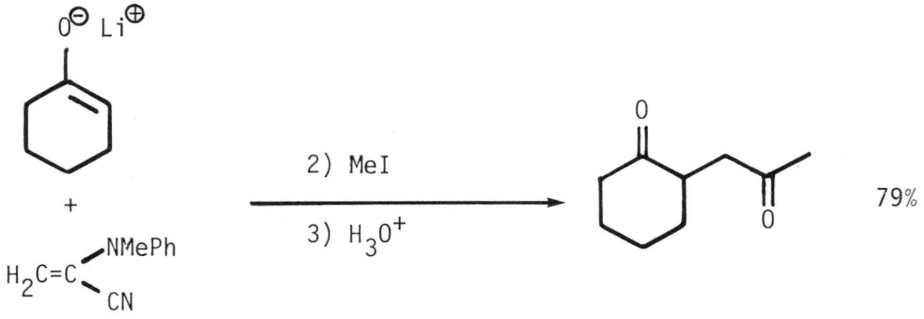

Synthesis, 413 (1980)

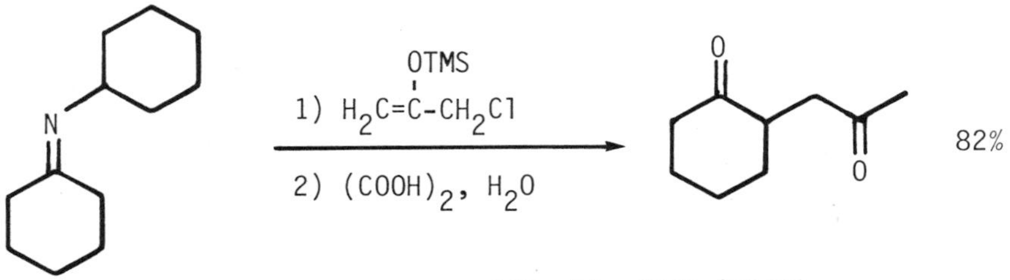

JOC, 46, 4631 (1981)

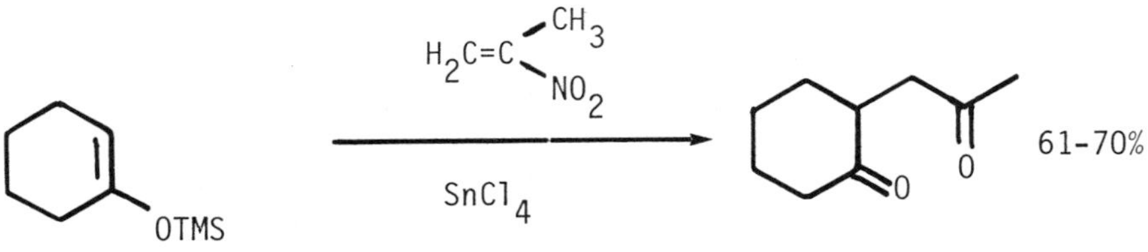

Org Syn, 60, 117 (1981)

2 [structure: cyclohexanone]

1) LDA   2) Cu(OTf)$_2$

or FeCl$_3$, DMF/THF

[structure: bicyclohexanone product]   45-73%

Chem Pharm Bull, 28, 262 (1980)
JOC, 45, 5408 (1980)

[structure: cyclohexene with OLi]

CH(OCH$_3$)$_3$

BF$_3$

[structure: cyclohexanone with CH(OMe)$_2$]   87%

Tetr Lett, 23, 3595 (1982)

[structure: cyclohexene with OSiMe$_3$ and vinyl]

1) CH$_3$Li, ZnCl$_2$

$\overset{O}{\overset{\|}{H-C}}$-CH$_2$SePh

2) Et$_3$N, MeSO$_2$Cl

[structure: bicyclic ketone with vinyl groups and H's]   74%

JOC, 47, 1632 (1982)

[structure: methylcyclohexenone]

1) LDA

2) [structure with OEt and ⊕Fp]

[structure: cyclohexenone with isopropenyl group]   93%

JOC, 46, 4103 (1981)

OTMS
OAc
ZnI_2
O

71%

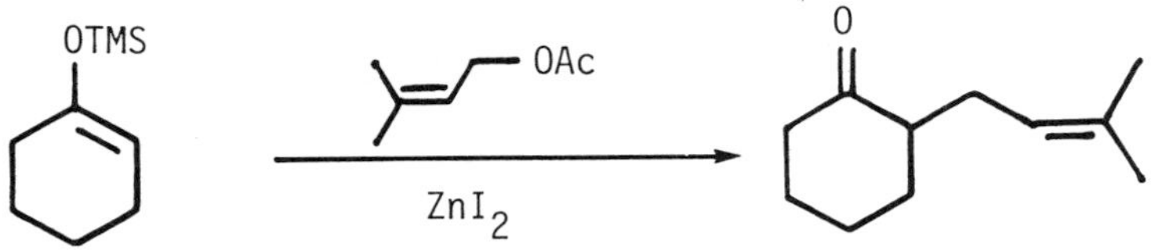

Synth Comm, 11, 217 (1981)

OLi
+
AcO
$Pd(dba)_2$
dppe, THF
O

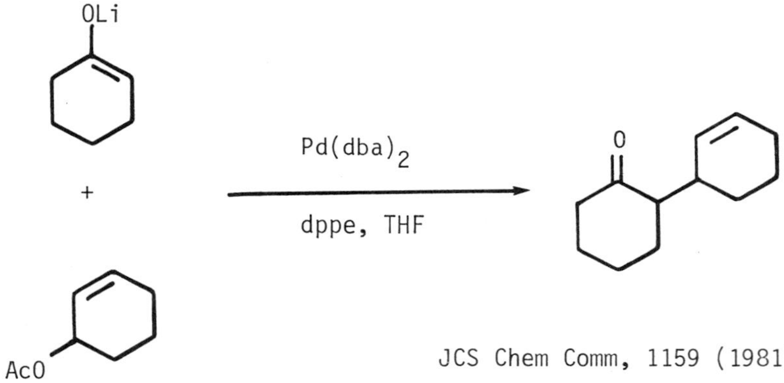

55%

JCS Chem Comm, 1159 (1981)

O
1) $KN(SiMe_3)_2$
2) BEt_3
3)
AcO
O

81%

JOC, 47, 3188 (1982)

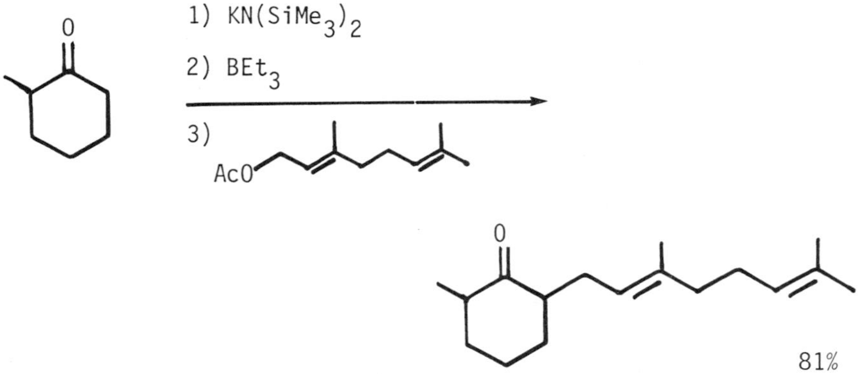

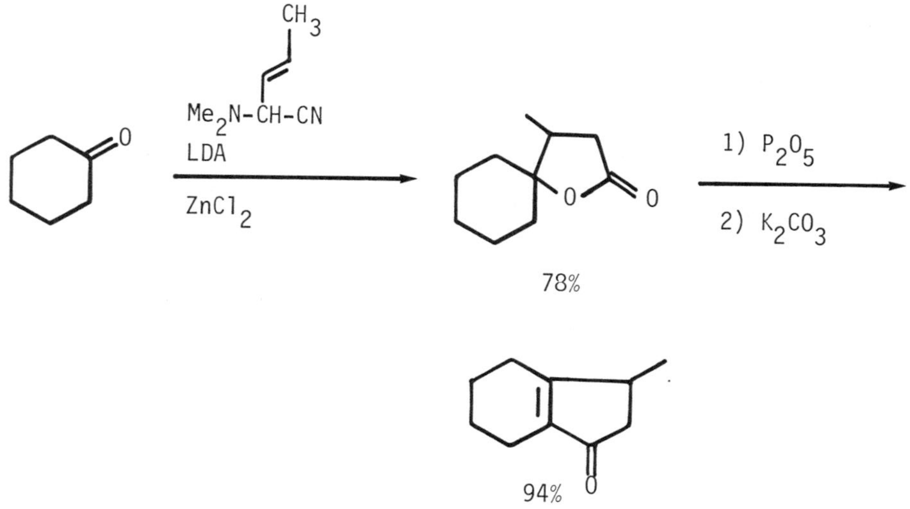

78%

94%

Tetr Lett, 21, 1205 (1980)

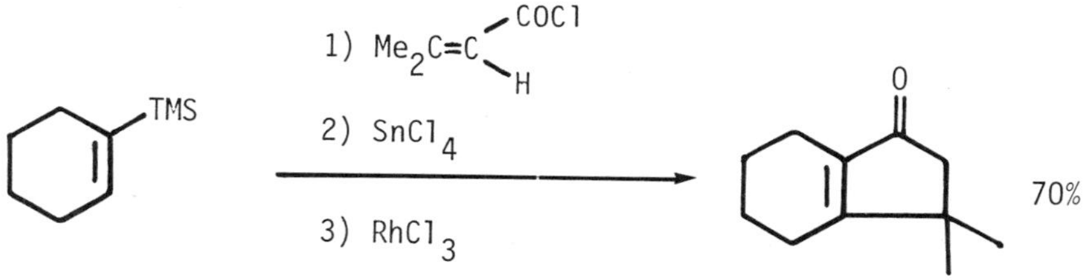

70%

JOC, 45, 3017 (1980)

70%

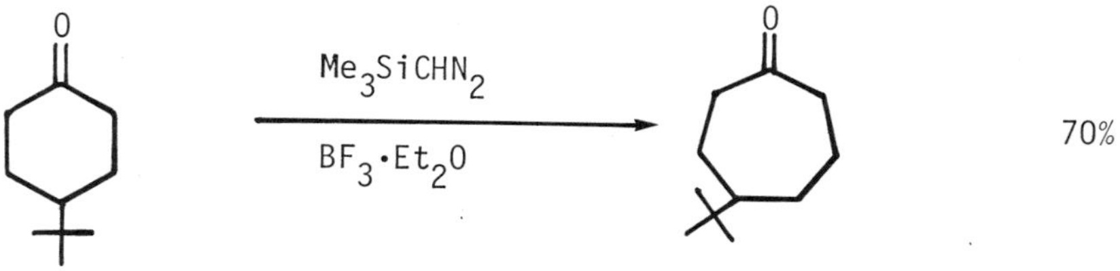

Chem Pharm Bull, 30, 119 (1982)
Tetr Lett, 21, 4619 (1980)

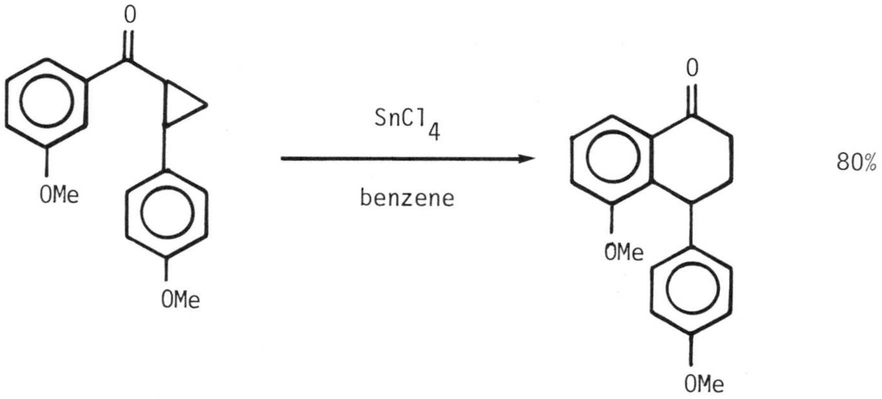

Tetr Lett, <u>21</u>, 1887 (1980)

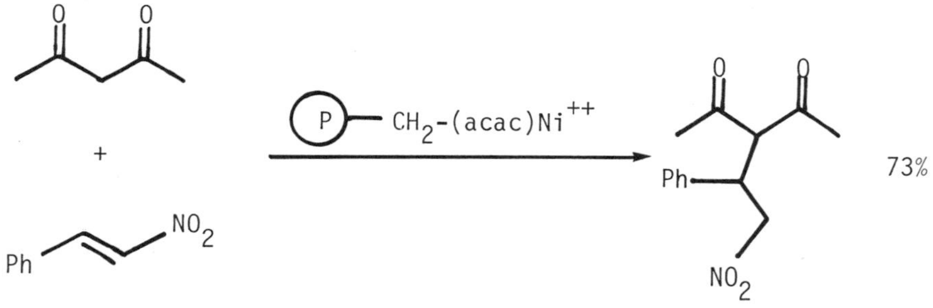

Synthesis, 467 (1982)

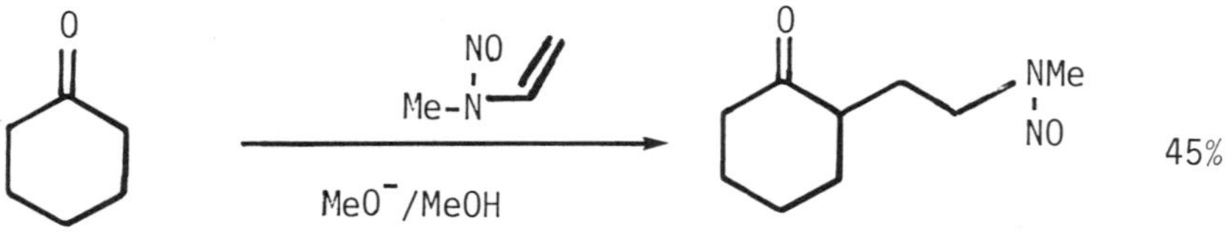

JOC, <u>45</u>, 2919 (1980)

OLi
|
Ph

+

O
‖
SePh

$C_{10}H_{21}$

THF →

Ph
|
O=C

$C_{10}H_{21}$

91%

JOC, 45, 2921 (1980)

(PhSeO)$_2$O →

S

O

89%

JCS Perkin I, 1650 (1980)

Ketones may also be alkylated and homologated via olefinic ketones (Section 374)

Related Methods: Aldehydes from Aldehydes (Section 49)

Section 178 Ketones from Nitriles

CN
|
Ph

$^{\ominus}$OH, O$_2$
―――――――――――
Et$_3$NBz Cl$^{\ominus}$
$^{\oplus}$

O
‖
Ph

84%

Synthesis, 1009 (1980)

CH$_3$CH$_2$CH$_2$CN

+

⌶ Br

$\xrightarrow[\text{2) NH}_4\text{Cl, H}_2\text{O}]{\text{1) Zn(Ag)}}$

[structure: O= ketone with allyl and propyl groups]   80%

Tetr Lett, <u>22</u>, 649 (1981)

[morpholine structure]
Ph-CH=CH-CHCN

$\xrightarrow[\text{3) H}_2\text{O} \quad \text{4) Cu(OAc)}_2\text{, EtOH}]{\text{1) LDA} \quad \text{2) EtBr}}$

Ph-CH=CH-C-Et   78%
(C=O)

Chem Lett, 1263 (1982)

Bz-CN

$\xrightarrow[\text{3) HCl}]{\text{1) HCl, CH}_3\text{OH} \quad \text{2)} \quad \text{Et}_3\text{N}}$

Bz-C-CH$_3$ 53%
(C=O)

Synthesis, 130 (1981)

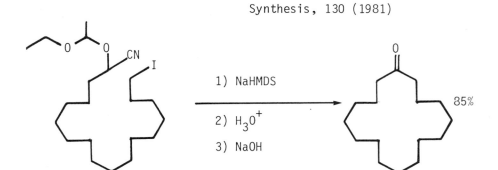

$\xrightarrow[\text{2) H}_3\text{O}^+ \quad \text{3) NaOH}]{\text{1) NaHMDS}}$

85%

Tetr Lett, <u>22</u>, 1359 (1981)

Section 179 Ketones from Olefins

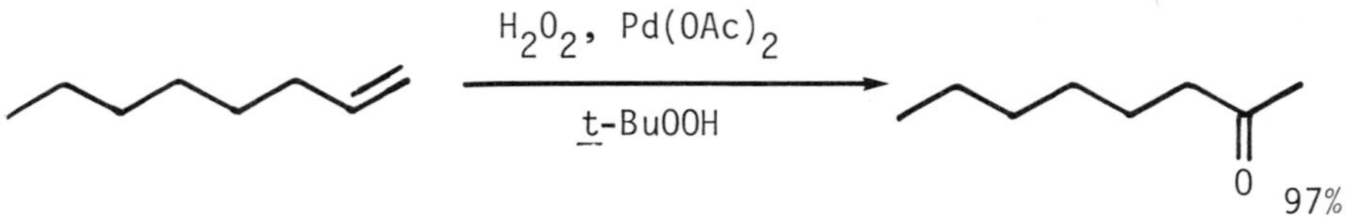
H_2O_2, $Pd(OAc)_2$
t-BuOOH
O
97%

JOC, 45, 5387 (1980)

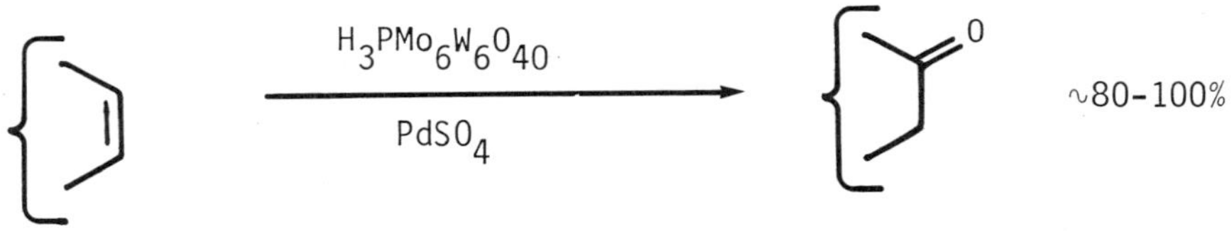
$H_3PMo_6W_6O_{40}$
$PdSO_4$
O
~80-100%

JCS Chem Comm, 1274 (1981)

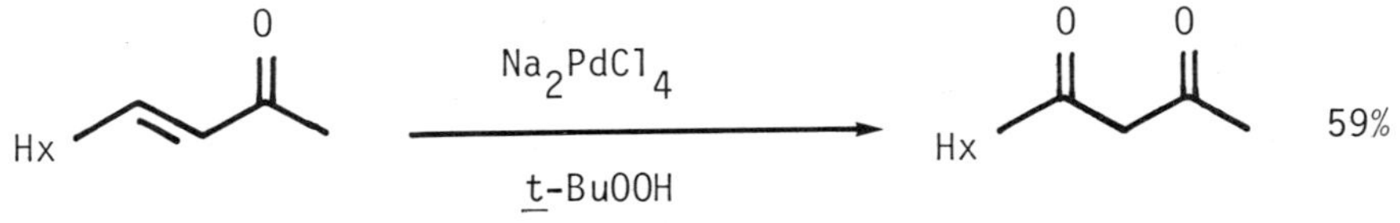
O
Hx
Na_2PdCl_4
t-BuOOH
O O
Hx
59%

Chem Lett, 257 (1980)

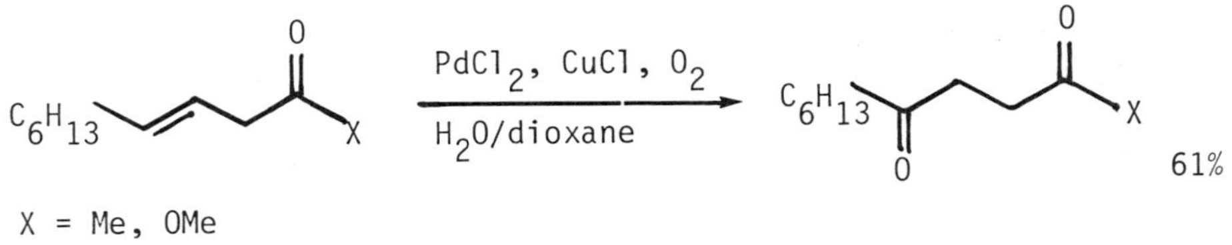
O
C_6H_{13}
X
$PdCl_2$, CuCl, O_2
H_2O/dioxane
O
C_6H_{13}
X
O
61%
X = Me, OMe

Chem Lett, 859 (1982)

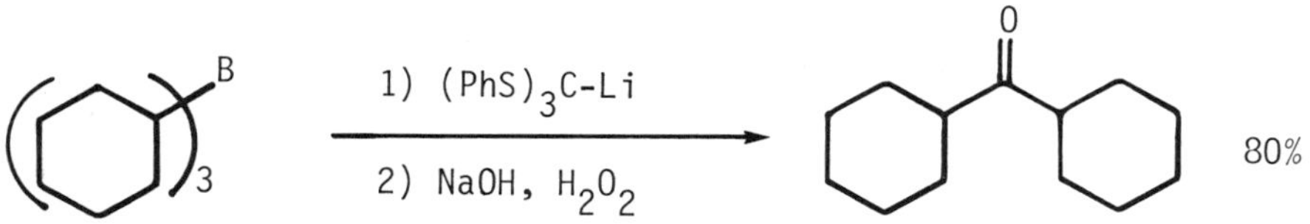
B
3
1) (PhS)3C-Li
2) NaOH, H2O2
O
80%

JCS Chem Comm, 1149 (1981)

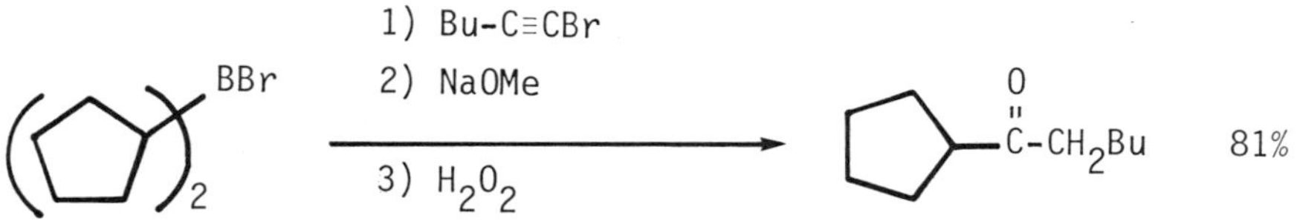
BBr
2
1) Bu-C≡CBr
2) NaOMe
3) H2O2
O
C-CH2Bu
81%

JOC, 47, 754 (1982)

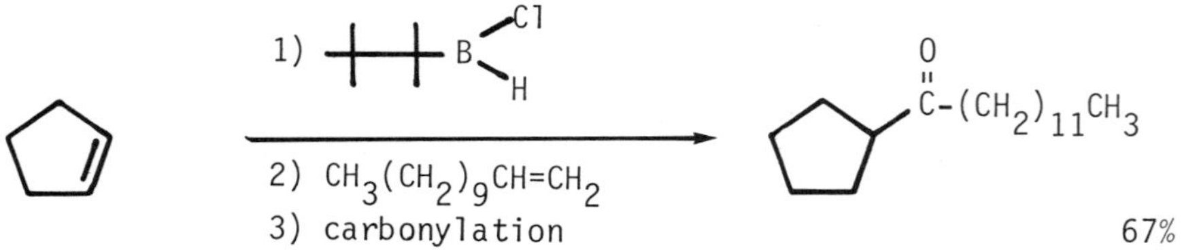
1)
B
Cl
H
2) CH3(CH2)9CH=CH2
3) carbonylation
O
C-(CH2)11CH3
67%

JOC, 45, 4540 and 4542 (1980)

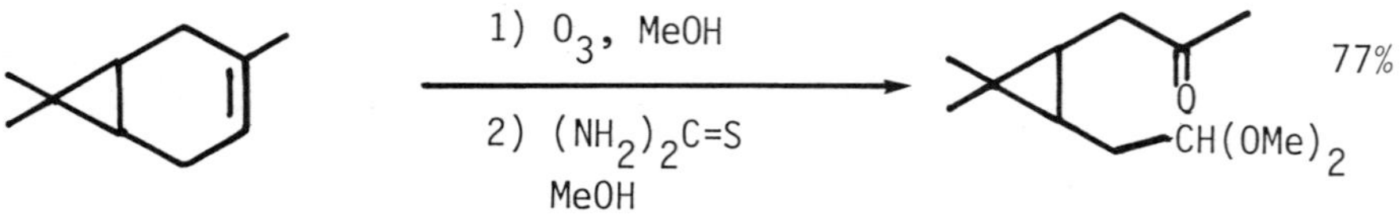
1) O3, MeOH
2) (NH2)2C=S
MeOH
O
CH(OMe)2
77%

Tetrahedron, 38, 3013 (1982)

$$h\nu, O_2$$
$$FeCl_3, \text{ pyridine}$$

54%

JOC, <u>46</u>, 509 (1981)

Review: "Ozonolysis -- A Modern Method in the Chemistry of
         Olefins"

                              Russ Chem Rev, <u>50</u>, 636 (1981)

See also Section 134 (Ethers and Epoxides from Olefins) and
Section 174 (Ketones from Ethers and Epoxides).

Section 180 <u>Ketones from Miscellaneous Compounds</u>

Conjugate reductions and reductive alkylations of enones are
listed in Section 74 (Alkyls from Olefins).

1) NaH

2) $KMnO_4$

91%

JOC, <u>47</u>, 4534 (1982)

80%

Synthesis, 44 and 662 (1980)

85%

Tetr Lett, 22, 5235 (1981)

$$Ph_2Hg \xrightarrow[\text{HMPT}]{CO, [Rh(CO)_2Cl]_2} Ph\text{-}\overset{O}{\underset{\|}{C}}\text{-}Ph \qquad 99\%$$

Bull Acad USSR Chem, 31, 211 (1982)

$$\underset{MeOC=CHOMe}{\overset{\ominus}{\underset{|}{BBu_3}}}$$

+

$EtOSO_2F$

$$\xrightarrow{\text{2) } H_3O^+} Bu\text{-}\overset{O}{\underset{\|}{C}}\text{-}CH_2Et \qquad 89\%$$

Chem Lett, 1059 (1981)

Angew Chem Int Ed, 20, 395 (1981)

JCS Chem Comm, 282 (1982)

Section 180A Protection of Ketones

See Section 362 for the formation of enol esters and Section 367
(Ether-Olefin) for the formation of enol ethers. Many of the
methods in Section 60A (Protection of Aldehydes) are also applicable to ketones.

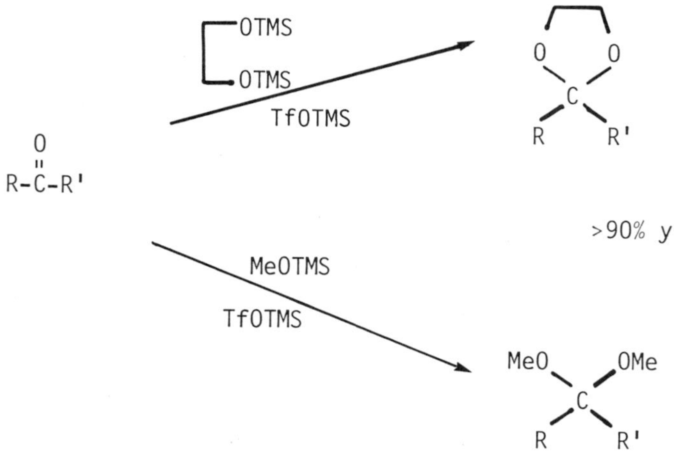

O
‖
R-C-R'

$\begin{array}{c}\text{—OTMS}\\\text{—OTMS}\end{array}$

TfOTMS

>90% yield

MeOTMS

TfOTMS

Tetr Lett, <u>21</u>, 1357 (1980)

HOCH₂CH₂OH

polyvinylpyridine HCl

resin

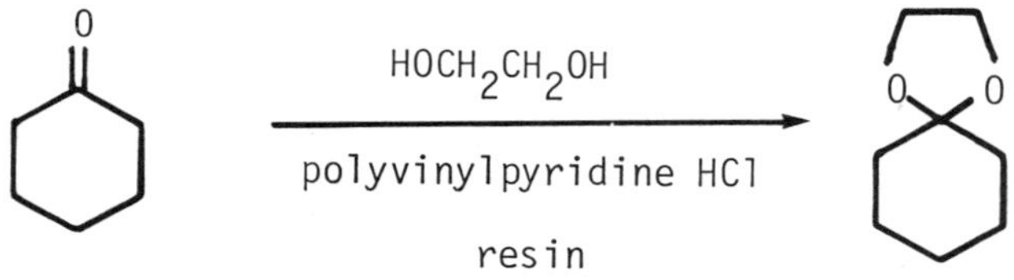

98%

Bull Chem Soc Japan, <u>54</u>, 309 (1981)

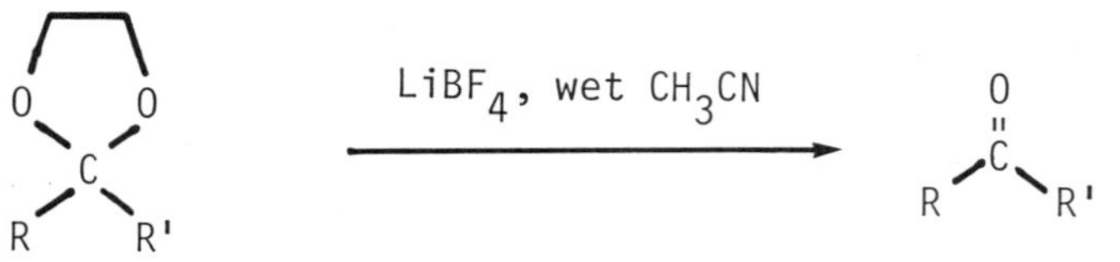

LiBF₄, wet CH₃CN

~90%

Also removes dimethyl and diethyl ketals.

Synth Comm, <u>12</u>, 267 (1982)

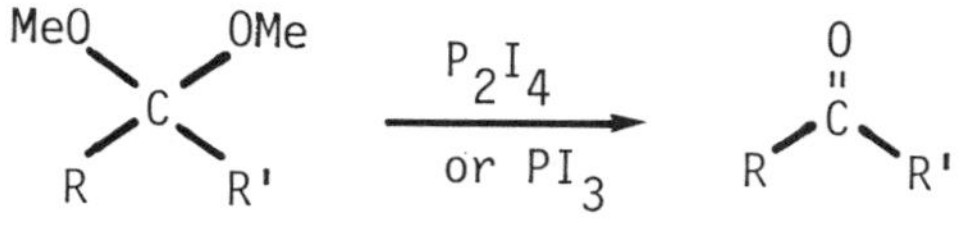

~70-90%

Angew Chem Int Ed, 19, 1006 (1980)

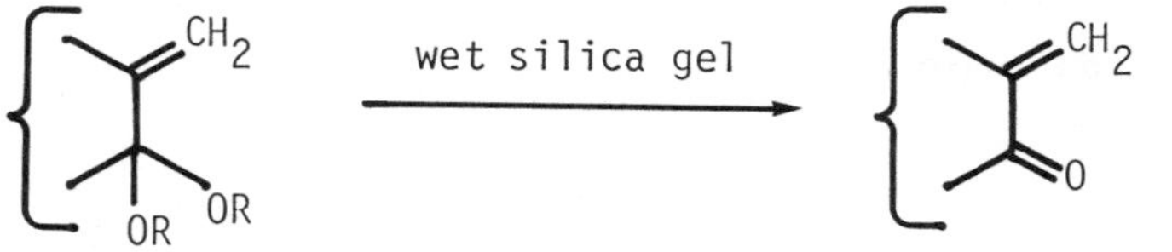

~90-100%

R = Me, Et, -(CH$_2$)$_2$

J Chem Res (S), 248 (1982)

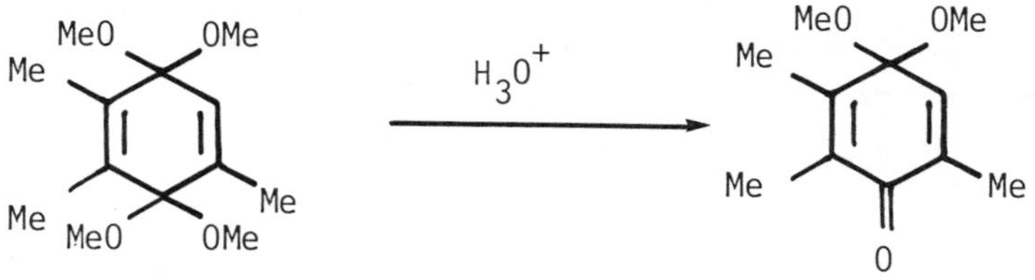

90%

JOC, 45, 3422 (1980)

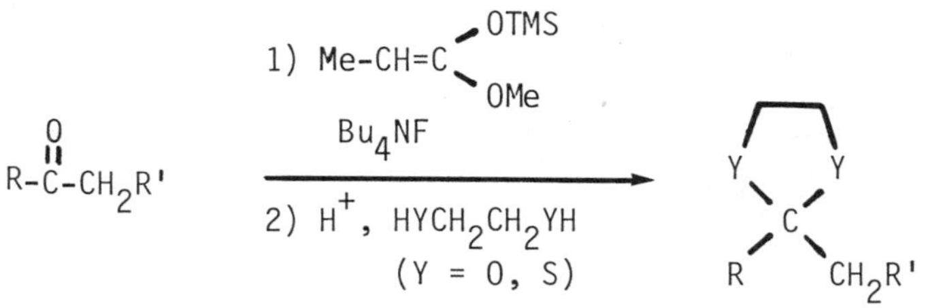

43-66%

Synthesis, 1089 (1982)

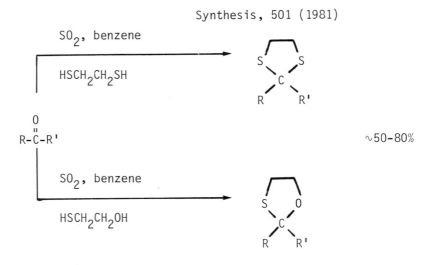

$$\underset{\substack{\text{O} \\ \text{R-C-R'}}}{} \xrightarrow[\text{Nafion-H, CCl}_4]{\text{HC(OMe)}_3} \underset{\substack{\text{R} \quad \text{R'}}}{\overset{\text{MeO} \quad \text{OMe}}{C}} \quad \sim 90\text{-}100\%$$

$$\xrightarrow[\text{Nafion-H, benzene}]{\text{HSCH}_2\text{CH}_2\text{SH}} \quad \sim 80\text{-}100\%$$

Synthesis, 282 (1981)

Review: "Methods for the Preparation of Acetals from Alcohols or Oxiranes and Carbonyl Compounds"

Synthesis, 501 (1981)

$$\xrightarrow[\text{HSCH}_2\text{CH}_2\text{SH}]{\text{SO}_2, \text{ benzene}}$$

$$\underset{\substack{\text{O} \\ \text{R-C-R'}}}{} \quad \sim 50\text{-}80\%$$

$$\xrightarrow[\text{HSCH}_2\text{CH}_2\text{OH}]{\text{SO}_2, \text{ benzene}}$$

Synthesis, 831 (1982)

$$\underset{R-\overset{\overset{\text{O}}{\|}}{C}-R'}{} \quad \xrightarrow[\text{AlCl}_3]{2R''SH} \quad \underset{R \quad \quad R'}{\overset{R''S \quad \quad SR''}{\diagdown C \diagup}}$$

~100%

R" = alkyl, dithiol

Tetr Lett, 21, 4225 (1980)

steroid $\xrightarrow[\text{CH}_2\text{Cl}_2/\text{MeOH}]{\text{HIO}_4}$

~98%

JCS Chem Comm, 886 (1980)

$\xrightarrow[\text{DMSO}]{\text{Me}_3\text{SiI (or Br)}}$ $\underset{R-\overset{\overset{\text{O}}{\|}}{C}-R'}{}$ 65-99%

R, R' = alkyl, aryl, cyclic

Synthesis, 965 (1982)

$\xrightarrow[\text{or MeS(SMe)}_2^{\oplus}\text{SbCl}_6^{\ominus}]{\text{HCl, H}_2\text{O/DMSO/dioxane}}$ $\underset{R-\overset{\overset{\text{O}}{\|}}{C}-R'}{}$ >95%

R, R' = alkyl, aryl

Synthesis, 679 (1982)

$$\underset{\substack{R \quad R'}}{\overset{\displaystyle S\text{---}S}{\diagdown C \diagup}} \quad \xrightarrow[\text{2) } H_2O]{\text{1) } Me_3O^{\oplus}\ ^{\ominus}BF_4} \quad \underset{R-\overset{\displaystyle O}{\overset{\|}{C}}-R'}{} \quad \sim 90\%$$

Synthesis, 135 (1981)

$$\underset{\substack{R \quad R'}}{\overset{\displaystyle S\text{---}S}{\diagdown C \diagup}} \quad \xrightarrow[\text{BF}_3\cdot\text{THF}]{PbO_2,\ H_2O} \quad \underset{R-\overset{\displaystyle O}{\overset{\|}{C}}-R'}{} \quad 80\text{-}96\%$$

Synthesis, 580 (1982)

$$\underset{\substack{R \quad R'}}{\overset{\displaystyle S\text{---}S}{\diagdown C \diagup}} \quad \xrightarrow[\text{35\% aq. HBF}_4]{HgO} \quad \underset{R-\overset{\displaystyle O}{\overset{\|}{C}}-R'}{} \quad \sim 80\text{-}100\%$$

Synthesis, 51 (1981)

$$\underset{\substack{R \quad R'}}{\overset{\displaystyle S\text{---}S}{\diagdown C \diagup}} \quad \xrightarrow{\text{\textcircled{P}}\text{---}(HgO\overset{\overset{\displaystyle O}{\|}}{C}CF_3)_2} \quad \underset{R-\overset{\displaystyle O}{\overset{\|}{C}}-R'}{} \quad 65\text{-}92\%$$

JOC, 47, 2212 (1982)

$$
\underset{\substack{R \quad R'}}{\overset{\substack{S \quad S}}{\Big\backslash C \Big/}} \xrightarrow{\text{(PhSeO)}_2\text{O}} \underset{R-\overset{O}{\overset{\|}{C}}-R'}{}
$$

JCS Perkin I, 1654 (1980)

$$
\underset{\substack{R \quad R'}}{\overset{\substack{S \quad S}}{\Big\backslash C \Big/}} \xrightarrow[\text{(p-Tol)}_3\text{N}]{\text{electrochemical oxidation}} \underset{R-\overset{O}{\overset{\|}{C}}-R'}{} \quad 85\text{-}97\%
$$

Tetr Lett, 21, 511 (1980)

$$
\underset{\substack{R \quad R'}}{\overset{\substack{S \quad S}}{\Big\backslash C \Big/}} \xrightarrow[\text{ether/H}_2\text{O}]{\text{ClSO}_2\text{F}} \underset{R-\overset{O}{\overset{\|}{C}}-R'}{} \quad 56\text{-}86\%
$$

Synthesis, 146 (1981)

$$
\underset{\substack{R \quad R'}}{\overset{\substack{S \quad S}}{\Big\backslash C \Big/}} \xrightarrow[\substack{\text{or trichloroisocyanuric acid,} \\ \text{AgNO}_3, \text{ MeCN/H}_2\text{O}}]{1)\ \text{HCl, 2)}\ \text{H}_2\text{O}_2} \underset{R-\overset{O}{\overset{\|}{C}}-R'}{} \quad 65\text{-}98\%
$$

Synthesis, 657 and 659 (1980)

$$
\underset{\underset{R-C-R'}{\overset{O}{\parallel}}}{} \quad \xrightarrow[\text{Bu}_2\text{SnCl}_2]{\text{R''NH}_2} \quad \underset{\underset{R}{\overset{N^{\diagup R''}}{\parallel}}{C}}{}{R'} \qquad >80\%
$$

Synth Comm, 12, 495 (1982)

$$
\xrightarrow{\text{NH}_2\text{OH}}
$$

Tetr Lett, 21, 1425 (1980)

$$
\xrightarrow[\text{Porasil, 194}^{\circ}]{\text{NH}_3, \text{O}_2}
$$

28%

JACS, 102, 1453 (1980)

$$
\underset{\underset{R}{\overset{N^{\diagup OH}}{\parallel}}{C}}{}{R'} \quad \xrightarrow[\text{H}_2\text{O}]{\text{Na}_2\text{S}_2\text{O}_4} \quad \underset{\underset{R-C-R'}{\overset{O}{\parallel}}}{}
$$

Aust J Chem, 32, 201 (1979)

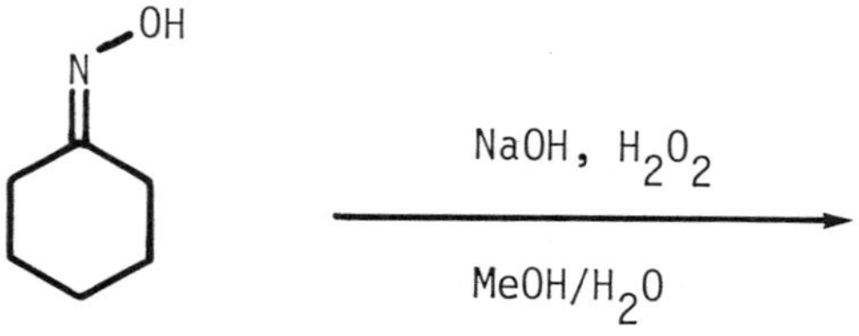

Synth Comm, <u>10</u>, 465 (1980)

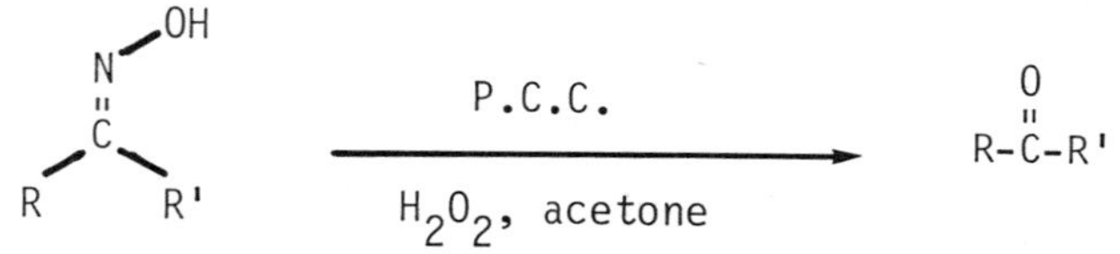

Synthesis, 125 (1980)

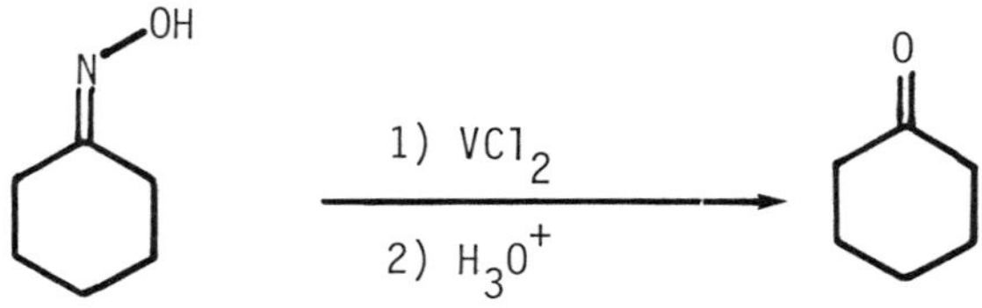

Synthesis, 220 (1980)

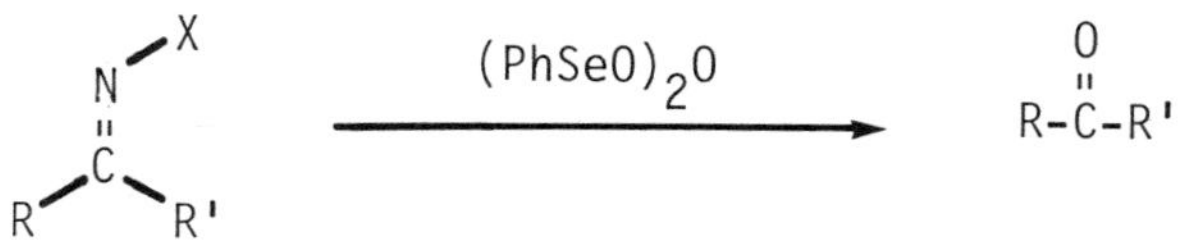

X = OH, NHPh   JCS Perkin I, 1212 (1980)

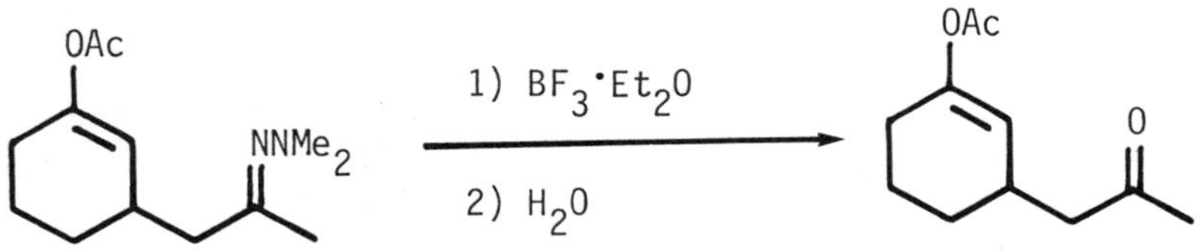

Synth Comm, 12, 15 (1982)

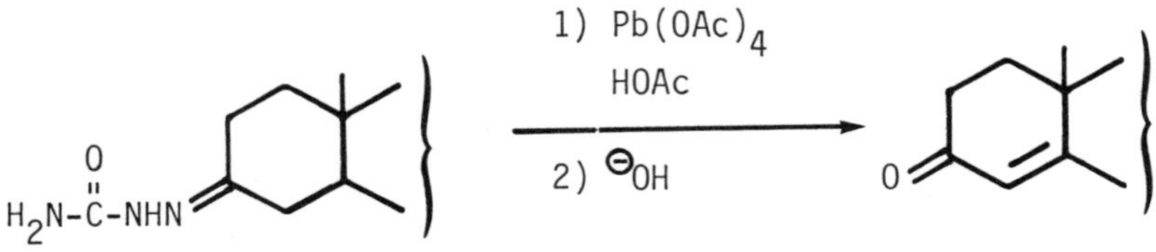

Tetr Lett, 21, 651 (1980)

CHAPTER 13
PREPARATION OF NITRILES

Section 181 Nitriles from Acetylenes

No additional examples.

Section 182 Nitriles from Carboxylic Acids and Acid Halides

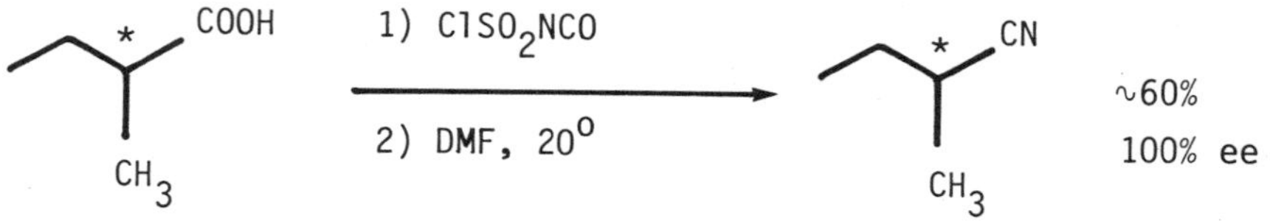
COOH
1) $ClSO_2NCO$
2) DMF, 20°
CH_3
CN
~60%
100% ee

Synth Comm, 12, 25 (1982)

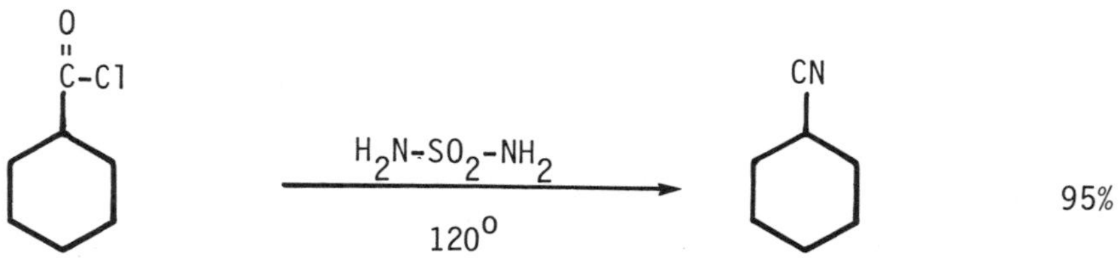
C-Cl
$H_2N-SO_2-NH_2$
120°
CN
95%

(Slightly lower yields starting from the acid)

Tetr Lett, 23, 1505 (1982)

Section 183 Nitriles from Alcohols

$$\underline{n}\text{-}C_8H_{17}\text{-OH} \quad \xrightarrow[\text{CCl}_4, \text{ MeCN}]{\text{Bu}_3\text{P, KCN}} \quad \underline{n}\text{-}C_8H_{17}\text{-CN} \qquad 76\%$$

Synthesis, 1007 (1980)

$$\underline{n}\text{-}C_7H_{15}\text{-CH}_2\text{OH} \quad \xrightarrow[\underline{\text{Cu}}, 300°]{\text{NH}_3} \quad \underline{n}\text{-}C_7H_{15}\text{-C}{\equiv}\text{N} \qquad 87\%$$

JOC, 46, 754 (1981)

NaCN, Me₃SiCl

NaI, DMF/MeCN

80%

JOC, 46, 2985 (1981)

Section 184 Nitriles from Aldehydes

$$C_7H_{15}\text{CHO} \quad \xrightarrow[\text{pyridine/toluene}]{\text{NH}_2\text{OH}\cdot\text{HCl}} \quad C_7H_{15}\text{-C}{\equiv}\text{N} \qquad 82\%$$

Synthesis, 190 (1982)

$$Ph-CHO \xrightarrow[\text{pyridine/HCl}]{\text{EtNO}_2} Ph-C\equiv N \qquad 80\%$$

Synthesis, 739 (1981)

$$CH_3(CH_2)_7CHO \xrightarrow{\text{diphenyl free sulfimide}} CH_3(CH_2)_7-C\equiv N \quad 78\%$$

Tetr Lett, 21, 761 (1980)

$$Ph-CHO \xrightarrow[\underline{t}\text{-BuOK, DME}]{\text{Ts-CH}_2\text{-NC}} Ph-CH_2CN \qquad 67\%$$

Synth Comm, 10, 399 (1980)

$$C_5H_{11}-CH=NOH \xrightarrow[\text{Et}_3N]{\text{ClSO}_2F} C_5H_{11}-C\equiv N \qquad 92\%$$

Synthesis, 659 (1980)

$$Ph-CHO \xrightarrow{\Delta} Ph-C\equiv N \qquad 98\%$$

Synthesis, 702 (1980)

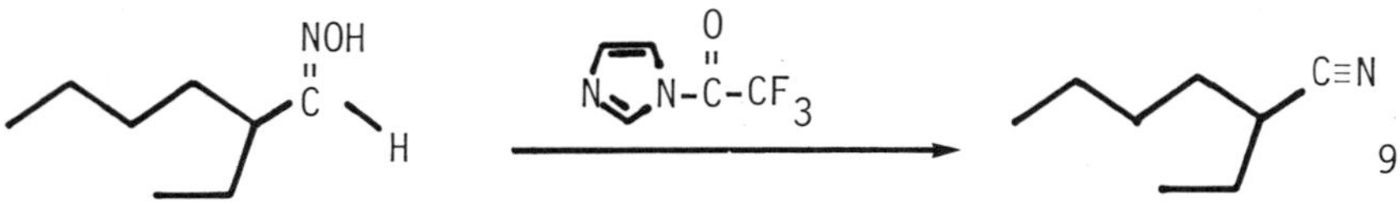

Bull Chem Soc Japan, 54, 1579 (1981)

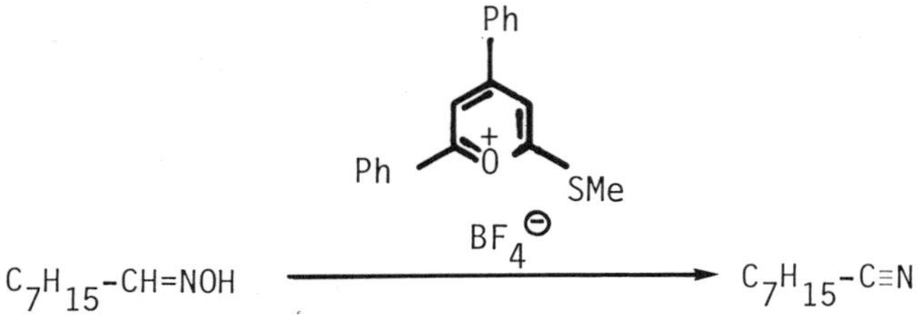

$C_7H_{15}-CH=NOH$ $C_7H_{15}-C≡N$ 86%

Synthesis, 1016 (1982)

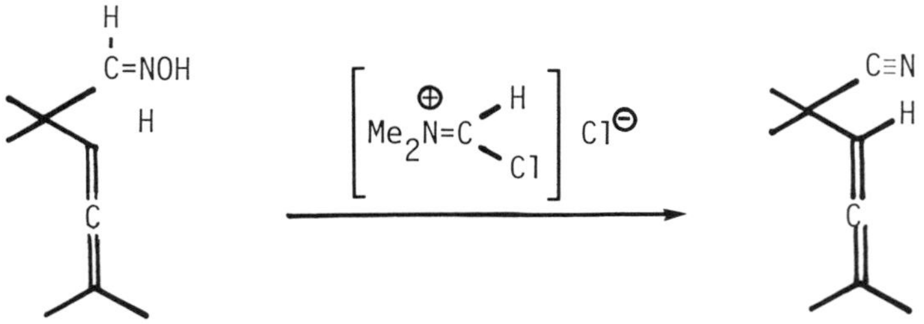

Tetr Lett, 22, 1599 (1981)

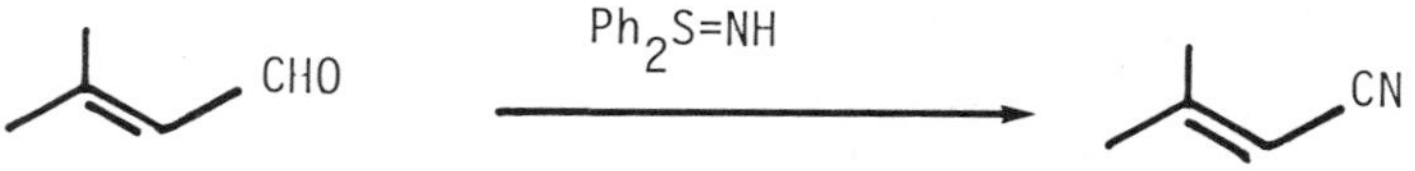

Synthesis, 1005 (1980)

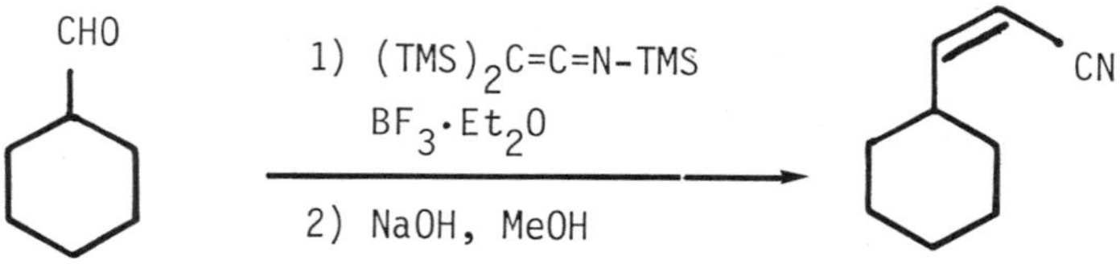

JCS Chem Comm, 56 (1982)

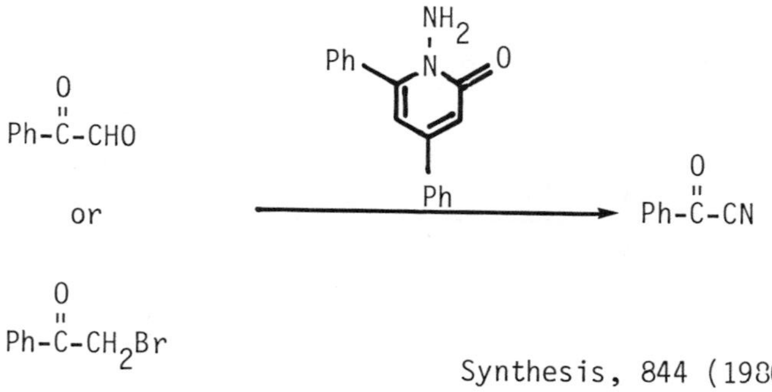

Synthesis, 844 (1980)

Section 185 <u>Nitriles from Alkyls, Methylenes, and Aryls</u>

No additional examples.

Section 186 Nitriles from Amides

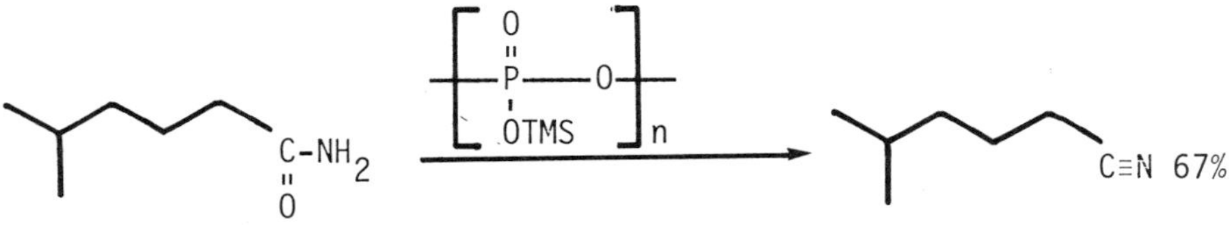

Synthesis, 591 (1982)

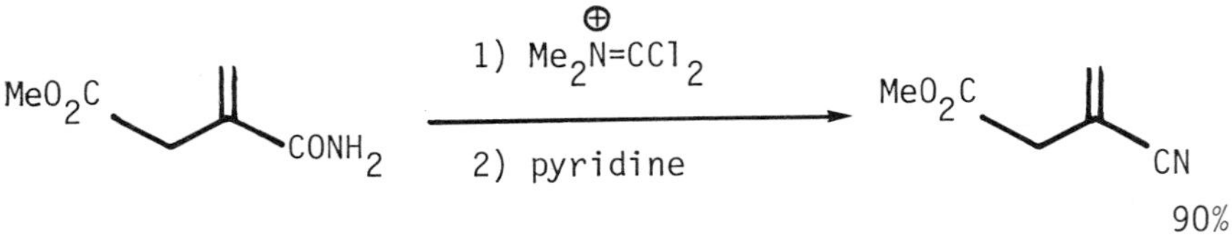

Synth Comm, 10, 479 (1980)

$$C_{11}H_{23}\text{-}\overset{S}{\overset{\|}{C}}\text{-}NH_2 \xrightarrow[\text{or } Bu_2SnO]{(Bu_3Sn)_2O} C_{11}H_{23}\text{-}C\equiv N \qquad 77\%$$

JOC, 47, 4594 (1982)

Section 187 Nitriles from Amines

$$C_9H_{19}\text{-}CH_2\text{-}NH_2 \xrightarrow[H_2O,\ KOH]{Ni(OH)_2\text{-anode}} C_9H_{19}\text{-}C\equiv N \qquad 91\%$$

Synthesis, 145 (1982)

$$Ph-CH_2NH_2 \xrightarrow[\substack{2) \ MeI \\ 3) \ NaCN, \ DME}]{} PhCH_2CN \qquad 80\%$$

Synthesis, 711 (1981)

Section 188 <u>Nitriles from Esters</u>

No additional examples.

Section 189 <u>Nitriles from Ethers and Epoxides</u>

No additional examples.

Section 190 <u>Nitriles from Halides</u>

$$\underline{n}-C_8H_{17}Br \xrightarrow[benzene]{\underset{\textcircled{P}}{\oplus} \ominus CN} \underline{n}-C_8H_{17}-CN \qquad 90\%$$

Uses a cationic resin loaded with cyanide.

Synthesis, 299 (1980)

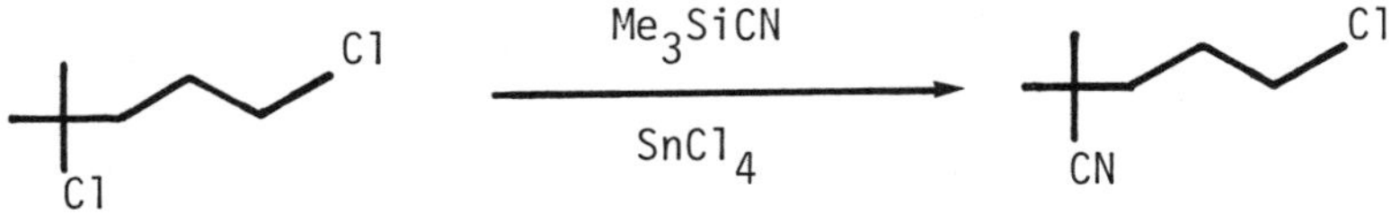

Angew Chem Int Ed, <u>20</u>, 1017 (1981)

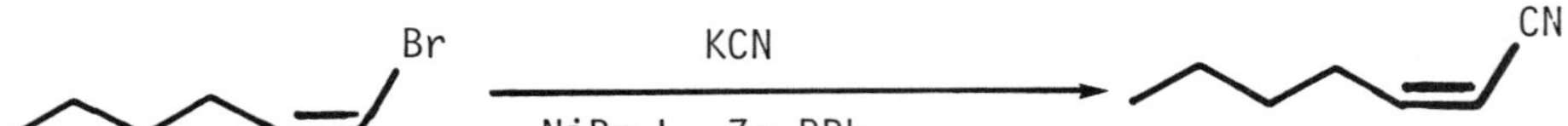

Chem Lett, 1565 (1982)

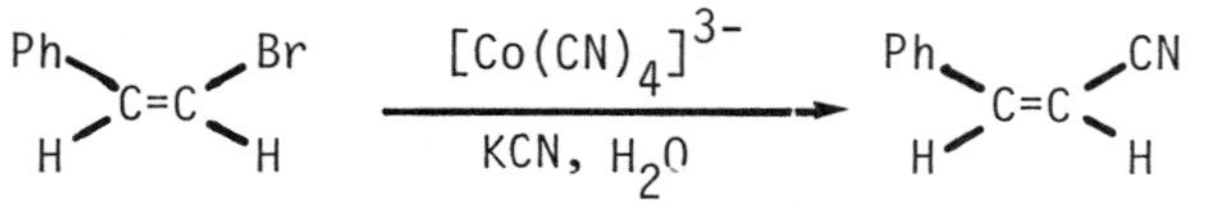

JACS, <u>104</u>, 1560 (1982)

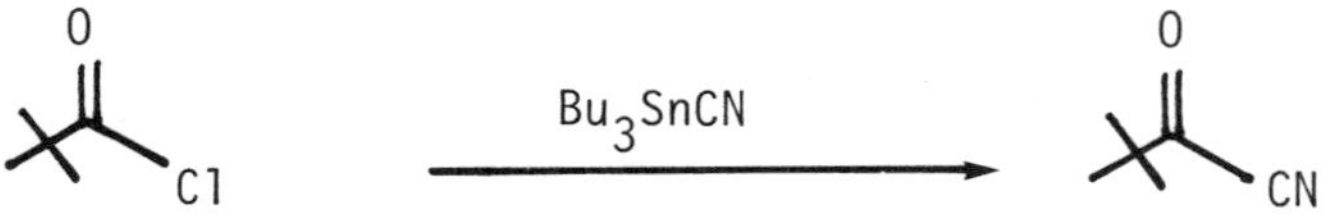

Tetr Lett, <u>21</u>, 2959 (1980)

Section 191 <u>Nitriles from Hydrides</u>

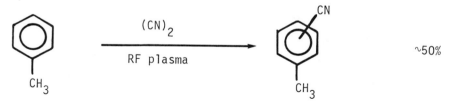

JACS, <u>102</u>, 7119 (1980)

Synth Comm, <u>10</u>, 495 (1980)

JCS Perkin I, 1132 (1980)

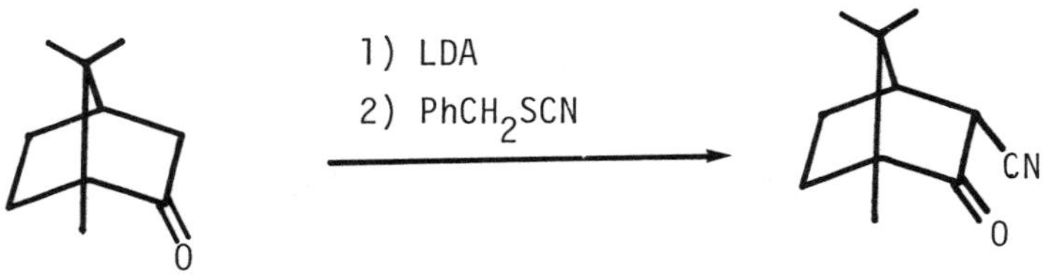

1) LDA

2) PhCH$_2$SCN

70%

Comptes Rendus, **291**, 179 (1980)

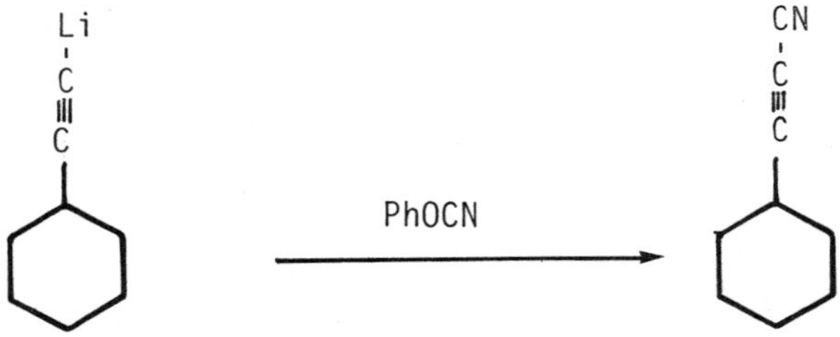

PhOCN

92%

Synthesis, 150 (1980)

Section 192 Nitriles from Ketones

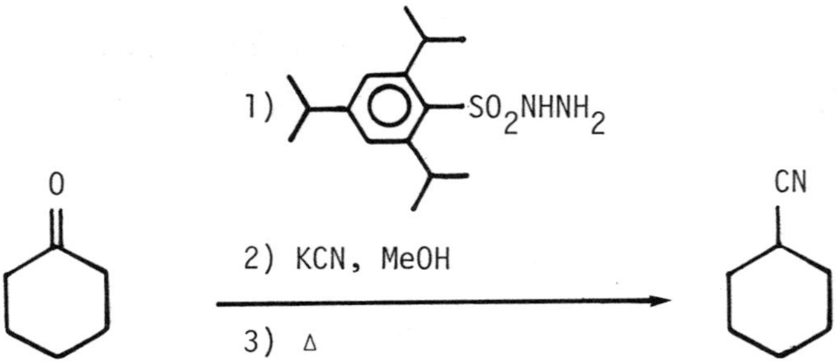

1) (2,4,6-triisopropylphenyl)-SO$_2$NHNH$_2$

2) KCN, MeOH

3) Δ

72%

JCS Perkin I, 1487 (1980)

1) TMSCN, ZnI_2

2) $POCl_3$, pyridine

82%

Chem Lett, 1427 (1979)

$Ph_2C=O$

1) Me_3SiCN, ZnI_2

2) H_3O^+, THF

$Ph_2C\underset{CN}{\overset{OH}{\diagdown}}$

Org Syn, <u>60</u>, 14 (1981)

Section 193 <u>Nitriles from Nitriles</u>

Conjugate reductions and Michael alkylations of olefinic nitriles
are found in Section 74 (Alkyls from Olefins).

$Ph-CH_2CN$

EtOH

RuH_2L_4

$Ph-\overset{Et}{\underset{|}{C}}HCN$

92%

Tetr Lett, <u>22</u>, 4107 (1981)

ArCHO
+
$Ar'CH_2CN$

1) ^-OH

2) $NaBH_4$, DMF

$ArCH_2\overset{}{\underset{Ar'}{C}}H-CN$

80-90%

JOC, <u>45</u>, 171 (1980)

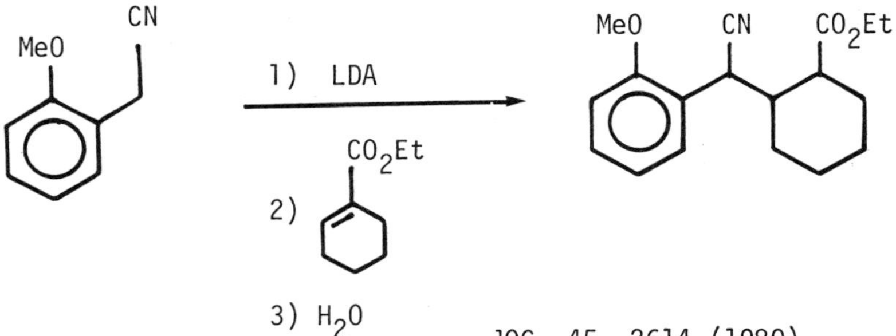

JOC, 45, 2614 (1980)

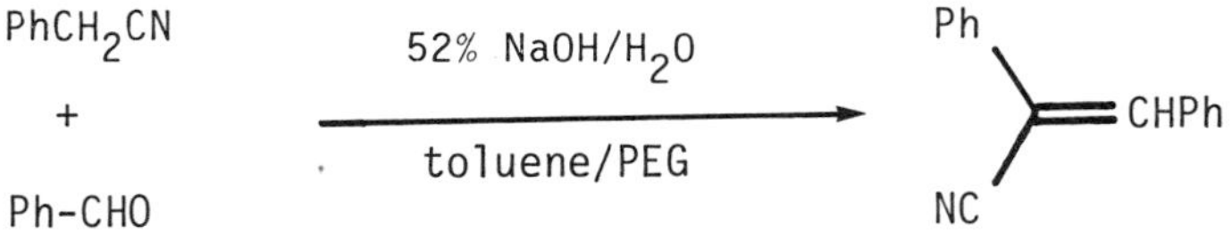

Synthesis, 913 (1981)

Section 194 Nitriles from Olefins

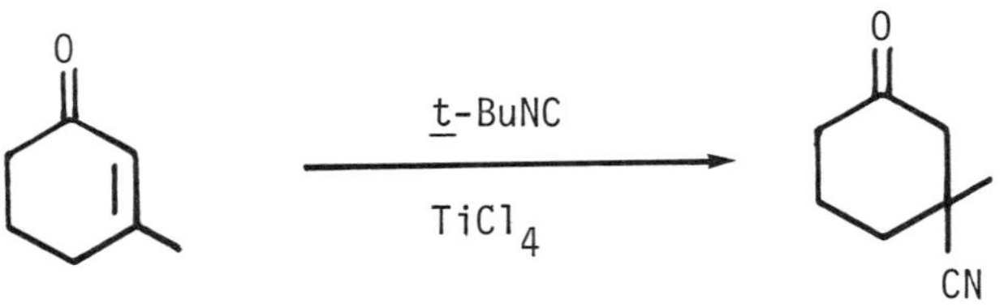

JACS, 104, 6449 (1982)

Section 195 <u>Nitriles from Miscellaneous Compounds</u>

No additional examples.

CHAPTER 14
PREPARATION OF OLEFINS

Section 196 <u>Olefins from Acetylenes</u>

$$Bu-C\equiv C-Bu \xrightarrow[\text{Ni on K/graphite}]{H_2}$$

Bu⌣Bu 98%

JCS Chem Comm, 540 (1981)

JOC, <u>46</u>, 5340 and 5344 (1981)

$$Ph-C\equiv C-Ph \xrightarrow[\text{MeOH, } \Delta]{Zn(Cu)}$$

Ph⌣Ph > 95%

Tetr Lett, <u>21</u>, 1069 (1980)

$$C_5H_{11}-C\equiv C-C_5H_{11} \xrightarrow[\text{NbCl}_5]{\text{NaAlH}_4}$$

$$\underset{C_5H_{11}}{\overset{H}{\diagdown}} C=C \underset{C_5H_{11}}{\overset{H}{\diagup}}$$ 62%

Chem Lett, 157 (1982)

BHBr·SMe$_2$ 1) Br-C≡C-Bu

2) NaOMe

3) CH$_3$COOH

76%

JOC, **47**, 754 (1982)
JOC, **47**, 3808 (1982)
Synthesis, 195 (1982)

Bu
B(Sia)$_2$

+

PhCH$_2$Br

NaOH

PdL$_4$

Bu
CH$_2$Ph

99%

Tetr Lett, **21**, 2865 (1980)
Bull Chem Soc Japan, **53**, 1670 (1980)

BBr$_2$·SMe$_2$

1) H-C≡C-Pr

2) NaOMe

3) I$_2$

74%

JOC, **47**, pages 171, 3806, and 5407 (1982)

Me
H-C-C≡CH
Cl

C$_8$H$_{17}$MgCl

Me
C=C=C
H
H
C$_8$H$_{17}$

61%

Tetr Lett, **21**, 5019 (1980)

$$\underset{\substack{| \\ \text{C} \\ ||| \\ \text{C} \\ |}}{\overset{\text{CH}_2\text{OSOMe}}{\underset{\text{CH}_2\text{OSOMe}}{}}} \xrightarrow[\text{THF}]{\text{[BuCuBr]MgCl·LiBr}}$$

80-95%

Rec Trav Chim Pays-Bas, <u>99</u>, 340 (1980)

$$\underset{\substack{| \\ \text{Br}}}{\overset{\text{Me}}{\text{Me-C-C}\equiv\text{C-H}}} \xrightarrow[\text{PdL}_4]{\text{Ph-ZnCl}} \underset{\text{Me}}{\overset{\text{Me}}{>}}\text{C=C=CHPh}$$

80-95%

Tetr Lett, <u>22</u>, 1451 (1981)

Review: "Carbometallation (C-metallation) of Alkynes: Stereo-

specific Synthesis of Alkenyl Derivatives"

Synthesis, 841 (1981)

Section 197 <u>Olefins from Carboxylic Acids</u>

$$\overset{\ominus}{\text{CHCOO}}{}^{\ominus}$$

+

CHO

$$\xrightarrow[\substack{\text{2) PhSO}_2\text{Cl, pyridine} \\ \text{3) } \Delta}]{}$$

~60%

JOC, <u>46</u>, 3359 (1981)

Section 198 <u>Olefins from Alcohols</u>

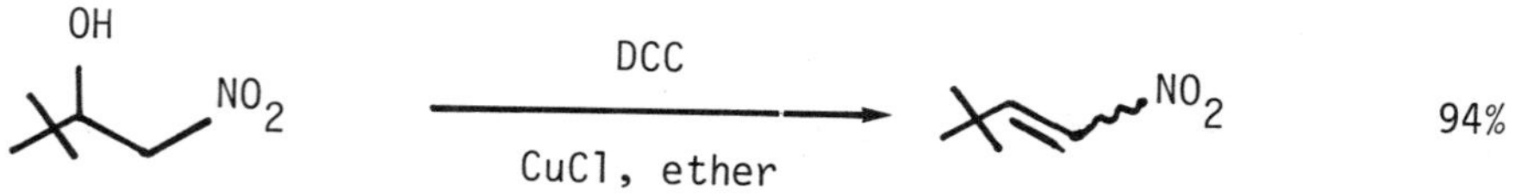

Synthesis, 1017 (1982)

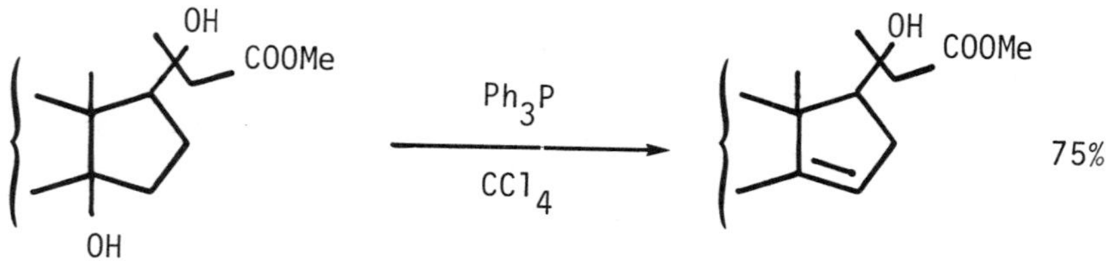

Z Chem, <u>20</u>, 372 (1980)

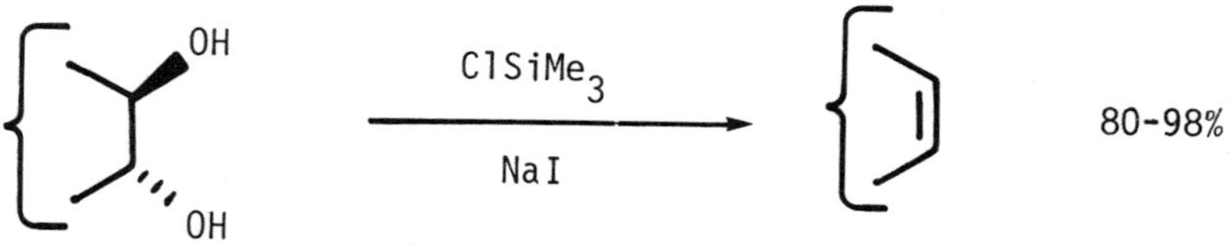

Tetr Lett, <u>23</u>, 1365 (1982)

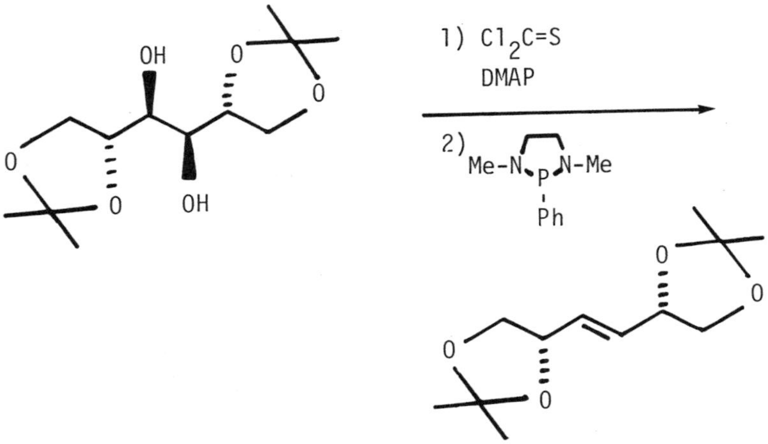

1) $Cl_2C=S$
   DMAP

2) Me-N⌐⌐N-Me
      P
      Ph

81%

Tetr Lett, 23, 1979 (1982)

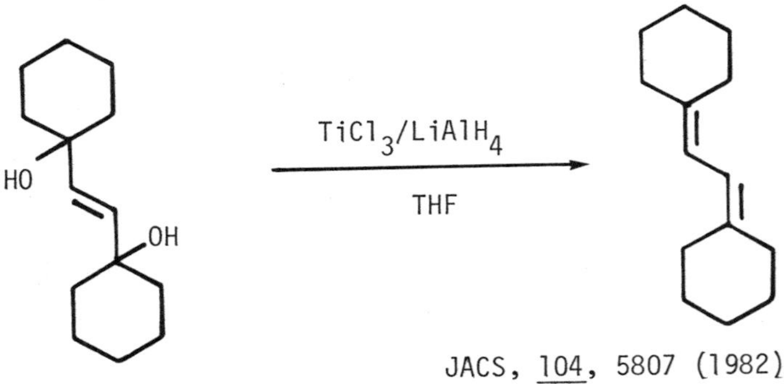

$TiCl_3/LiAlH_4$

THF

75%

JACS, 104, 5807 (1982)

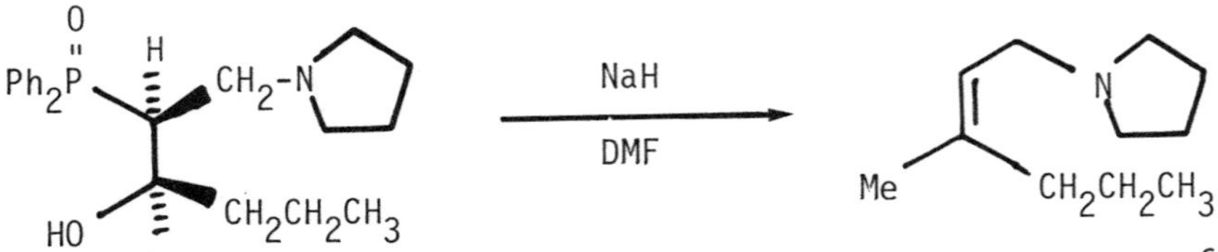

NaH

DMF

63%

Tetr Lett, 23, 4505 (1982)

Section 199 <u>Olefins from Aldehydes</u>

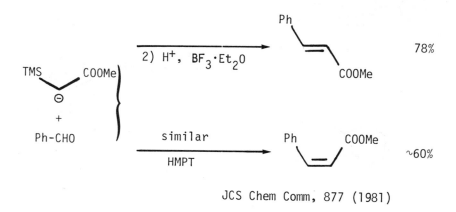

JCS Chem Comm, 877 (1981)

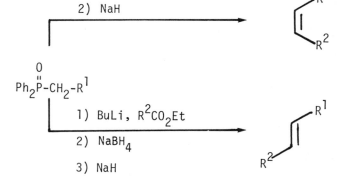

JCS Chem Comm, 100 (1981)

Tetr Lett, $\underline{21}$, 1375 (1980)

Tetr Lett, $\underline{21}$, 4021 (1980)

Angew Chem Int Ed, $\underline{21}$, 776 (1982)

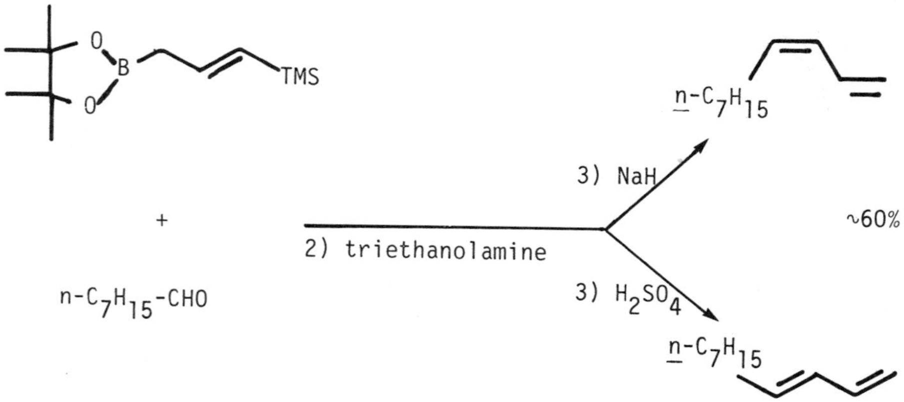

Tetr Lett, 22, 2751 (1981)

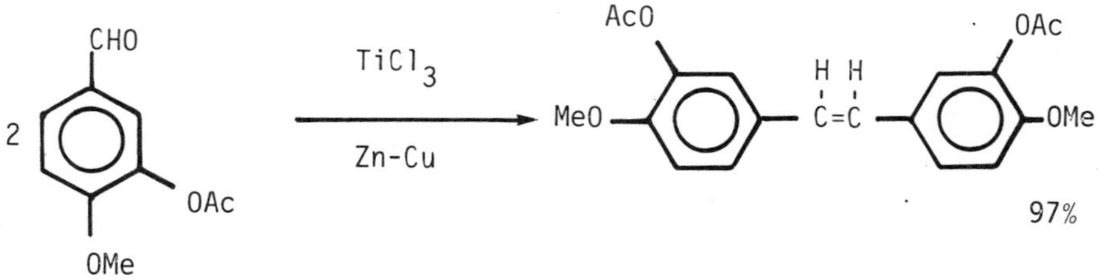

JOC, 46, 4292 (1981)

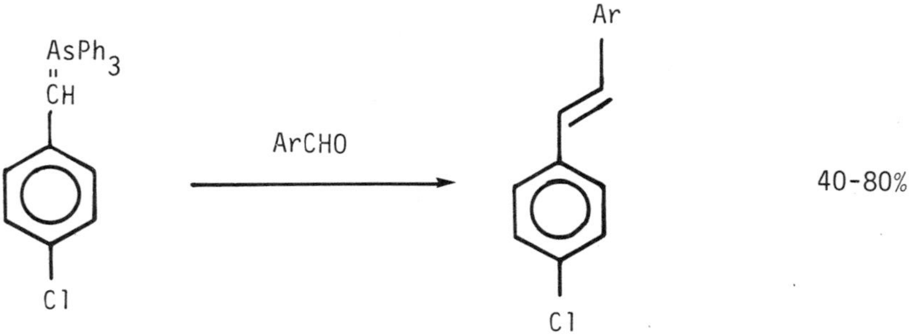

Indian J Chem, 18B, 359 (1979)

J Chem Research (S), 106 (1981)

Synthesis, 647 (1980)

PhCH$_2$CH$_2$CHO

\+

CH$_3$CCH$_3$ (O)

$\xrightarrow[\text{ether}]{\text{LiI}}$

PhCH$_2$CH$_2$—CH=CH—C(=O)—CH$_3$

75%

JCS Chem Comm, 486 (1980)

Cl—⟨C$_6$H$_4$⟩—CHO

\+

(RO)$_2$P=O
  |
 $\ominus$CHCN

P—⟨ ⟩—⟨C$_6$H$_4$⟩—CH$_2$ $\overset{\oplus}{N}$Me$_3$

$\longrightarrow$

Cl—⟨C$_6$H$_4$⟩—CH=CHCN

97%

JCS Perkin I, 2516 (1980)

Related Methods: "Olefins from Ketones (Section 207)

Section 200 <u>Olefins from Alkyls, Metyylenes, and Aryls</u>

This section contains dehydrogenations to form olefins and unsaturated ketones, esters, and amides. It also includes the conversion of aromatic rings to olefins. Hydrogenation of aryls to alkanes and dehydrogenations to form aryls are included in Section 74 (Alkyls, Methylenes, and Aryls from Olefins).

Also works with ketones.

Tetr Lett, <u>23</u>, 2105 (1982)

1) NaH
2) Se
3) MeI
4) H_2O_2

83%

Tetr Lett, <u>22</u>, 3043 (1981)

1) PhSeCl, pyr.

2) H_2O_2, CH_2Cl_2

85%

JOC, <u>46</u>, 2920 (1981)

1) Pd^{II}, THF

2) DBN

100%

Aust J Chem, <u>33</u>, 1537 (1980)

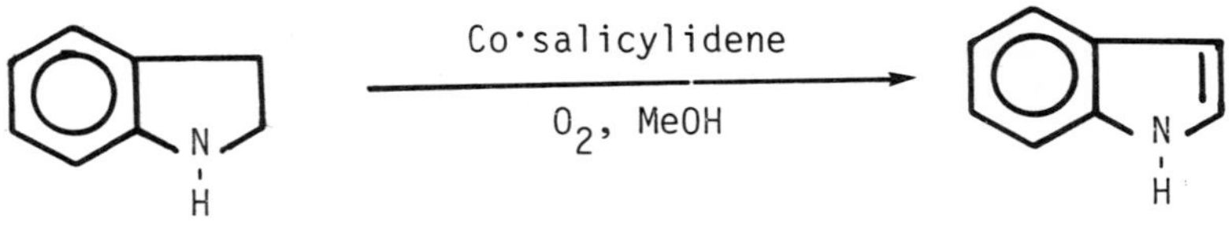

Chem Lett, 1287 (1980)

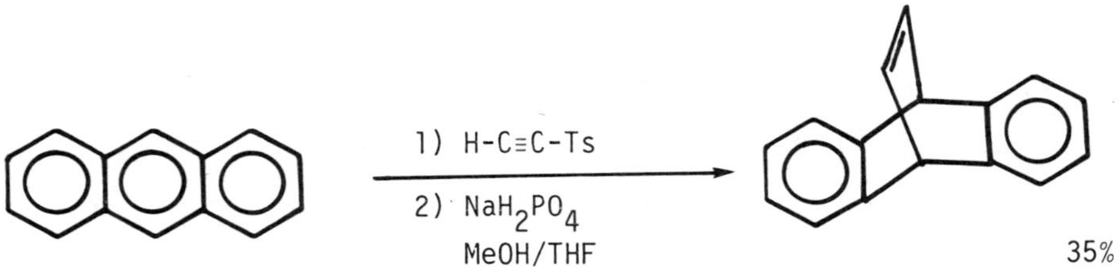

35%

JCS Chem Comm, 639 (1980)

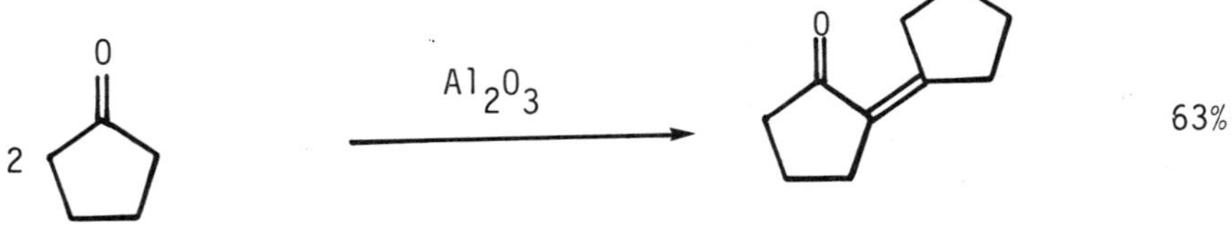

63%

Synthesis, 60 (1982)

~80-90%

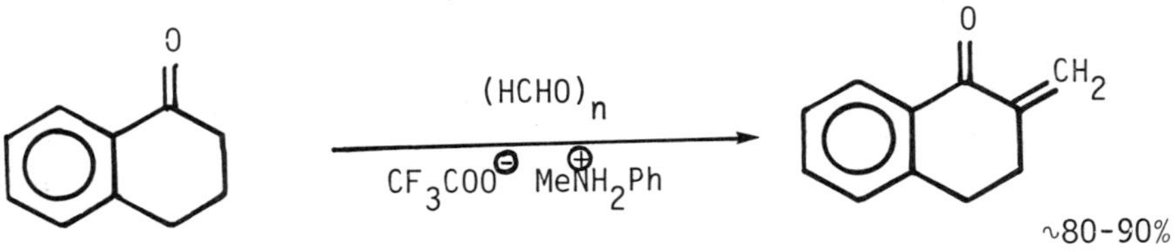

Org Syn, 60, 88 (1981)

1) 2 LDA
2) (MeOCSS)$_2$
3) ZnCl$_2$
4) BuCHO

65%

(E̲ isomer is formed in 58% yield if ZnCl$_2$ is omitted.)

Chem Lett, 595 (1980)

H$_2$C(OMe)$_2$

POCl$_3$/NaOAc

70%

Synthesis, 34 (1982)

Related methods: Alkyls and Aryls from Alkyls and Aryls (Section 65) Alkyls and Aryls from Olefins (Section 74)

Section 201 Olefins from Amides

N(SO$_2$—⬡—NO$_2$)$_2$ 180°

90%

Tetr Lett, 22, 199 (1981)

Section 202 <u>Olefins from Amines</u>

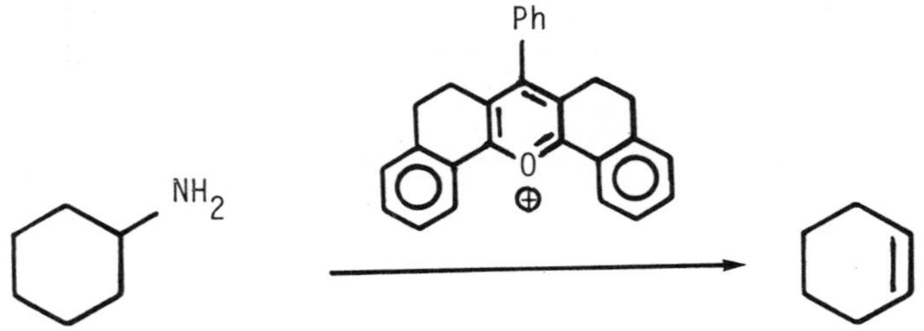

An alternative to the Hofmann elimination.

JCS Perkin I, 2347 (1982)
JCS Chem Comm, 96 (1981)

Section 203 <u>Olefins from Esters</u>

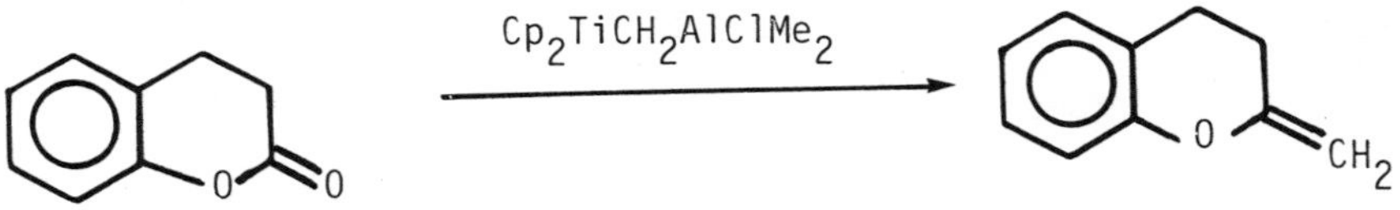

$Cp_2TiCH_2AlClMe_2$

85%

JACS, <u>102</u>, 3270 (1980)

1) DIBAL-H
2) $Ph_3P=CH_2$

3) H^+

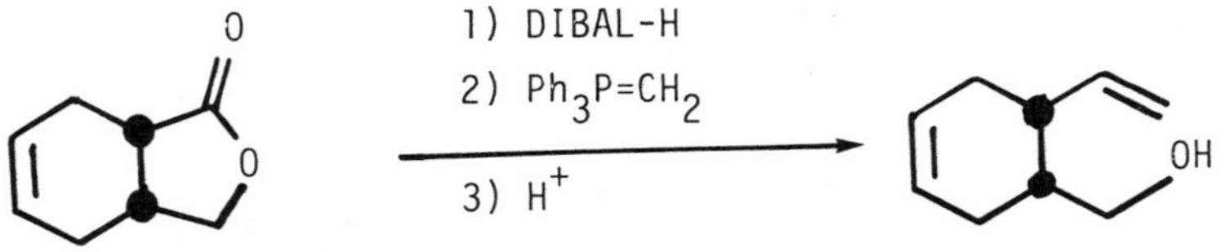

55%

Synthesis, 1015 (1980)

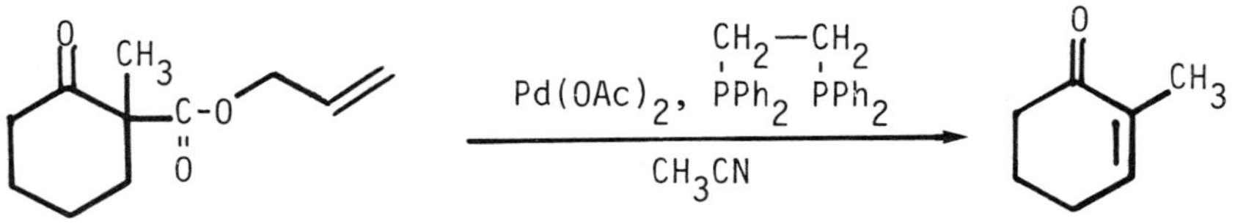

JACS, 104, 5844 (1982)

Section 204 Olefins from Ethers and Epoxides

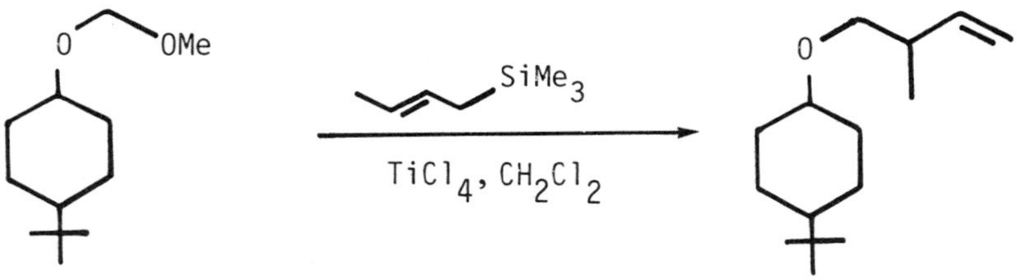

JOC, 47, 2496 (1982)

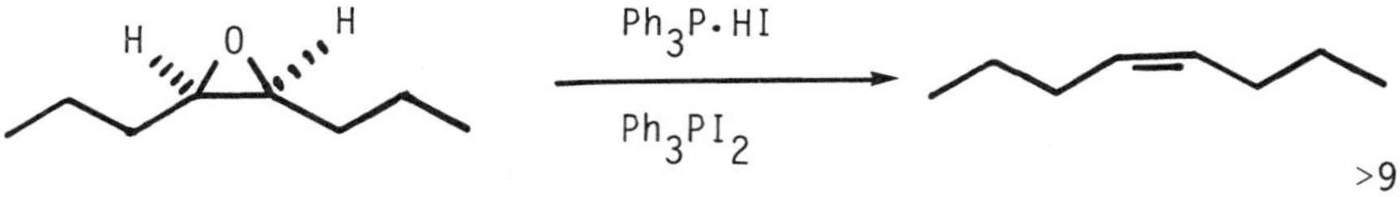

Synthesis, 828 (1980)

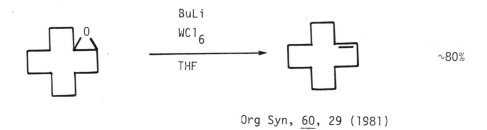

$$\underset{\text{THF}}{\overset{\text{Li}}{\longrightarrow}}$$

96%

Tetr Lett, <u>21</u>, 1173 (1980)

$$\underset{\text{NaI}}{\overset{\text{Me}_3\text{SiI}}{\longrightarrow}}$$

92%

Tetr Lett, <u>22</u>, 3551 (1981)

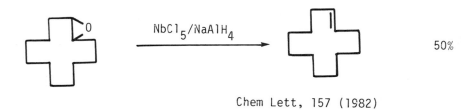

$$\underset{\text{THF}}{\overset{\substack{\text{BuLi} \\ \text{WCl}_6}}{\longrightarrow}}$$

~80%

Org Syn, <u>60</u>, 29 (1981)

$$\overset{\text{NbCl}_5/\text{NaAlH}_4}{\longrightarrow}$$

50%

Chem Lett, 157 (1982)

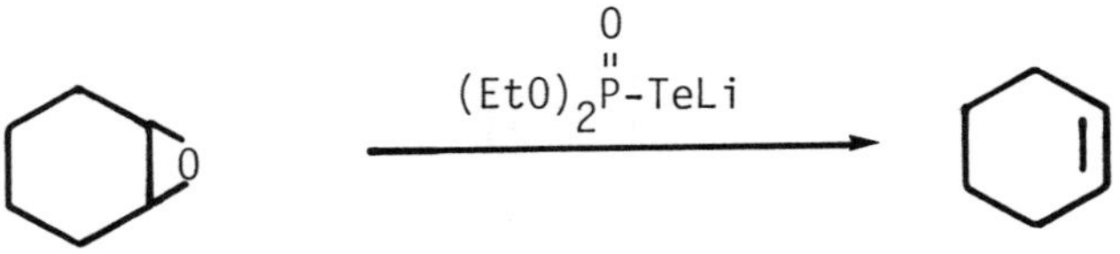

JOC, <u>45</u>, 2347 (1980)

88%

Section 205 Olefins from Halides and Sulfonates

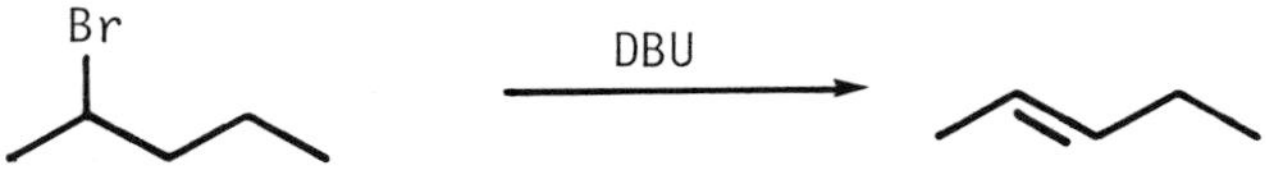

JOC, <u>47</u>, 1944 (1982)

90%

JOC, <u>47</u>, 4358 (1982)

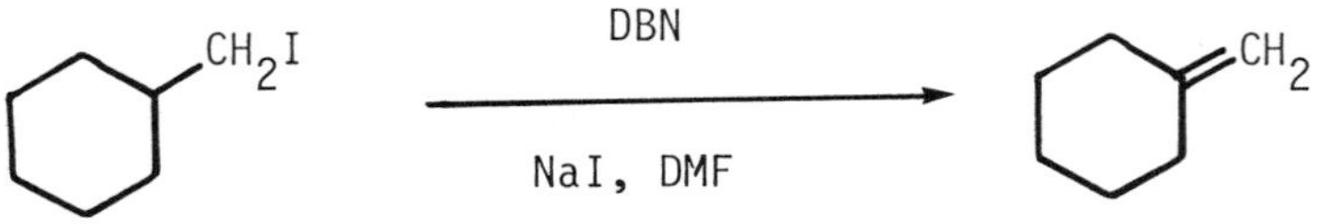

60%

JOC, <u>47</u>, 4358 (1982)

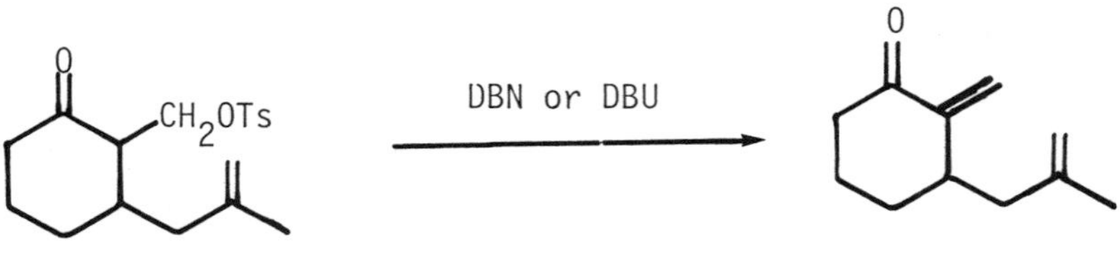

95%

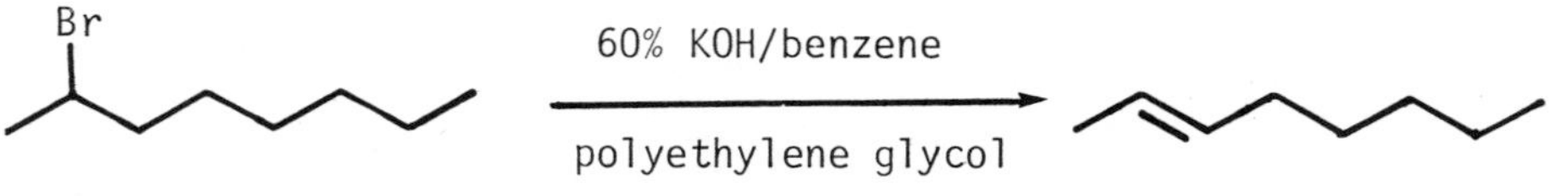
Br
60% KOH/benzene
polyethylene glycol

80%

JOC, 47, 2493 (1982)

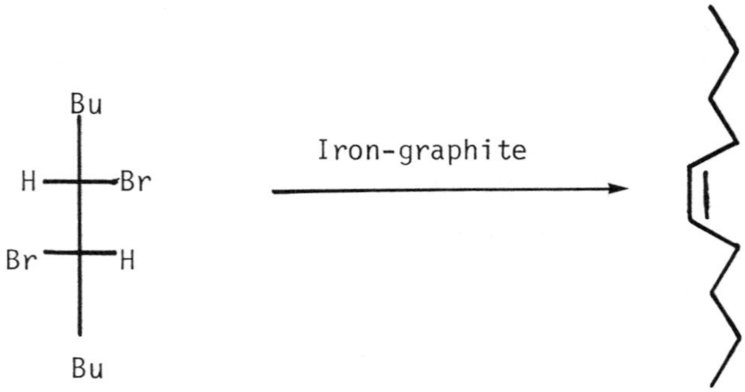
Bu
H
Br
Br
H
Bu
Iron-graphite

90%

JOC, 47, 876 (1982)

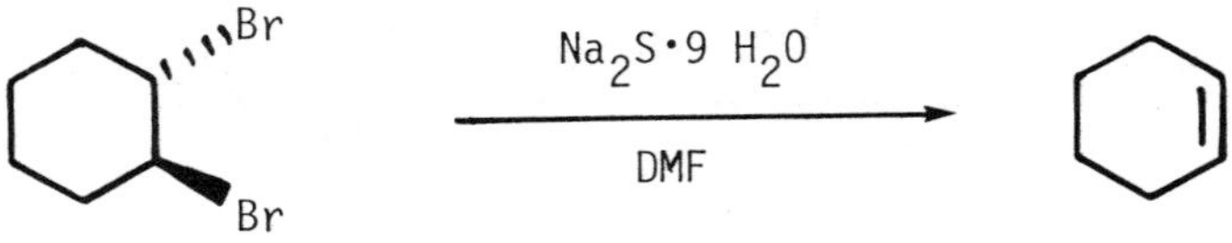
Br
Br
Na2S·9 H2O
DMF

76%

Synthesis, 879 (1981)

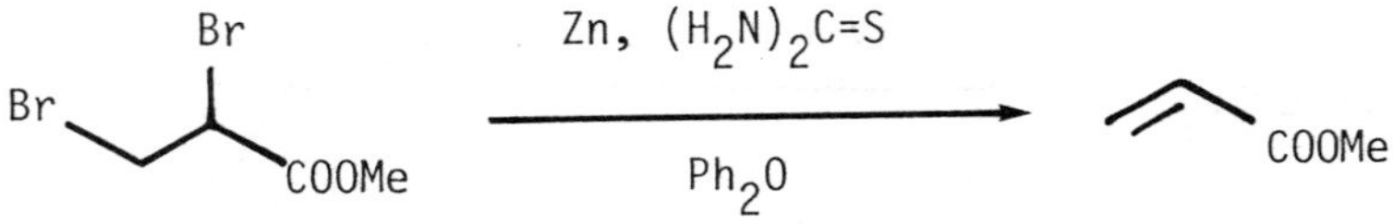
Br
Br
COOMe
Zn, (H2N)2C=S
Ph2O
COOMe

75%

Synth Comm, 11, 901 (1981)

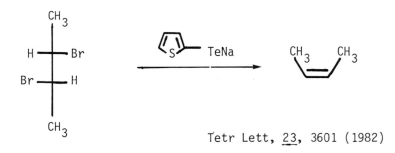

Synthesis, 1025 (1982)

Tetr Lett, $\underline{21}$, 1877 (1980)

Tetr Lett, $\underline{23}$, 3601 (1982)

Tetr Lett, $\underline{23}$, 4215 (1982)

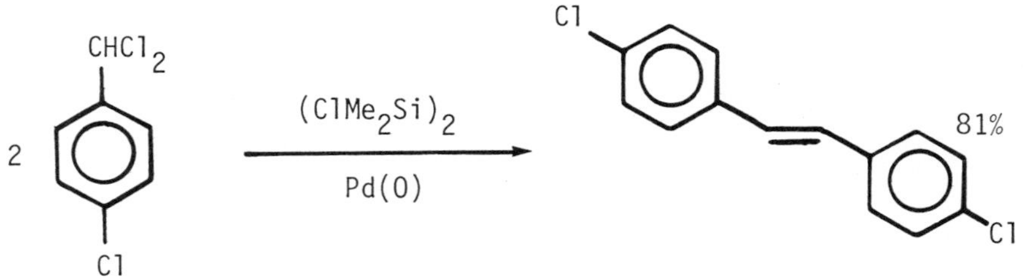

Chem Lett, 613 (1982)

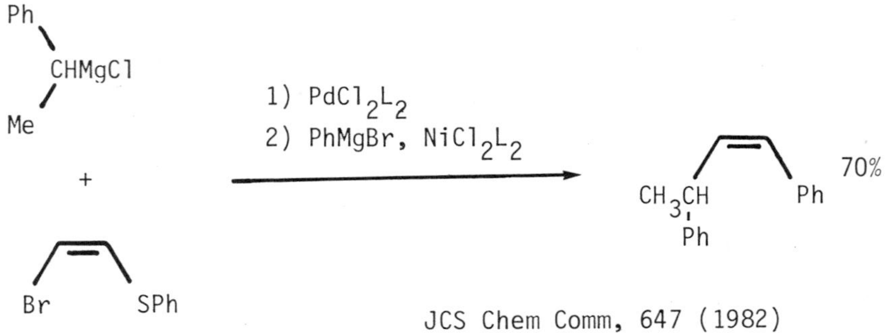

JCS Chem Comm, 647 (1982)

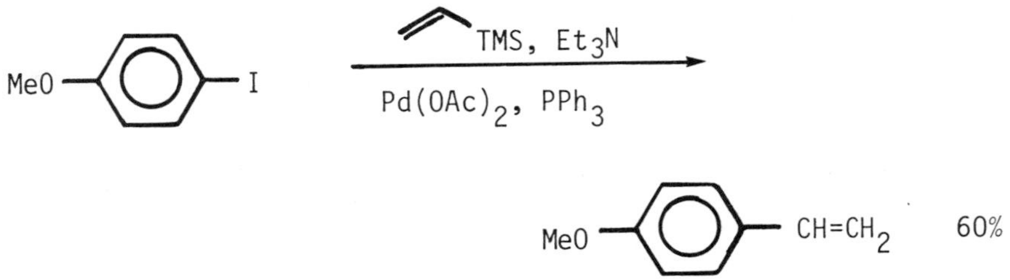

Chem Lett, 1993 (1982)

Tetr Lett, <u>22</u>, 127 (1981)

Review: "Palladium-Catalyzed Vinylation of Organic Halides"

Org React, <u>27</u>, 345 (1982)

Tetr Lett, <u>22</u>, 959 (1981)
Tetr Lett, <u>23</u>, 1591 (1982)

$$\left(\begin{array}{c} H \\ Bu \end{array} C=C \begin{array}{c} H \\ H \end{array}\right)_2 CuLi$$

+

$PhSO_2CH=CHMe$

$\xrightarrow{\text{2) } H^+ \quad \text{3) Na-Hg}}$

$$\begin{array}{c} Me \\ Me-CH \end{array} C=C \begin{array}{c} Bu \\ H \end{array}$$

JCS Chem Comm, 523 (1981)

PhMeCHMgCl

+

$CH_2=CHBr$

$\xrightarrow{\begin{array}{c}\text{chiral, polymer-bound}\\ \text{ligand} \cdot NiCl_2\end{array}}$

$$\begin{array}{c} Me \\ Ph \end{array} C-CH=CH_2 \quad H$$

93%

48% ee(S)

Tetr Lett, <u>21</u>, 4623 (1980)

$2\ \underline{n}-C_5H_{11}MgBr \xrightarrow{\begin{array}{c}\text{1) } Me_3SiCH_2COOEt\\ \text{2) } H_2SO_4/THF\end{array}} CH_2=C(\underline{n}-C_5H_{11})_2$ 78%

Tetr Lett, <u>23</u>, 1035 (1982)

$\underline{n}-C_9H_{19}-CH_2I \xrightarrow{\begin{array}{c}\text{1) } LiCHBr_2\\ \text{2) } BuLi\end{array}} \underline{n}-C_9H_{19}-CH=CH_2$ 80%

$Pr-CH_2I \xrightarrow{\begin{array}{c}\text{1) } PhCBr_2Li\\ \text{2) } BuLi\end{array}} Pr-CH=CH-Ph$ 74%

Comptes Rendus, <u>294</u>, 37 (1982)

1) ClTMS

2) $\diagup\!\!\!\!\diagdown\!\!\!\!\diagup$ I

$\overset{H}{\underset{Me}{\diagdown}}C=PPh_3$ $\xrightarrow{\hspace{3cm}}$ $\overset{CH_3}{\diagdown}C=CH-C_8H_{17}$ ~80%

3) CsF

4) $C_8H_{17}CHO$

Angew Chem Int Ed, <u>21</u>, 545 (1982)

Section 206 <u>Olefins from Hydrides</u>

This section contains examples of the conversion R-H → R-C=C. For

conversions of methylenes to olefins (R-CH$_2$-R' → R-C-R'), see Sec-

C

tion 200.

1) LDA

$\xrightarrow{\hspace{3cm}}$ 93%

OEt

2) $\overset{}{\underset{\oplus Fp}{}}$

JOC, <u>46</u>, 4103 (1981)

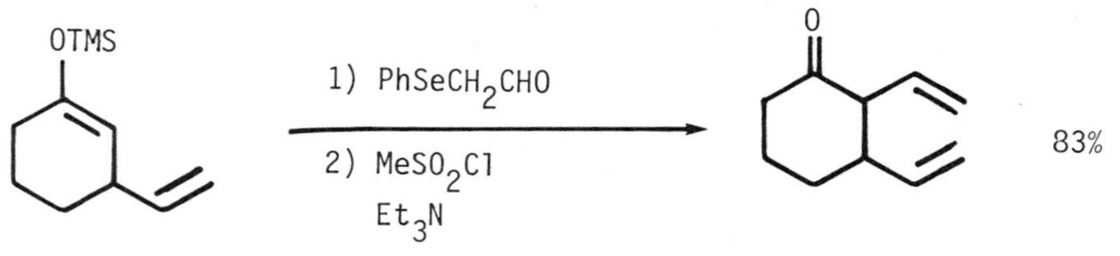

JCS Chem Comm, 434 (1981)

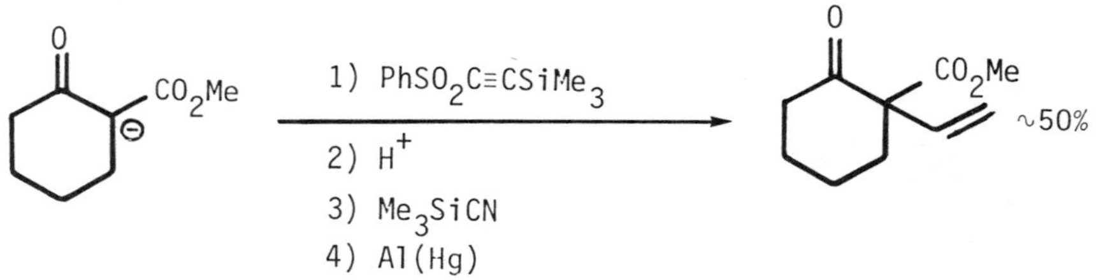

JOC, 47, 4713 (1982)

Section 207 Olefins from Ketones

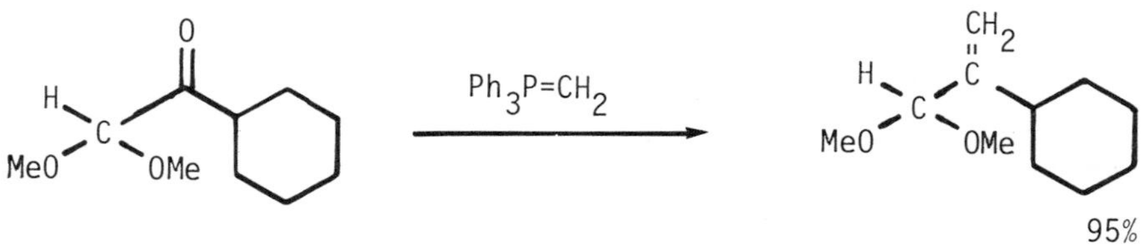

J Chem Research (S), 248 (1982)

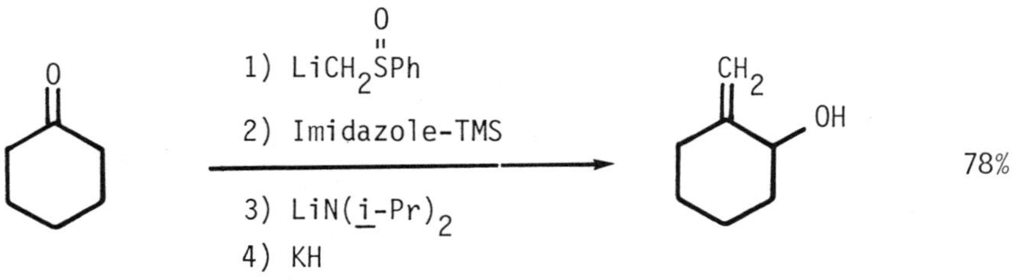
1) LiCH2SPh
2) Imidazole-TMS
3) LiN(i-Pr)2
4) KH
CH2
OH
78%

Synthesis, 640 (1980)

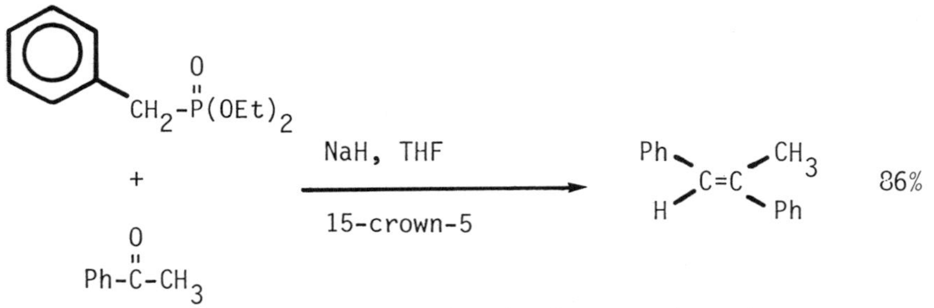
CH2-P(OEt)2
+
Ph-C-CH3
NaH, THF
15-crown-5
Ph
CH3
C=C
H
Ph
86%

Synthesis, 117 (1981)

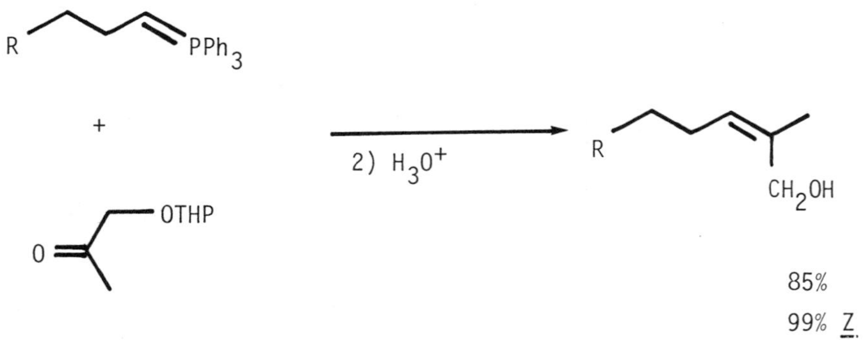
R
PPh3
+
OTHP
2) H3O+
R
CH2OH
85%
99% Z

JOC, 45, 4260 (1980)

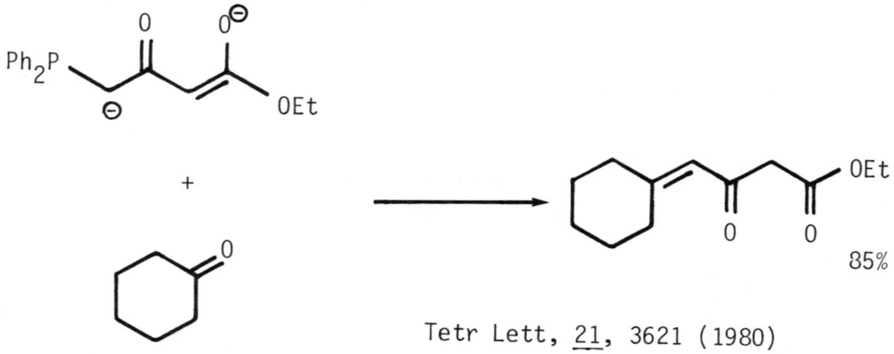

Tetr Lett, 21, 3621 (1980)

Review: "Cycloalkenes by Intramolecular Wittig Reactions"

Tetrahedron, 36, 1717 (1980)

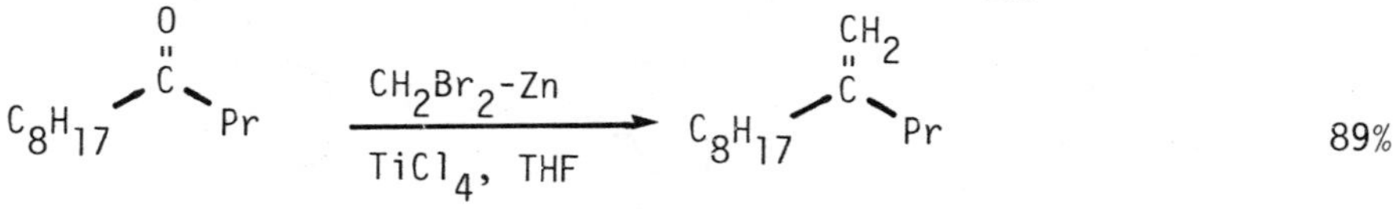

89%

Bull Chem Soc Japan, 53, 1698 (1980)

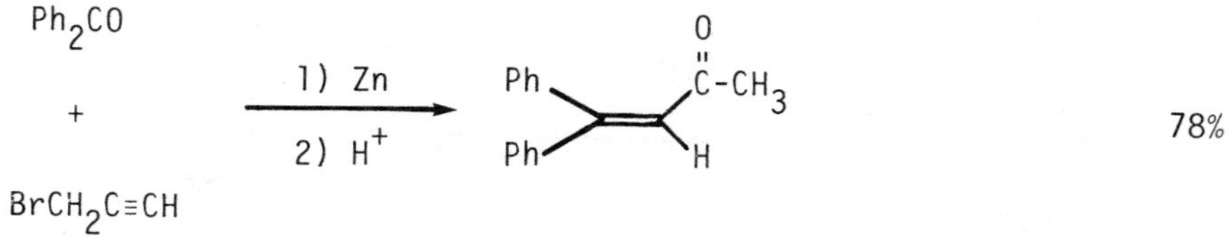

78%

Synth Comm, 10, 637 (1980)

$$
\begin{array}{c}
\text{S} \\
\underset{\text{Li}}{\overset{\|}{\text{OC-S-CH-COOEt}}}
\end{array}
$$

+

2) H_3O^+

→

H
$\overset{\text{H}}{\underset{}{\text{C-COOEt}}}$

81%

Bull Chem Soc Japan, 52, 3619 (1979)

C_5H_{11} H

H $Al(\underline{i}\text{-Bu})_2$

1) Cl_2TiCp_2

2)

→

$CH_2C_5H_{11}$

61%

Chem Lett, 429 (1982)

1) $Me_3SiCH_2CH_2Li$

2) $BF_3 \cdot AcOH$

→

H

65%

JOC, 47, 1983 (1982)

1) NH$_2$NHSO$_2$Ar
   MeCN-HCl

2) sec-BuLi

3) BuBr

Org Syn, 61, 141 (1983)

OTMS

PhMgBr

———————————

Ni(acac)$_2$

Ph

73%

Tetr Lett, 21, 3915 (1980)

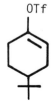

OTf

Bu$_2$CuLi

———————————

Bu

100%

Tetr Lett, 21, 4313 (1980)

Related methods: Olefins from Aldehydes (Section 199).

Section 208 Olefins from Nitriles

No additional examples.

Section 209 Olefins from Olefins

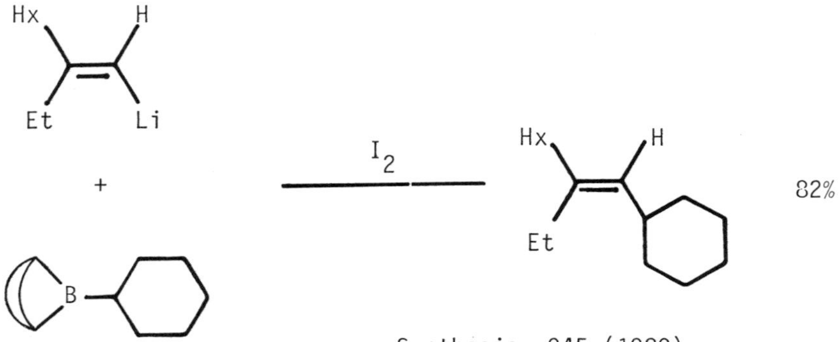

Hx
H
Et
Li
+
B
I_2
Hx
H
Et
82%

Synthesis, 945 (1980)

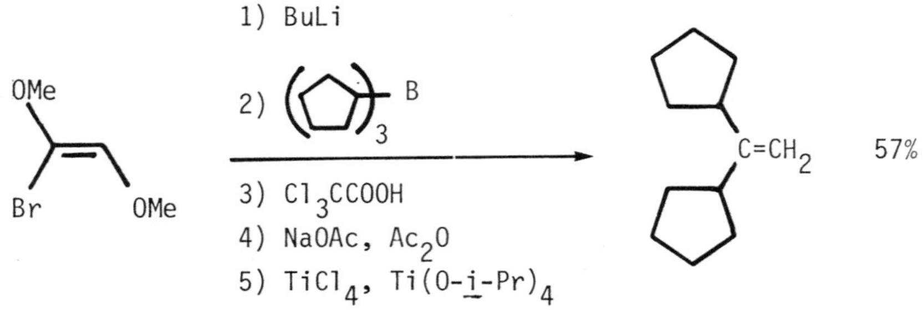

OMe
Br
OMe
1) BuLi
2) B
3
3) Cl_3CCOOH
4) NaOAc, Ac_2O
5) $TiCl_4$, $Ti(O\text{-}\underline{i}\text{-}Pr)_4$
$C{=}CH_2$
57%

Chem Lett, 591 (1980)

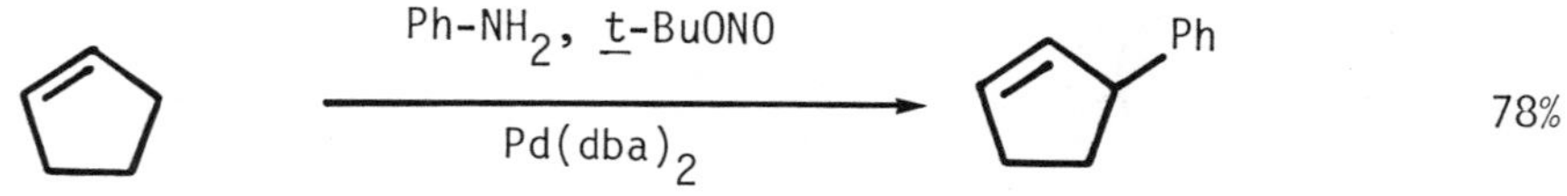
Ph-NH2, t-BuONO
Pd(dba)2
Ph
78%

JOC, 46, 4885 (1981)

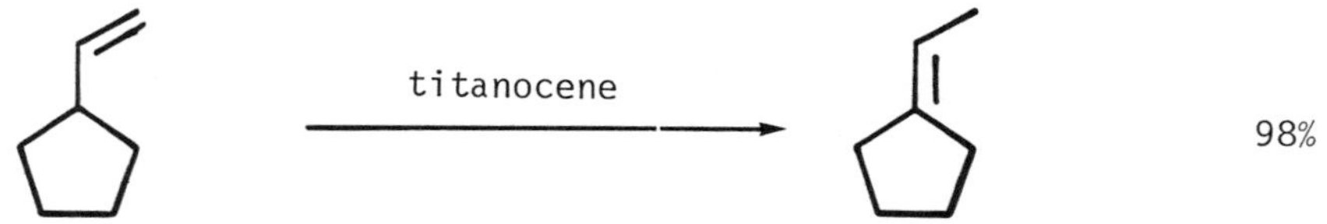
titanocene
98%

Synthesis, 53 (1982)

Review: "The 1,5-Shift Reaction"

Russ Chem Rev, 50, 666 (1981)

Section 210 Olefins from Miscellaneous Compounds

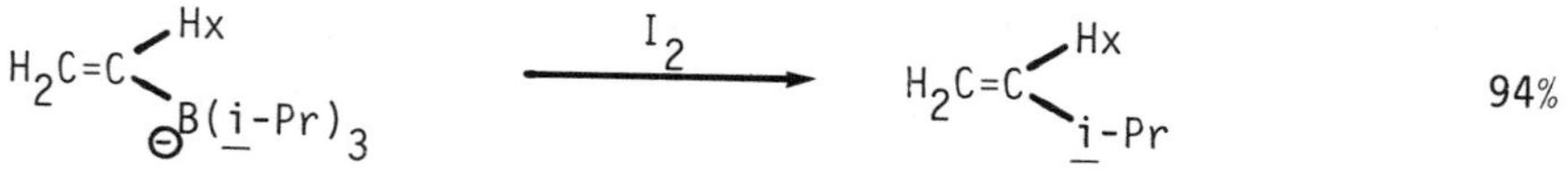
H2C=C
Hx
B(i-Pr)3
I2
H2C=C
Hx
i-Pr
94%

Tetr Lett, 21, 945 (1980)

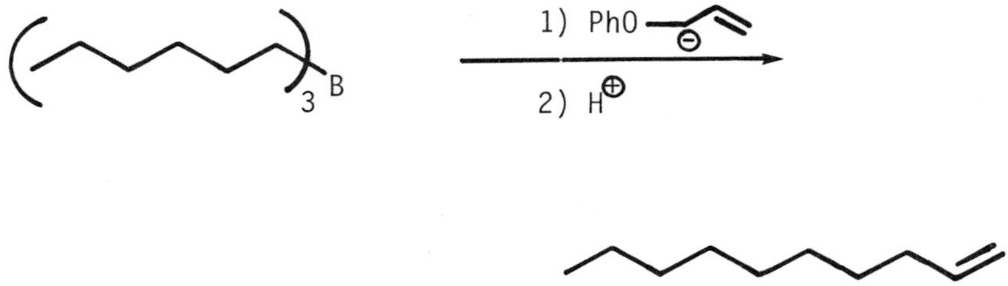

Synth Comm, 12, 813 (1982)

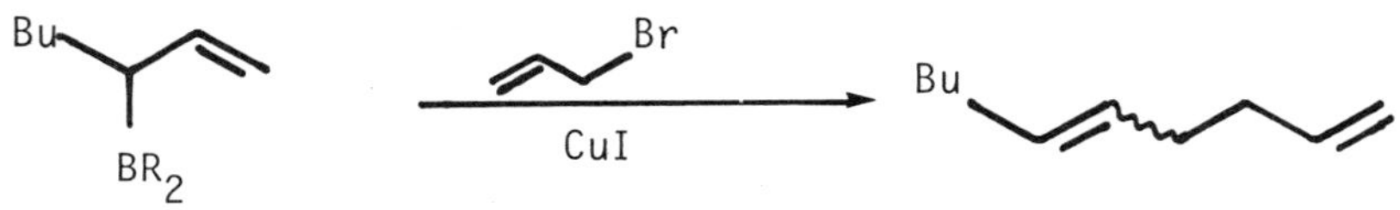

Synth Comm, 12, 813 (1982)

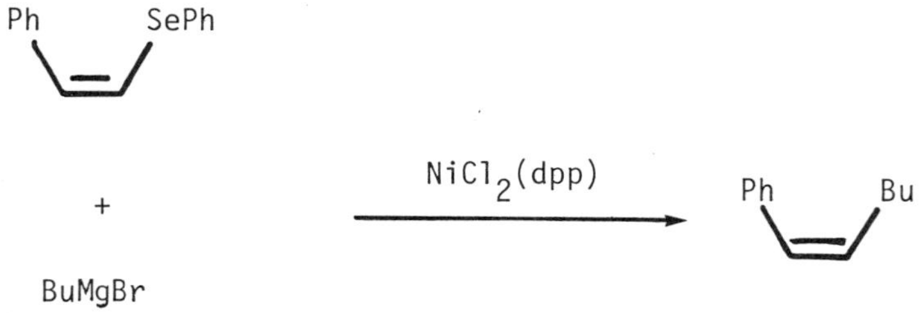

Tetr Lett, 21, 87 (1980)

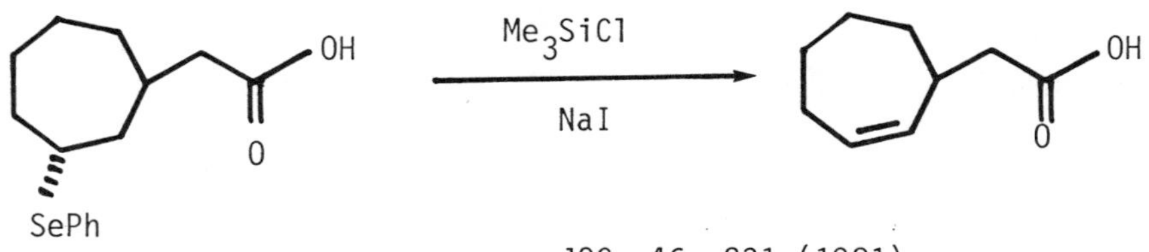

JOC, <u>46</u>, 231 (1981)

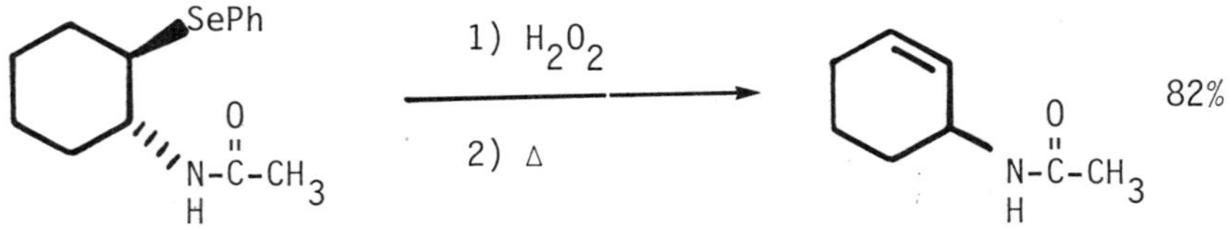

82%

JCS Chem Comm, 546 (1981)

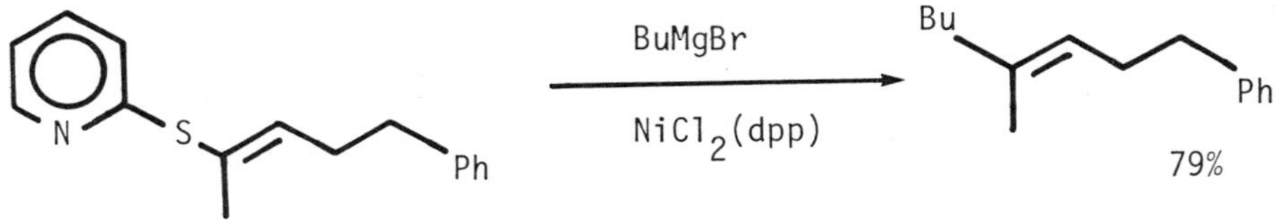

79%

Chem Lett, 1209 (1980.)

$$\eta\text{-C}_5\text{H}_5\text{Fe(CO)}_2\underset{\underset{\text{Hx}}{|}}{\text{CHCH}_2\text{Hx}} \xrightarrow[\text{2) NaI, acetone}]{\text{1) Ph}_3\text{C}^{\oplus}\;\overset{\ominus}{\text{BF}}_4} \text{HxCH=CHHx} \qquad 88\%$$

JOC, <u>45</u>, 291 (1980)

Section 210A <u>Protection of Olefins</u>

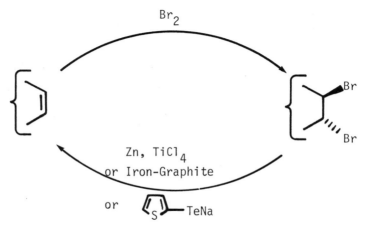

Br$_2$

Zn, TiCl$_4$
or Iron-Graphite

or [thiophene]─TeNa

Synthesis, 1025 (1982)
JOC, <u>47</u>, 876 (1982)
Tetr Lett, <u>23</u>, 3601 (1982)

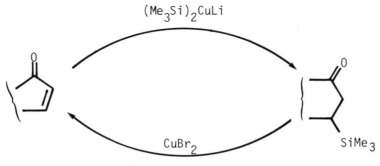

(Me$_3$Si)$_2$CuLi

CuBr$_2$

SiMe$_3$

JCS Perkin I, 2520 (1981)

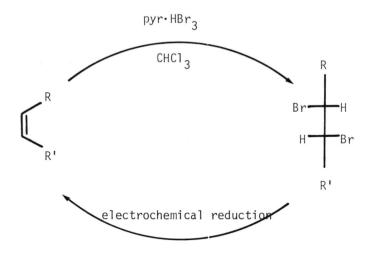

Tetr Lett, _22_, 623 (1981)

Use of (diene)$Fe(CO)_3$ complexes in the synthesis of insect phero-
mones. The $Fe(CO)_3$ complex locks the 1,3-diene in the $\underline{E}$ config-
uration, while other synthetic transformations take place.
Cleaved by $Et_3N{\rightarrow}O$.

JCS Chem Comm, 373 (1981)

CHAPTER 15

PREPARATION OF
DIFUNCTIONAL COMPOUNDS

Section 300 <u>Acetylene - Acetylene</u>

No additional examples.

Section 301 <u>Acetylene - Carboxylic Acid</u>

$I(CH_2)_9COOH$

$+$ $\xrightarrow{\text{HMPA}}$ $HC{\equiv}C(CH_2)_9COOH$ 96%

$LiC{\equiv}CH{\cdot}EDA$

Synth Comm, <u>10</u>, 653 (1980)

Section 302 <u>Acetylene - Alcohol</u>

1) $H_2C=C=C{\overset{\displaystyle H}{\underset{\displaystyle B(OH)_2}{\Big\langle}}}$

    (+)dialkyl tartrate

56%

92% ee

JACS, <u>104</u>, 7667 (1982)

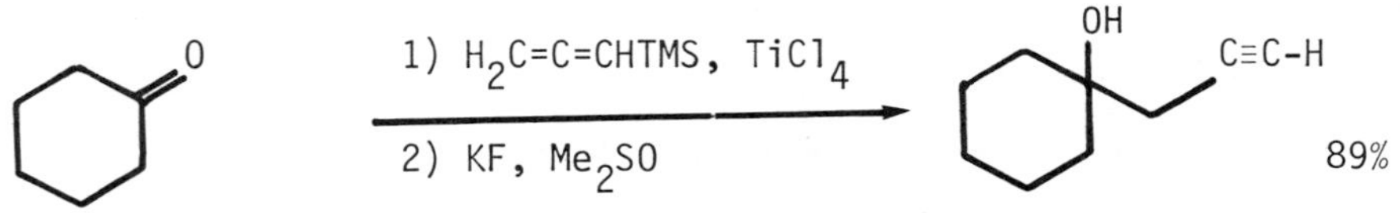
O
1) H2C=C=CHTMS, TiCl4
2) KF, Me2SO
OH
C≡C-H
89%

JOC, 45, 3925 (1980)

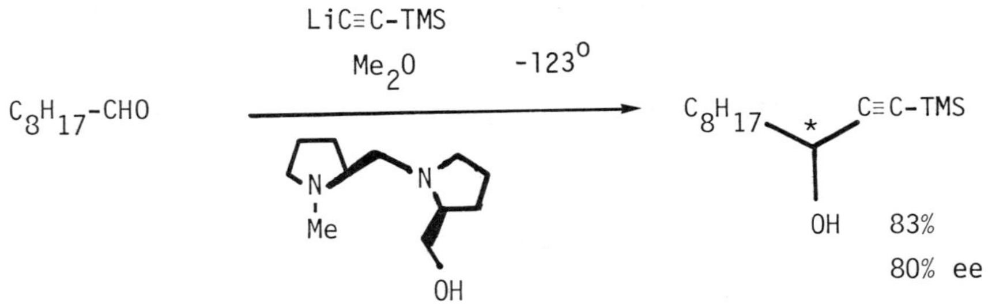
LiC≡C-TMS
Me2O -123°
C8H17-CHO
N
Me
N
OH
C8H17
*
C≡C-TMS
OH
83%
80% ee

Chem Lett, 255 (1980)

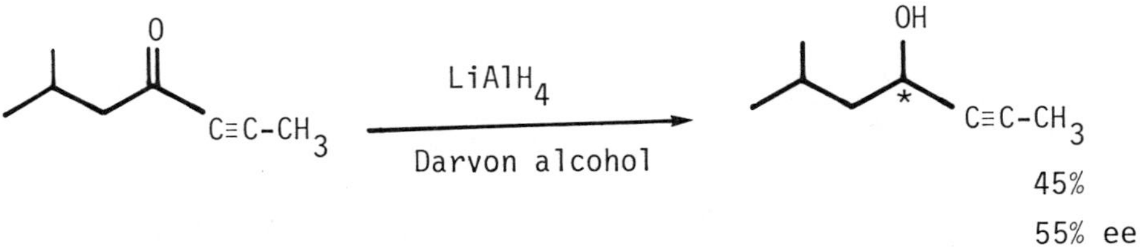
O
C≡C-CH3
LiAlH4
Darvon alcohol
OH
*
C≡C-CH3
45%
55% ee

JOC, 45, 582 (1980)

Using a chiral binaphthyl, Tetr Lett, 22, 247 (1981)

Section 303 <u>Acetylene - Aldehyde</u>

$$CH_2-CH-CH(OEt)_2 \quad \xrightarrow[H_2O/pentane]{\overset{\oplus \ominus}{Bu_4N \ OH}} \quad H-C\equiv C-CH(OEt)_2 \qquad 67\%$$
$$\underset{Br \ \ Br}{}$$

Org Syn, <u>59</u>, 10 (1980)

Section 304 <u>Acetylene - Amide</u>

No additional examples.

Section 305 <u>Acetylene - Amine</u>

$$Ph_2NH \quad \xrightarrow[NaOH/K_2CO_3, \ PTC]{BrCH_2C\equiv C-CH_2CH_3} \quad Ph_2N-CH_2C\equiv C-CH_2CH_3 \qquad 70\%$$

JOC (USSR), <u>18</u>, 848 (1982)

$$Hx-CH=CF_2 \quad \xrightarrow[Et_2O]{LDA} \quad Hx-C\equiv C-N(\underline{i}-Pr)_2 \qquad 70\%$$

Chem Lett, 935 (1980)

$$\begin{array}{c} O \\ \parallel \\ CH_3(CH_2)_{14}C-Cl \end{array}$$

+

$$\begin{array}{c} SnBu_3 \\ | \\ C \\ ||| \\ C \\ | \\ NPh_2 \end{array}$$

$$\xrightarrow{\hspace{2cm}}$$

$$\begin{array}{c} O \\ \parallel \\ C-(CH_2)_{14}CH_3 \\ | \\ C \\ ||| \\ C \\ | \\ NPh_2 \end{array} \qquad 88\%$$

Liebigs Ann Chem, 1907 (1981)

Section 306 Acetylene - Ester

$$C_5H_{11}-C\equiv CH \xrightarrow[\text{NaOAc, MeOH}]{\substack{CO \\ PdCl_2, \; CuCl_2}} C_5H_{11}-C\equiv C-CO_2Me \qquad 74\%$$

Tetr Lett, 21, 849 (1980)

$$Oct-C\equiv C-\overset{O}{\overset{\|}{C}}CH_2CH_2\overset{O}{\overset{\|}{C}}-OMe \xrightarrow[\text{2) NaOH, MeOH}]{\text{1) }\alpha\text{-pinylborane}}$$

Oct-C≡C

70-75%

85-90% ee

JOC, 46, 4107 (1981)

Section 307 Acetylene - Ether

$$H_2C=C=C\overset{I}{\underset{OMe}{\big\langle}}$$

+

(BuCuBr)MgCl·LiBr

$$\xrightarrow{\text{THF/HMPT}} Bu-CH_2-C\equiv C-OMe \qquad 80\%$$

JOC, 45, 1158 (1980)

Et-C≡C-Li   1) THF

+

2) CH_3I, THF/DMSO

OCH_3

C≡C-Et   92%

Synthesis, 459 (1981)

Section 308 <u>Acetylene - Halide</u>

Synth Comm, <u>10</u>, 345 (1980)

$$H-C\equiv C-CH_2OH \xrightarrow[\text{HMPT}]{Bu_3PI_2} H-C\equiv C-CH_2I \qquad 60\%$$

Aust J Chem, <u>35</u>, 517 (1982)

Section 309 <u>Acetylene - Ketone</u>

$$Ph-\overset{O}{\overset{\|}{C}}-CH_2-\overset{O}{\overset{\|}{C}}-CH_3 \xrightarrow{KF, Et_2NCF_2CHFCF_3} Ph-C\equiv C-\overset{O}{\overset{\|}{C}}-CH_3 \qquad 72\%$$

Chem Lett, 1327 (1980)

Bull Acad USSR Chem, <u>30</u>, 918 (1981)

Ph-I

          $\xrightarrow[\text{CO}]{\text{PdCl}_2\text{(dppf)}}$ $\overset{\overset{\text{O}}{\|}}{\text{Ph-C}}\text{-C}\equiv\text{C-Ph}$ ~85%

+

H-C≡C-Ph

JCS Chem Comm, 333 (1981)

$\text{Me}_2\text{CH-}\overset{\overset{\text{O}}{\|}}{\text{C}}\text{-Cl}$ $\xrightarrow[\text{(Ph}_3\text{P)}_2\text{PdCl}_2]{\text{Bu}_3\text{SnC}\equiv\text{C-CH(OEt)}_2}$ $\text{Me}_2\text{CH-}\overset{\overset{\text{O}}{\|}}{\text{C}}\text{-C}\equiv\text{C-CH(OEt)}_2$ 70%

JOC, <u>47</u>, 2549 (1982)

+

AcBr

70%

Bull Acad USSR Chem, <u>29</u>, 418 (1980)

Bu-C≡CAlMe$_2$

1) Ni(acac)$_2$
   DIBAL-H

2)

3) H$_3$O$^{\oplus}$

71%

JOC, <u>45</u>, 3053 (1980)

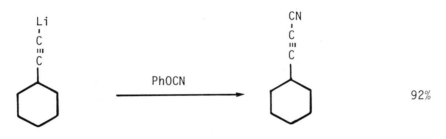

Liebigs Ann Chem, 1907 (1981)

Section 310 Acetylene - Nitrile

PhOCN

92%

Synthesis, 150 (1980)

Section 311 Acetylene - Olefin

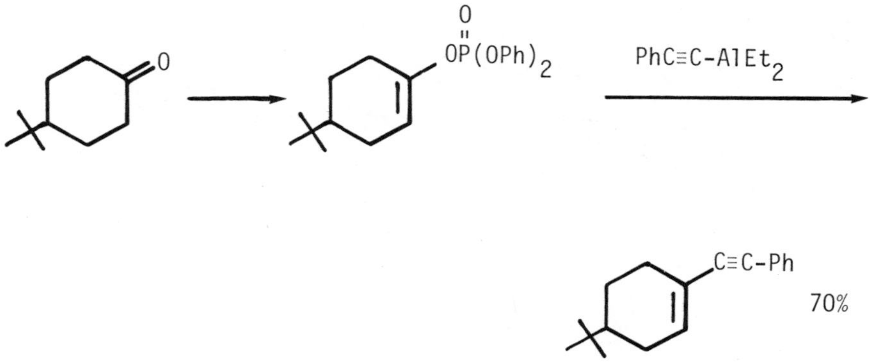
O
OP(OPh)2
PhC≡C-AlEt2
C≡C-Ph
70%

Tetr Lett, 21, 2531 (1980)

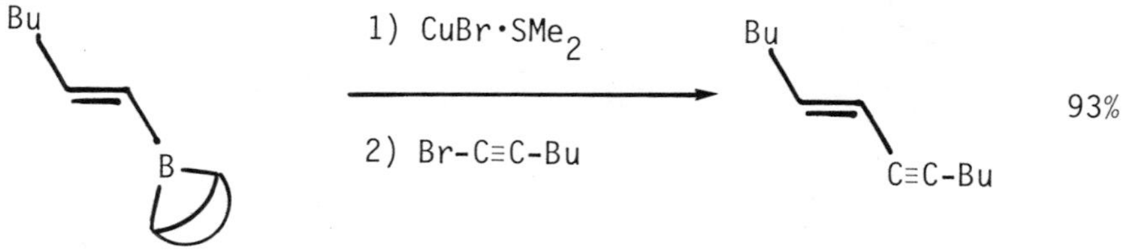
Bu
B
1) CuBr·SMe2
2) Br-C≡C-Bu
Bu
C≡C-Bu
93%

JOC, 46, 645 (1981)

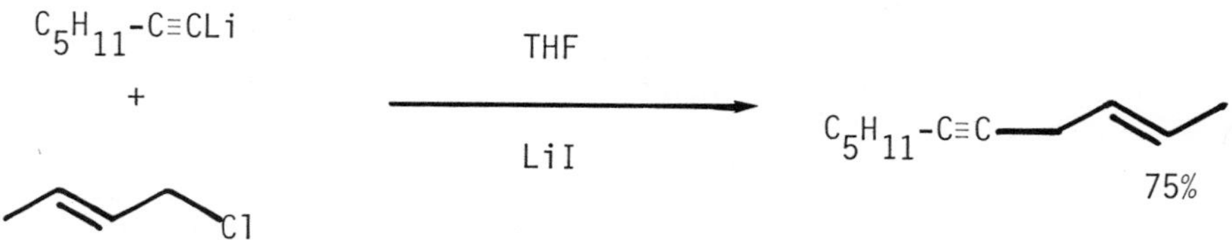
C5H11-C≡CLi
+
Cl
THF
LiI
C5H11-C≡C
75%

Chem Lett, 669 (1980)

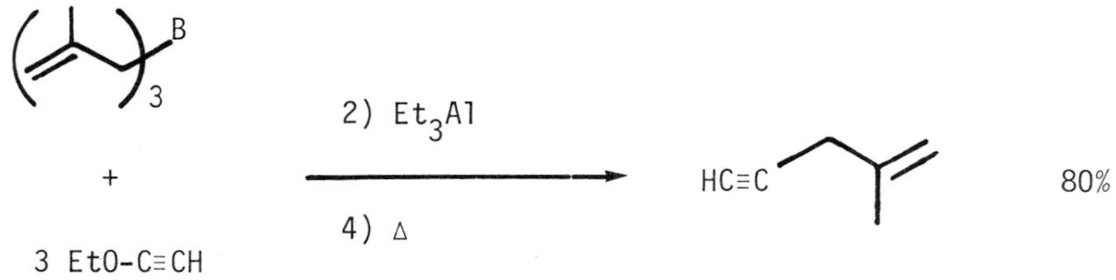

$$\left(\diagdown\diagup \right)_3 B$$

+

3 EtO-C≡CH

2) Et$_3$Al

4) Δ

HC≡C ⟶⟨⟩ 80%

Synthesis, 904 (1980)

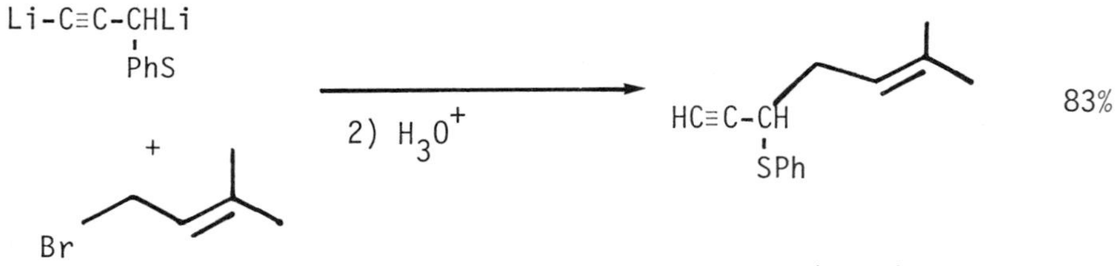

Li-C≡C-CHLi
    PhS

+

Br ⟶⟨⟩

2) H$_3$O$^+$

HC≡C-CH
      SPh   83%

JOC, 46, 5041 (1981)

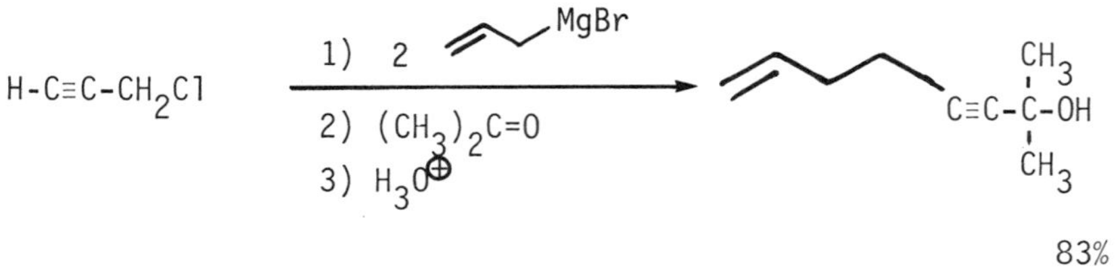

H-C≡C-CH$_2$Cl

1) 2 ⟶ MgBr

2) (CH$_3$)$_2$C=O

3) H$_3$O$^\oplus$

⟶ C≡C-C-OH with CH$_3$, CH$_3$

83%

Angew Chem Int Ed, 21, 286 (1982)

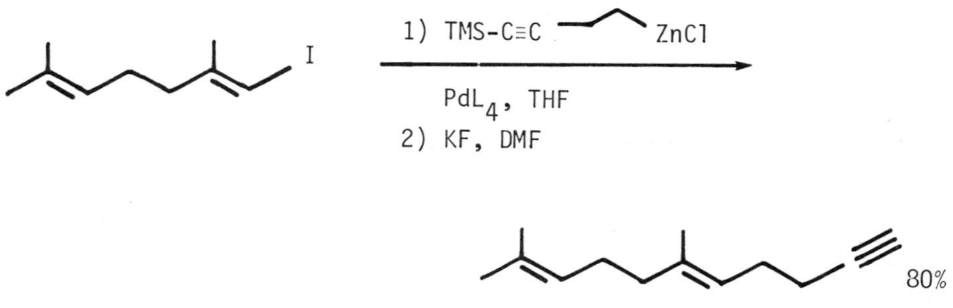

1) TMS-C≡C ⟍⟋ ZnCl

PdL$_4$, THF

2) KF, DMF

80%

JACS, 102, 3298 (1980)

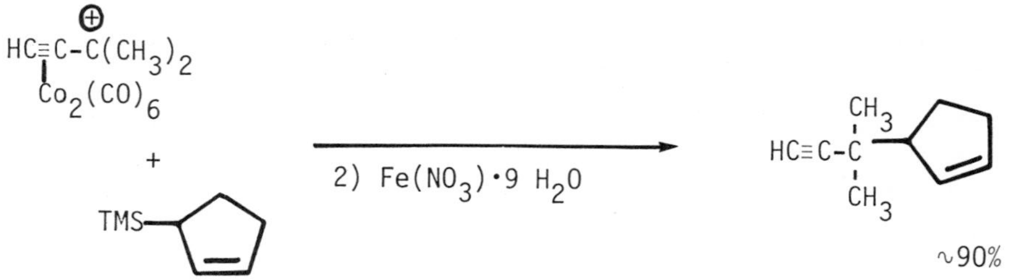

HC≡C-C(CH$_3$)$_2$
⊕
Co$_2$(CO)$_6$

+

TMS-

2) Fe(NO$_3$)·9 H$_2$O

CH$_3$
HC≡C-C
CH$_3$

∿90%

Tetr Lett, 21, 1595 (1980)

Section 312 Carboxylic Acid - Carboxylic Acid

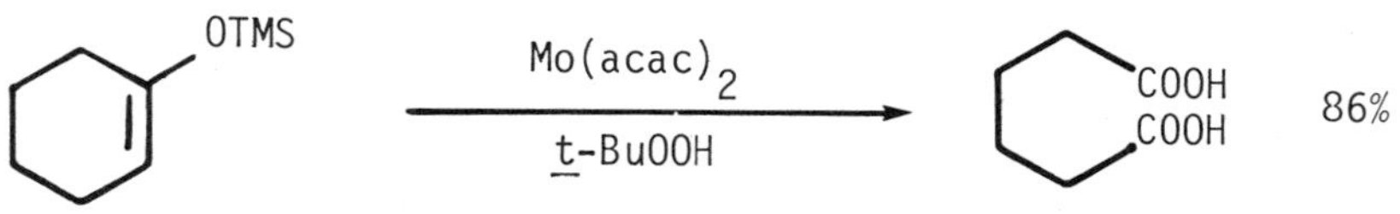

OTMS

Mo(acac)$_2$
─────────
t-BuOOH

COOH
COOH

86%

Tetr Lett, 22, 2595 (1981)

LiCH-COOLi
  |
  Bu

   + $\xrightarrow{\quad\quad\quad\quad}$ Et
                       |

$\xrightarrow[\text{2) } H_3O^+]{}$ HOOC-CH-CH-COOH   60%
                     |
                     Bu

Br-CH-COOLi
  |
  Et

Synthesis, 710 (1980)

$\underset{\underset{\displaystyle Cl-\overset{O}{\overset{\|}{C}}-(CH_2)_n-\overset{O}{\overset{\|}{C}}-Cl}{}}{}$ $\xrightarrow[\substack{\text{3) HCl, heat}}]{\substack{\text{1) Meldrum's acid, DMAP} \\ \text{2) NaCNBH}_3\text{, HOAc/THF}}}$ $HO-\overset{O}{\overset{\|}{C}}-(CH_2)_{n+4}-\overset{O}{\overset{\|}{C}}-OH$

~50% overall

Synth Comm, 12, 19 (1982)

Section 313 Carboxylic Acid - Alcohol

Ph-CH$_2$COOMe $\xrightarrow[\substack{\text{2) KOH, H}_2\text{O} \\ \text{3) H}^+}]{\text{1) PhI(OAc)}_2}$ Ph-CH-COOH   50 %
                                         |
                                       OH

Tetr Lett, 22, 2747 (1981)

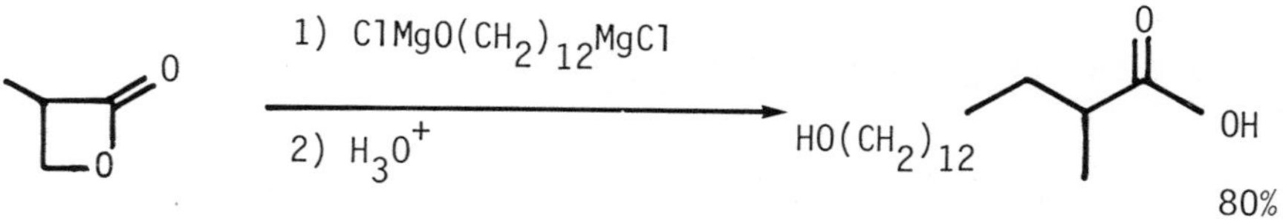
O
O
1) ClMgO(CH2)12MgCl
2) H3O+
HO(CH2)12
O
OH
80%

Chem Lett, 569 (1982)

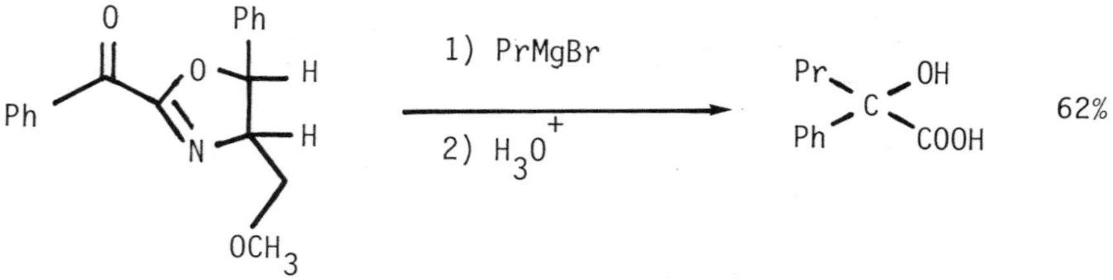
O
Ph
Ph
O
H
N
H
OCH3
1) PrMgBr
2) H3O+
Pr
OH
C
Ph
COOH
62%

JOC, 45, 2785 (1980)

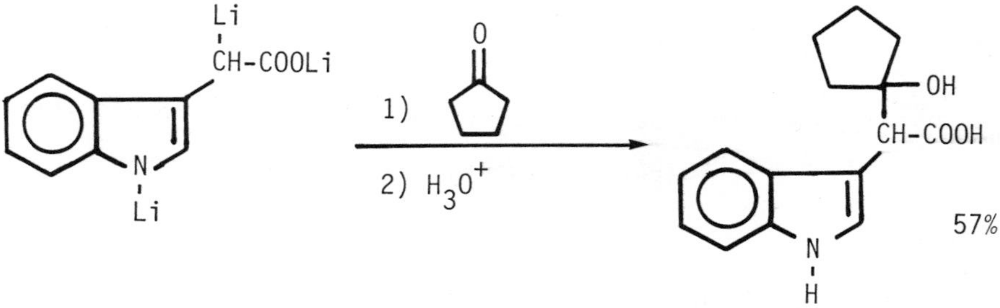
Li
CH-COOLi
N
Li
O
1)
2) H3O+
OH
CH-COOH
N
H
57%

JOC, 45, 447 (1980)

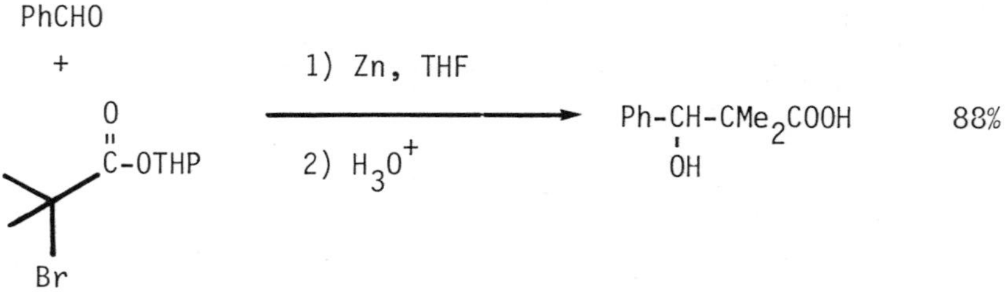
PhCHO
+
O
C-OTHP
Br
1) Zn, THF
2) H3O+
Ph-CH-CMe2COOH
OH
88%

Bull Soc Chim France II, 145 (1980)

Section 314 Carboxylic Acid - Aldehyde

No additional examples.

Section 315 Carboxylic Acid - Amide

$$RCH=C\begin{smallmatrix} NHAc \\ \\ COOH \end{smallmatrix} \xrightarrow[\text{chiral Rh catalyst}]{H_2} RCH_2-\overset{*}{\underset{COOH}{CH}}\overset{NHAc}{}$$

JOC, 45, 5187 (1980)

Chem Ber, 113, 2323 (1980)

JACS, 102, 988 (1980)

JACS, 103, 2273 (1981)

JOC, 46, 2954 and 2960 (1981)

Synthesis, 76 (1981)

Related methods: Section 316 (Acid-Amine); Section 344 (Amide-

Ester); Section 351 (Amine - Ester)

Section 316 Carboxylic Acid - Amine

$$\begin{smallmatrix} H_2N \\ \\ H_3C \end{smallmatrix}C(COO^{\ominus})_2 \xrightarrow{\text{chiral Co(III) complex}} H_3C-\overset{NH_2}{\underset{*}{CH}}-COO^{\ominus}$$

88% yield

66% ee

JACS, 103, 2459 (1981)

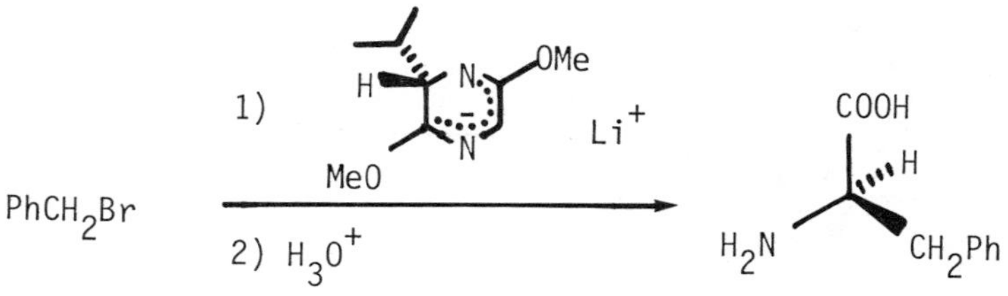

~60%

91% ee

Angew Chem Int Ed, 20, 798 (1981)

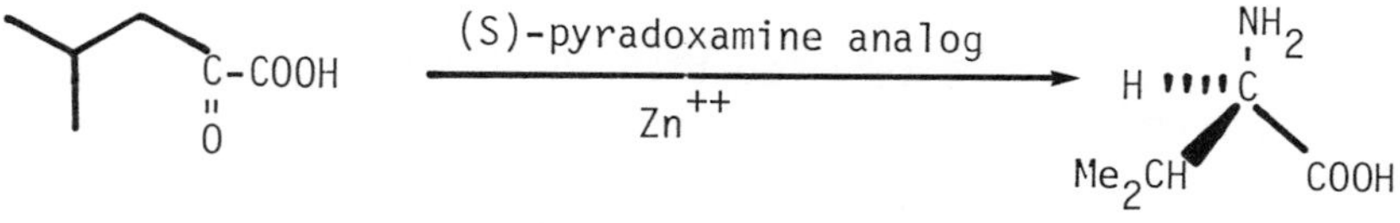

68%

96% ee

Chem Lett, 1765 and 1769 (1982)

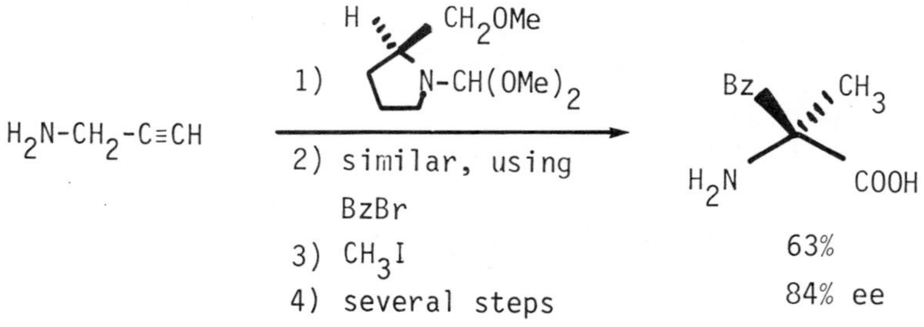

63%

84% ee

Angew Chem Int Ed, 19, 725 (1980)

Tetrahedron, <u>36</u>, 2961 (1980)

Related methods: Section 315 (Acid-Amide); Section 344 (Amide-
Ester); Section 351 (Amine-Ester)

Section 317 <u>Carboxylic Acid - Ester</u>

No additional examples.

Section 318 <u>Carboxylic Acid - Ether, Epoxide</u>

No additional examples.

Section 319 <u>Carboxylic Acid - Alkyl Halide</u>

Org Syn, <u>59</u>, 20 (1980)

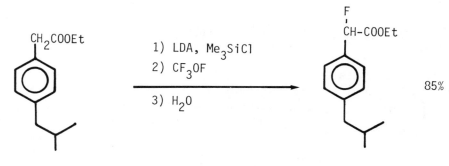

JACS, <u>102</u>, 4845 (1980)

Synthesis, 963 (1982)

Chem Lett, 1255 (1982)

Section 320 <u>Carboxylic Acid - Ketone</u>

Tetr Lett, <u>22</u>, 2595 (1981)

Chem Lett, 1505 (1982)

Tetr Lett, <u>23</u>, 271 (1982)

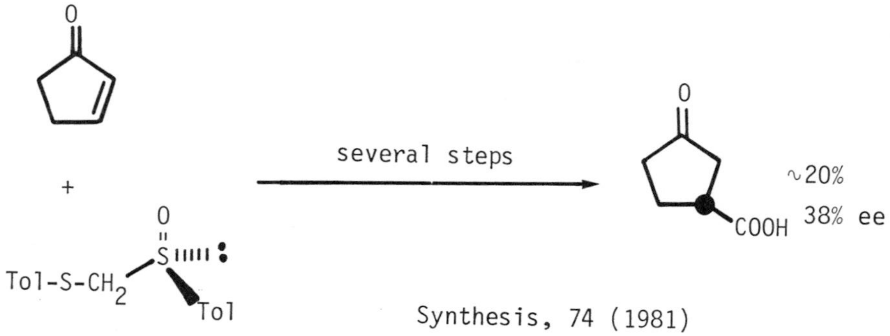
O
+
O
Tol-S-CH2
Tol
several steps
O
COOH
~20%
38% ee

Synthesis, 74 (1981)

Also via: Ketoesters (Section 360)

Section 321 Carboxylic Acid - Nitrile

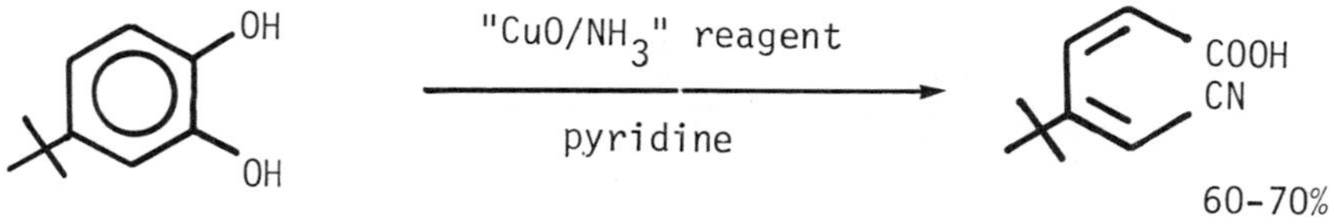
OH
OH
"CuO/NH3" reagent
pyridine
COOH
CN
60-70%

JOC, 45, 2737 (1980)

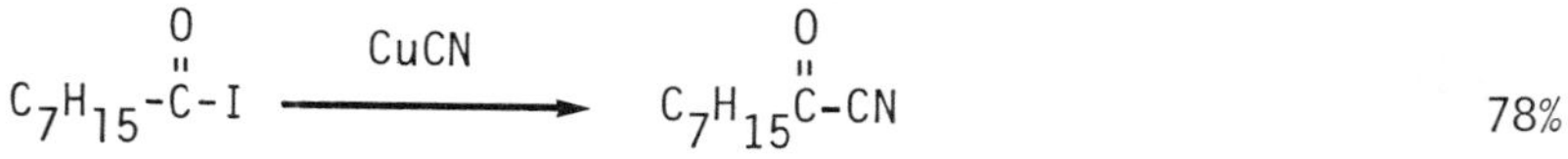
O
C7H15-C-I
CuCN
O
C7H15C-CN
78%

Angew Chem Int Ed, 21, 83 (1982)

$$\text{Bu}_3\text{SnCN}$$

92%

Tetr Lett, <u>21</u>, 2959 (1980)

$$\text{Ph-I} \xrightarrow[\text{PhPdIL}_2]{\text{CO, KCN}} \text{Ph}\overset{\text{O}}{\overset{\|}{\text{C}}}\text{-CN}$$

83%

Bull Chem Soc Japan, <u>54</u>, 637 (1981)

$$\text{Ph-}\overset{\text{O}}{\overset{\|}{\text{C}}}\text{-CHO}$$

or

$$\text{Ph-}\overset{\text{O}}{\overset{\|}{\text{C}}}\text{-CH}_2\text{Br}$$

$$\text{Ph-}\overset{\text{O}}{\overset{\|}{\text{C}}}\text{-CN}$$

97%

Synthesis, 844 (1980)

Review: "The Chemistry of Acyl Cyanides"

Angew Chem Int Ed, <u>21</u>, 36 (1982)

Also via: Section 361 (Cyanoesters)

Section 322 <u>Carboxylic Acid - Olefin</u>

Ph-CHO → (MeO)$_2$PH, NaOMe / ClCH$_2$COOH, MeOH → Ph-CH=CH-COOH 89%

JOC, <u>46</u>, 2514 (1981)

CH$_2$=CH-COOH
+
Ph$_4$Sn

→ H$_2$PtCl$_6$ / CH$_3$COOH → Ph-CH=CH-COOH 86%

Bull Acad USSR Chem, <u>30</u>, 2211 (1982)

→ sodium chlorite / NaH$_2$PO$_4$, H$_2$O/<u>t</u>-BuOH → COOH 90%

Tetrahedron, <u>37</u>, 2091 (1981)

TMS-CH$_2$-C-TMS
(O)

→ 1) LDA, MeI 2) LDA, BuCHO 3) H$^+$ → COOH ∿70%
Bu   Me

JACS, <u>103</u>. 6217 (1981)

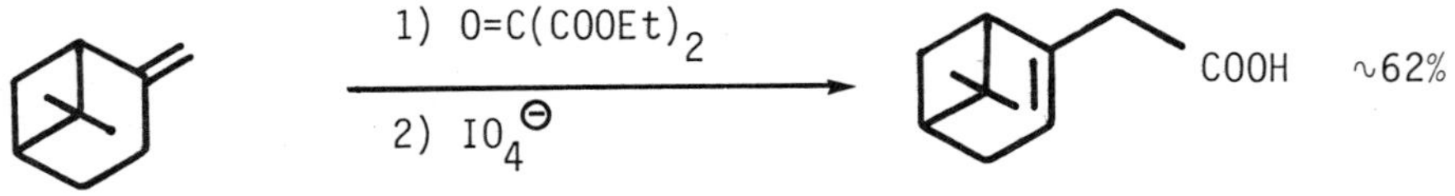

1) O=C(COOEt)$_2$

2) IO$_4^{\ominus}$

COOH   ~62%

JACS, 102, 2473 (1980)

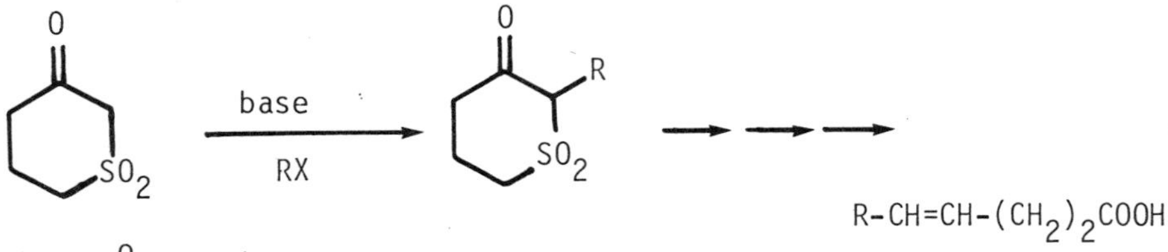

base

RX

(R = 1° alkyl)

R-CH=CH-(CH$_2$)$_2$COOH

Chem Ber, 114, 909 (1981)

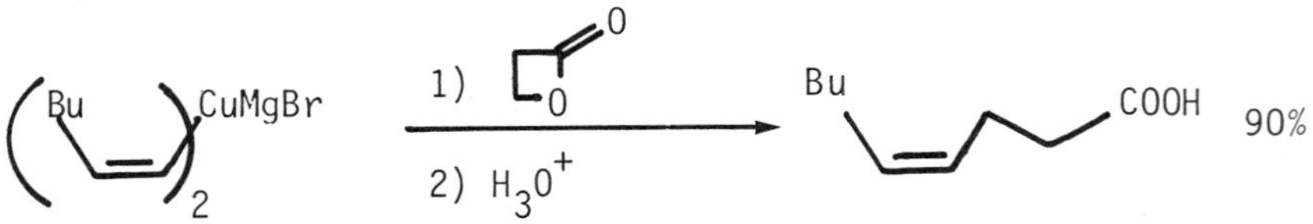

1)

2) H$_3$O$^+$

Bu   COOH   90%

Chem Lett, 1123 (1980)

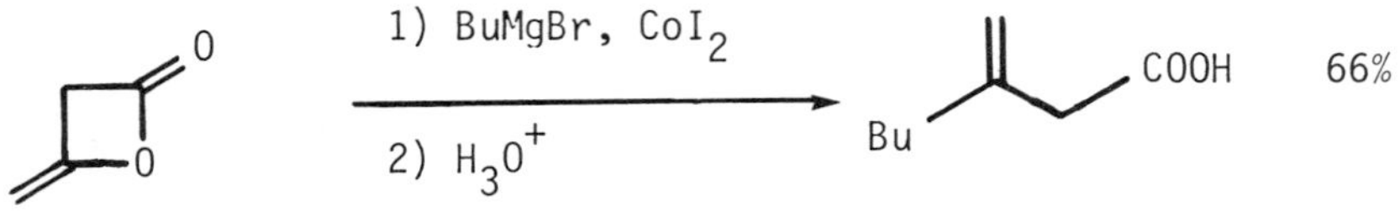

1) BuMgBr, CoI$_2$

2) H$_3$O$^+$

Bu   COOH   66%

Bull Chem Soc Japan, 55, 3555 (1982)

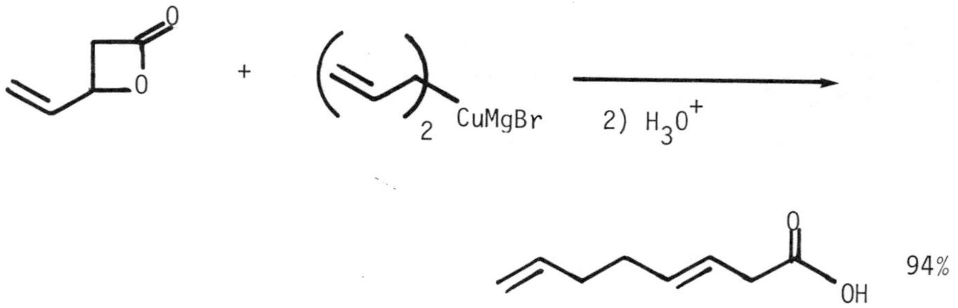

94%

Tetr Lett, _22_, 1817 (1981)

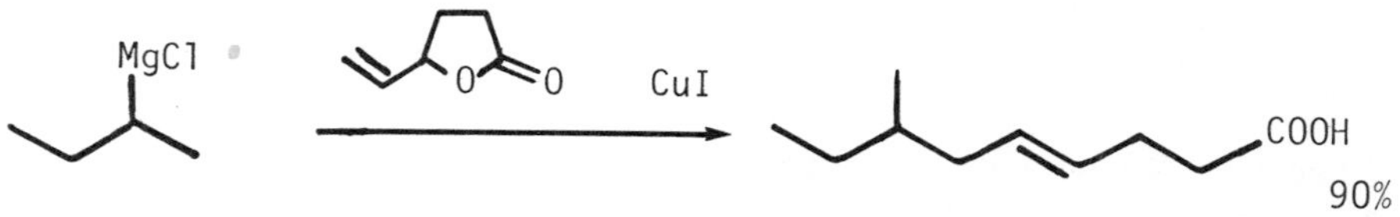

90%

Tetr Lett, _23_, 3583 (1982)

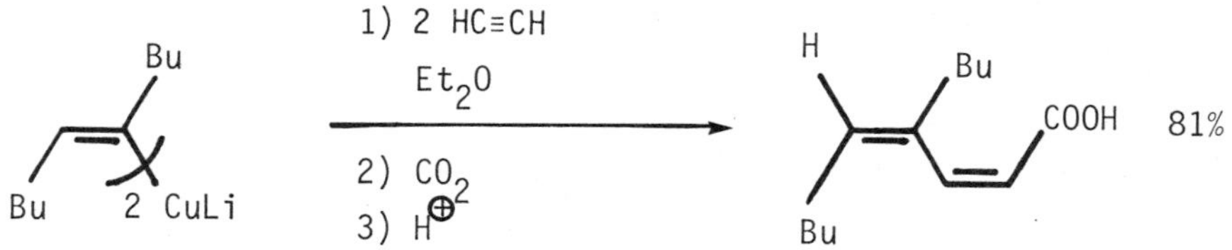

81%

Tetr Lett, _23_, 5151 (1982)

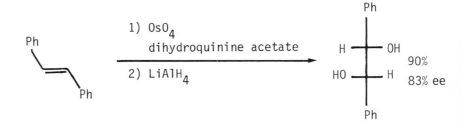

Chem Lett, 1151 (1982)

Also via: Hydroxy acids (Section 313); Olefinic amides (Section 349); Olefinic esters (Section 362); Olefinic nitriles (Section 376)

Section 323 Alcohol - Alcohol

$$
\begin{array}{c}
\text{Ph} \\
\diagdown \\
\diagup \\
\text{Ph}
\end{array}
\xrightarrow[\text{2) LiAlH}_4]{\begin{array}{c}\text{1) OsO}_4 \\ \text{dihydroquinine acetate}\end{array}}
\begin{array}{c}
\text{Ph} \\
\text{H} \!-\!\!\!\!|\!-\! \text{OH} \\
\text{HO} \!-\!\!\!\!|\!-\! \text{H} \\
\text{Ph}
\end{array}
\quad
\begin{array}{c}90\% \\ 83\% \text{ ee}\end{array}
$$

JACS, 102, 4263 (1980)

OsO$_4$/Et$_3$NO

pyridine

CH$_2$CH$_2$OMEM

CH$_2$CH$_2$MEM

OH

OH

Tetr Lett, 21, 449 (1980)

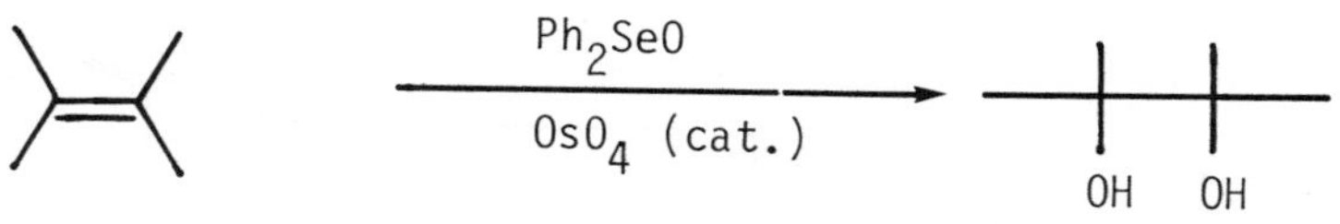
Ph2SeO
OsO4 (cat.)
OH OH

98%

Tetr Lett, 22, 2051 (1981)

Review: "Osmium Tetroxide Cis-Hydroxylation of Unsaturated Substrates"

Chem Rev, 80, 187 (1980)

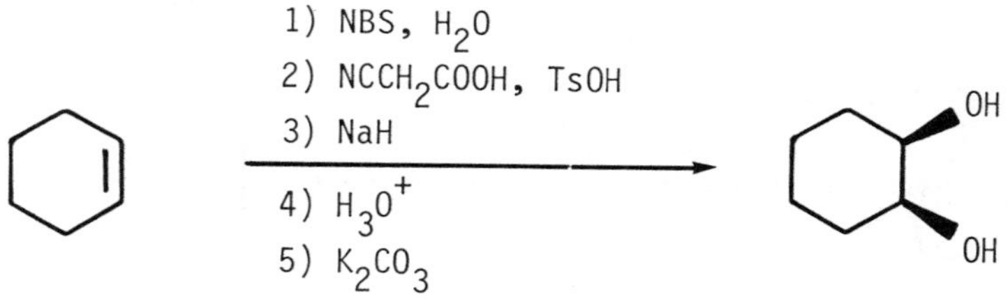
1) NBS, H2O
2) NCCH2COOH, TsOH
3) NaH
4) H3O+
5) K2CO3
OH
OH

~70%

Tetr Lett, 23, 4217 (1982)

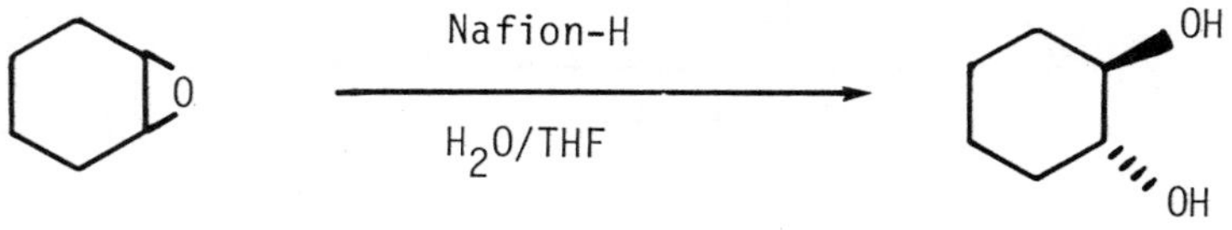
O
Nafion-H
H2O/THF
OH
OH

73%

Synthesis, 280 (1981)

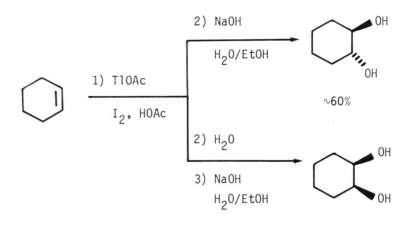

Org Syn, <u>59</u>, 169 (1980)

$$\text{LiEt}_3\text{BH} \longrightarrow \quad 94\%$$

JOC, <u>45</u>, 1 (1980)

Ph-C≡C-CH₃

1) 2 BuLi
2) BH₃·THF
3) NaOH, H₂O₂

70%

Tetrahedron, <u>36</u>, 299 (1980)

2 Ph-CHO $\xrightarrow{\begin{array}{c}\text{TiCl}_3,\ \text{NaOH}\\ \underline{\text{or}}\ \text{Fe(CO)}_5,\ \text{pyridine}\end{array}}$ Ph-C—C-Ph　　88%

$$\begin{array}{cc}\text{OH} & \text{OH}\\ | & |\\ \text{Ph-C} & \text{C-Ph}\\ | & |\\ \text{H} & \text{H}\end{array}$$

Chem Lett, 1141 (1980)

Tetr Lett, **23**, 3517 (1982)

$$\underset{\text{Ph-}\overset{\text{O}}{\overset{||}{\text{C}}}\text{-COOH}}{}$$

+

$$\text{CH}_3\text{-}\overset{\text{O}}{\overset{||}{\text{C}}}\text{-CH}_3$$

$\xrightarrow{\text{TiCl}_3,\ \text{H}_2\text{O}}$

$$\begin{array}{cc}\text{COOH} & \text{CH}_3\\ | & |\\ \text{Ph-C} & \text{C-CH}_3\\ | & |\\ \text{OH} & \text{OH}\end{array}$$

85%

JOC, **47**, 2852 (1982)

$\xrightarrow{\text{Ce-I}_2}$

95%

Tetr Lett, **23**, 1353 (1982)

$\xrightarrow{\text{BrMg(CH}_2)_4\text{MgBr}}$

72%

JOC, **45**, 1828 (1980)

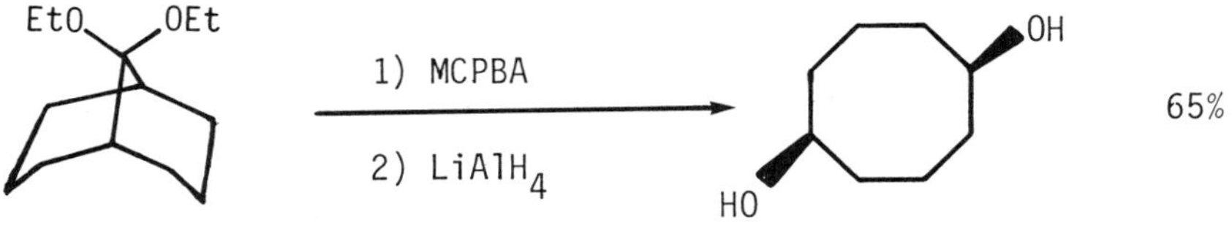

65%

JACS, 104, 1769 (1982)

Review: "Advances in Stereochemical Control: The 1,2- and 1,3-
Diol Systems"

Aldrichimica Acta, 15, 47 (1982)

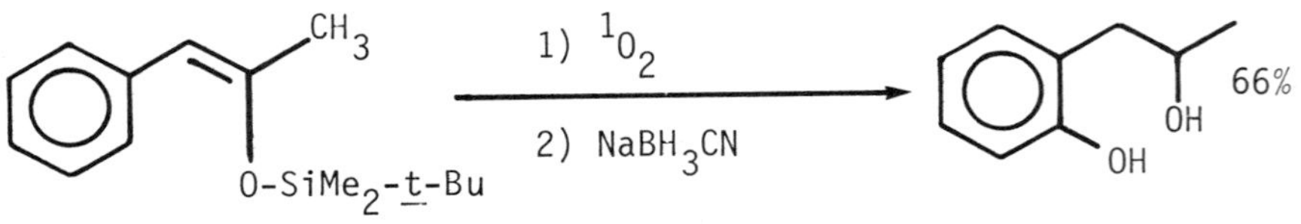

66%

Tetr Lett, 23, 1717 (1982)

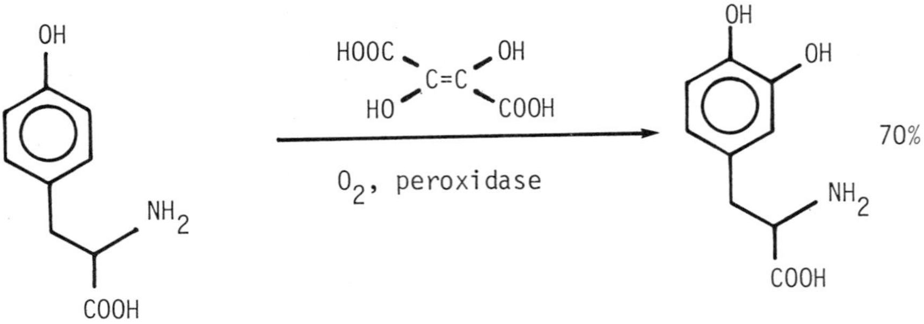

70%

JACS, 103, 6263 (1981)

Also via: Hydroxyesters (Section 327); Diesters (Section 357)

Section 324 Alcohol - Aldehyde

$$\text{(OH)} \quad (CH_2)_8CH_2OH \xrightarrow[\text{benzene}]{RuCl_2L_3} \text{(OH)} \quad (CH_2)_8CHO \qquad 89\%$$

Tetr Lett, 22, 1605 (1981)

$$\overset{O}{\bigcirc}(CH_2)_6COOH \xrightarrow[ErCl_3]{NaBH_4} \overset{OH}{\bigcirc}(CH_2)_6COOH \qquad 75\%$$
CHO   CHO

Tetr Lett, 22, 4077 (1981)

1) t-BuNH₂
   (protects the aldehyde)
2) Li(t-BuO)₃AlH

3) H₂O
4) basic alumina

83%

Tetrahedron, 38, 1827 (1982)

Review: "The Directed Aldol Reaction"

Org React, 28, 203 (1982)

Related methods: Alcohol - Ketone (Section 33)

Section 325 Alcohol - Amide

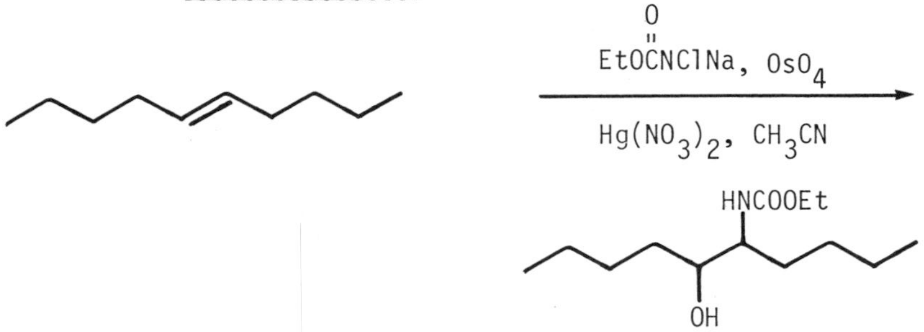

JOC, 45, 2710 (1980)

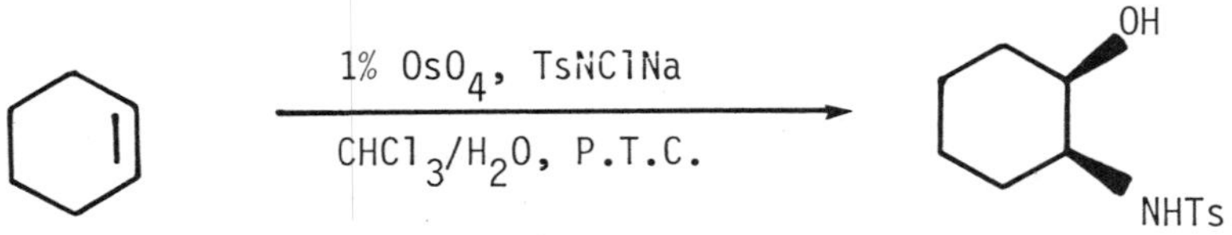

~80%

Org Syn, 61, 85 (1983)

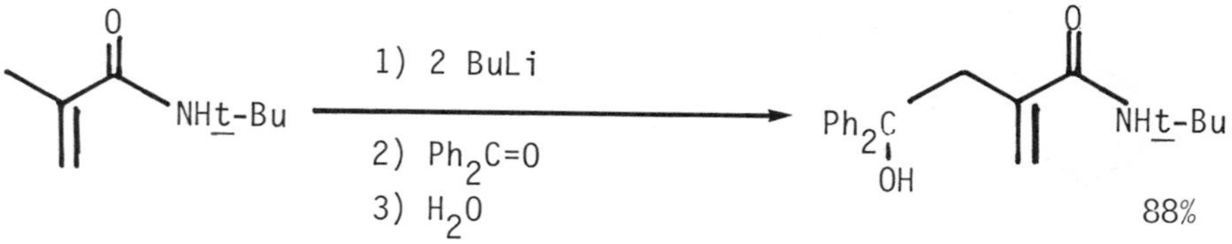

1) 2 BuLi

2) Ph$_2$C=O

3) H$_2$O

88%

JOC, 45, 4257 (1980)

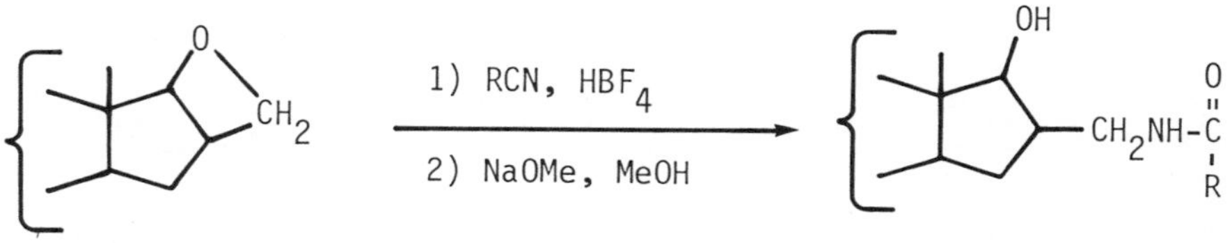

1) RCN, HBF$_4$

2) NaOMe, MeOH

Tetr Lett, 22, 341 (1981)

Section 326 Alcohol - Amine

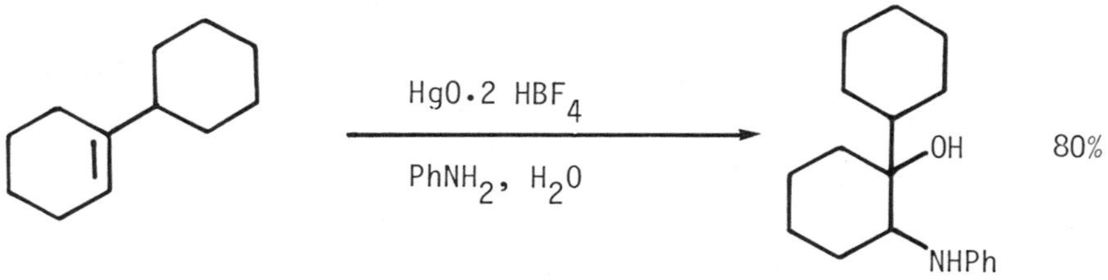

HgO·2 HBF$_4$

PhNH$_2$, H$_2$O

80%

Synthesis, 376 (1981)

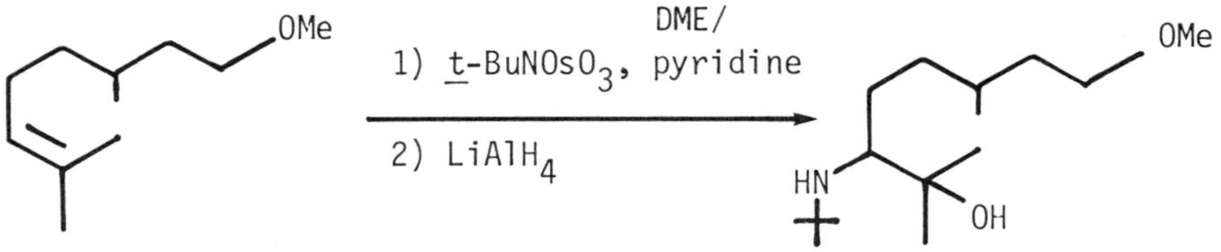
OMe
DME/
1) t-BuNOsO3, pyridine
2) LiAlH4
OMe
HN
OH

JOC, 45, 2257 (1980)

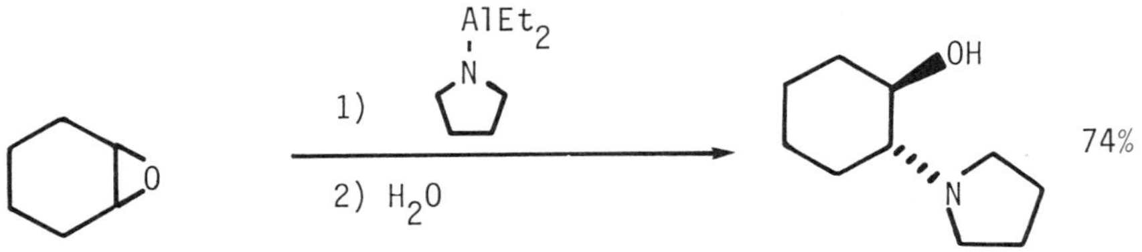
AlEt2
N
1)
2) H2O
O
OH
N
74%

Tetr Lett, 22, 195 (1981)

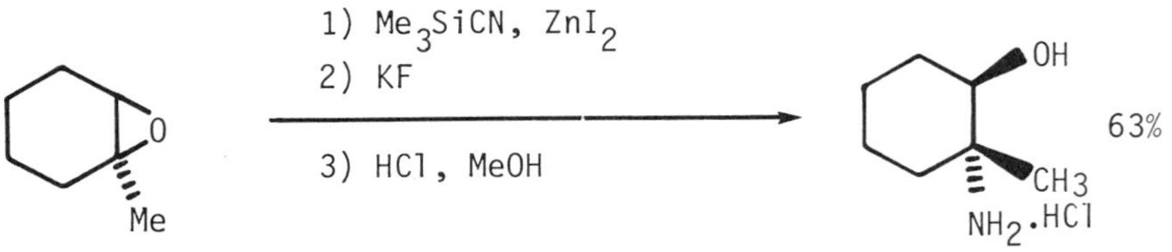
1) Me3SiCN, ZnI2
2) KF
3) HCl, MeOH
O
Me
OH
CH3
NH2·HCl
63%

JACS, 104, 5849 (1982)

Review: "Stereoselective Synthesis of Diastereomeric Amino Alcohols from Chiral Aminocarbonyl Compounds by Reduction or by Addition of Organometallic Reagents"

Synthesis, 605 (1982)

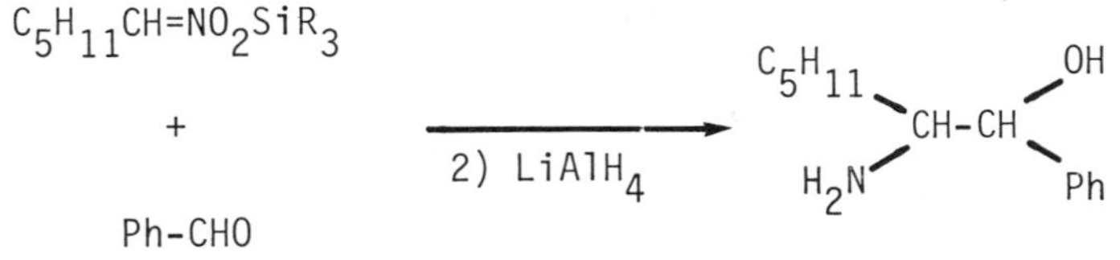
$C_5H_{11}CH=NO_2SiR_3$
+
Ph-CHO
2) $LiAlH_4$
C_5H_{11}
OH
CH-CH
H_2N
Ph

64%

Helv Chim Acta, 64, 2264 (1981)

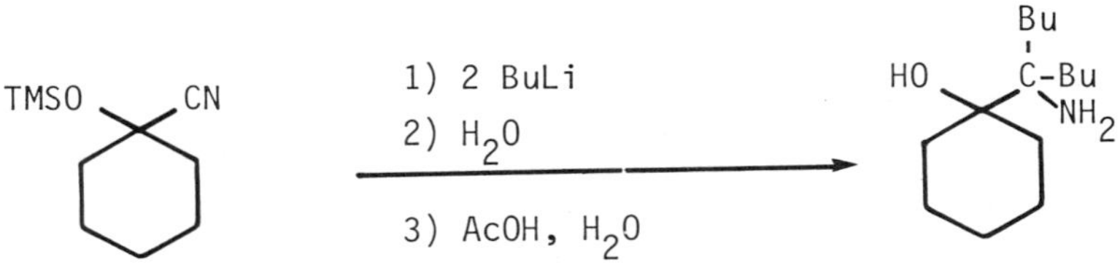
TMSO
CN
1) 2 BuLi
2) H_2O
3) AcOH, H_2O
Bu
HO
C-Bu
NH_2

92%

Synthesis, 270 (1981)

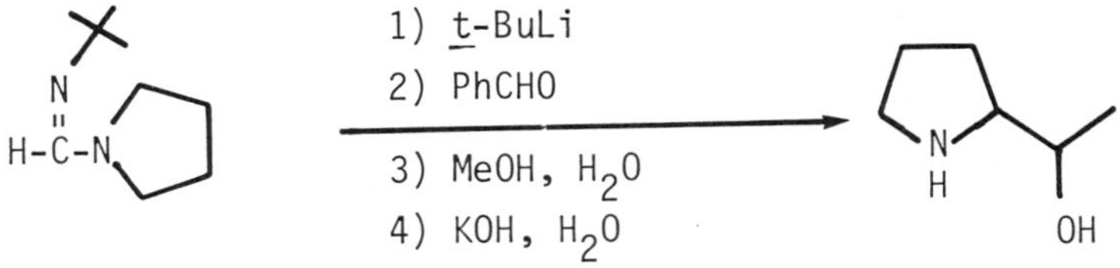
N
H-C-N
1) t-BuLi
2) PhCHO
3) MeOH, H_2O
4) KOH, H_2O
N
H
Ph
OH

57%

JACS, 102, 7125 (1980)

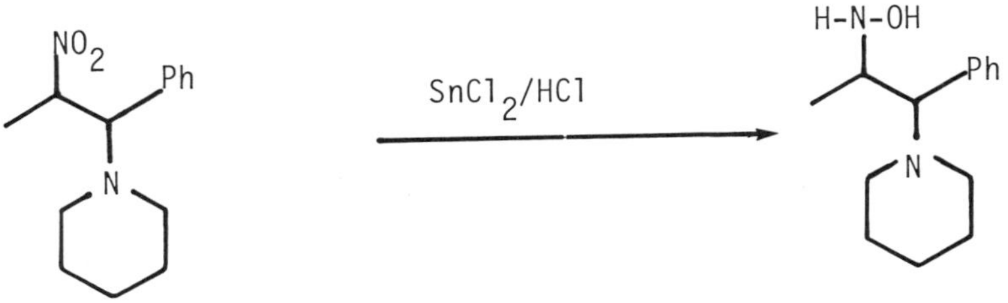
NO_2
Ph
N
$SnCl_2/HCl$
H-N-OH
Ph
N

90%

Synthesis, 599 (1982)

Section 327 <u>Alcohol - Ester</u>

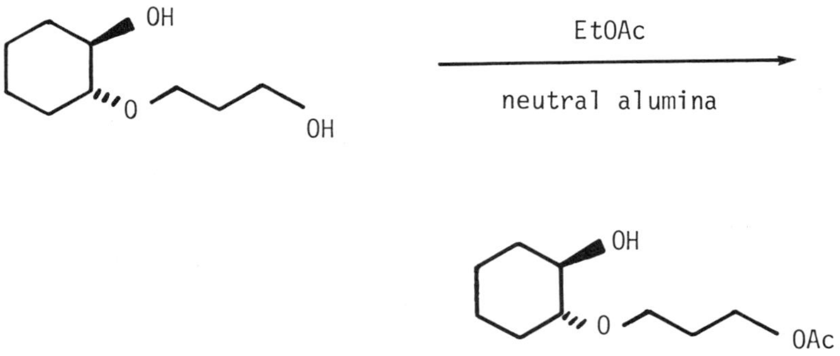

EtOAc

neutral alumina

71%

Synthesis, 789 (1981)
Tetr Lett, <u>22</u>, 5003 and 5007 (1981)

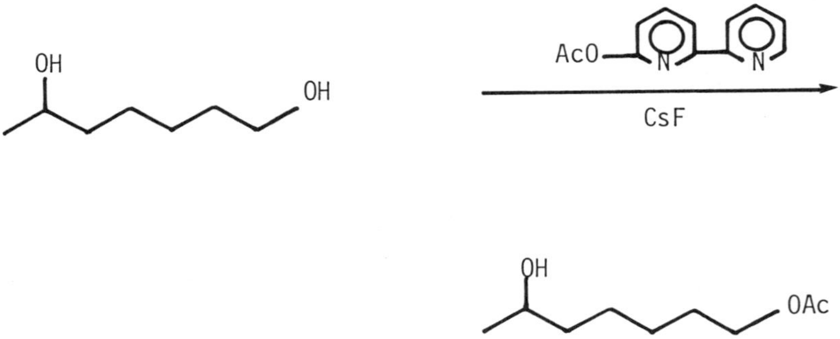

CsF

81%

Chem Lett, 563 (1981)

$$\text{CH}_2\text{OH} \text{ (para-OH benzene)} \xrightarrow[\text{BF}_3\cdot\text{Et}_2\text{O}]{\text{Ac}_2\text{O}} \text{CH}_2\text{OAc} \text{ (para-OH benzene)}$$

68%

Chem Pharm Bull, <u>29</u>, 3202 (1981)

$$\underset{\displaystyle Ph-\overset{\displaystyle O}{\overset{\|}{C}}-COOEt}{} \xrightarrow{\text{chiral dihydropyridine}} \underset{\displaystyle Ph-\overset{\displaystyle OH}{\overset{|}{C}H}-COOEt}{\overset{\displaystyle *}{}}$$

80%

93% ee

JACS, <u>103</u>, 2091 (1981)

JACS, <u>103</u>, 4613 (1981)

$$\text{CH}_3-\overset{O}{\overset{\|}{C}}-(\text{CH}_2)_3-\overset{O}{\overset{\|}{C}}-\text{OMe} \xrightarrow[\text{ether}]{\text{LiAlH}_4 \cdot \text{SiO}_2} \text{CH}_3-\overset{OH}{\overset{|}{C}H}(\text{CH}_2)_3-\overset{O}{\overset{\|}{C}}-\text{OMe}$$

84%

Tetr Lett, <u>23</u>, 4585 (1982)

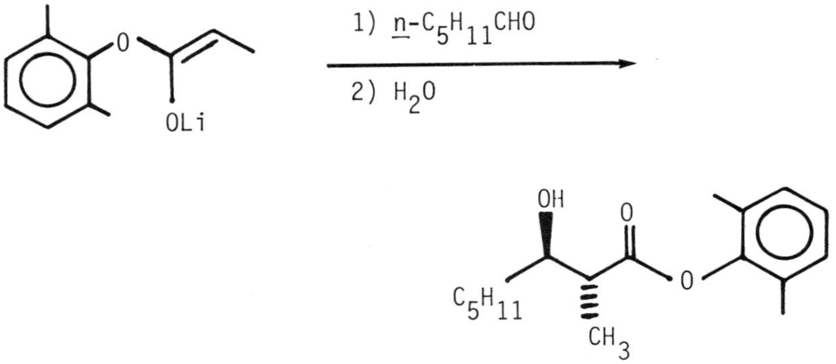

91%

JOC, **45**, 1727 (1980)

95%

80:20 erythro:threo

Chem Lett, 161 (1982)

84%

Synth Comm, **11**, 505 (1981)

Also via: Hydroxyacids (Section 313)

Section 328 Alcohol - Ether, Epoxide

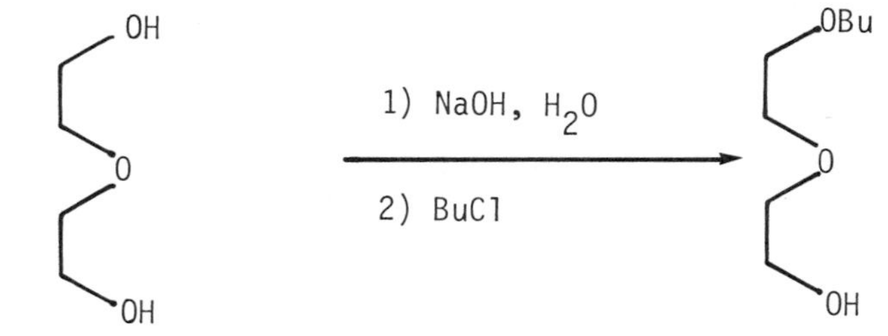
OH
O
OH
1) NaOH, H_2O
2) BuCl
OBu
O
OH

70%

JOC, 45, 1095 (1980)

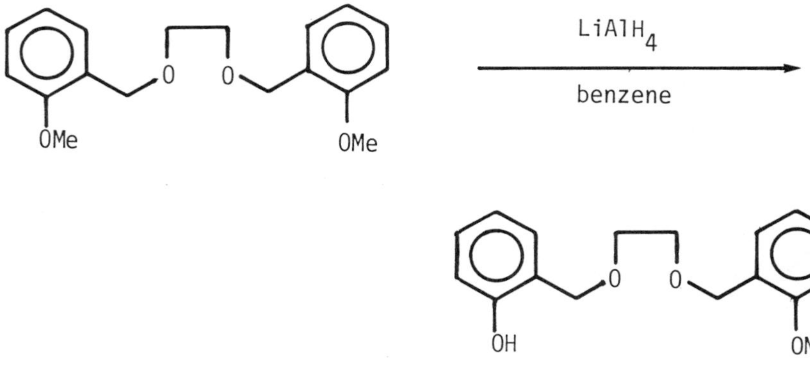
O
O
OMe
OMe
$LiAlH_4$
benzene
OH
OMe

67%

JCS Chem Comm, 507 (1980)

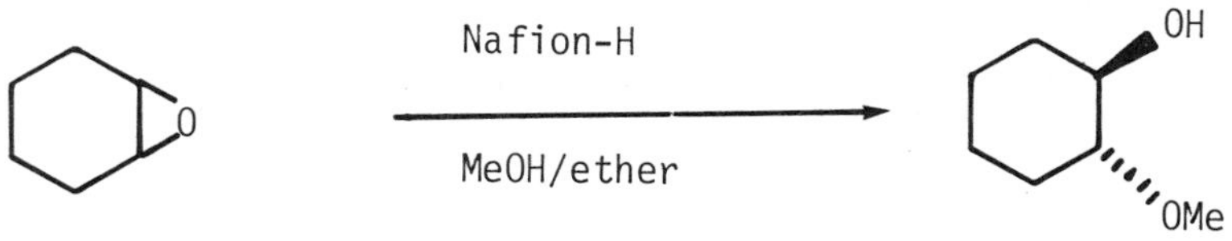
O
Nafion-H
MeOH/ether
OH
OMe

74%

Synthesis, 280 (1981)

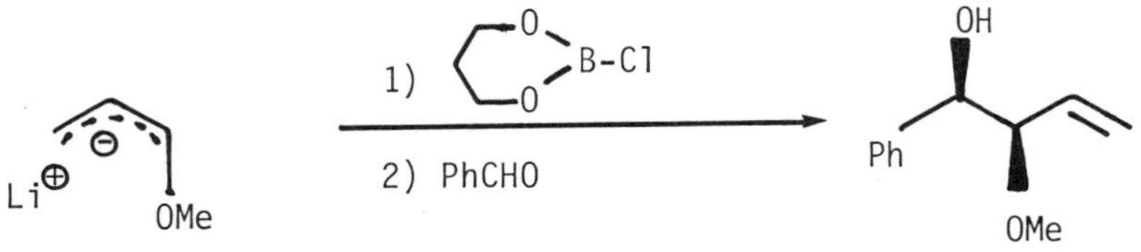

JOC, 47, 2498 (1982)

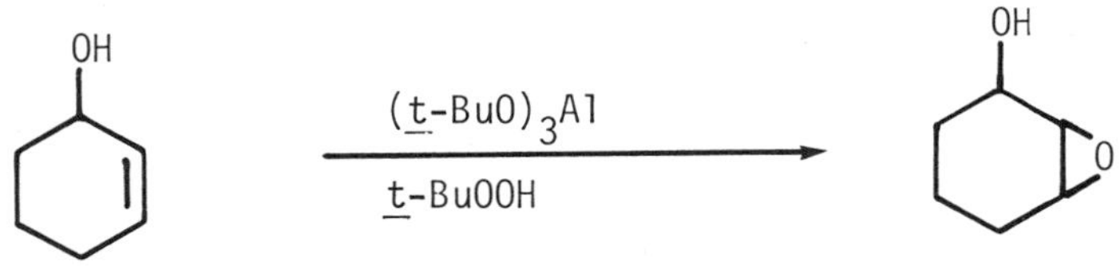

Tetr Lett, 21, 1657 (1980)

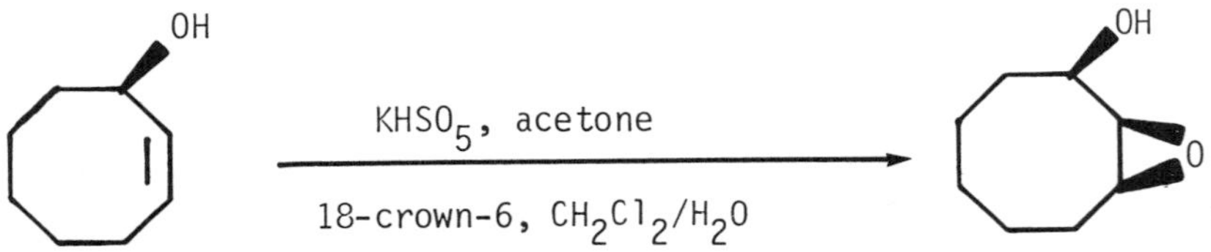

JOC, 47, 2670 (1982)

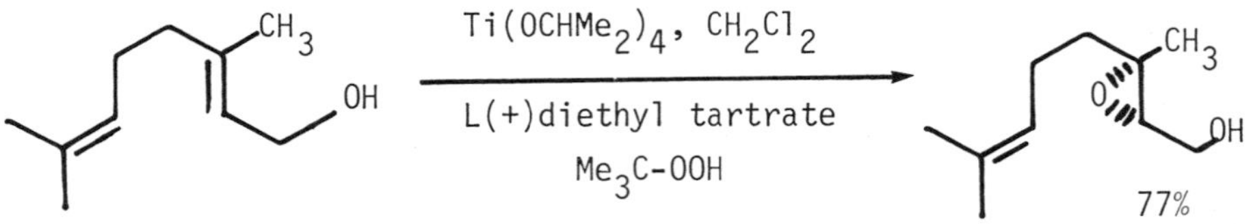

JACS, 102, 5974 (1980)

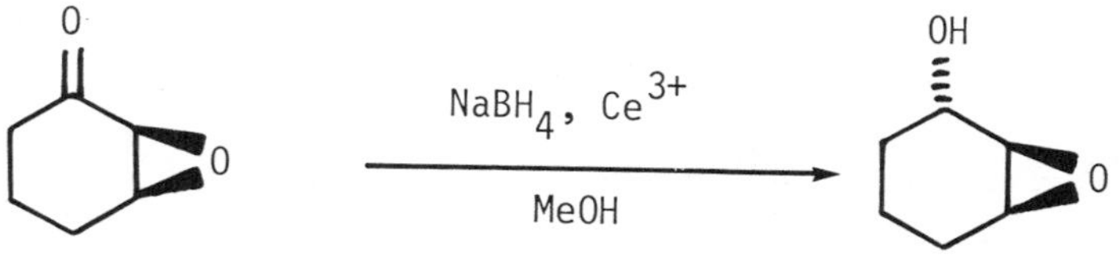

O
O
$NaBH_4$, Ce^{3+}
MeOH
OH
O

85%
98% trans

Synth Comm, 10, 623 (1980)

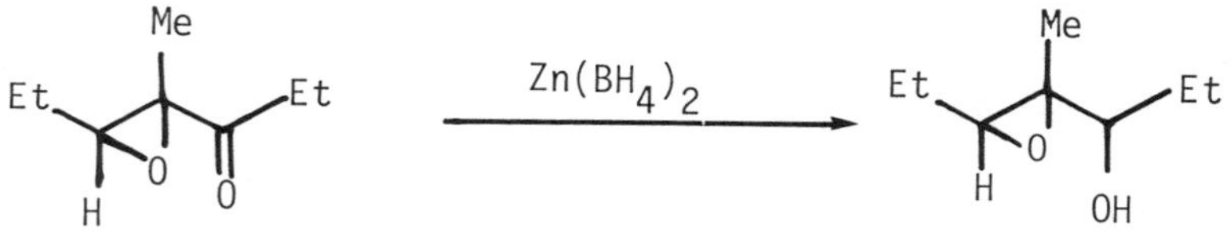

Me
Et
Et
O
H
O
$Zn(BH_4)_2$
Me
Et
Et
O
H
OH

83%
99% erythro

Tetr Lett, 22, 4723 (1981)

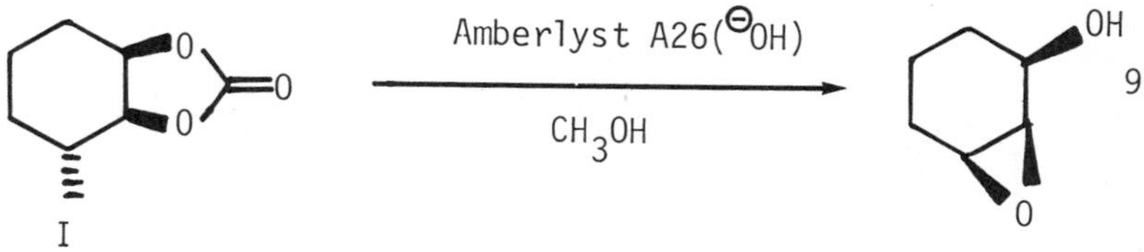

O
O
O
I
Amberlyst A26(⊖OH)
CH_3OH
OH
O

90%

JOC, 47, 4626 (1982)

Section 329 Alcohol - Halide

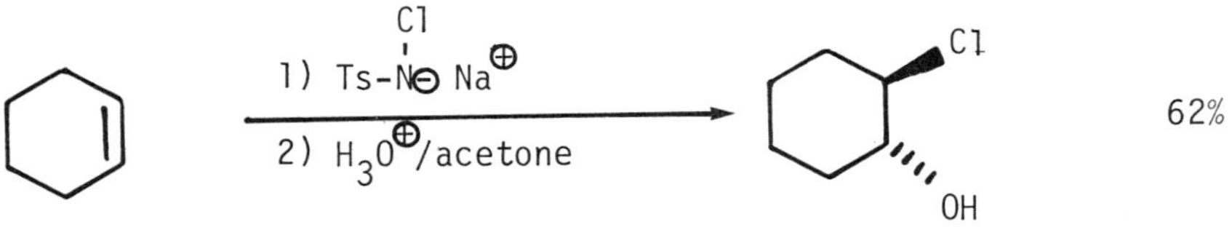

Synthesis, 362 (1981)

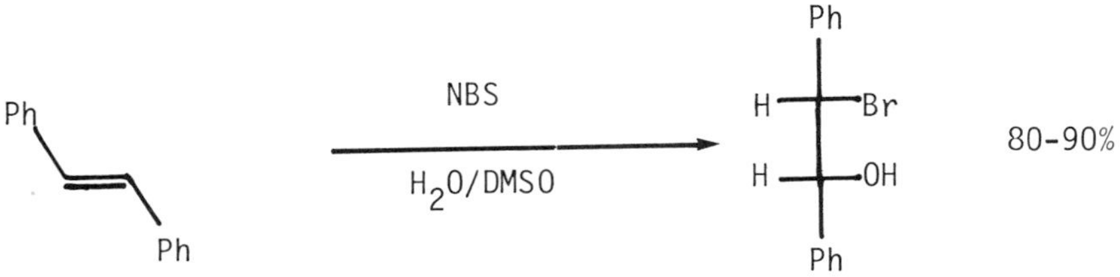

Org Syn, 59, 16 (1980)

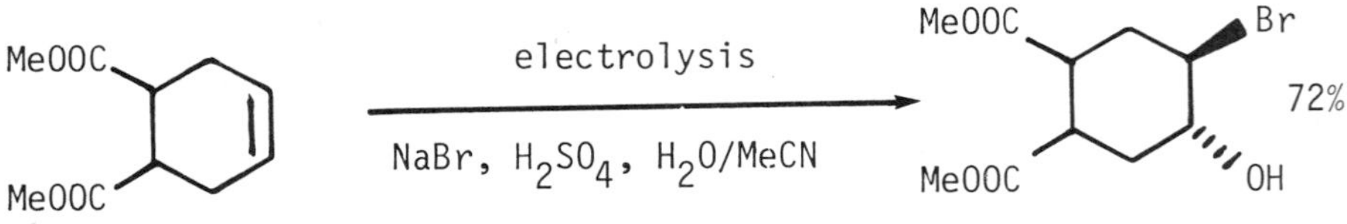

JOC, 46, 3312 (1981)

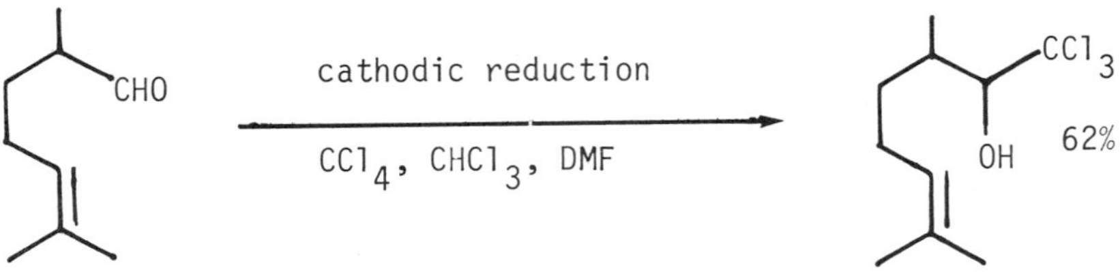

Tetr Lett, 23, 1609 (1982)

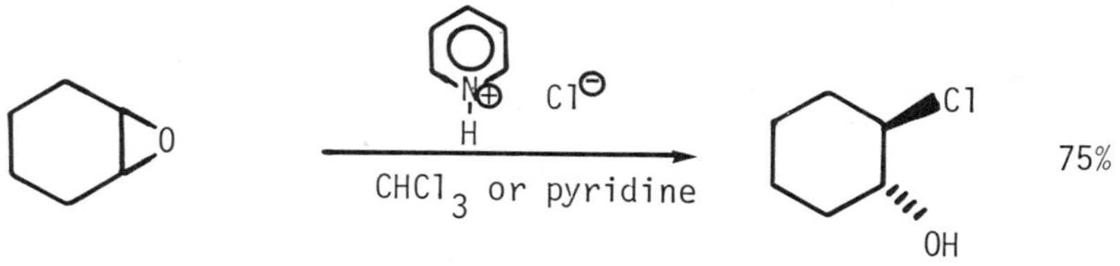

Synth Comm, <u>11</u>, 287 (1981)

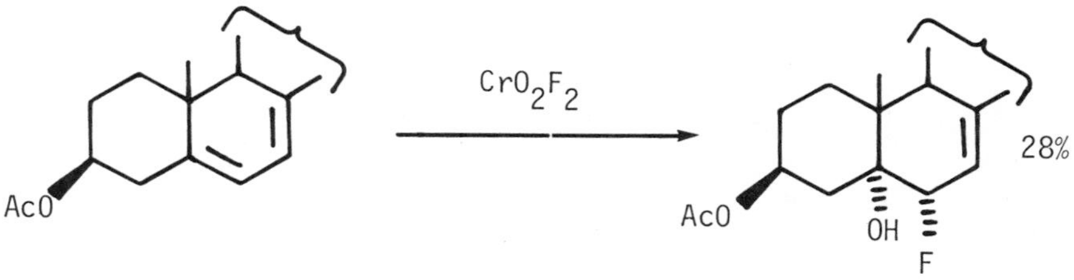

JCS Perkin I, 639 (1980)

Section 330 Alcohol - Ketone

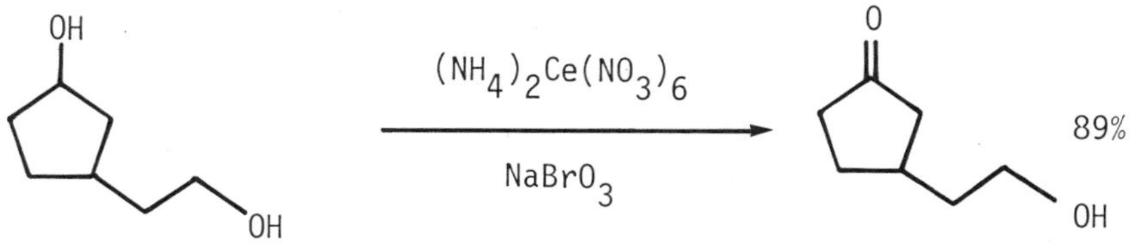

Tetr Lett, <u>23</u>, 539 (1982)

NaOCl, HOAc

85%

Tetr Lett, 23, 4647 (1982)

Ni(OH)$_2$ electrode

t-BuOH, KOH

78%

Tetrahedron, 38, 3299 (1982)

1) Me$_3$SiI, HMDS
2) OsO$_4$, NMMO
3) H$_2$O

98%

Tetr Lett, 22, 607 (1981)

$$Ph-\overset{O}{\underset{||}{C}}-CH_3 \xrightarrow[{^-OH, \ CH_3OH}]{\begin{array}{c}PhI=O \\ or \ PhI(OAc)_2\end{array}} Ph-\overset{O}{\underset{||}{C}}-CH_2OH$$

60%

Tetr Lett, 22, 1283 (1981)

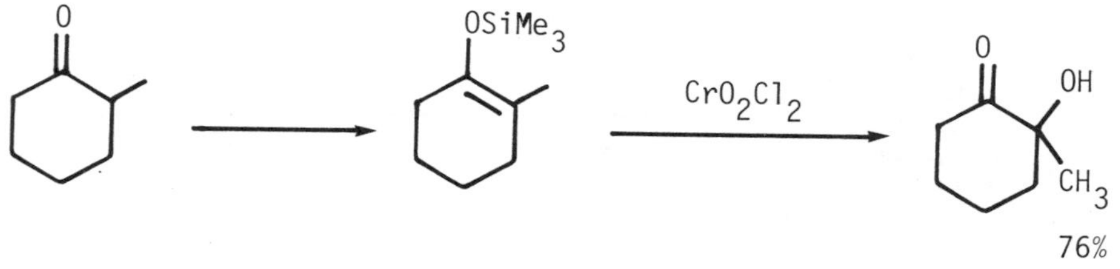

Tetr Lett, <u>23</u>, 2917 (1982)

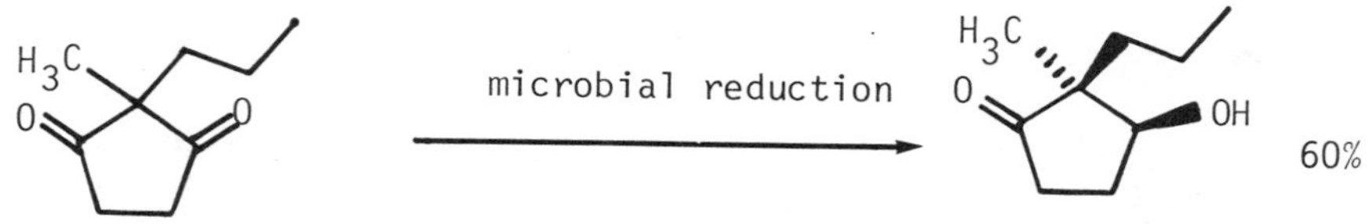

JOC, <u>47</u>, 2820 (1982)

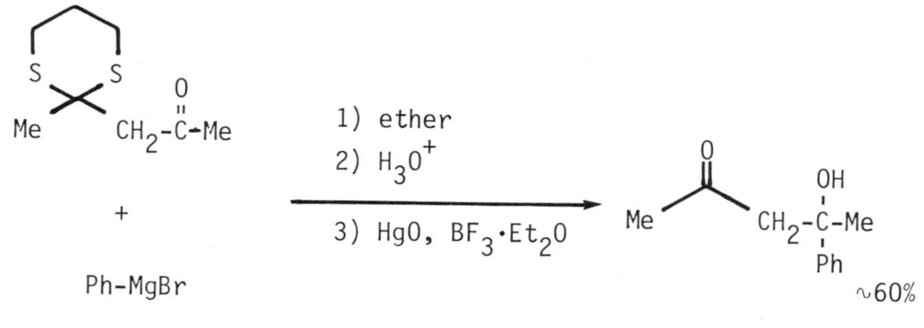

Synthesis, 561 (1980)

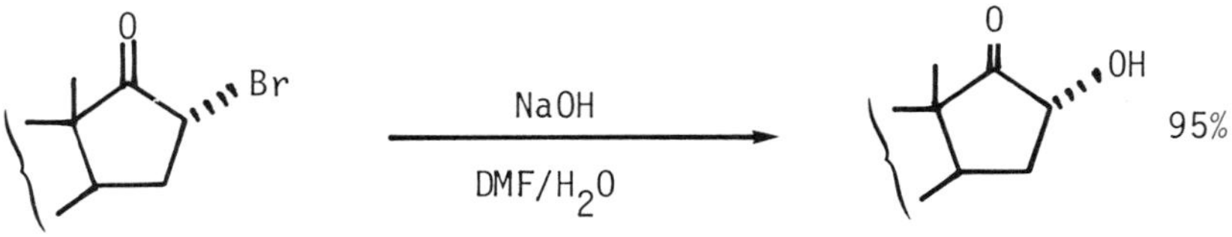

JOC, <u>47</u>, 4024 (1982)

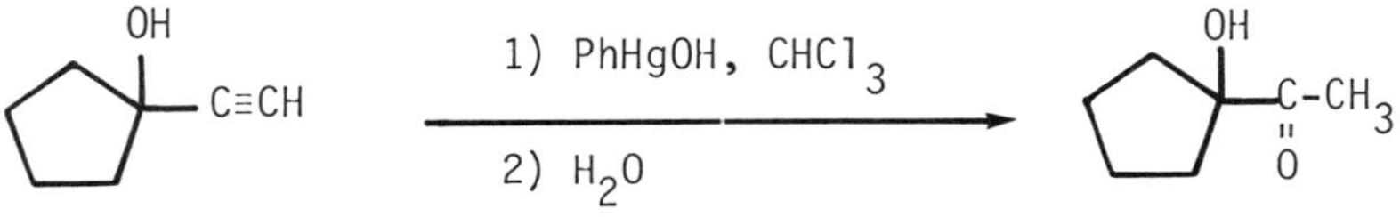

JOC, <u>47</u>, 3331 (1982)

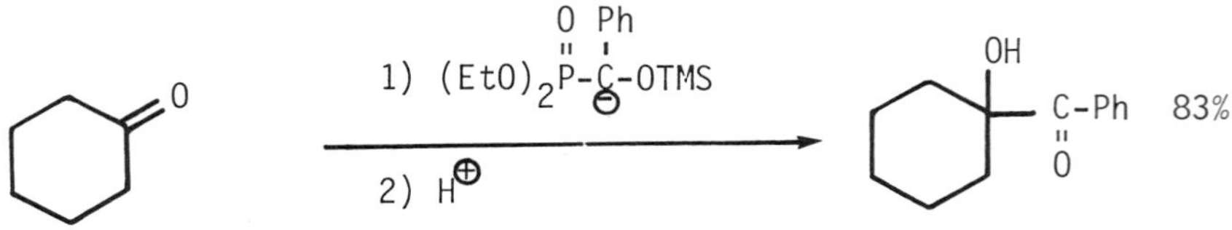

Tetr Lett, <u>21</u>, 1017 (1980)

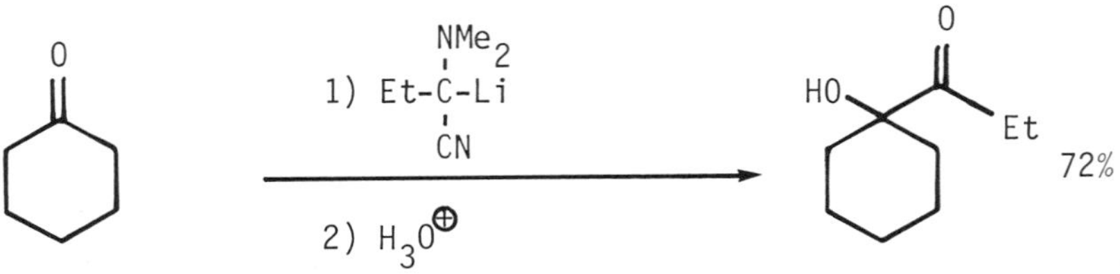

Chem Lett, 71 (1980)

$Et_2AlCl-Zn$

CuBr, THF

PhCHO

97%

Bull Chem Soc Japan, $\underline{53}$, 3301 (1980)

1) $(Et_2N)_3S^{\oplus}$ $TMSF_2^{\ominus}$

2) $\displaystyle \diagup\hspace{-0.5em}\diagdown$ — CHO

3) H_2O

67%

(100% erythro)

JACS, $\underline{103}$, 2106 (1981)

$\underline{i}$-Pr_2NLi

Et_3B, THF

2) PhCHO

90%

<u>Threo</u> product is favored by a 97/3 ratio.

Tetr Lett, $\underline{23}$, 2387 (1982)

1) $PhB(OH)_2$

2) $C_8H_{17}CHO$

3) H_2O_2

>90%

Chem Lett, 509 (1982)

Chem Lett, 1291 (1982)

Bull Chem Soc Japan, 53, 174 (1980)

Aldol reactions of boron enolates with aldehydes to form erythro

alcohols: JACS, 102, 4548 (1980)
            JACS, 103, 1566 (1981)
            JACS, 103, 2127 (1981)
            JACS, 103, 3099 (1981)
            JACS, 103, 3229 (1981)

Using Zr enolates:

            JACS, 103, 2876 (1981)

Erythro or threo selectivity:

            JACS, 103, 4278 (1981)

92:8 erythro:threo

Chem Lett, 467 (1982)

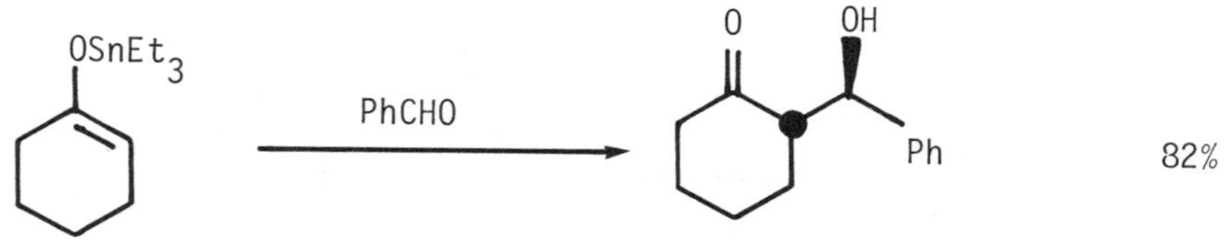

ALCOHOL - KETONE

JCS Chem Comm, 162 (1981)

Tetr Lett, 23, 627 (1982)

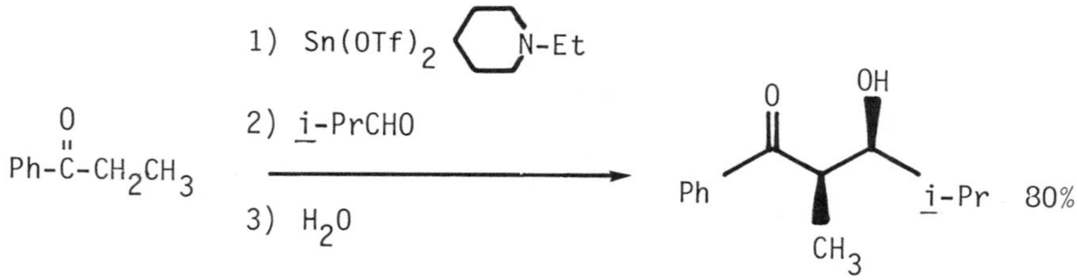

91:9 erythro:threo

Chem Lett, 353 (1982)

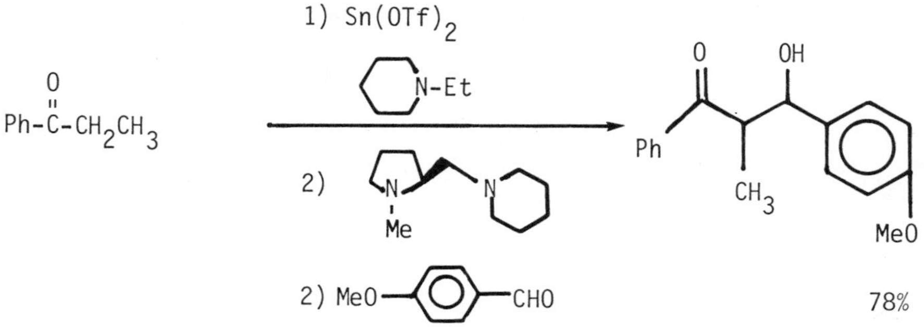

Chem Lett, 1441 (1982)

Chem Lett, 1459 (1982)

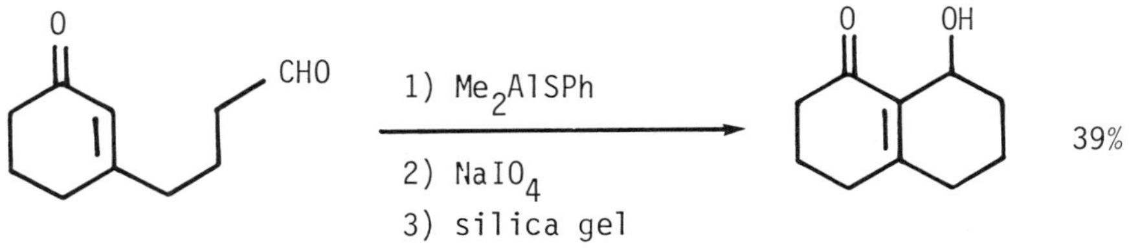

39%

Tetr Lett, <u>21</u>, 361 (1980)

Bull Chem Soc Japan, <u>54</u>, 274 (1981)

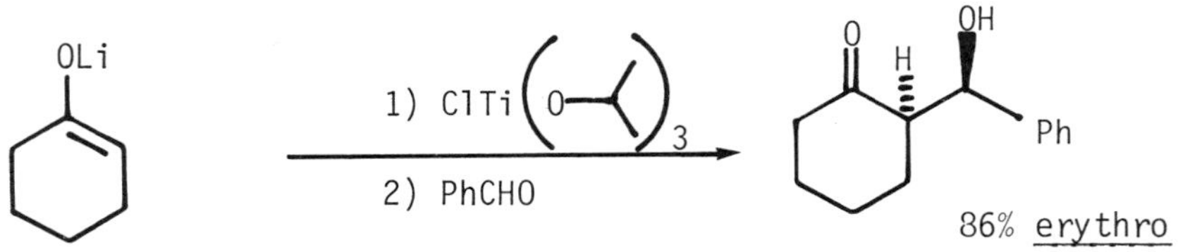

86% <u>erythro</u>

Tetr Lett, <u>22</u>, 4691 (1981)

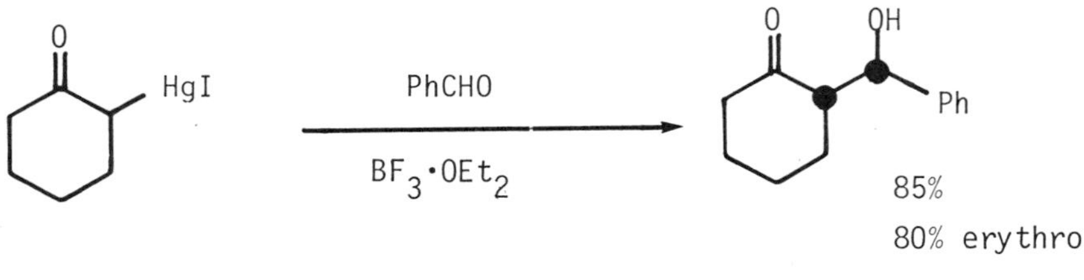

85%

80% erythro

JACS, <u>104</u>, 2323 (1982)

Review: "The Directed Aldol Reaction"

Org React, <u>28</u>, 203 (1982)

CH₃CH=C(Et)(OZrClCp₂)

$$CH_3CH=C \overset{Et}{\underset{OZrClCp_2}{\diagdown}}$$

+

Ph-CHO

2) H₂O

→

Ph — CH(OH) — CH(CH₃) — C(=O) — Et

90%

83% erythro

Tetr Lett, <u>21</u>, 4607 (1980)

(1-lithio-cyclohexen-1-ol lithium enolate)

1) cyclohexanone

2) H₂O

→

2-(1-hydroxycyclohexyl)cyclohexanone

64%

JACS, <u>104</u>, 1777 (1982)

cyclohexanone

+

NC—CH(O-EE)—CH=CH₂

(EE = ethoxyethyl)

LDA

THF

→

1-(1-hydroxycyclohexyl)prop-2-en-1-one

72%

JOC, <u>45</u>, 395 (1980)

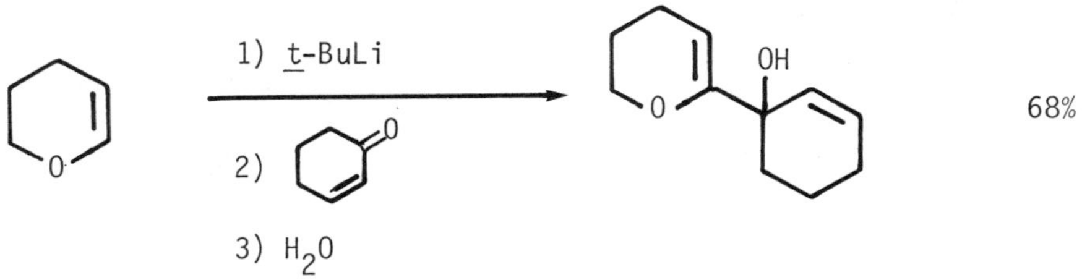
1) t-BuLi
2)
3) H2O
O
OH
68%

Tetrahedron, 37, 3997 (1981)

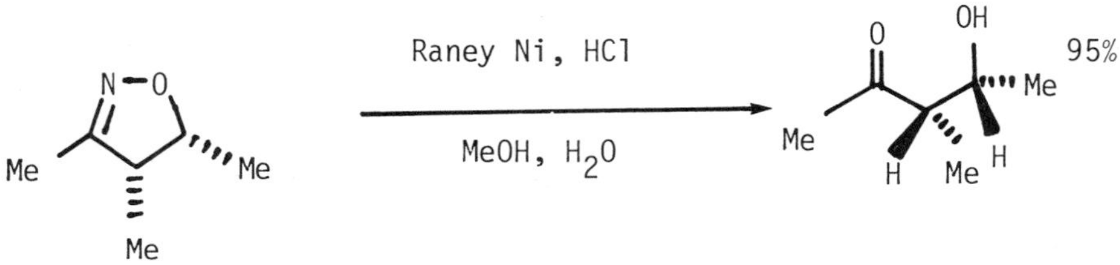
Raney Ni, HCl
MeOH, H2O
N—O
Me
95%
OH
H

Tetr Lett, 23, 3123 (1982)

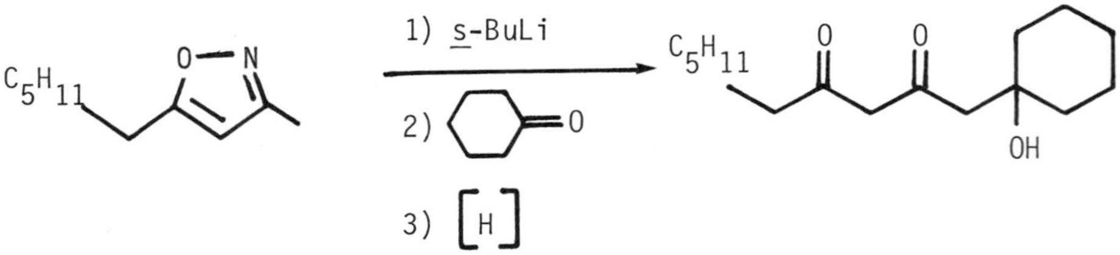
1) s-BuLi
2)
3) [H]
C5H11
O—N
OH

Tetr Lett, 22, 3699 (1981)

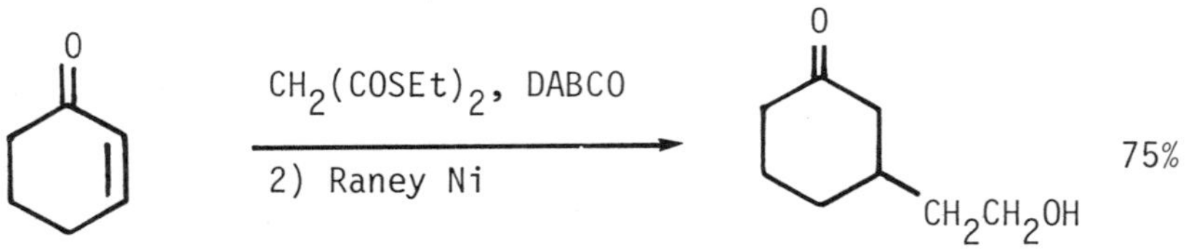
CH2(COSEt)2, DABCO
2) Raney Ni
CH2CH2OH
75%

Can J Chem, 60, 94 (1982)

Section 331 <u>Alcohol - Nitrile</u>

$Ph_2C=O$ $\xrightarrow[\text{2) } H_3O^+, \text{ THF}]{\text{1) } Me_3SiCN, \text{ } ZnI_2}$ $Ph_2C\overset{OH}{\underset{CN}{\diagdown}}$

Org Syn, <u>60</u>, 14 (1981)

$\xrightarrow[\text{3) } NH_4Cl]{\text{1) LDA} \quad \text{2)}}$ 72%

Synth Comm, <u>10</u>, 49 (1980)

CH_3CH_2CN

+

$Ph-CHO$

$\xrightarrow[\text{i-Pr}_2NEt]{Bu_2BOTf}$ $Ph-CH-\overset{CH_3}{\underset{OH}{CH}}-CN$ 90%

Chem Lett, 1401 (1982)

$N\equiv C-CH_2-\overset{O}{\overset{\|}{C}}-SBz$ $\xrightarrow[\text{3) } NaBH_4]{\text{1) NaH, i-BuX} \quad \text{2) NaH, PhCH}_2X}$ $N\equiv C-\overset{i-Bu}{\underset{CH_2Ph}{C}}-CH_2OH$ 76%

Tetr Lett, <u>23</u>, 3151 (1982)

Section 332 Alcohol - Olefin

Allylic and benzylic hydroxylation (C=C-CH → C=C-C-OH, etc.) is
listed in Section 41 (Alcohols and Phenols from Hydrides).

JCS Chem Comm, 121 (1981)
Tetr Lett, 22, 675 (1981)

Reagent	Reference
(i-Bu)$_3$Al	JOC, 47, 4640 (1982)
Ph$_2$SiH$_2$, L$_3$RhCl	Organometallics, 1, 1390 (1982)
NaBH$_4$-CeCl$_3$	JACS, 103, 5454 (1981)
t-BuNH$_2$·BH$_3$	Tetr Lett, 21, 693 and 697 (1980)
ZnBH$_4$ (erythro product)	Tetr Lett, 21, 1641 (1980)
LiAlH$_4$, N-methylephedrine	JCS Chem Comm, 1026 (1980)
(optically active product)	Chem Lett, 981 (1980)
(i-Bu)$_2$Al-O—	Bull Chem Soc Japan, 54, 3033 (1981)

(gives stereoselectivity in reaction with PG intermediate)

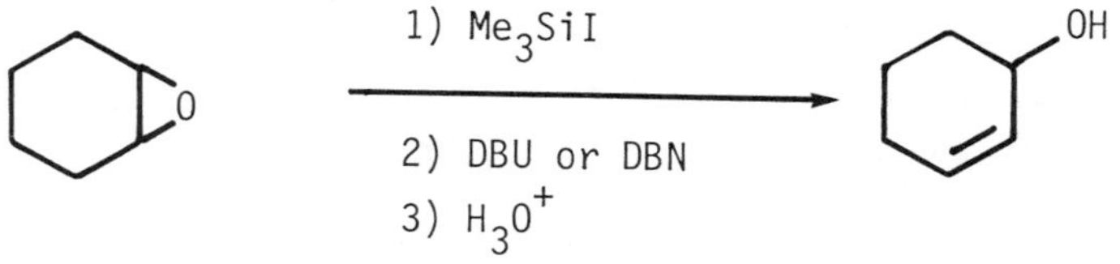

1) Me$_3$SiI

2) DBU or DBN

3) H$_3$O$^+$

81%

Tetr Lett, <u>21</u>, 2329 (1980)

JOC, <u>45</u>, 924 and 2579 (1980)

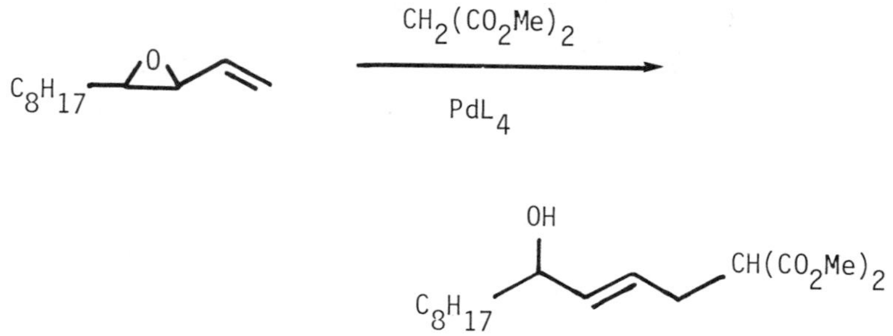

C$_8$H$_{17}$

CH$_2$(CO$_2$Me)$_2$

PdL$_4$

OH

C$_8$H$_{17}$ CH(CO$_2$Me)$_2$ 84%

Tetr Lett, <u>22</u>, 2575 (1981)

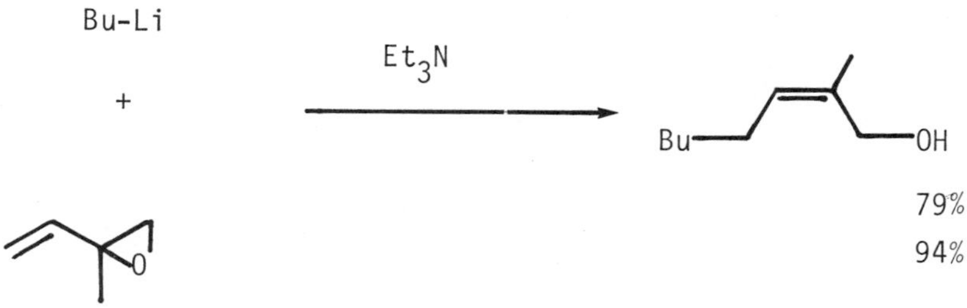

Bu-Li

+

Et$_3$N

Bu OH

79%

94% <u>Z</u>

Tetr Lett, <u>22</u>, 577 (1981)

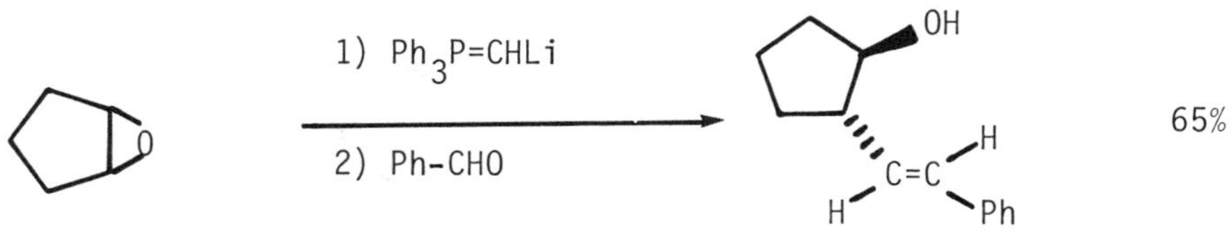
1) $Ph_3P=CHLi$
2) Ph-CHO
O
OH
H
C=C
H
Ph
65%

JACS, 104, 4724 (1982)

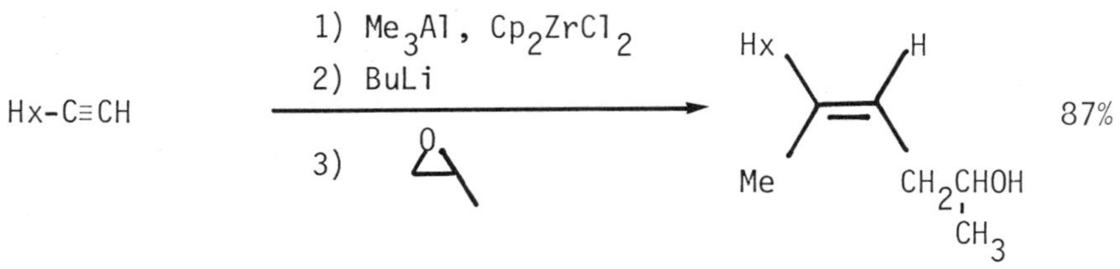
Hx-C≡CH
1) Me_3Al, Cp_2ZrCl_2
2) BuLi
3)
O
Hx
H
Me
CH_2CHOH
CH_3
87%

Synthesis, 1034 (1980)

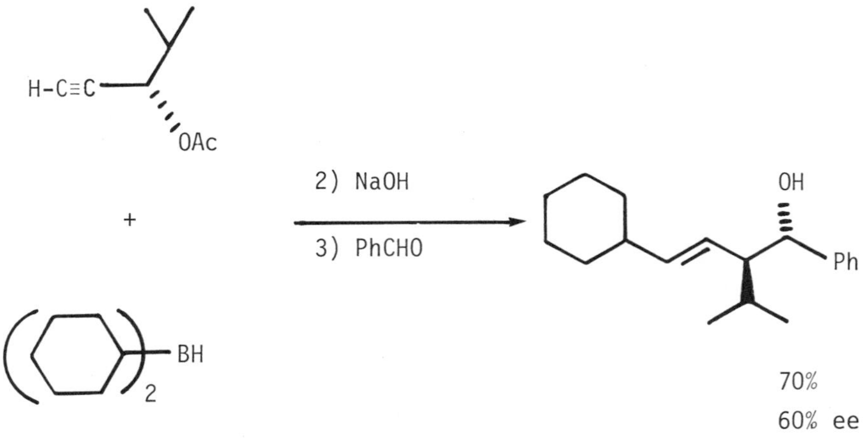
H-C≡C
OAc
+
BH
2
2) NaOH
3) PhCHO
OH
Ph
70%
60% ee

JACS, 104, 2330 (1982)

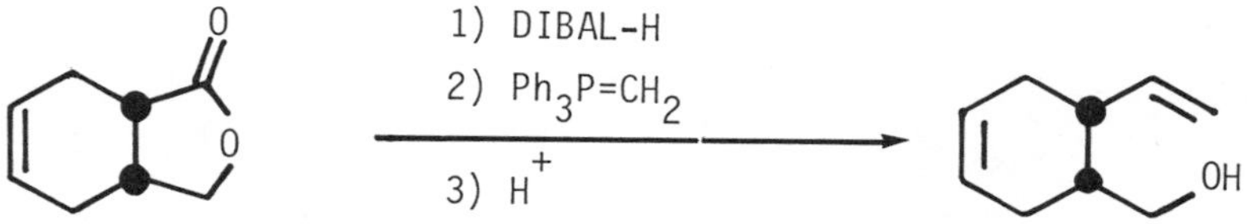
O
O
1) DIBAL-H
2) Ph3P=CH2
3) H+
OH

55%

Synthesis, 1015 (1980)

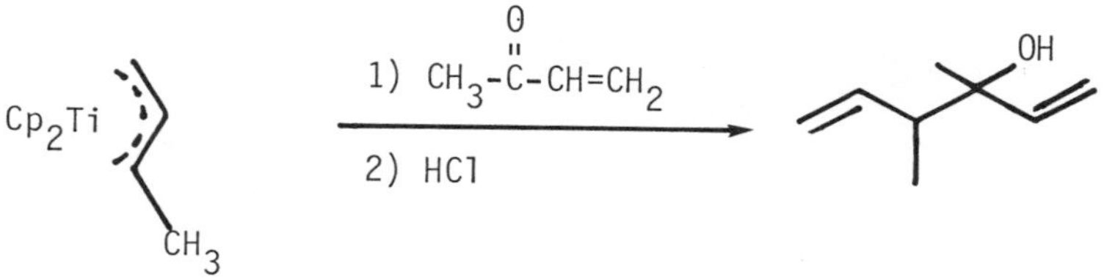
Cp2Ti
CH3
O
1) CH3-C-CH=CH2
2) HCl
OH

91%

Tetr Lett, 22, 243 (1981)

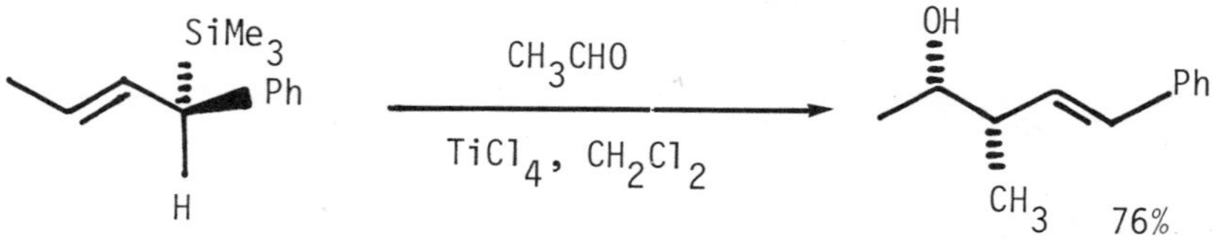
SiMe3
Ph
H
CH3CHO
TiCl4, CH2Cl2
OH
Ph
CH3

76%

86% ee

JACS, 104, 4963 (1982)

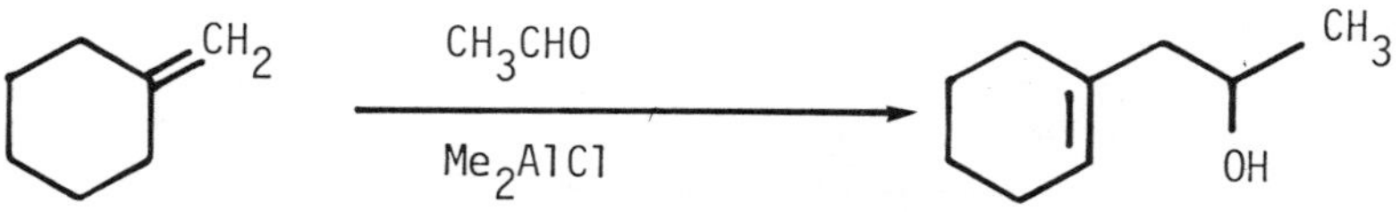
CH2
CH3CHO
Me2AlCl
CH3
OH

91%

Tetr Lett, 21, 1815 (1980)

JACS, 104, 555 (1982)

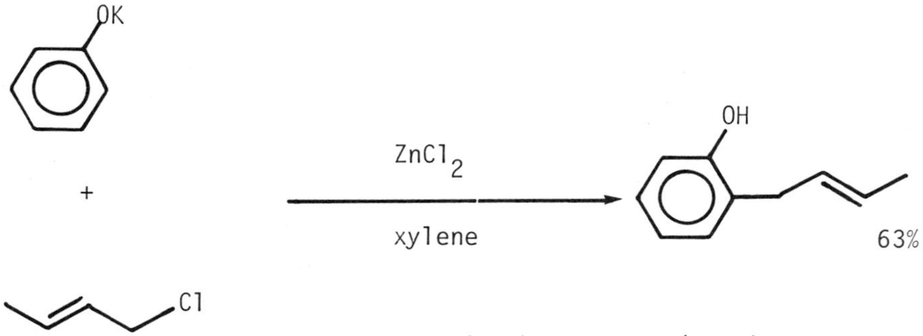

Synthesis, 310 (1981)

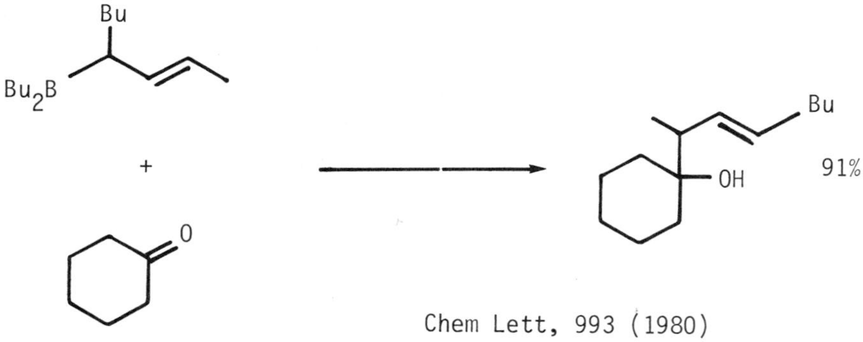

Chem Lett, 993 (1980)

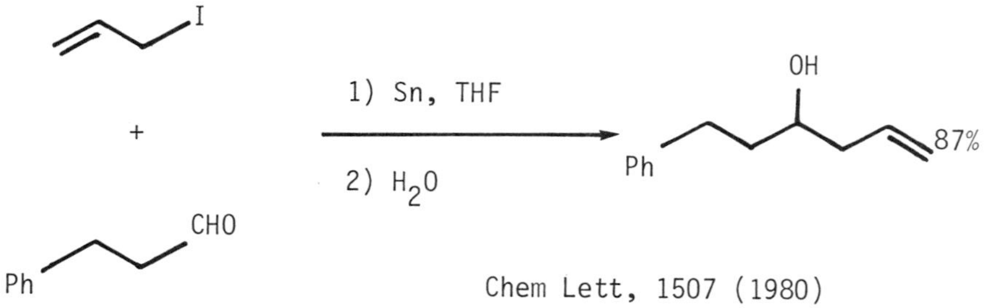

Chem Lett, 1507 (1980)
Chem Lett, 1527 (1981)

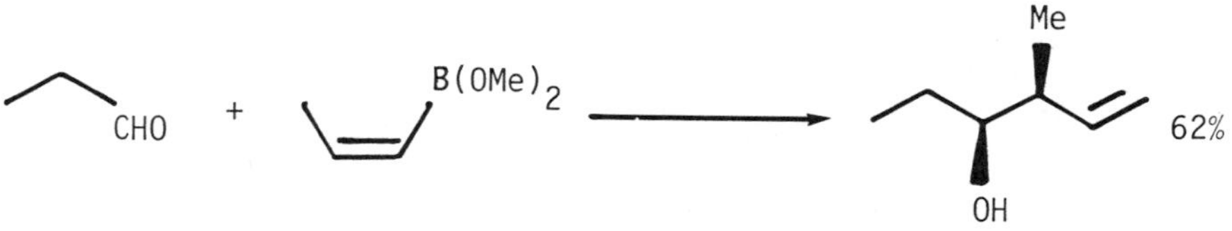

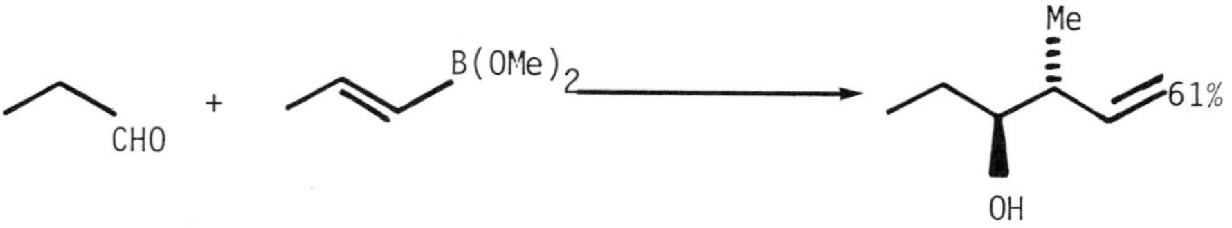

Helv Chim Acta, 65, 1258 (1982)

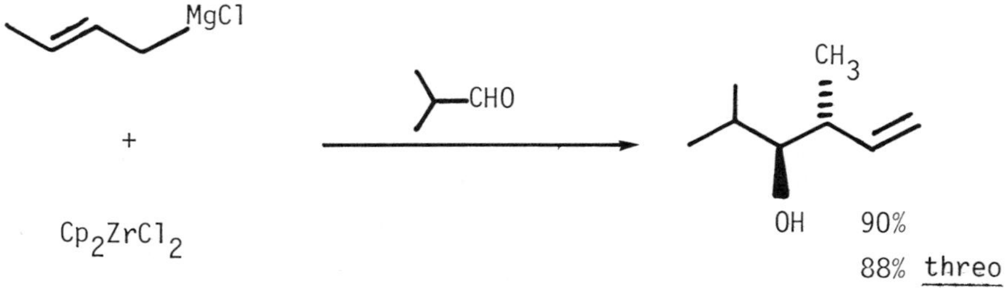

Tetr Lett, 22, 2895 (1981)

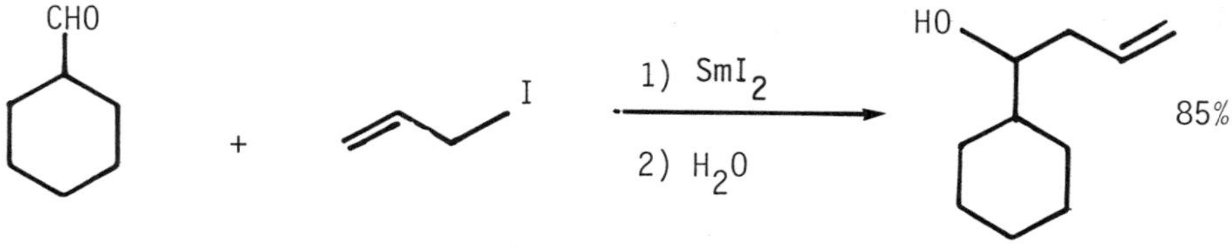

Tetr Lett, 23, 3497 (1982)

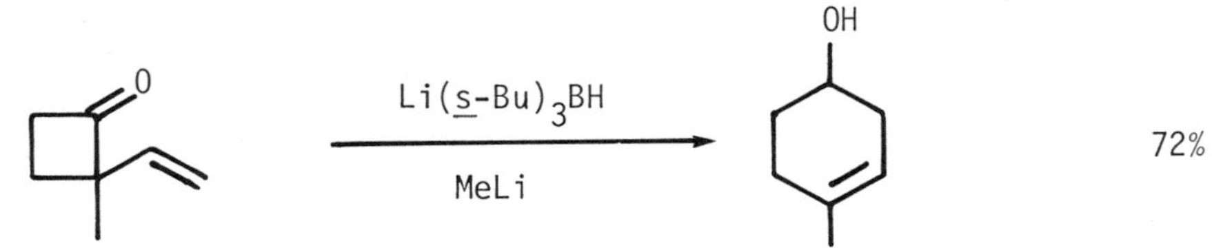

Tetrahedron, 37, 3943 (1981)

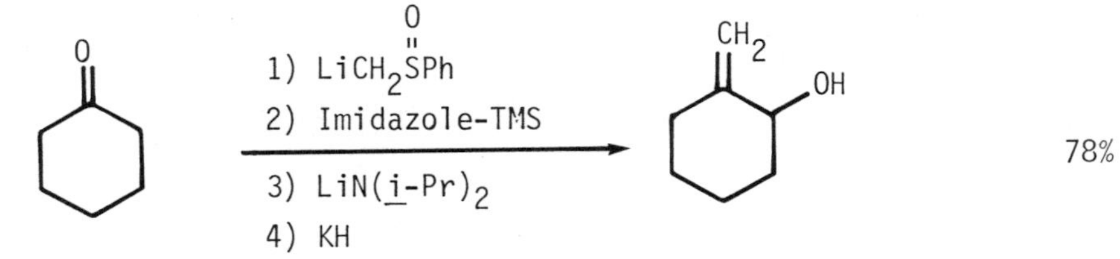

Synthesis, 640 (1980)

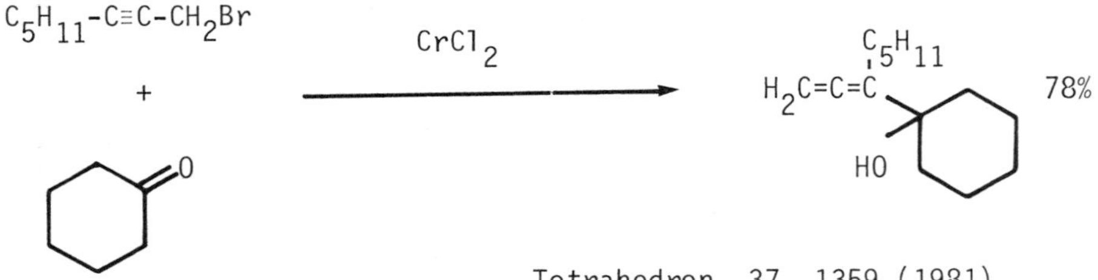

$C_5H_{11}-C{\equiv}C-CH_2Br$

Tetrahedron, 37, 1359 (1981)

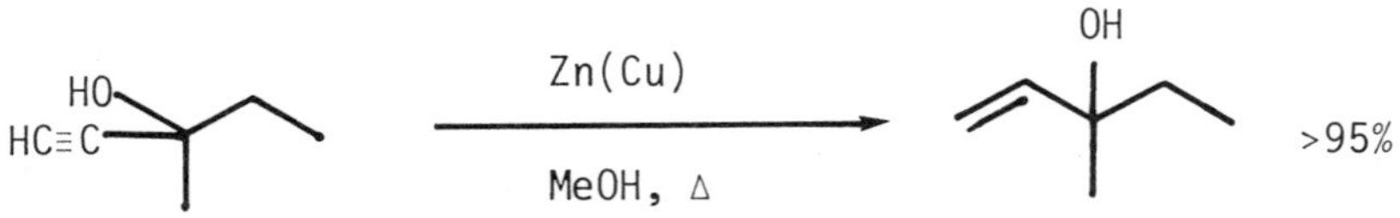

Tetr Lett, 21, 1069 (1980)

Also via:　Acetylenes - Alcohols (Section 302)

Section 333 <u>Aldehyde - Aldehyde</u>

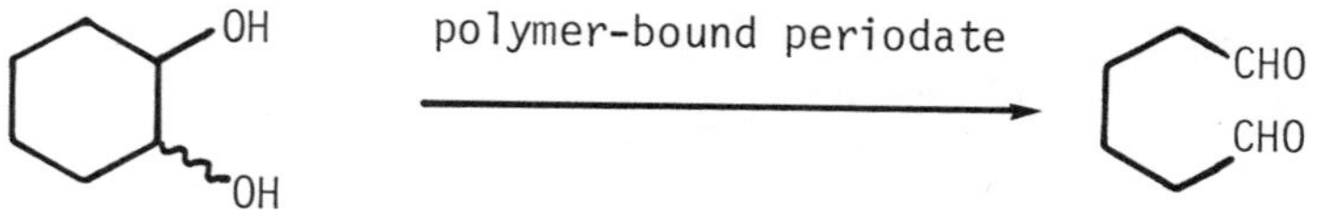

polymer-bound periodate

90%

JCS Perkin I, 509 (1982)

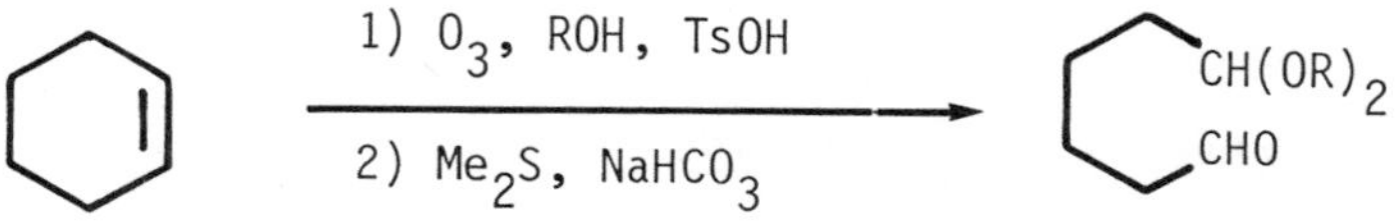

1) O_3, ROH, TsOH

2) Me_2S, $NaHCO_3$

93%

Tetr Lett, <u>23</u>, 3867 (1982)

Section 334 <u>Aldehyde - Amide</u>

No additional examples.

Section 335 <u>Aldehyde - Amine</u>

No additional examples.

Section 336 Aldehyde - Ester

1) O_3, ROH, $NaHCO_3$
2) Ac_2O, Et_3N

→ CHO / COOR 96%

1) O_3, ROH, TsOH
2) Ac_2O, Et_3N

→ $CH(OR)_2$ / COOR 83%

Tetr Lett, <u>23</u>, 3867 (1982)

Section 337 Aldehyde - Ether, Epoxide

No additional examples.

Section 338 Aldehyde - Halide

CHO

Amberlyst A-26 ($ICl_2^{\ominus}$)

CCl_4

→ CHO / Cl 84%

JCS Chem Comm, 1278 (1980)

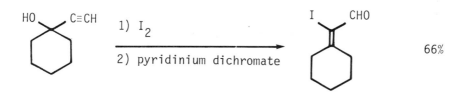

CHO

electrogenerated I$^{\oplus}$

TFA

CHO

I

74%

Acta Chem Scand B, <u>34</u>, 47 (1980)

Bu-CBr$_2$Li

+

1)

2) H$_3$O$^+$ Bu-CBr$_2$-CHO   78%

O
‖
H-C-OMe   May also be used to form α-monohalo aldehydes.

Synthesis, 644 (1980)

HO   C≡CH

1) I$_2$

2) pyridinium dichromate

I   CHO

66%

Tetr Lett, <u>22</u>, 1041 (1981)

Section 339 Aldehyde - Ketone

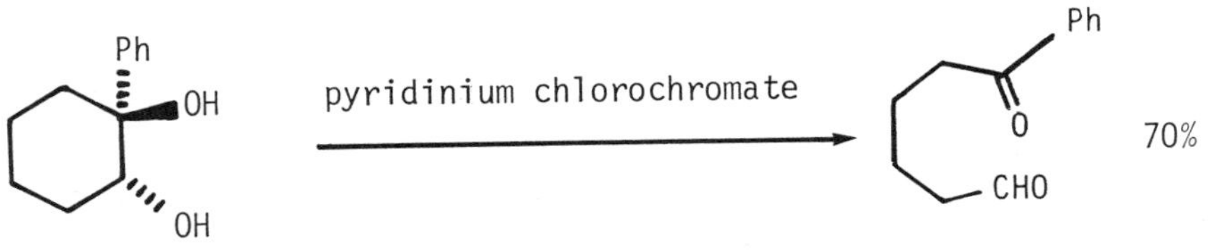
Ph
OH
OH
pyridinium chlorochromate
Ph
O
CHO
70%

Synth Comm, 12, 833 (1982)

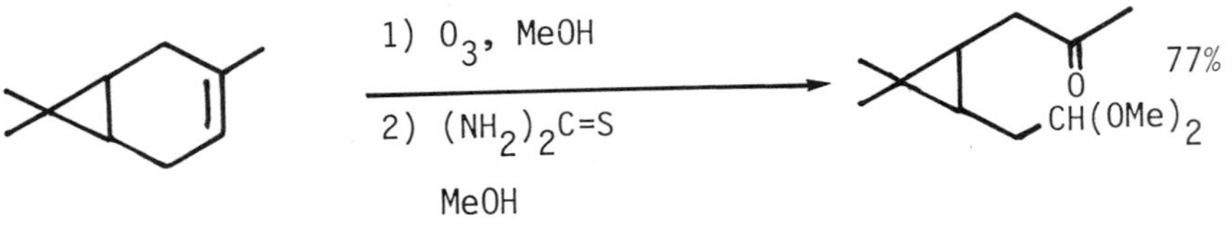
1) O3, MeOH
2) (NH2)2C=S
MeOH
O
CH(OMe)2
77%

Tetrahedron, 38, 3013 (1982)

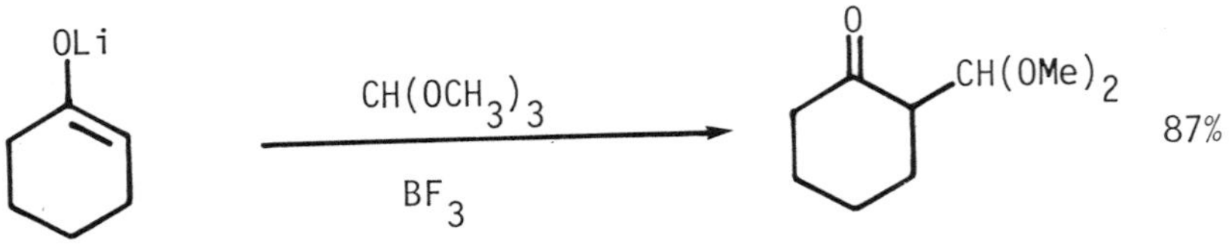
OLi
CH(OCH3)3
BF3
O
CH(OMe)2
87%

Tetr Lett, 23, 3595 (1982)

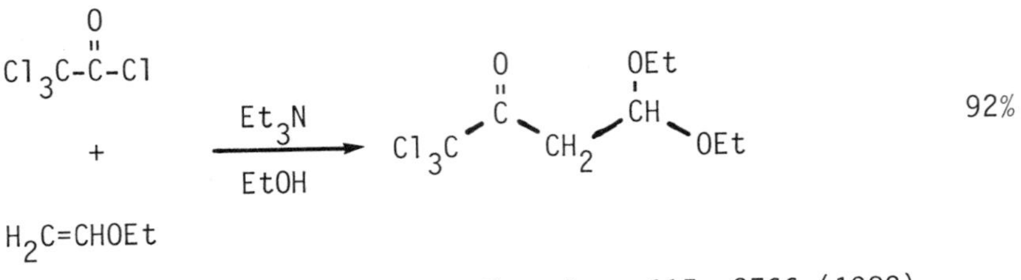
O
Cl3C-C-Cl
+
H2C=CHOEt
Et3N
EtOH
O
C
Cl3C
CH2
OEt
CH
OEt
92%

Chem Ber, 115, 2766 (1982)

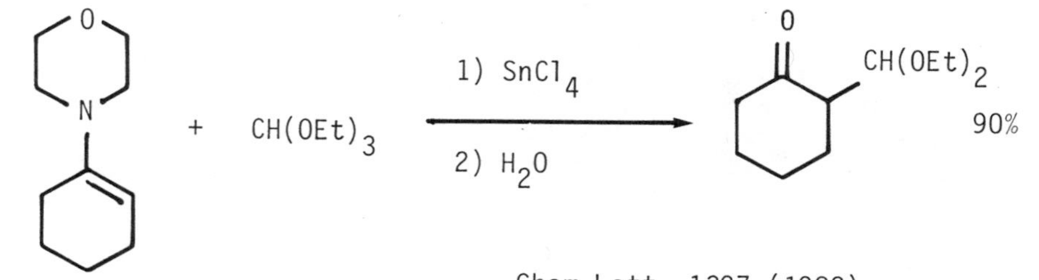
O
N
+ CH(OEt)3
1) SnCl4
2) H2O
O
CH(OEt)2
90%

Chem Lett, 1307 (1982)

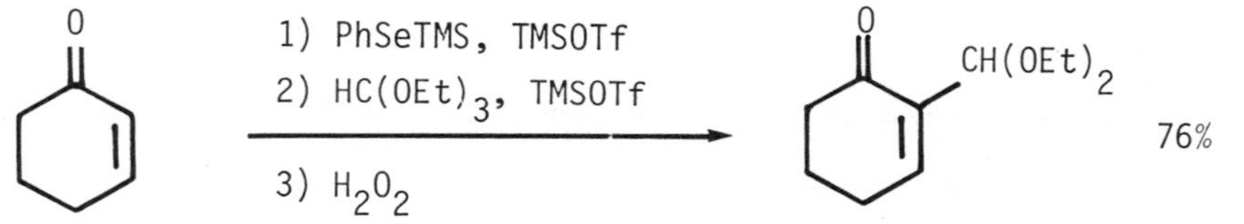
O
1) PhSeTMS, TMSOTf
2) HC(OEt)3, TMSOTf
3) H2O2
O
CH(OEt)2
76%

Tetr Lett, 22, 1809 (1981)

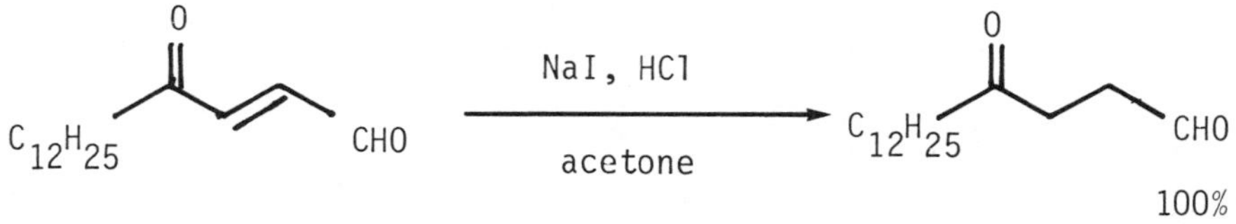
O
C12H25
CHO
NaI, HCl
acetone
O
C12H25
CHO
100%

Synthesis, 245 (1980)

Section 340 Aldehyde - Nitrile

1) NaH, t-BuOH
2) KMnO$_4$
3) Na$_2$S$_2$O$_5$, H$_2$SO$_4$

81%

JOC, 47, 4534 (1982)

Section 341 Aldehyde - Olefin

For the oxidation of allylic alcohols to olefinic aldehydes see also Section 48 (Aldehydes from Alcohols).

1) $\left(\begin{array}{c} \text{N} \end{array} \text{Se} \right)_2$ + Br$_2$

2) NaIO$_4$

90%

Tetr Lett, 23, 2105 (1982)

1) PhSeCl, pyr.

2) H$_2$O$_2$, CH$_2$Cl$_2$

85%

JOC, 46, 2920 (1981)

$H_2C=C=C$ 〈 OMe / TMS

1) BuLi
2) BuBr
3) Bu_4NF
4) $H_3O^{\oplus}$

Bu ⌇ CHO

Tetr Lett, 21, 3987 (1980)

(structure: 2-methyl-1-bromopropene) Br

+

(structure) $CH(OEt)_2$

1) (piperidine, N-H), $Pd(OAc)_2$,

$P(tol)_3$

2) H_2O, $(COOH)_2$

(structure) CHO

JOC, 46, 1061 (1981)

PhMgBr

+

EtO-CH=C-CHO
        |
        CH_3

⟶

Ph-CH=C-CHO   52%
        |
        CH_3

Angew Chem Int Ed, 19, 816 (1980)

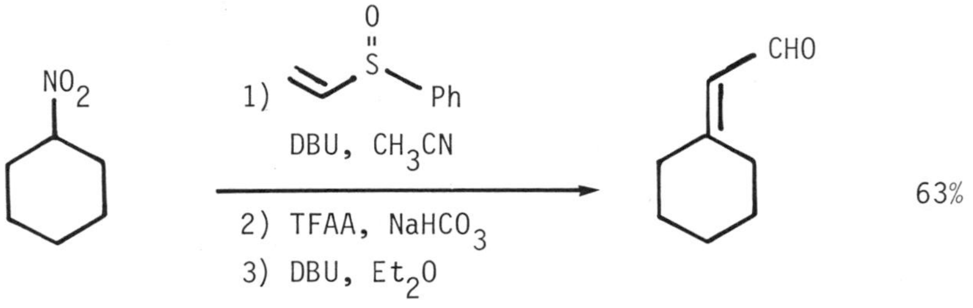

JOC, <u>47</u>, 5017 (1982)

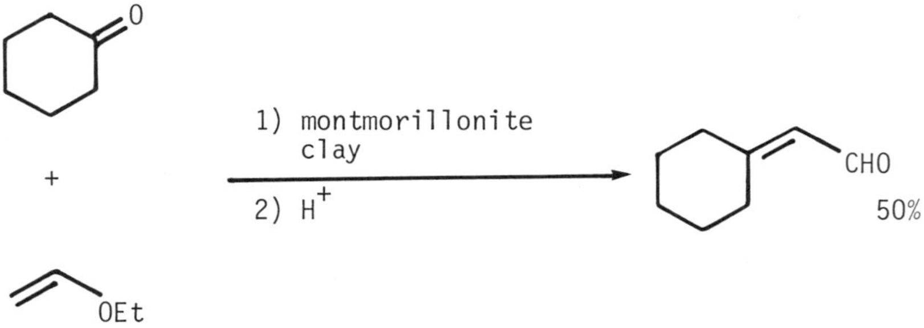

Synthesis, 137 (1981)

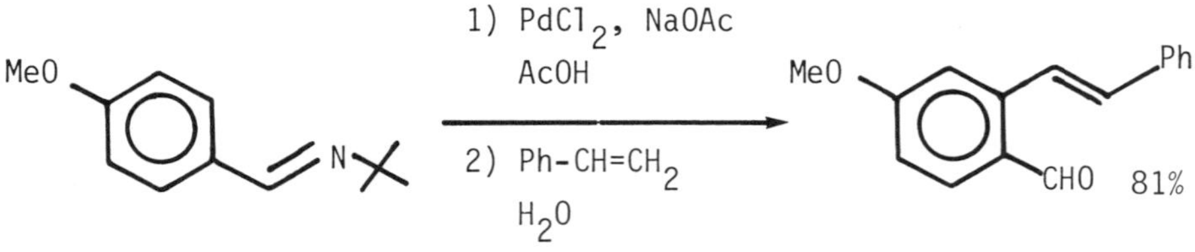

Tetr Lett, <u>23</u>, 1957 (1982)

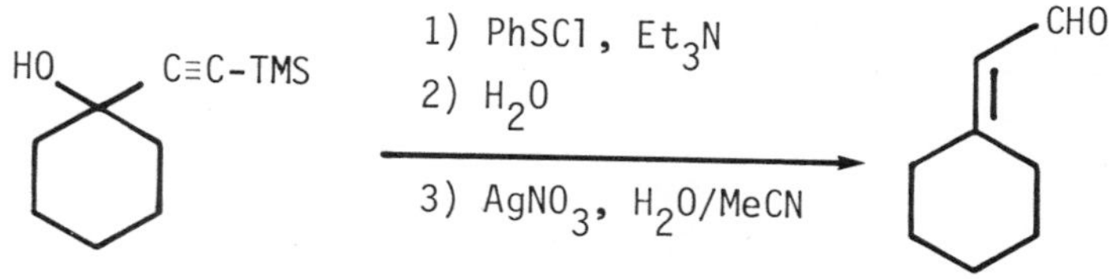

Tetr Lett, _22_, 2021 (1981)

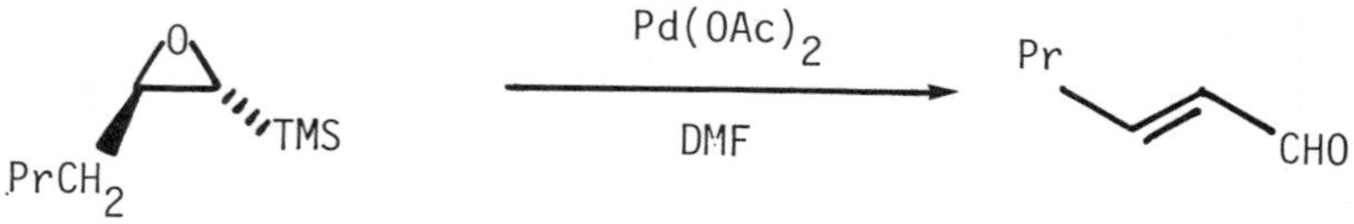

Chem Lett, 1997 (1982)

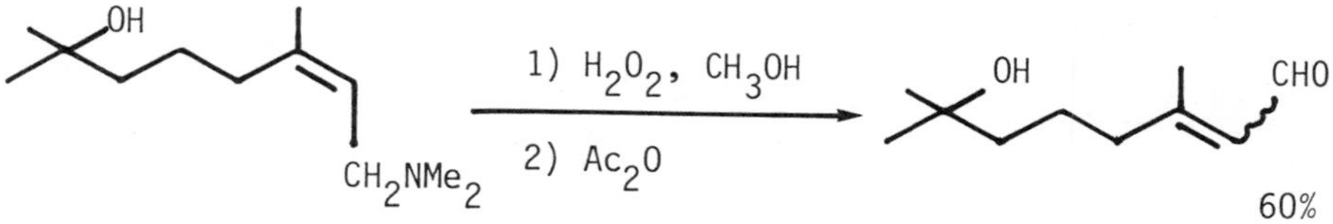

Chem Lett, 1987 (1982)

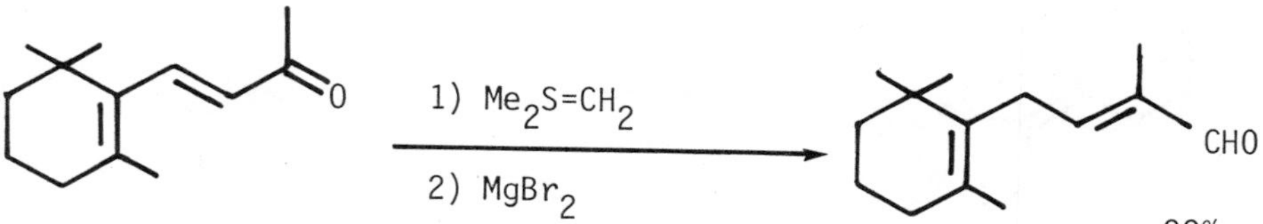

Helv Chim Acta, _63_, 1665 (1980)

Also via: β-Hydroxyaldehydes (Section 324)

Section 342 Amide - Amide

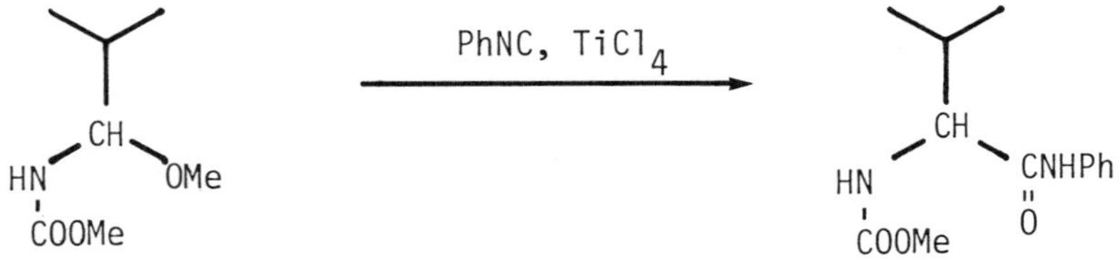

PhNC, TiCl$_4$ → 82%

Tetr Lett, 22, 2411 (1981)

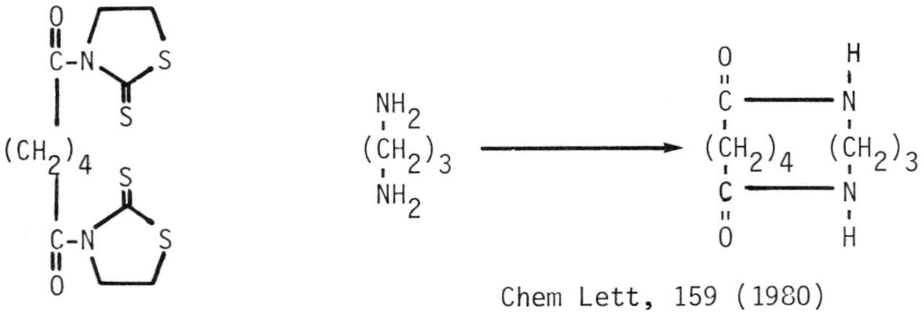

89%

Chem Lett, 159 (1980)

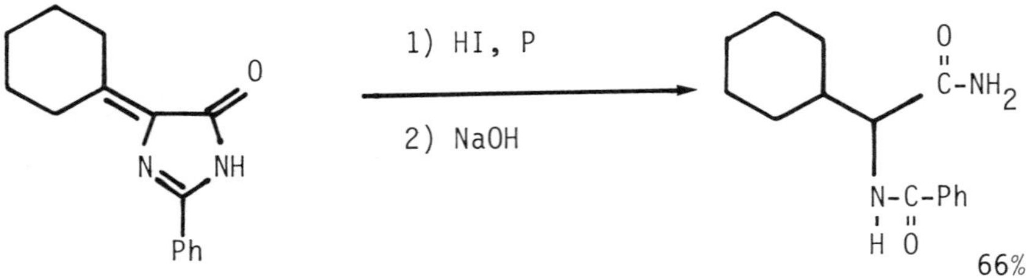

1) HI, P

2) NaOH

66%

Indian J Chem, 19B, 70 (1980)

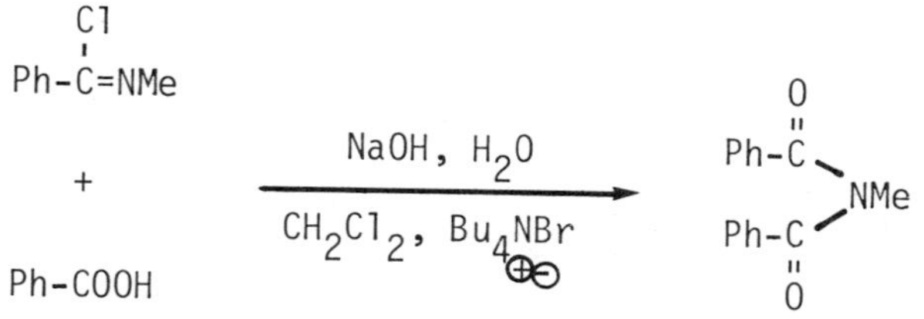

Synthesis, 114 (1980)

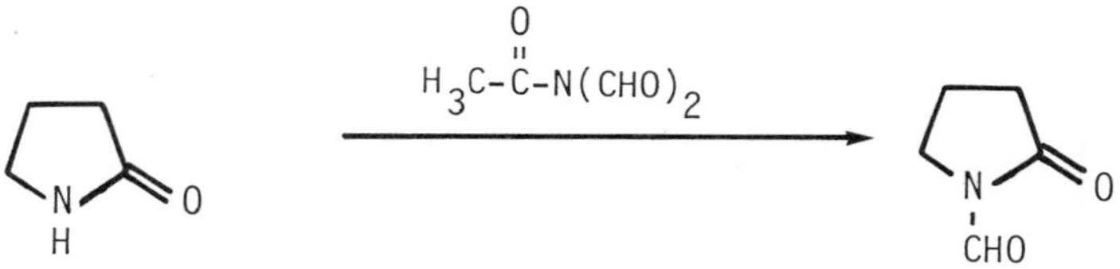

Synthesis, 264 (1982)

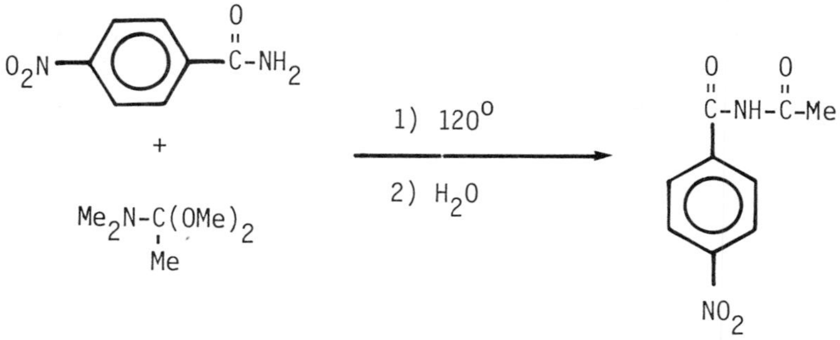

Synthesis, 119 (1980)

Also via: Dicarboxylic acids (Section 312); Diamines (Section 350)

Section 343 Amide - Amine

Synthesis, 1092 (1982)

Section 344 Amide - Ester

Angew Chem Int Ed, 21, 203 (1982)

Angew Chem Int Ed, 21, 203 (1982)

$$
\begin{array}{c}
O \\
\| \\
P(OEt)_2 \\
| \\
Z\text{-}NH\text{-}CH\text{-}COOEt
\end{array}
\xrightarrow[B^{\ominus}]{\text{-CHO}}
\begin{array}{c}
H \\
C \\
\| \\
Z\text{-}NH\text{-}C\text{-}COOEt
\end{array}
\qquad 80\%
$$

Angew Chem Int Ed, <u>21</u>, 776 (1982)

$$
\text{(NH)}
\xrightarrow{\text{MeO} \quad O \quad OEt}
\left(\text{N-}\overset{O}{\overset{\|}{C}}\text{-OEt}\right)
\qquad 77\%
$$

JOC, <u>45</u>, 4519 (1980)

Related methods: Section 315 (Acid-Amide); Section 316 (Acid-
Amine); Section 351 (Amine-Ester)

Section 345 <u>Amide - Ether</u>

$$
\xrightarrow[-2e^{\ominus}]{CH_3\text{-}OH}
\qquad 88\%
$$

Synthesis, 315 (1980)

Ph-O-CH$_2$COOH

+

(2-methylaniline structure with NH$_2$ and CH$_3$)

→

Ph-O-CH$_2$ C=O, NH (with 2-methylphenyl, CH$_3$)   92%

Synthesis, 547 (1980)

Section 346 Amide - Halide

No additional examples.

Section 347 Amide - Ketone

(CH$_3$)$_2$CH-C≡C-NMe$_2$ $\xrightarrow[\text{RuCl}_2\text{L}_3]{\text{Ph-I=O}}$ (CH$_3$)$_2$CH-C(=O)-C(=O)-NMe$_2$ 44%

Tetr Lett, 23, 3661 (1982)

PhBr + Et$_2$NH $\xrightarrow[\text{PdCl}_2(\text{PMePh}_2)_2]{\text{CO}}$ Ph-C(=O)-C(=O)-NEt$_2$ 81%

Tetr Lett, 23, 3383 (1982)

Tetr Lett, $\underline{23}$, 1201 (1982)

Tetr Lett, $\underline{21}$, 2033 (1980)

Tetr Lett, $\underline{23}$, 5531 (1982)

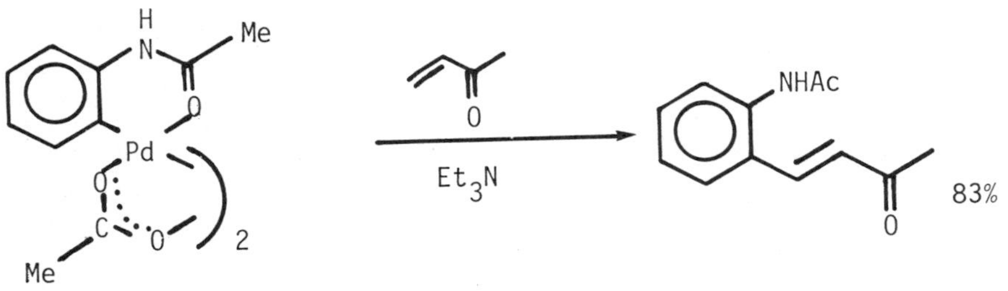
H
N
Me
O
Pd
O
C
O
2
Me
O
Et_3N
NHAc
O
83%

JOC, 46, 4416 (1981)

Section 348 Amide - Nitrile

No additional examples.

Section 349 Amide - Olefin

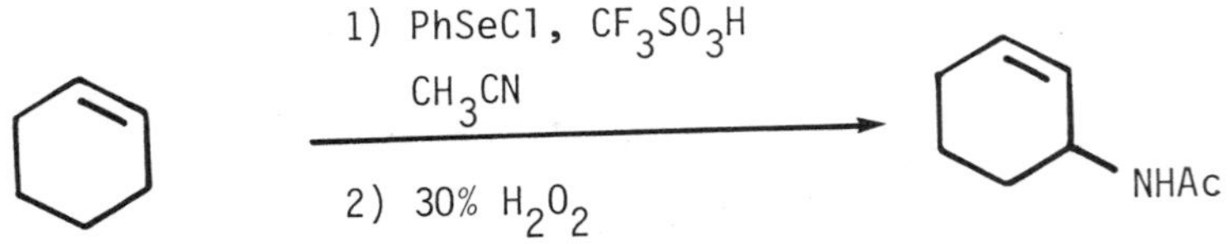
1) PhSeCl, CF_3SO_3H
CH_3CN
2) 30% H_2O_2
NHAc

JOC, 46, 4727 (1981)

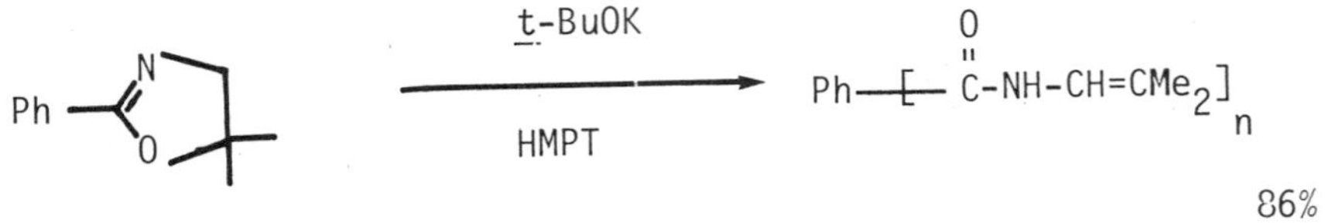
Ph
N
O
t-BuOK
HMPT
Ph-[-C(O)-NH-CH=CMe_2]$_n$
86%

Angew Chem Int Ed, 21, 630 (1982)

Also via: Olefinic acids (Section 322)

Section 350 Amine - Amine

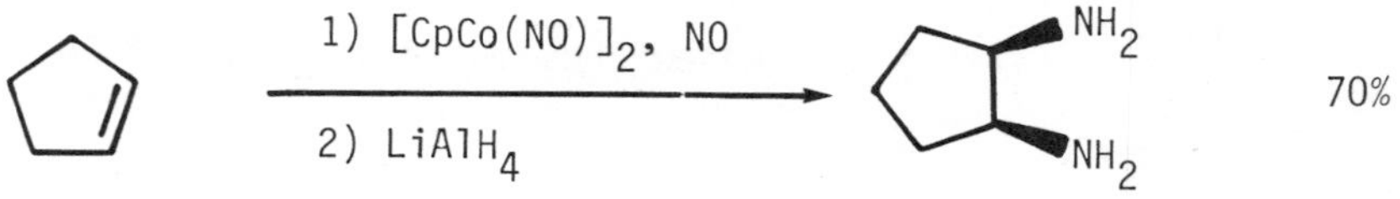

70%

JACS, 102, 5676 (1980)

Section 351 Amine - Ester

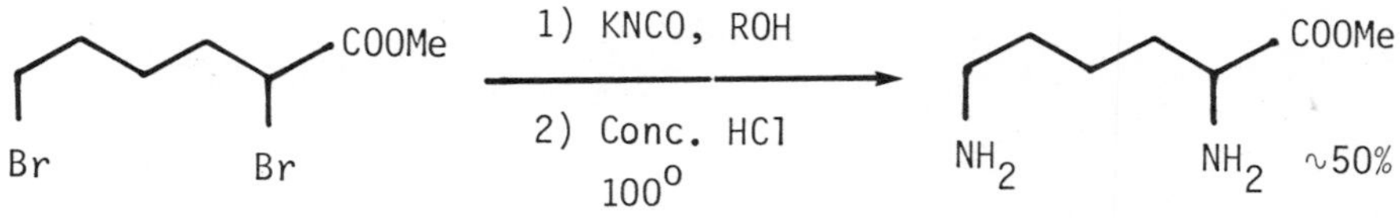

Chem Ber, 114, 173 (1981)

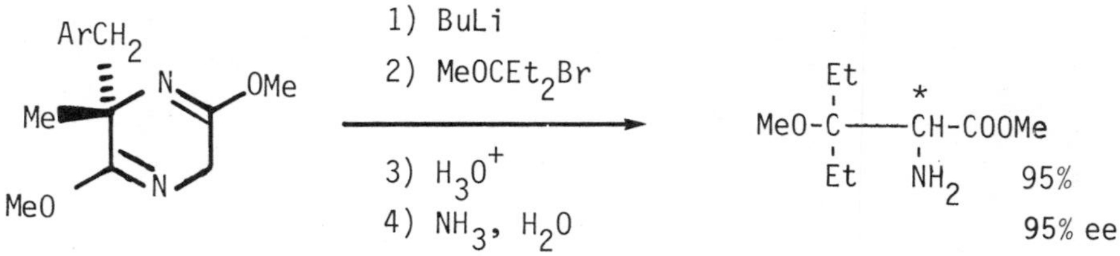

$$MeO-\underset{\underset{Et}{|}}{\overset{\overset{Et}{|}}{C}}-\overset{*}{C}H-COOMe$$

NH₂   95%

95% ee

Synthesis, 966 (1981)
Synthesis, 861 and 870 (1982)

Angew Chem Int Ed, <u>19</u>, 212 (1980)

$CH_3-CH-COOEt$
  |
  $N=CH-\langle\bigcirc\rangle-Cl$

Tetr Lett, <u>23</u>, 4259 (1982)

$$\underset{\text{O}}{\overset{\text{O}}{t\text{-BuOC}}}-CH_2MgCl$$

+

$\underset{\text{Ph}}{BuO-CH-NMe_2}$

→

$\underset{\text{Ph}}{\overset{\text{O}}{t\text{-BuOC}}-CH_2-CH-NMe_2}$ 76%

Bull Soc Chim France II, 395 (1982)

Me_3N

+

$H_2C(COOEt)_2$

electrolysis
────────────→

$Me_2NCH_2CH(COOEt)_2$ 53%

Tetrahedron, <u>37</u>, 2297 (1981)

Review: "Methods of Synthesis and Properties of β-Dimethylamino-
             ethyl and Choline Esters of Amino Acids and Peptides"

                                    Russ Chem Rev, 50, 1151 (1981)

Related methods: Section 315 (Acid-Amide); Section 316 (Acid-
                   Amine); Section 344 (Amide-Ester)

Section 352 Amine - Ether

                                                                    71%

                        Synthesis, 376 (1981)

                                                                    63%

                   Chem Ber, 115, 2635 (1982)

Section 353 Amine - Halide

HF

93%

JOC, 45, 5328 and 5333 (1980)

Review: "The Synthesis of β-Halogenated Enamines"

Org Prep Proc Int, 13, 241 (1981)

Section 354 Amine - Ketone

$Me_2N=CH_2^{\oplus}$

or $BuOCH_2NMe_2$, TMSI

79%

Tetr Lett, 21, 805 (1980)
Tetr Lett, 23, 547 (1982)

$Ph-\overset{O}{\overset{\|}{C}}-CH_2CH_3$

1) CH_2I_2,

2) NaOH, H_2O

$Ph-\overset{O}{\overset{\|}{C}}-CH-CH_3$ 60%

BCS Japan, 55, 1331 (1982)

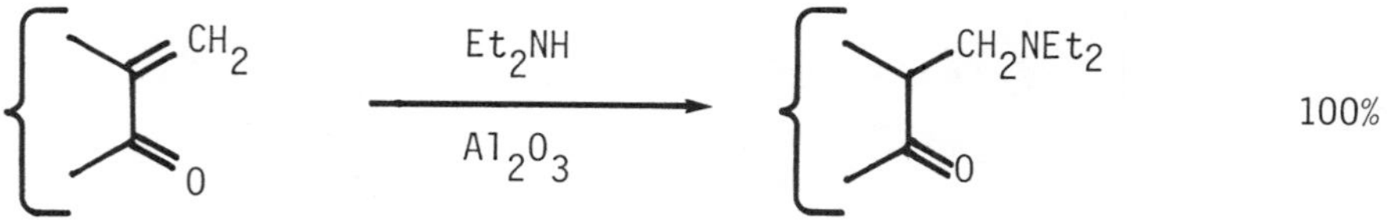

Tetr Lett, <u>21</u>, 809 (1980)

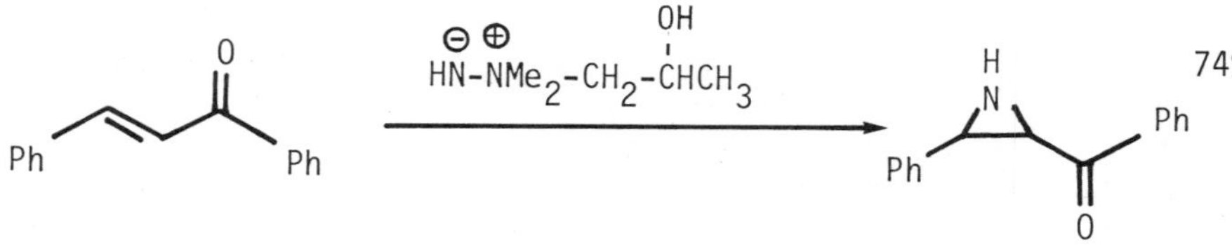

Synthesis, 650 (1980)

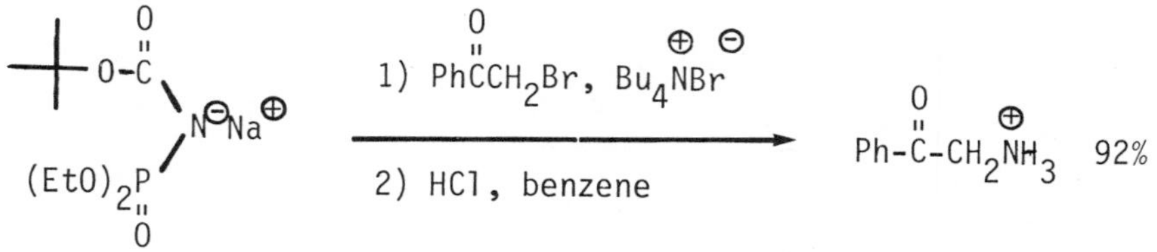

Synthesis, 922 (1982)

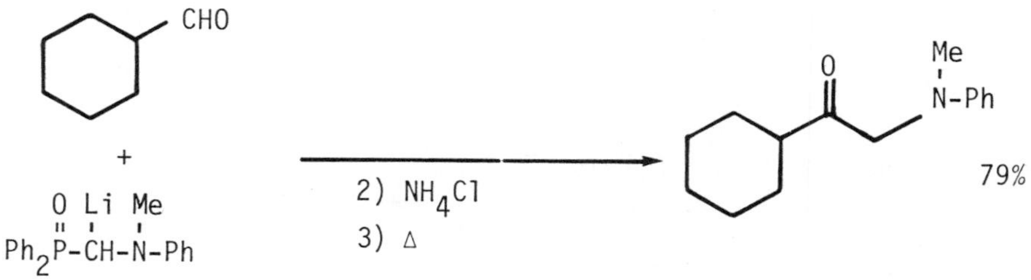

Tetr Lett, <u>22</u>, 2799 (1981)

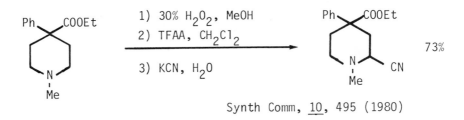

$$CH_3-CH-CN \atop NMe_2$$

1) LDA, THF

2) ⬡—CHO

3) Δ

72%

Tetr Lett, 23, 639 (1982)

Review: "Amino Acids as the Amine Component in the Mannich Reaction"

Russ Chem Rev, 51, 387 (1982)

Section 355 Amine - Nitrile

1) 30% H_2O_2, MeOH
2) TFAA, CH_2Cl_2

3) KCN, H_2O

73%

Synth Comm, 10, 495 (1980)

Section 356 Amine - Olefin

$$(MeO)_2 \overset{O}{\overset{\|}{P}} CHN_2$$

t-BuOK, Et_2NH

69%

Tetr Lett, 21, 2041 (1980)

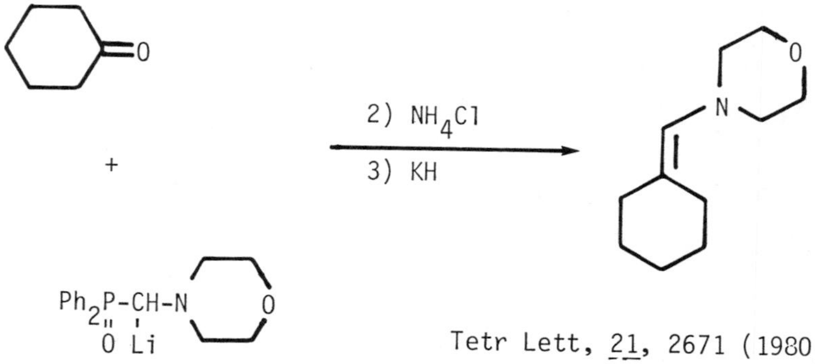
O
+
2) NH_4Cl
3) KH
N
O
Ph_2P-CH-N
O
O Li

92%

Tetr Lett, 21, 2671 (1980)

Review: "Enamines: Recent Advances in Synthetic, Spectroscopic, Mechanistic, and Stereochemical Aspects."

Tetrahedron, 38, 1975 (1982)
Tetrahedron, 38, 3363 (1982)

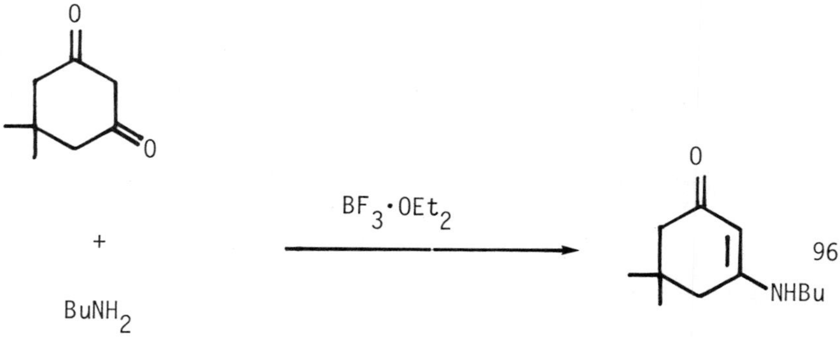
O
O
+
$BuNH_2$
$BF_3 \cdot OEt_2$
O
NHBu

96%

Synthesis, 880 (1981)

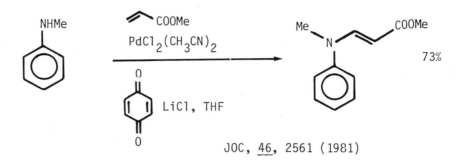

JOC, <u>46</u>, 2561 (1981)

Tetr Lett, <u>21</u>, 3763 (1980)

Nouveau J Chem, <u>4</u>, 727 (1980)

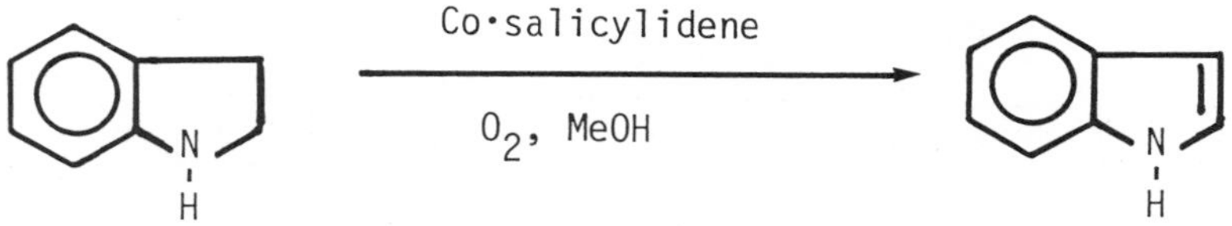

Chem Lett, 1287 (1980)

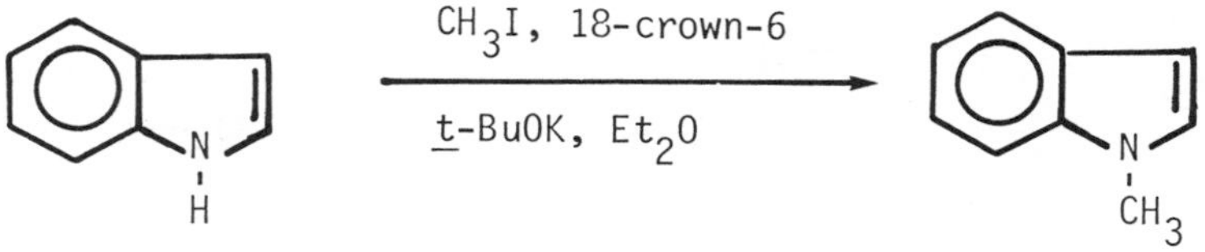

87%

JOC, 45, 3172 (1980)

Section 357 Ester - Ester

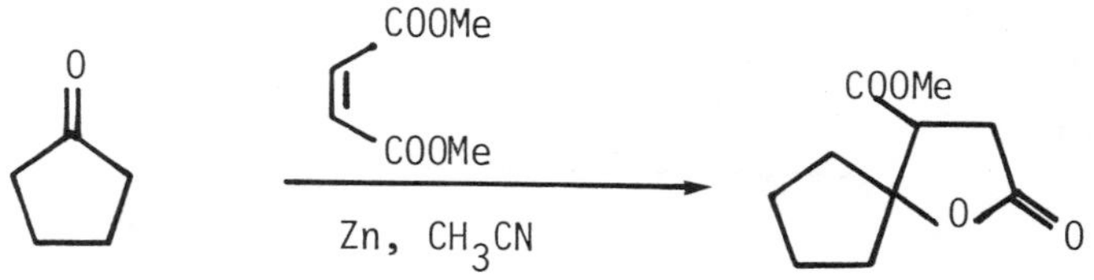

66%

Chem Lett, 1217 (1981)

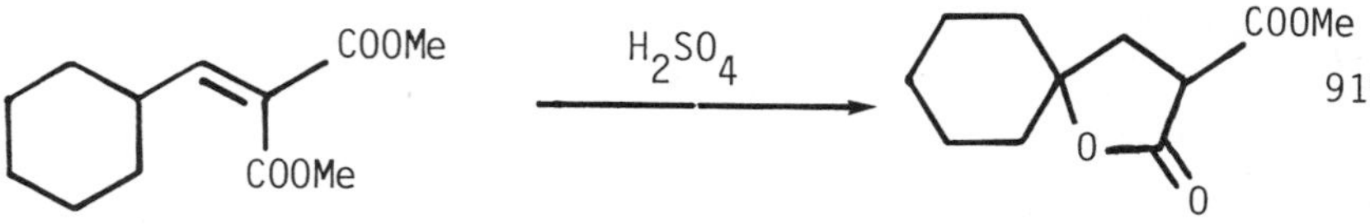

91%

Synth Comm, 11, 35 (1981)

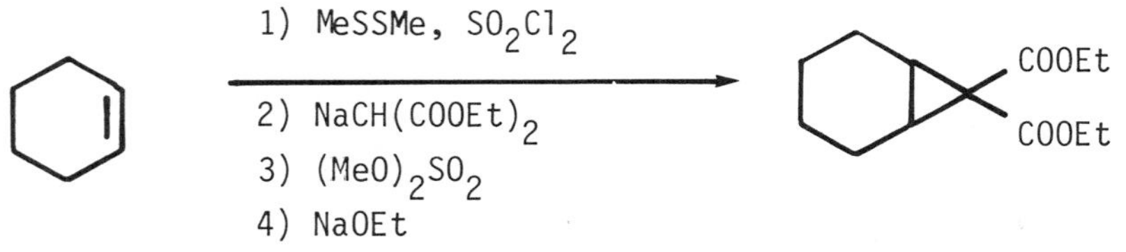

1) MeSSMe, SO_2Cl_2

2) NaCH(COOEt)$_2$
3) (MeO)$_2SO_2$
4) NaOEt

COOEt   62%
COOEt

Synthesis, 690 (1980)

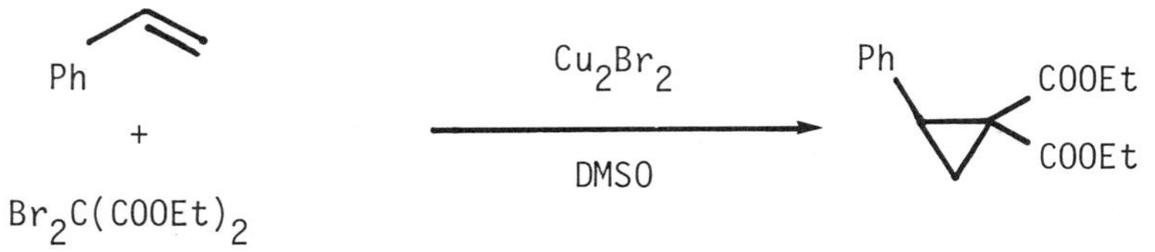

Ph

+

Br$_2$C(COOEt)$_2$

Cu_2Br_2

DMSO

Ph   COOEt
COOEt   71%

Bull Chem Soc Japan, 54, 2539 (1981)

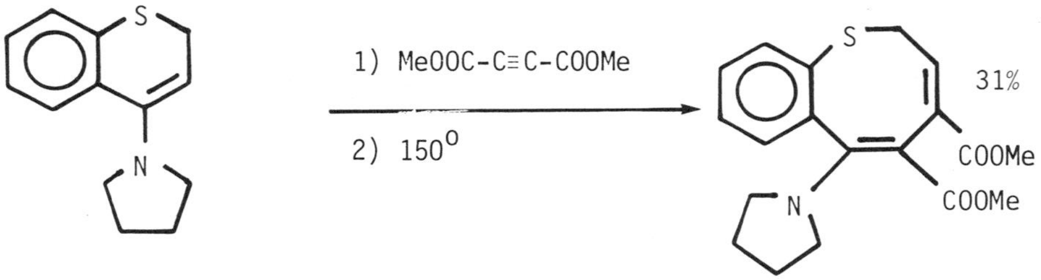

1) MeOOC-C≡C-COOMe

2) 150°

31%

COOMe
COOMe

Acta Chem Scand B, 435 (1982)

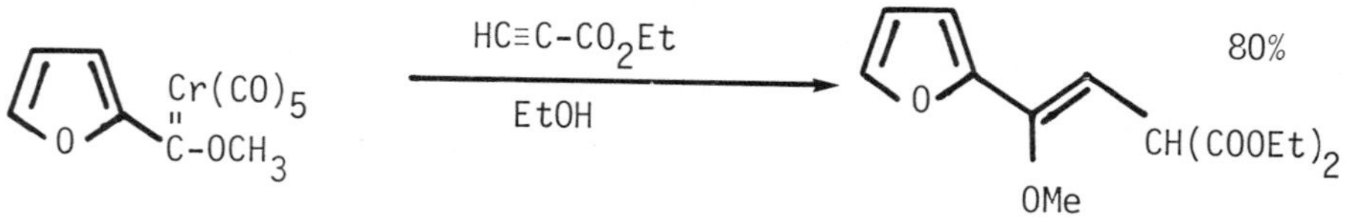

80%

Tetr Lett, 23, 3765 (1982)

Also via: Dicarboxylic acids (Section 312); Hydroxyesters (Section
327)' Diols (Section 323)

Section 358 Ester - Ether

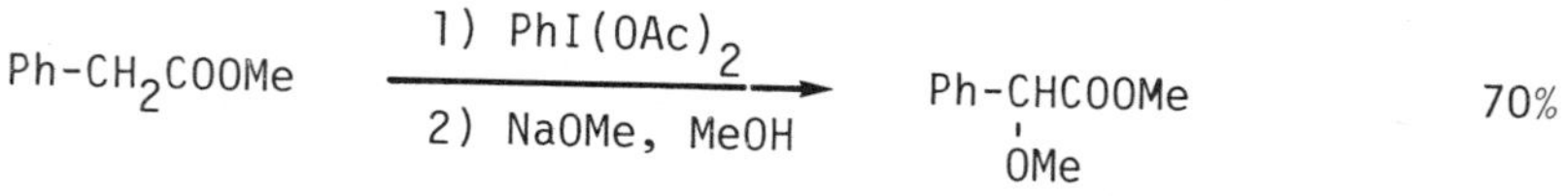

70%

Tetr Lett, 22, 2747 (1981)

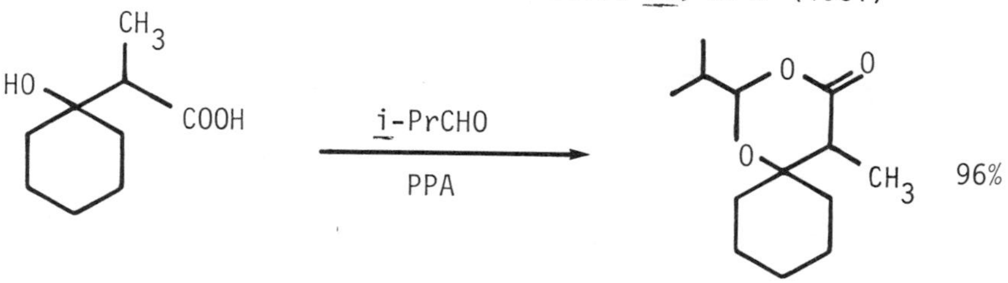

96%

Aust J Chem, 33, 685 (1980)

Section 359 Ester - Halide

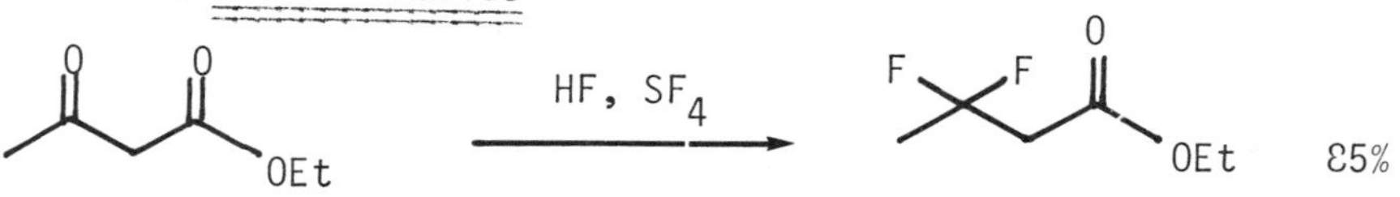

85%

JOC (USSR), 17, 1260 (1981)

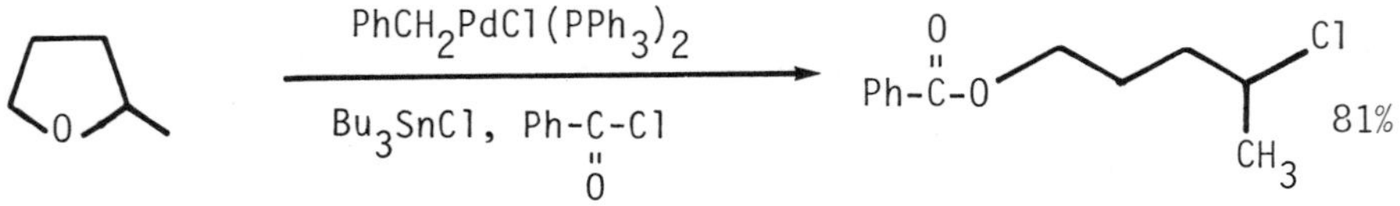
PhCH2PdCl(PPh3)2
Bu3SnCl, Ph-C-Cl
O
O
Ph-C-O
Cl
CH3
81%

JOC, 47, 1215 (1982)

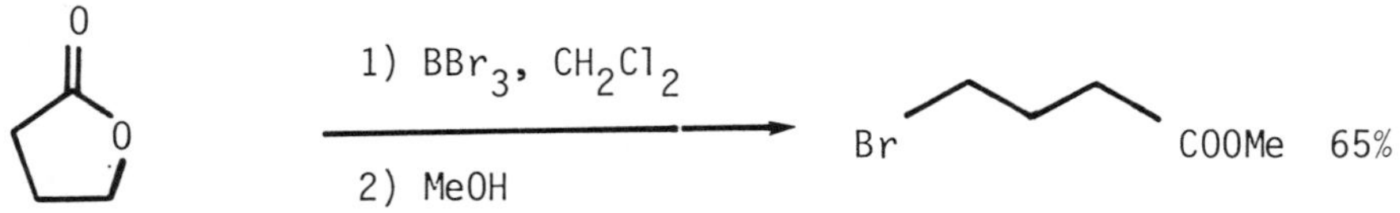
O
O
1) BBr3, CH2Cl2
2) MeOH
Br
COOMe
65%

Synthesis, 963 (1982)

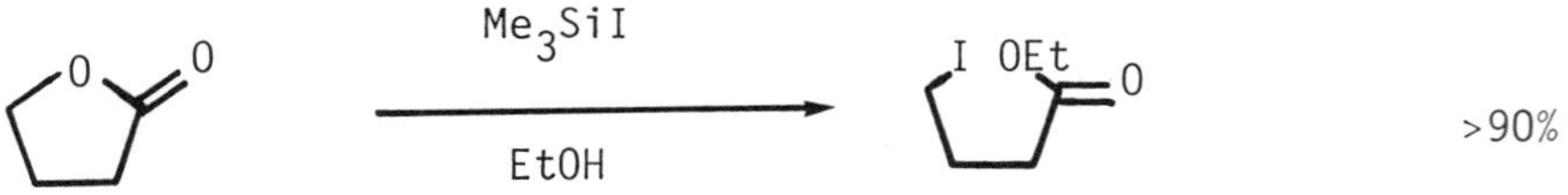
O
O
Me3SiI
EtOH
I OEt
O
>90%

Synth Comm, 11, 763 (1981)

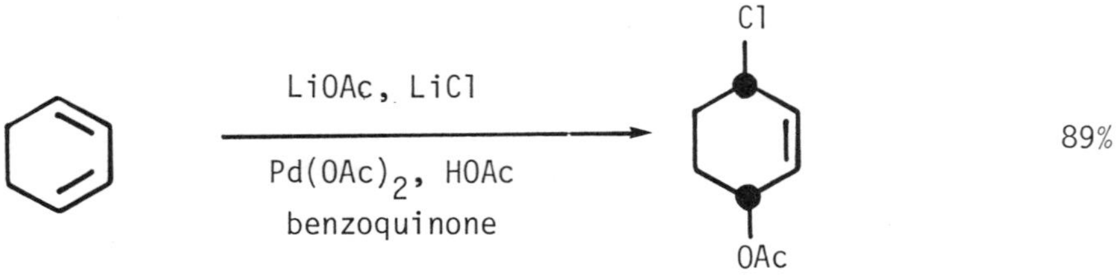
LiOAc, LiCl
Pd(OAc)2, HOAc
benzoquinone
Cl
OAc
89%

Tetr Lett, 23, 1617 (1982)

81%

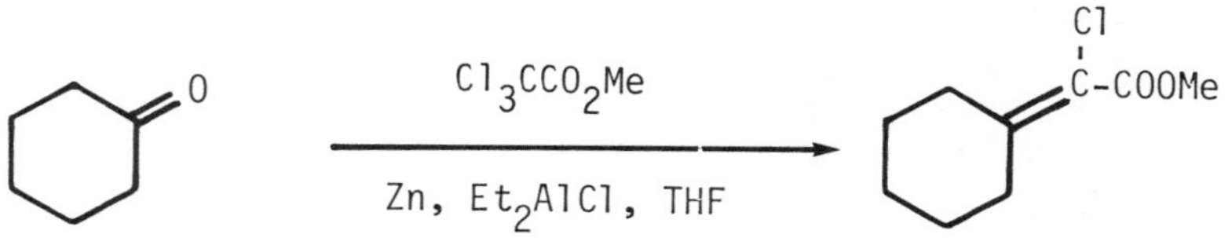

Bull Chem Soc Japan, 53, 1698 (1980)

90%

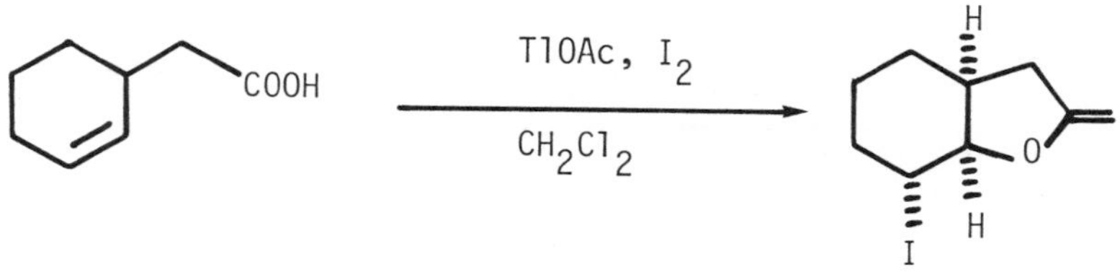

Aust J Chem, 32, 2793 (1979)

60-90%

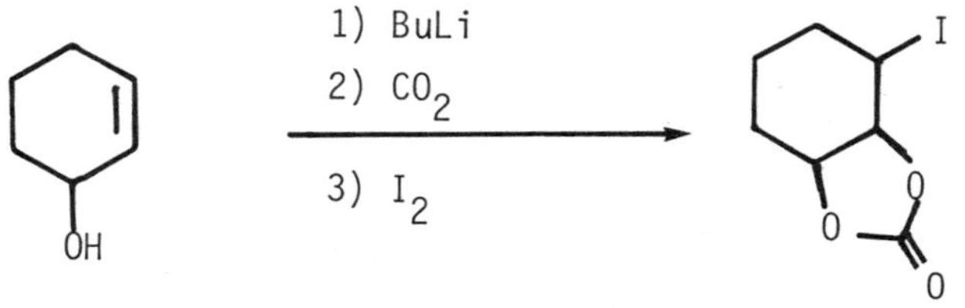

JCS Chem Comm, 465 (1981)

$$C_2F_5\overset{O}{\overset{\|}{C}}\text{-OCl} \xrightarrow[\text{NaF}]{CF_2=CH_2,\ ClF} C_2F_5\overset{O}{\overset{\|}{C}}\text{-OCF}_2CH_2Cl$$

80%

JOC, 45, 1214 (1980)

Also via: Haloacids (Section 319); Halohydrins (Section 329)

Section 360 <u>Ester - Ketone</u>

$$\text{(i-Pr)}-C\equiv C-OMe \xrightarrow[\text{RuCl}_2\text{L}_3]{\text{Ph-I=O}} \text{(i-Pr)}-\overset{O}{\overset{||}{C}}-\overset{O}{\overset{||}{C}}-OMe \qquad 60\%$$

Tetr Lett, <u>23</u>, 3661 (1982)

BuMgBr

+

$$\text{EtO}-\overset{O}{\overset{||}{C}}-\overset{O}{\overset{||}{C}}-OEt \xrightarrow[\text{2) H}_3\text{O}^+]{\text{THF}} \text{Bu}-\overset{O}{\overset{||}{C}}-\overset{O}{\overset{||}{C}}-OEt \qquad 75\%$$

Synth Comm, <u>11</u>, 943 (1981)

Ph-MgBr

+

imidazole-$\overset{O}{\overset{||}{N-C}}-\overset{O}{\overset{||}{C}}-OEt \longrightarrow \text{Ph}-\overset{O}{\overset{||}{C}}-\overset{O}{\overset{||}{C}}-OEt \qquad 72\%$

JOC, <u>46</u>, 211 (1981)

$$\overset{N_3}{\underset{|}{\text{PhCH}_2\text{CHCO}_2\text{Et}}} \xrightarrow[\text{2) HCl, H}_2\text{O}]{\text{1) LiOEt}} \text{PhCH}_2\overset{O}{\overset{||}{C}}-CO_2\text{Et} \qquad 94\%$$

JOC, <u>45</u>, 4952 (1980)

$$(\text{EtO})_3\text{CCN} \xrightarrow[\text{2) H}_3\text{O}^+]{\text{1) BuLi}} \text{Bu}-\overset{O}{\overset{||}{C}}-\overset{O}{\overset{||}{C}}-OEt$$

Tetr Lett, <u>22</u>, 1509 (1981)

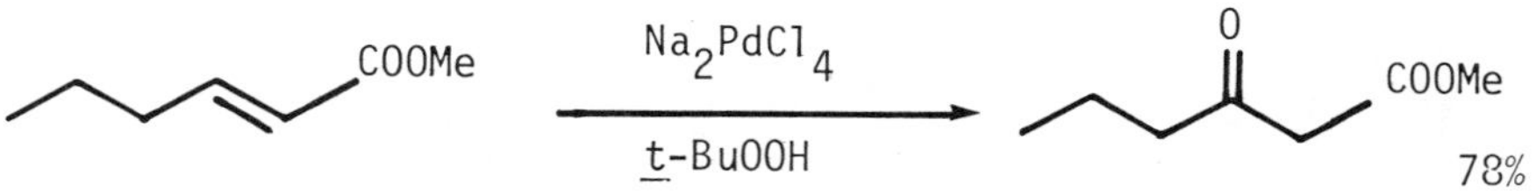

Chem Lett, 257 (1980)

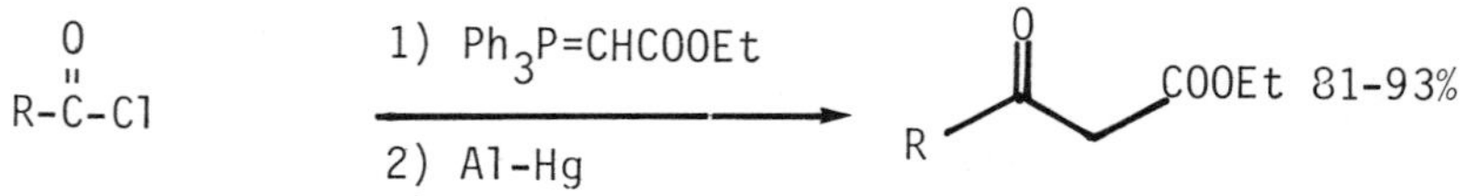

JOC, 47, 4955 and 4963 (1982)

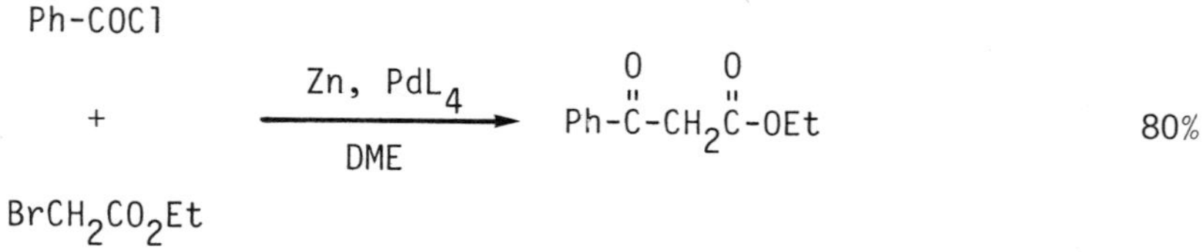

Ph-COCl

+

BrCH$_2$CO$_2$Et

$\xrightarrow[\text{DME}]{\text{Zn, PdL}_4}$

$\underset{\text{O}}{\overset{\text{O}}{\text{Ph-C-CH}_2\text{C-OEt}}}$ 80%

Chem Lett, 1559 (1982)

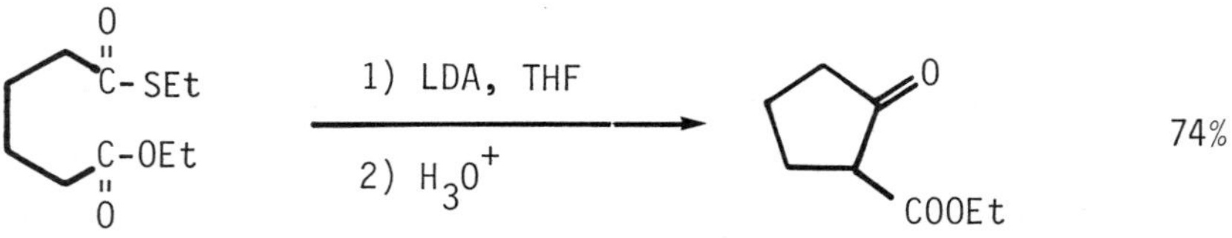

Tetr Lett, 22, 1353 (1981)

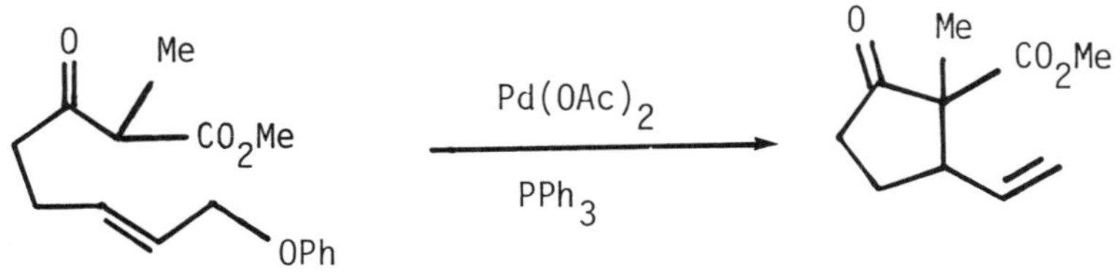

Tetr Lett, 21, 1475 (1980)

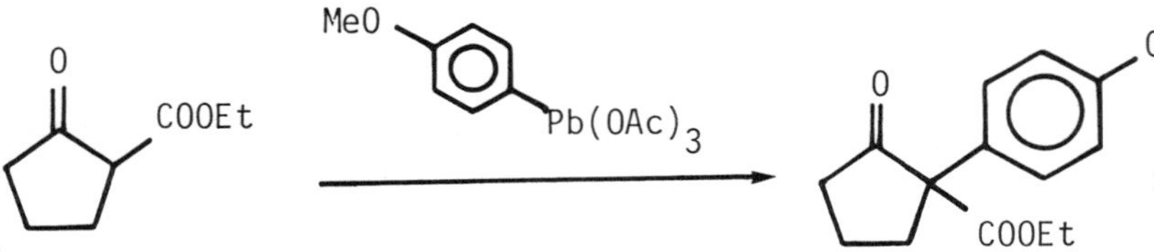

Aust J Chem, 33, 113 (1980)

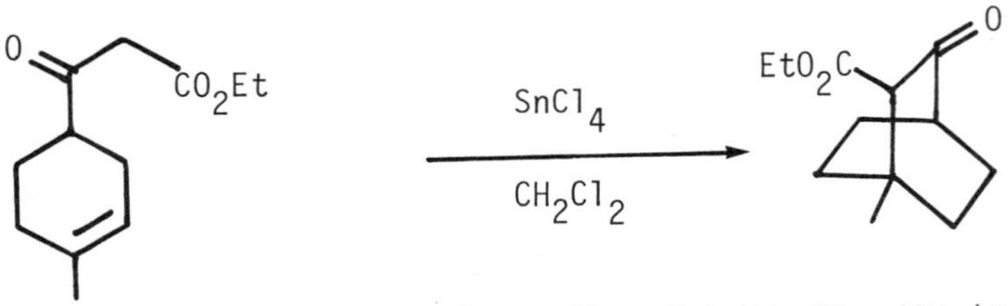

Angew Chem Int Ed, 20, 687 (1981)

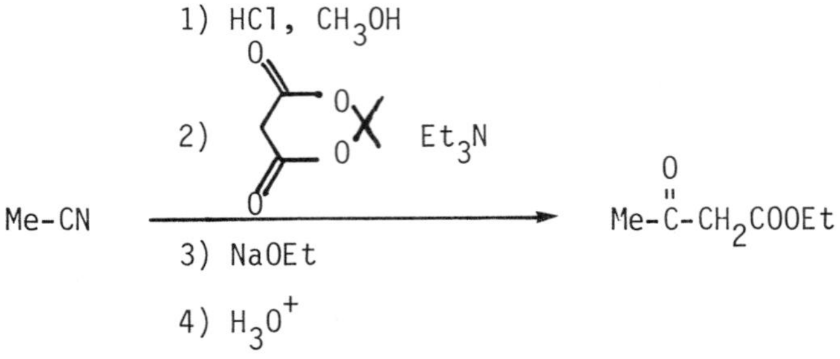

1) HCl, CH$_3$OH

2) [structure] Et$_3$N

Me-CN → Me-C-CH$_2$COOEt 85%

3) NaOEt

4) H$_3$O$^+$

Synthesis, 130 (1981)

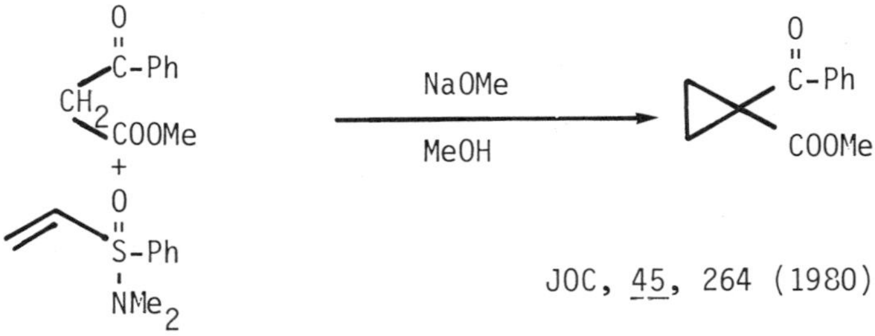

JOC, 45, 264 (1980)

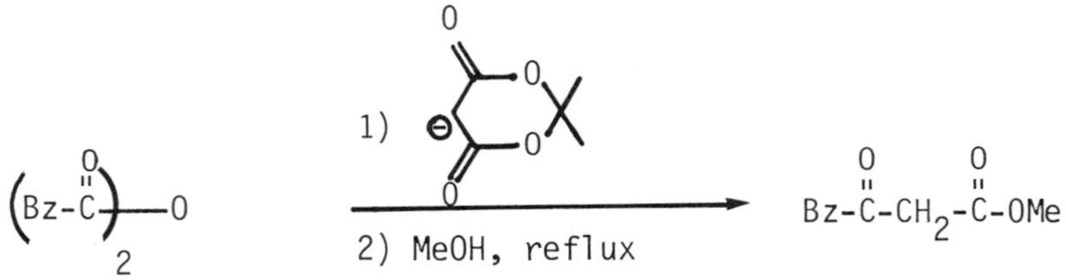

Synthesis, 451 (1982)

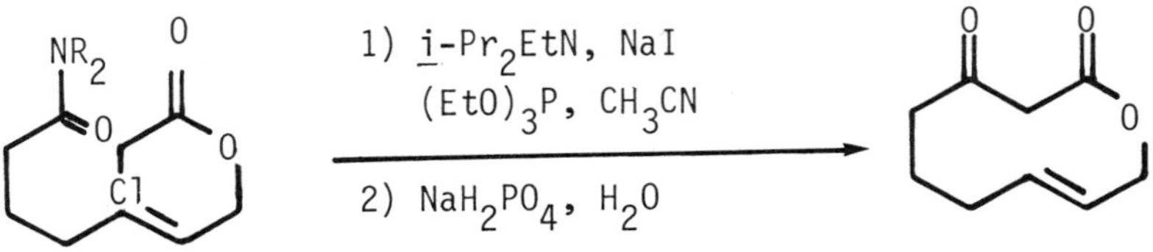

JOC, 45, 1868 (1980)

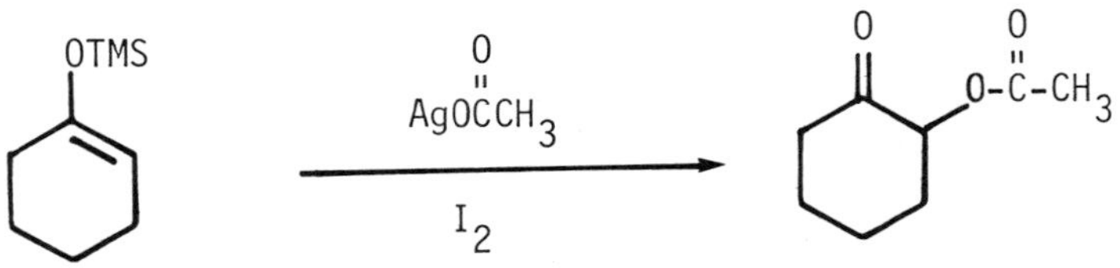

JOC, 46, 2717 (1981)

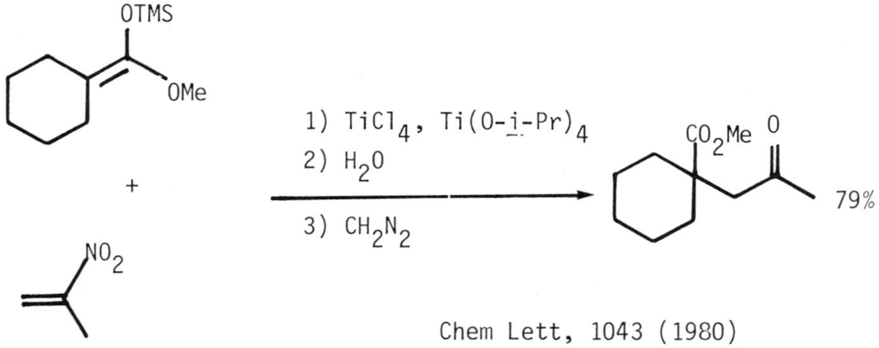
OTMS
OMe
+
NO_2
1) $TiCl_4$, Ti(O-i-Pr)$_4$
2) H_2O
3) CH_2N_2
CO_2Me
O
79%

Chem Lett, 1043 (1980)

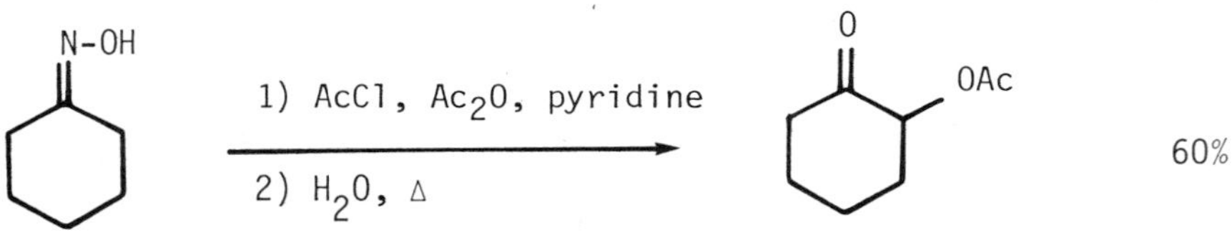
N-OH
1) AcCl, Ac_2O, pyridine
2) H_2O, Δ
O
OAc
60%

Synthesis, 223 (1981)

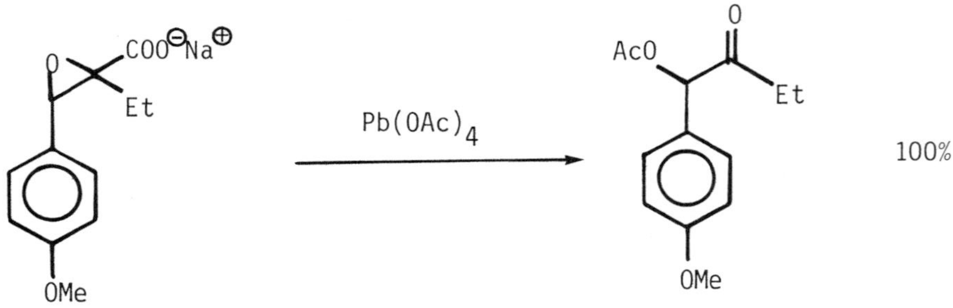
$COO^{\ominus}Na^{\oplus}$
O
Et
OMe
$Pb(OAc)_4$
O
AcO
Et
OMe
100%

Chem Lett, 879 (1982)

Also via: Ketoacids (Section 320); Hydroxyketones (Section 330)

Section 361 Ester - Nitrile

Tetr Lett, <u>23</u>, 4927 (1982)

JOC, <u>45</u>, 2614 (1980)

Angew Chem Int Ed, <u>21</u>, 130 (1982)

Section 362 Ester - Olefin

This section contains syntheses of enol esters and esters of unsaturated acids.

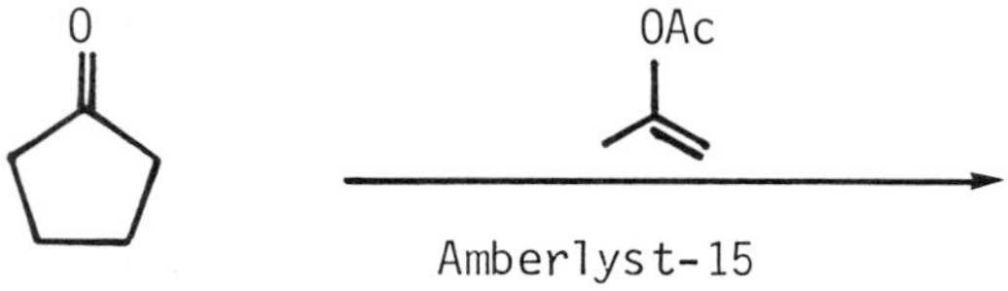

Indian J Chem, 21B, 358 (1982)

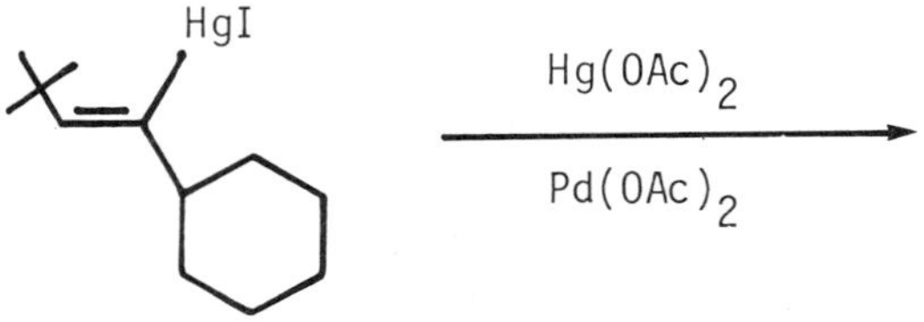

JACS, 102, 1966 (1980)

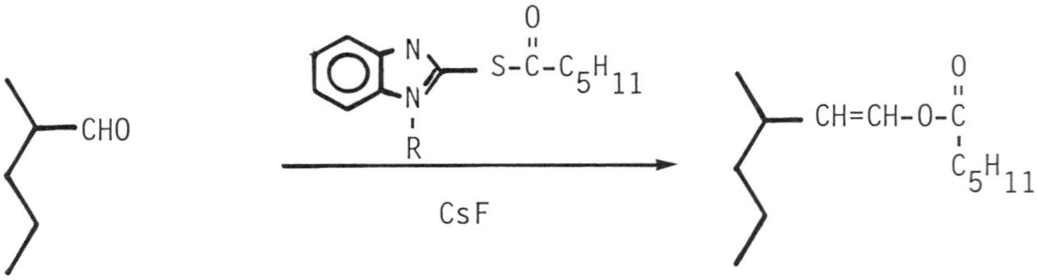

Chem Lett, 529 (1980)

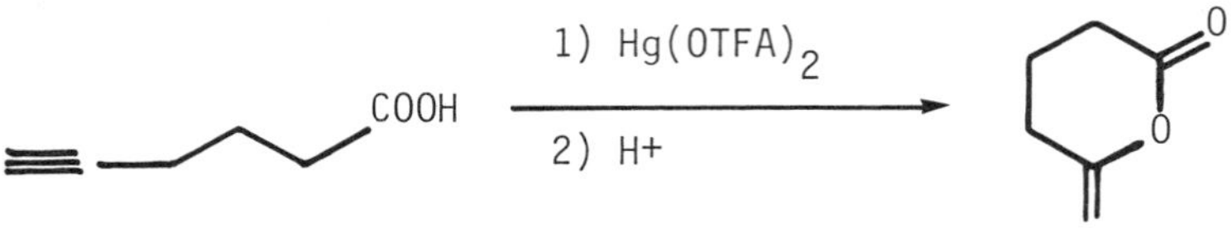

JACS, 103, 5459 (1981)

Related methods: Protection of Aldehydes (Section 60A); Protection

of Ketones (Section 180A)

$$\text{TMS} \diagdown \overset{\ominus}{} \diagup \text{COOMe} \xrightarrow[\text{2) H}^{\oplus}, \text{ BF}_3 \cdot \text{Et}_2\text{O}]{} \overset{\text{Ph}}{\diagdown} \diagup_{\text{COOMe}} \qquad 78\%$$

+

Ph-CHO $\xrightarrow[\text{HMPT}]{\text{similar}}$ Ph COOMe $\sim$60%

JCS Chem Comm, 877 (1981)

CO, MeOH

PdCl$_2$, NaOAc   95%

Chem Lett, 879 (1981)

MeO-C-OTMS

1) :C(Cl)(CH$_3$)

2) Et$_3$N, MeOH   87%

Synthesis, 58 (1982)

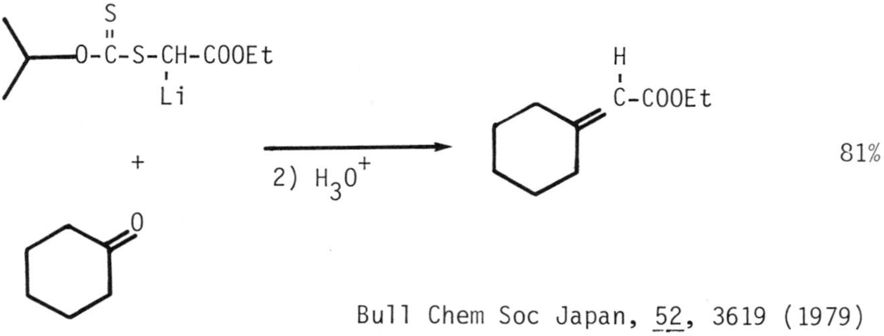

81%

Bull Chem Soc Japan, 52, 3619 (1979)

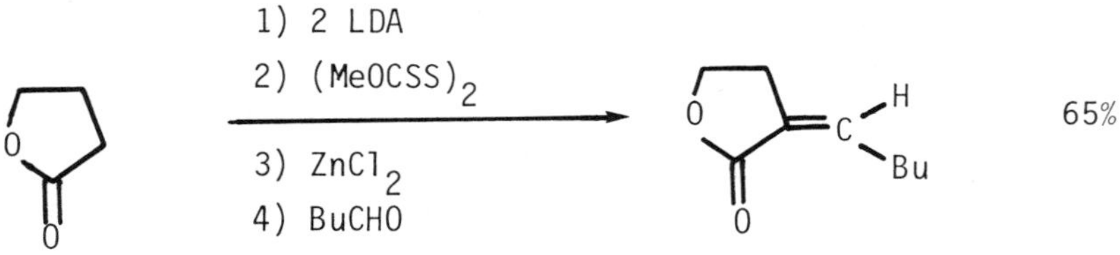

65%

(E isomer is formed in 58% yield if ZnCl$_2$ is omitted.)

Chem Lett, 595 (1980)

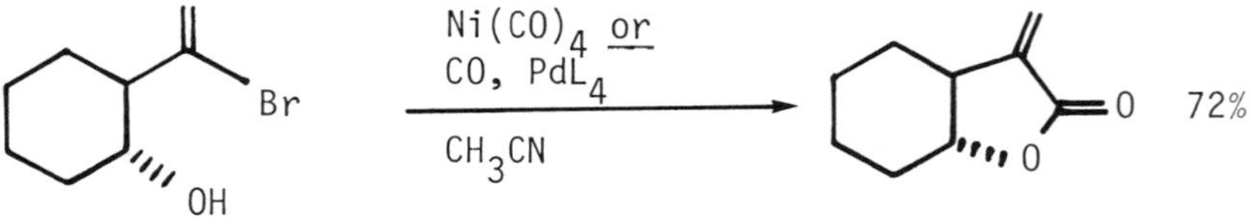

72%

JOC, 46, 1723 (1981)
JOC, 47, 3630 (1982)

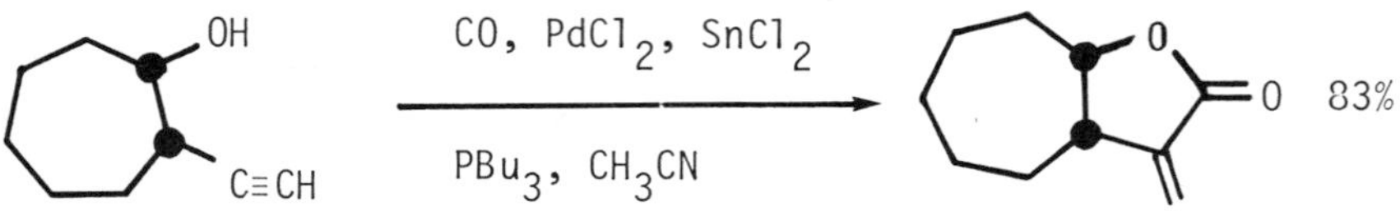

83%

JACS, 103, 7520 (1981)

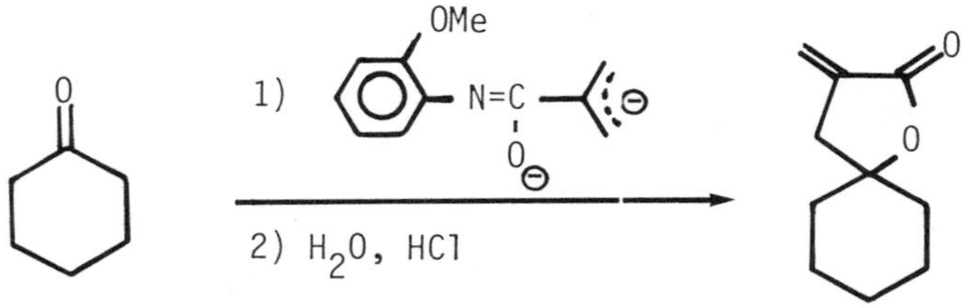

1)

OMe

N=C⟨ ⟩⊖

O⊖

2) H$_2$O, HCl

~50%

Chem Lett, 1567 (1980)

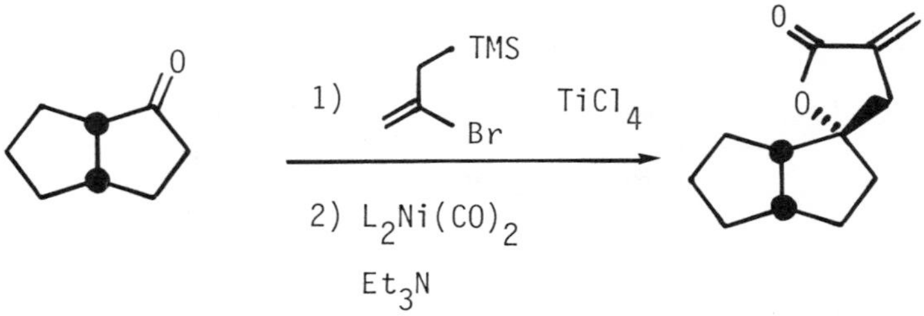

1)

TMS

Br

TiCl$_4$

2) L$_2$Ni(CO)$_2$

Et$_3$N

88%

JACS, 104, 6879 (1982)

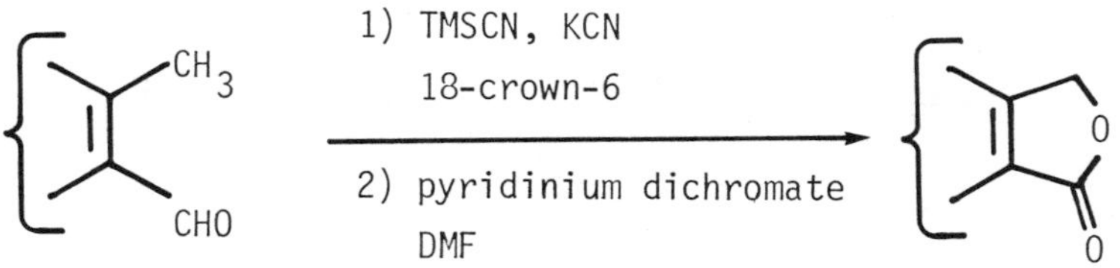

1) TMSCN, KCN
   18-crown-6

2) pyridinium dichromate
   DMF

~70%

Tetr Lett, 21, 731 (1980)

1) NaH

   SPh

2) Me-CH-Cl

Et-CH(COOEt)$_2$

3) NaBr, HMPT

MeCH=C⟨COOEt / Et⟩

73%

Synthesis, 131 (1982)

$$\text{Tetr Lett, } \underline{23}, 3683 \ (1982)$$

$$\text{JOC, } \underline{45}, 4135 \ (1980)$$

$$\text{JOC, } \underline{45}, 264 \ (1980)$$

$$\text{JCS Chem Comm, } 821 \ (1982)$$

Also via: Acetylenic esters (Section 306); Olefinic acids (Section
           322); β-Hydroxyesters (Section 327)

Section 363 Ether - Ether

See Section 60A (Protection of Aldehydes) and Section 180A (Protection of Ketones) for reactions involving the formation of acetals and ketals.

No additional examples.

Section 364 Ether - Halide

Bull Chem Soc Japan, 53, 219 (1980)

Tetr Lett, 21, 2005 (1980)

$$H_2C=C \overset{CH_2OEt}{\underset{CH_3}{\diagdown}} \quad \xrightarrow[\text{MeOH}]{\text{NBS}} \quad BrCH_2-\overset{CH_3}{\underset{OMe}{\overset{|}{C}}}-CH_2OEt \qquad 62\%$$

Z Chem, <u>20</u>, 209 (1980)

$(F_3C)_2C=O$

+

$F_2C=CFCF_2OSO_2F$

$\xrightarrow{\text{KF}}$ $(F_3C)_2CFOCF_2CF=CF_2$

JACS, <u>103</u>, 5598 (1981)

Section 365   Ether, Epoxide - Ketone

$\xrightarrow[\text{HC(OMe)}_3, \text{ CH}_3\text{OH}]{\text{Tl(NO}_3)_3}$

30-50%

JOC, <u>46</u>, 3326 (1981)

$\xrightarrow[\text{HgO}]{\underline{t}\text{-BuOCl, PdL}_4}$

64%

Chem Lett, 369 (1982)

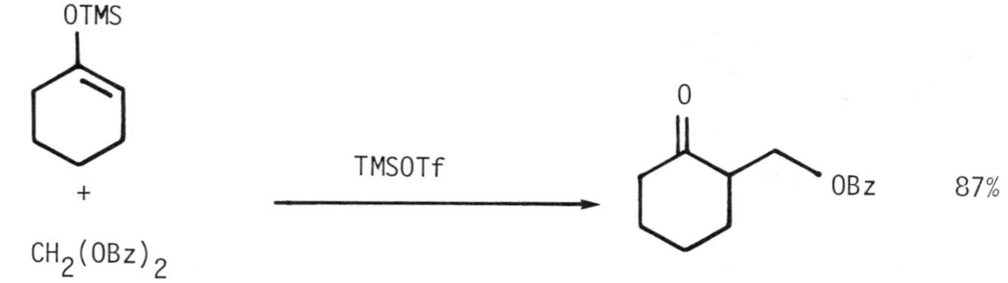

Tetr Lett, 21, 2527 (1980)

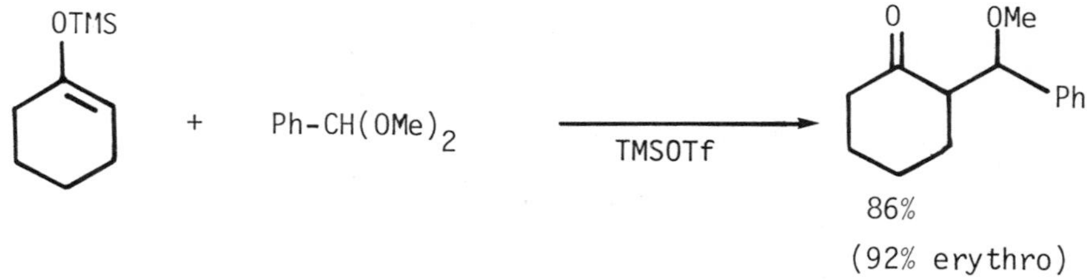

86%

(92% erythro)

JACS, 102, 3248 (1980)

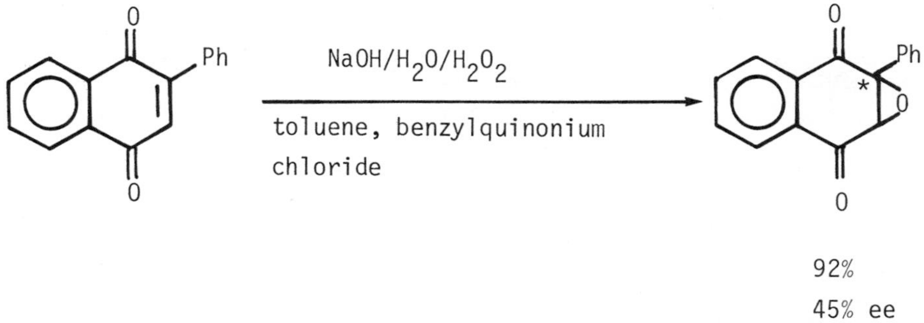

92%

45% ee

JOC, 45, 2498 (1980)

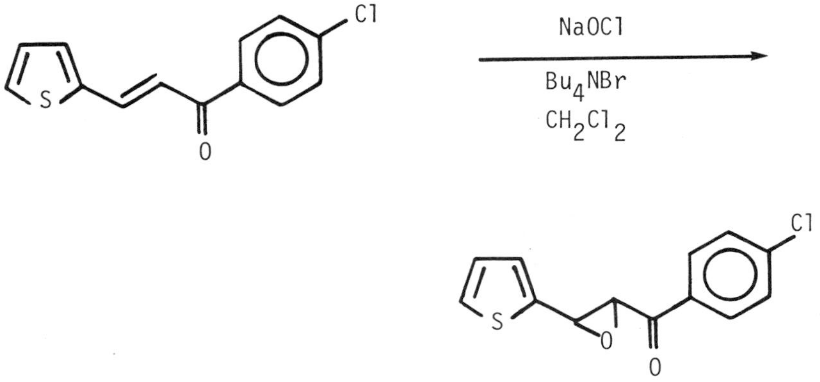

NaOCl

Bu$_4$NBr

CH$_2$Cl$_2$

82%

Gazz Chim Ital, 110, 267 (1980)

Section 366 Ether, Epoxide - Nitrile

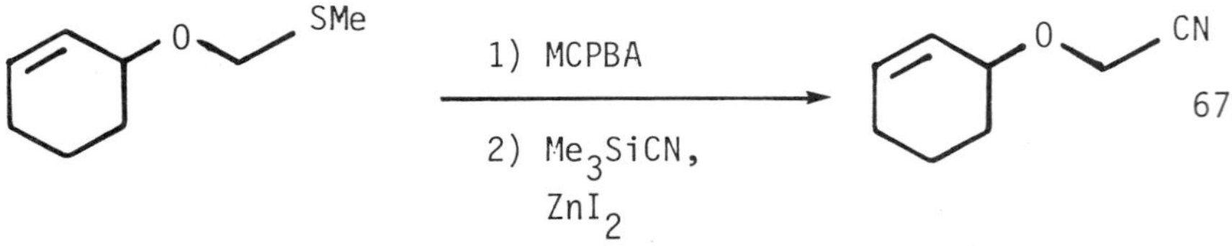

1) MCPBA

2) Me$_3$SiCN, ZnI$_2$

67%

Tetr Lett, 22, 4925 (1981)

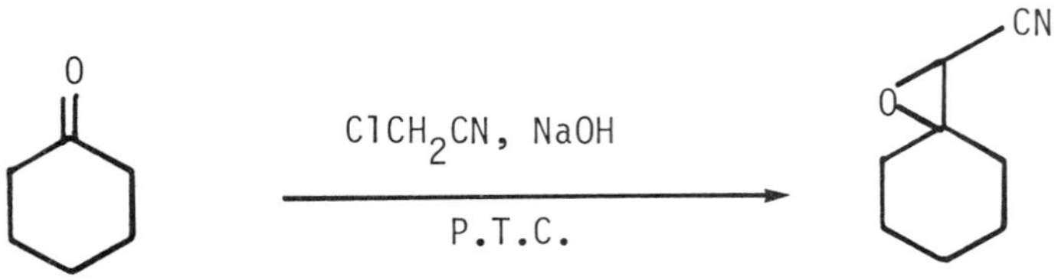

ClCH$_2$CN, NaOH

P.T.C.

76%

Bull Chem Soc Japan, 53, 1463 (1980)

Section 367 Ether - Olefin

Related methods: Protection of Ketones (Section 180A).

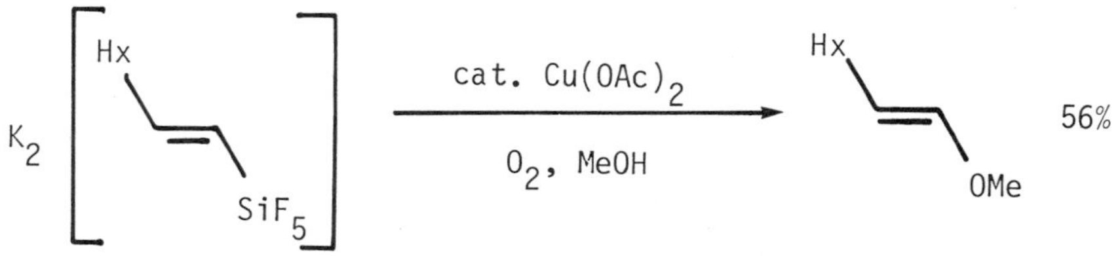

Tetr Lett, 21, 4105 (1980)

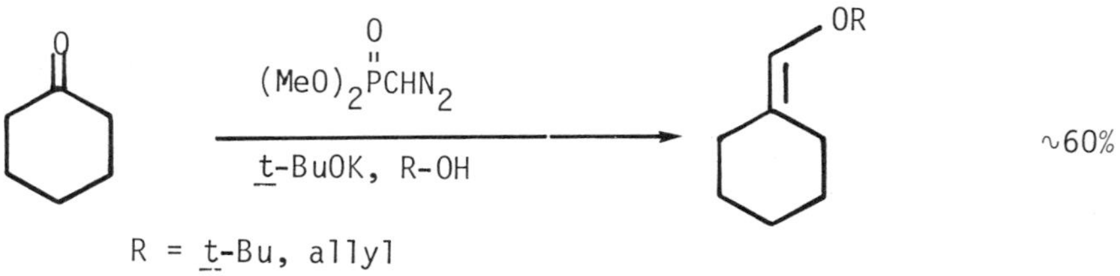

~60%

R = t-Bu, allyl

Tetr Lett, 21, 2041 (1980)
Tetr Lett, 21, 5003 (1980)

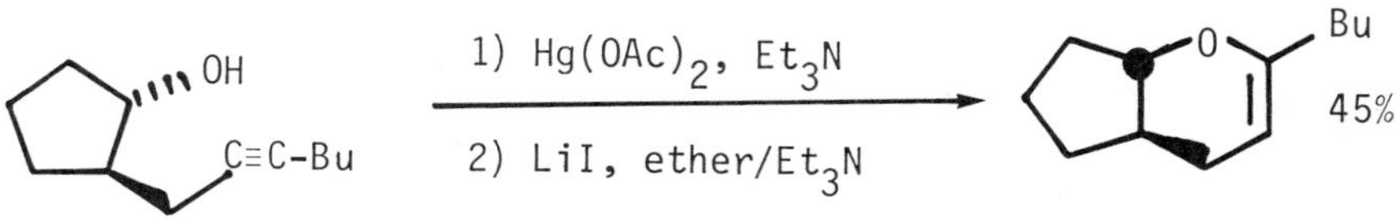

JACS, 104, 5842 (1982)

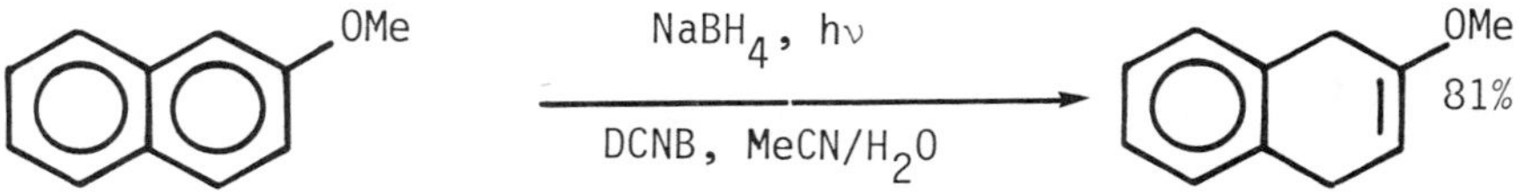

JOC, 46, 788 (1981)

Organometallics, 1, 1467 (1982)
Tetr Lett, 23, 323 (1982)

JOC (USSR), 18, 395 (1982)

Tetr Lett, 22, 745 (1981)

1) ClCH=CHCH$_2$TMS
   AlCl$_3$

2) NaBH$_4$, MeOH
3) NaOH

Tetr Lett, 21, 4369 (1980)

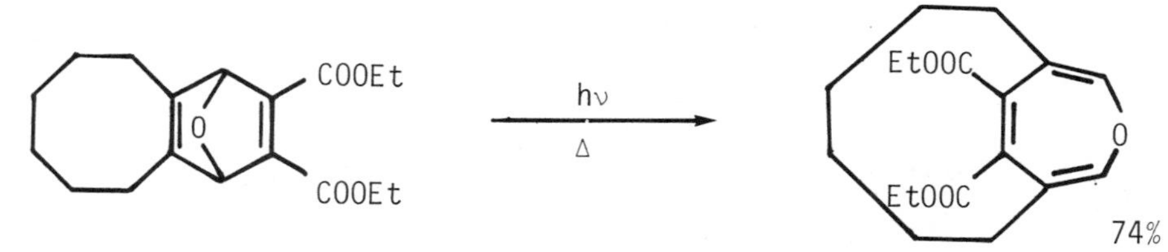

Chem Ber, <u>114</u>, 3725 (1981)

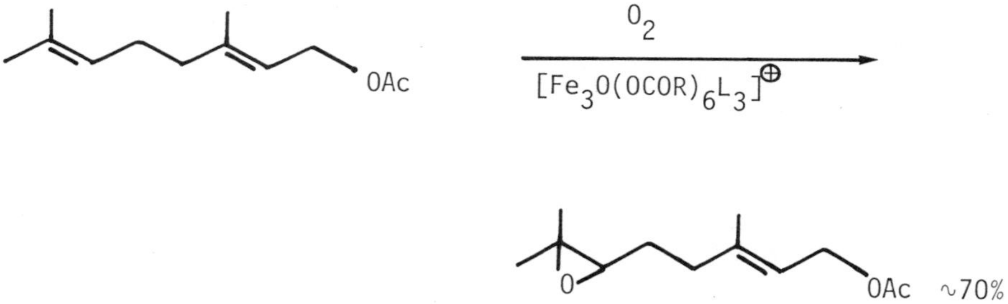

JACS, <u>104</u>, 6450 (1982)

Section 368 <u>Halide - Halide</u>

Halocyclopropanations are found in Section 74 (Alkyls from Olefins).

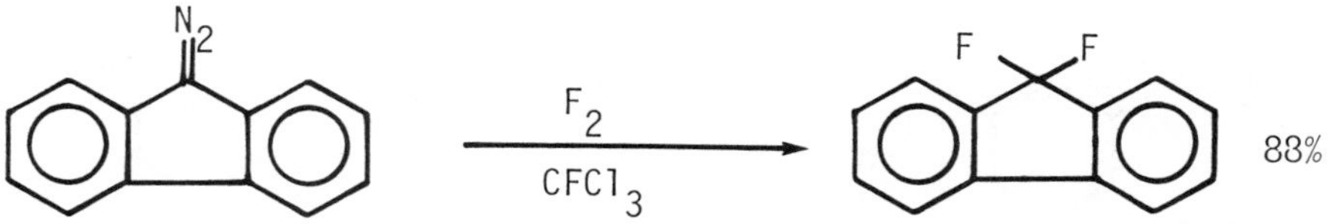

JOC, <u>46</u>, 3917 (1981)

1) LiCH(TePh)$_2$
2) Br$_2$, NaBr

R-X $\xrightarrow{\hspace{3cm}}$ R-CHBr$_2$ 76-85%

R = 1^0 alkyl

Chem Lett, 1081 (1982)

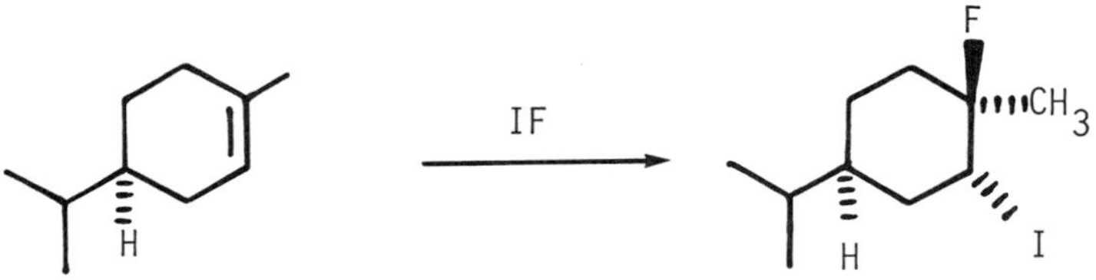

IF

45%

Tetr Lett, 21, 4543 (1980)

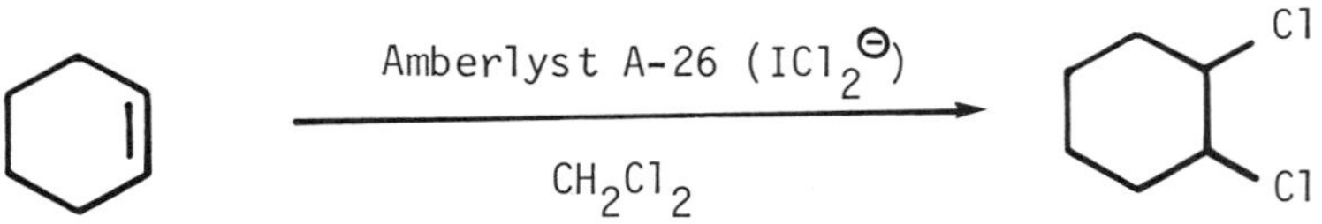

Amberlyst A-26 (ICl$_2^{\ominus}$)

$\xrightarrow{\hspace{3cm}}$

CH$_2$Cl$_2$

75%

JCS Chem Comm, 1278 (1980)

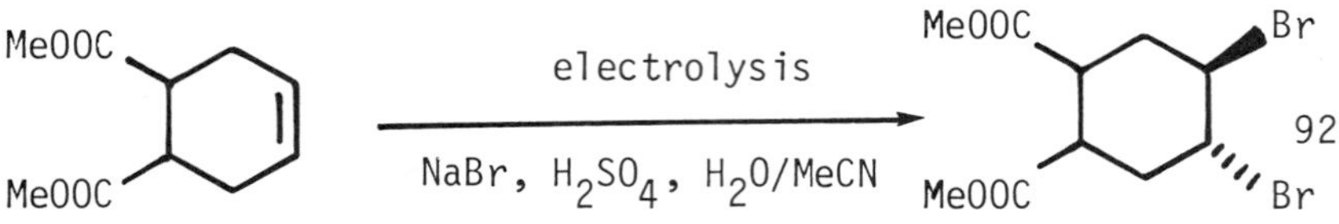

electrolysis

$\xrightarrow{\hspace{3cm}}$

NaBr, H$_2$SO$_4$, H$_2$O/MeCN

92%

JOC, 46, 3312 (1981)

Bu-CH=CH$_2$ $\xrightarrow[\text{AIBN}]{\text{CH}_2\text{ClI}}$ Bu-CH-CH$_2$-CH$_2$ 81%
　　　　　　　　　　　　　　　　　　　　|　　　|
　　　　　　　　　　　　　　　　　　　　I　　　Cl

Bull Chem Soc Japan, <u>53</u>, 770 (1980)

Chem Lett, 141 (1982)

$\xrightarrow{\text{Me}_3\text{SiBr}}$ Br-(CH$_2$)$_3$-CH
　　　　　　　　　　　　　　　　　　　　　CH$_2$OAc
　　　　　　　　　　　　　　　　　　　　　|
　　　　　　　　　　　　　　　　　　　　　Br   87%

Synthesis, 383 (1981)

$\xrightarrow[\text{CH}_3\text{CN}]{\text{CuBr}_2, \text{Me}_3\text{CONO}}$ 99%

JOC, <u>45</u>, 2570 (1980)

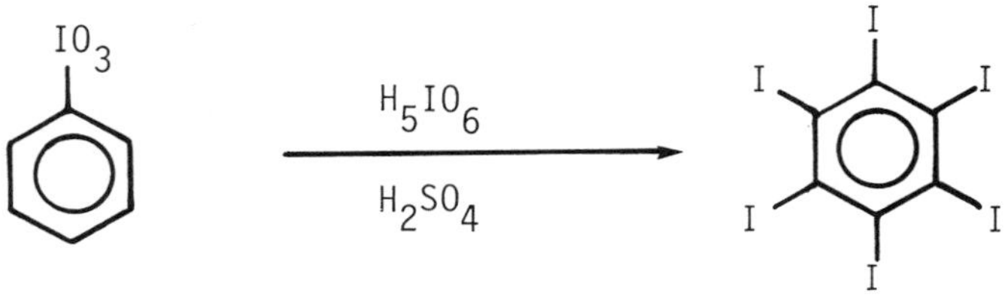
IO_3
H_5IO_6
H_2SO_4
I
I
I
I
I
I

48%

JOC, 47, 4770 (1982)

Section 369 Halide - Ketone

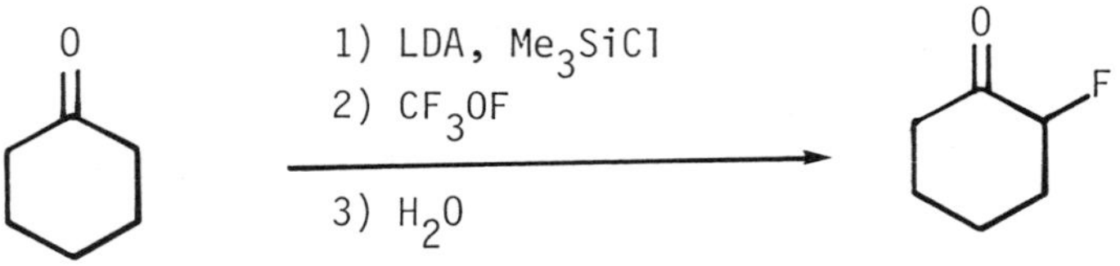
O
1) LDA, Me_3SiCl
2) CF_3OF
3) H_2O
O
F

74%

JACS, 102, 4845 (1980)

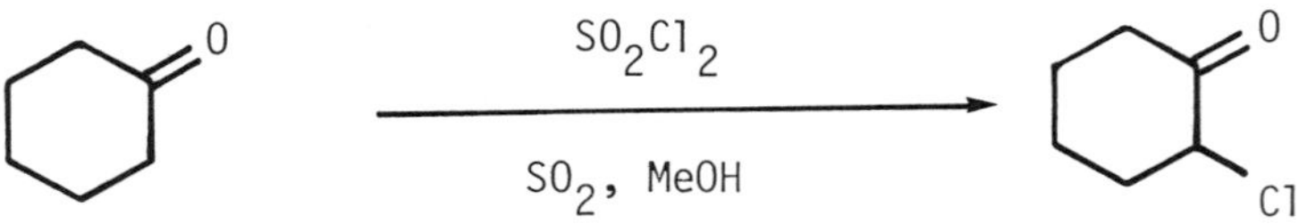
O
SO_2Cl_2
SO_2, MeOH
O
Cl

95%

JOC, 46, 4486 (1981)

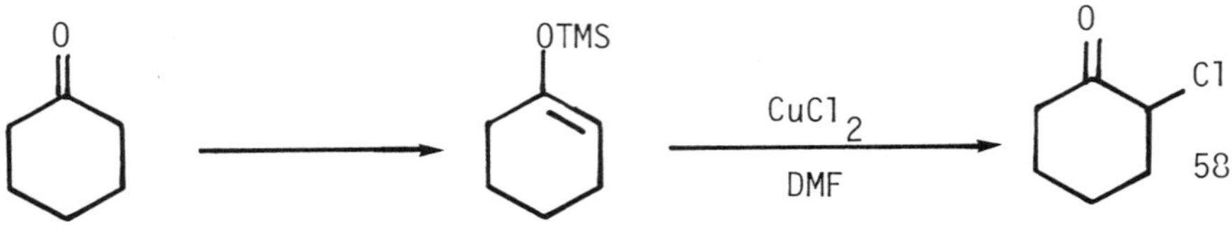
O
OTMS
$CuCl_2$
DMF
O
Cl

58%

JOC, 45, 2022 (1980)

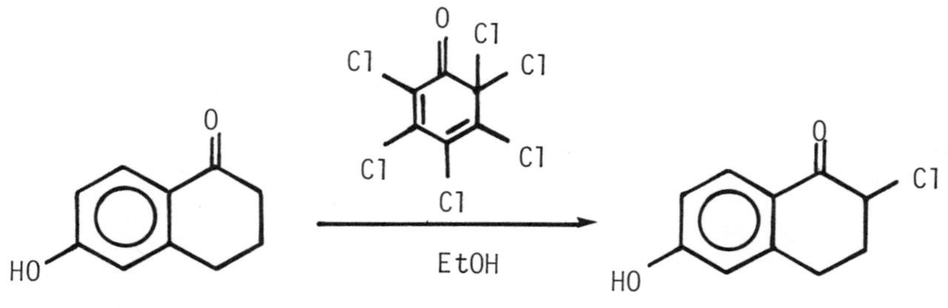

Synthesis, 1018 (1982)

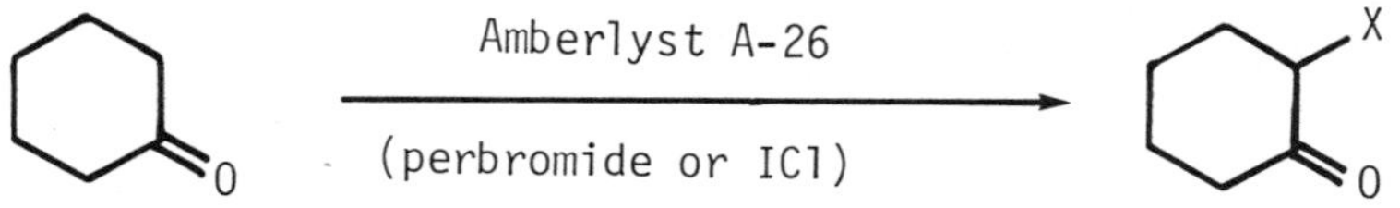

45% (X = Cl)
90% (X = Br)

JCS Chem Comm, 1278 (1980)
Synthesis, 143 (1980)

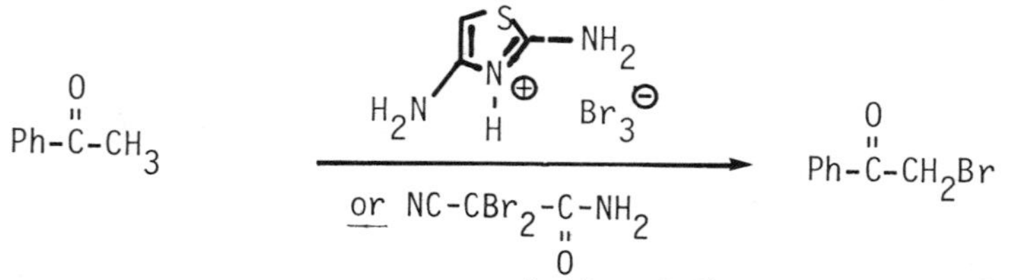

$$Ph-\overset{O}{\overset{\|}{C}}-CH_3$$

or NC-CBr$_2$-$\overset{O}{\overset{\|}{C}}$-NH$_2$

$$Ph-\overset{O}{\overset{\|}{C}}-CH_2Br \qquad 95\%$$

Indian J Chem, 17B, 305 (1979)
Synthesis, 487 (1980)

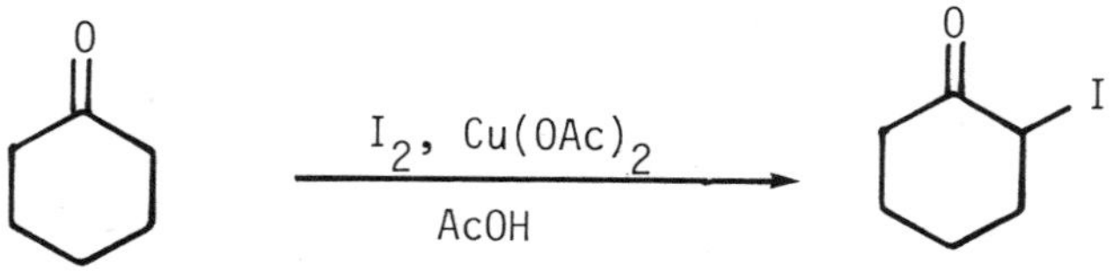

$$\frac{I_2, \ Cu(OAc)_2}{AcOH}$$

100%

Synthesis, 312 (1981)

Synthesis, 1021 (1982)

JOC, <u>46</u>, 509 (1981)

Tetr Lett, <u>21</u>, 4521 (1980)

Tetr Lett, <u>22</u>, 1429 (1981)

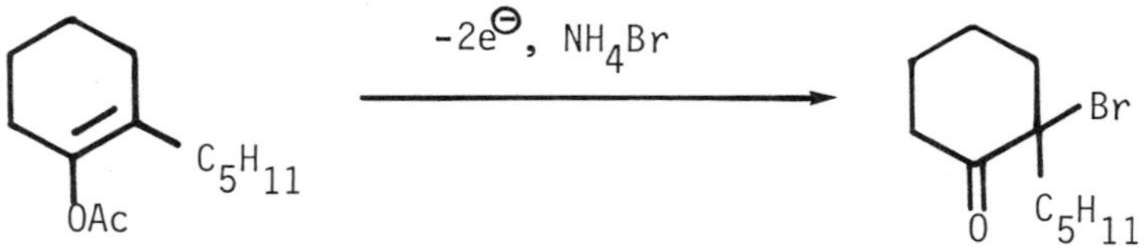

$-2e^{\ominus}$, NH_4Br
OAc
C_5H_{11}
Br
O
C_5H_{11}

83%

JOC, 45, 2731 (1980)

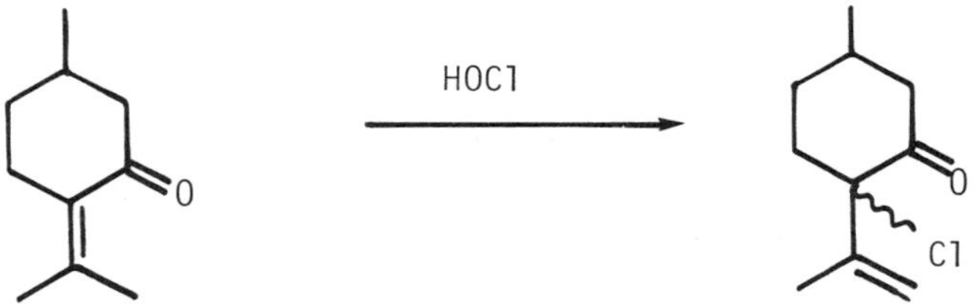

HOCl
O
O
Cl

75%

Tetr Lett, 22, 5019 (1981)

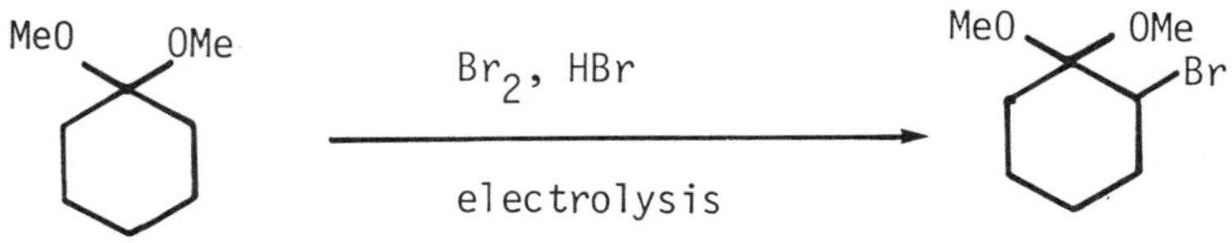

MeO
OMe
Br_2, HBr
electrolysis
MeO
OMe
Br

68%

Synth Comm, 10, 821 (1980)

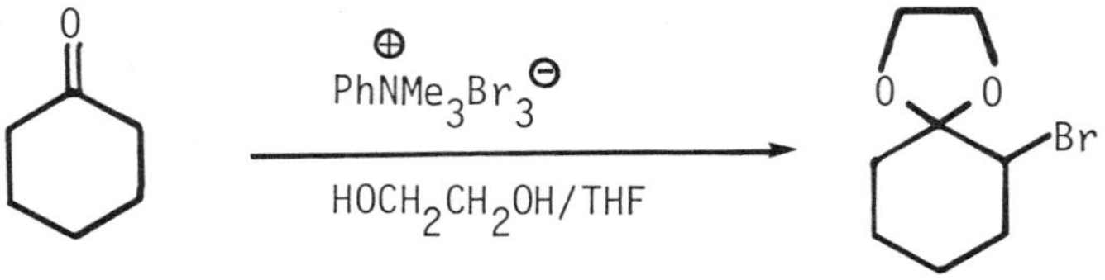

O
$PhN^{\oplus}Me_3Br_3^{\ominus}$
$HOCH_2CH_2OH/THF$
O
O
Br

89%

Synthesis, 309 (1982)

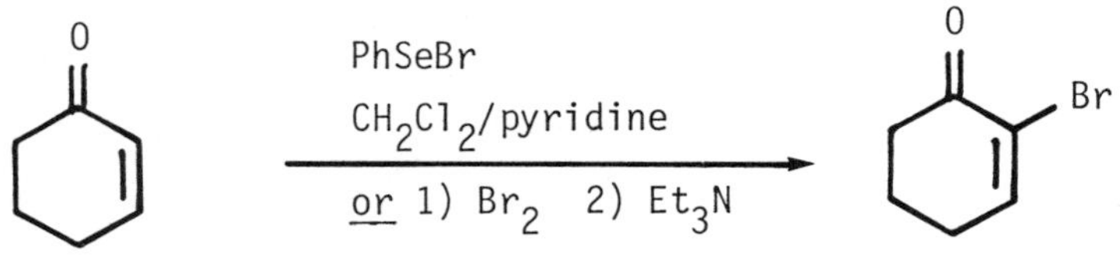
PhSeBr
CH_2Cl_2/pyridine
or 1) Br_2 2) Et_3N
Br

100%

Tetr Lett, 22, 3301 (1981)
JOC, 47, 5088 (1982)

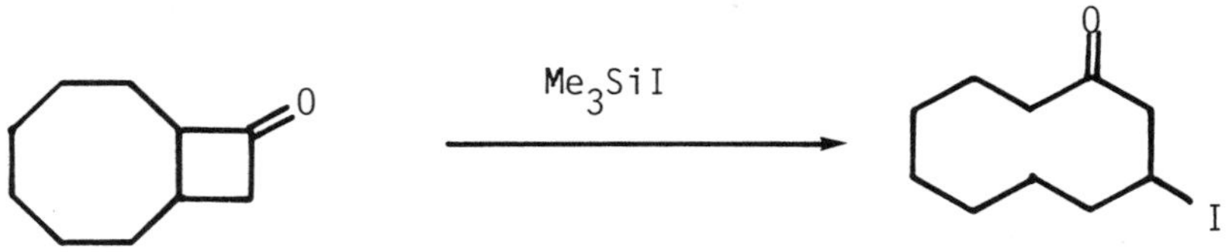
Me_3SiI
I

95%

Tetr Lett, 21, 2639 (1980)

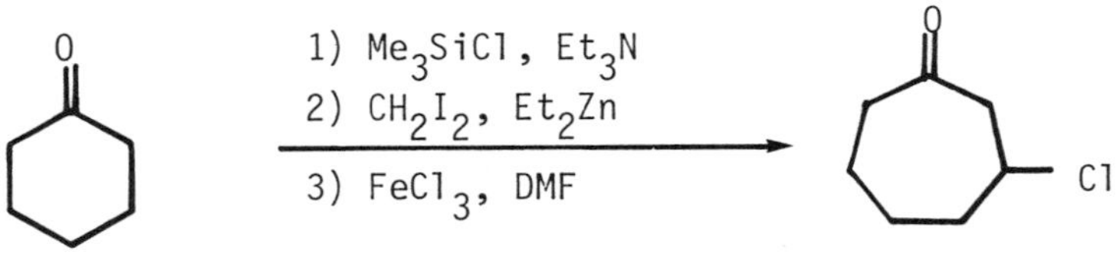
1) Me_3SiCl, Et_3N
2) CH_2I_2, Et_2Zn
3) $FeCl_3$, DMF
Cl

>80%

Org Syn, 59, 113 (1980)

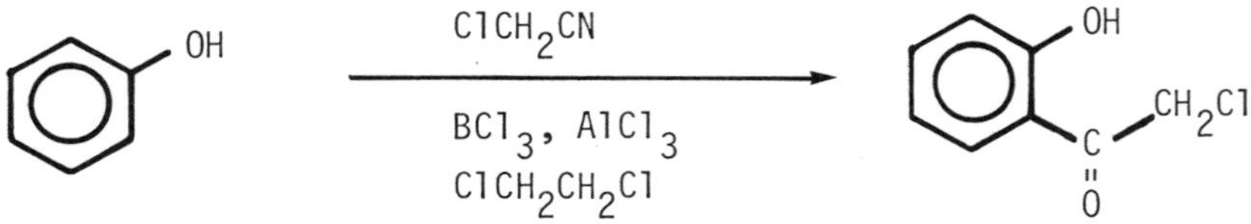
OH
$ClCH_2CN$
BCl_3, $AlCl_3$
$ClCH_2CH_2Cl$
OH
CH_2Cl

85%

JOC, 46, 189 (1981)

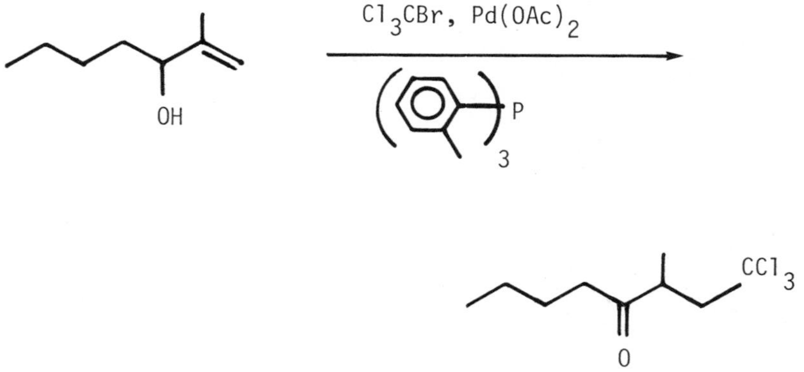

80%

Chem Lett, 1605 (1981)

Review: "Fluorine-Containing β-Diketones"

Russ Chem Rev, 50, 180 (1981)

Section 370 Halide - Nitrile

No additional examples.

Section 371 Halide - Olefin

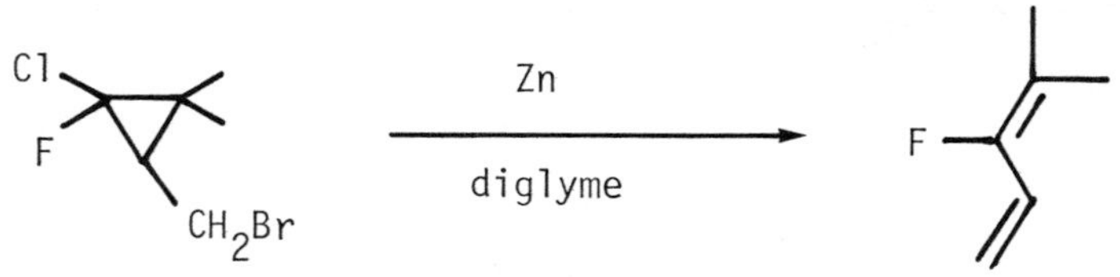

80%

Helv Chim Acta, 63, 1236 (1980)

$\underline{n}\text{-}C_8H_{17}MgCl$

Cl Cl
(alkene)

$\xrightarrow{\quad NiL_4 \quad}$

$\underline{n}\text{-}C_8H_{17}$ Cl

65%

Tetr Lett, <u>22</u>, 315 (1981)

$H\text{-}C\equiv C\text{-}CH_2TMS$

+

Bu-CHO

$\xrightarrow{\quad TiCl_4 \quad}$ $BuCH=CH\text{-}C=CH_2$
                                        |
                                        Cl

72%

Tetr Lett, <u>22</u>, 453 (1981)

$Ph\text{-}\overset{O}{\overset{||}{C}}\text{-}CH_3$

$\xrightarrow{\quad \text{(catechol)}PCl_3 \quad}$

$\underset{Cl}{\overset{Ph}{\diagup}}C=CH_2$

64%

Z Chem, <u>22</u>, 126 (1982)

NaOMe

MeOH

93%

NaOMe

MeOH

74%

Synthesis, 999 (1981)

$$Hx-C\equiv CH \xrightarrow[CH_2Cl_2]{Bu_4N^{\oplus} \ HBr_2^{\ominus}} \quad \begin{array}{c} Hx \\ \diagdown \\ Br \diagup \end{array} C=CH_2 \qquad 80\%$$

Synthesis, 805 (1980)

$$\xrightarrow[\text{2) Zn, HOAc}]{\text{1) Br}_2\text{CHLi}} \qquad 68\%$$

Tetr Lett, 22, 3745 (1981)

$$\xrightarrow{Ph_3P=CHBr} \qquad 72\%$$

77% Z

Tetr Lett, 21, 4021 (1980)

$$HC\equiv C-(CH_2)_8COOMe \xrightarrow[\substack{\text{2) H}_2\text{O} \\ \text{3) ICl, NaOAc}}]{\text{1) catecholborane}} \quad \begin{array}{c} H \\ \diagdown \\ I \diagup \end{array} C=CH(CH_2)_8COOMe$$

70%

Synth Comm, 11, 247 (1981)

$$\left[\begin{array}{c} Me \\ | \\ C \\ ||| \\ C \\ | \\ Me \end{array} Fe^* - \right]^{\oplus} \xrightarrow[\text{2) I}_2]{\text{1) LiPh}_2\text{Cu}} \quad \begin{array}{c} Me \diagdown \underset{||}{C} \diagup Ph \\ I \diagup \underset{}{C} \diagdown Me \end{array} \qquad \sim 70\%$$

Fe* = CpFe(CO)(PPh₃)   JACS, 102, 5923 (1980)

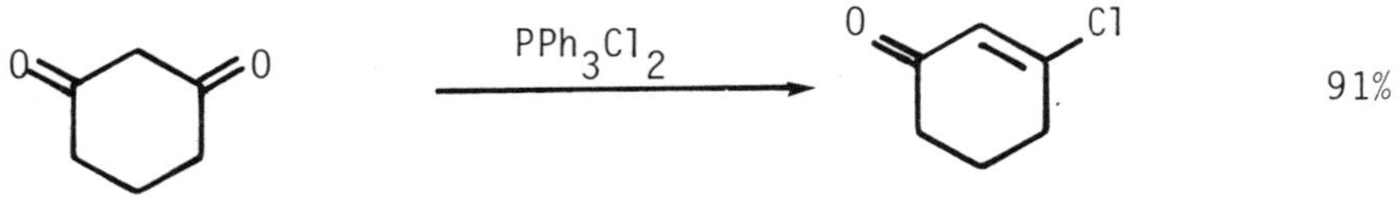

91%

Can J Chem, $\underline{60}$, 210 (1982)

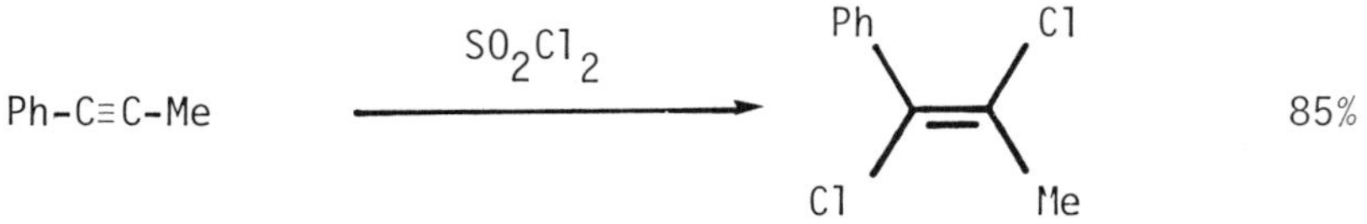

Ph-C≡C-Me

85%

Bull Chem Soc Japan, $\underline{54}$, 2843 (1981)

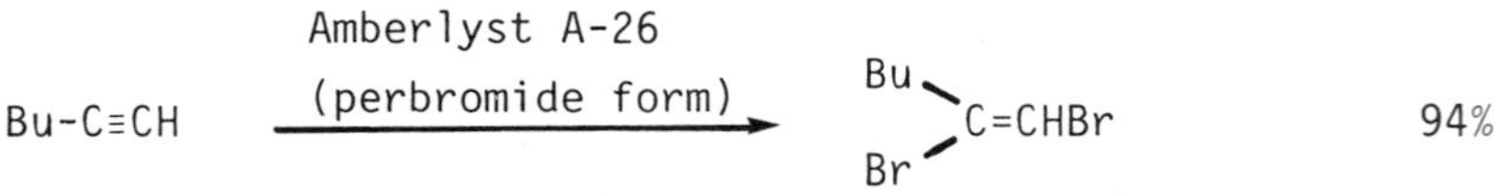

Bu-C≡CH

94%

Synthesis, 143 (1980)

73%

80%

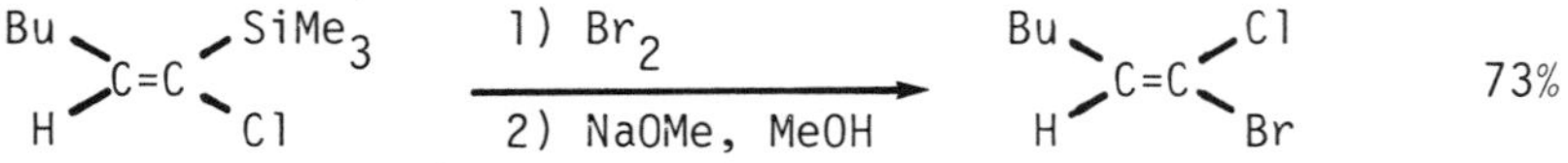

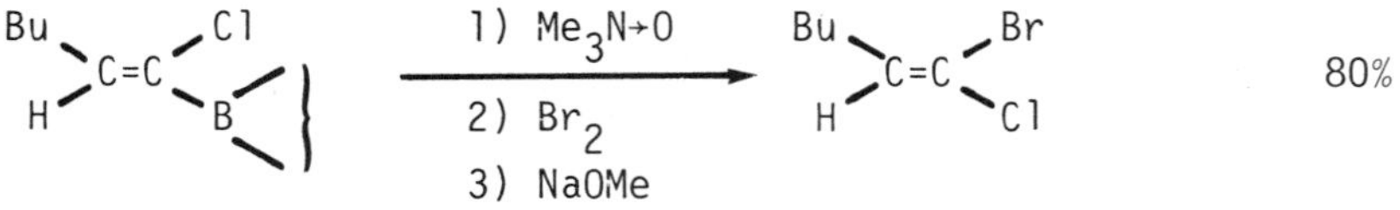

Synthesis, 127 (1982)

Angew Chem Int Ed, 19, 138 (1980)

JOC, 46, 824 (1981)

1) LDA

2) BuBr

81%

JOC, 46, 1504 (1981)

JOC, 45, 2566 (1980)

Section 372 <u>Ketone - Ketone</u>

$$Ph-C \equiv C-C_5H_{11} \xrightarrow[RuCl_2L_3]{PhIO} \underset{Ph}{\overset{O}{\|}} \overset{O}{\underset{\|}{}} C_5H_{11} \quad 72\%$$

Helv Chim Acta, <u>64</u>, 2531 (1981)

$$Ph-C \equiv C-Ph \xrightarrow[CHCl_3/H_2O]{2 \ PhI(OCOCF_3)_2} \overset{O \ O}{\underset{\| \ \|}{}} Ph-C-C-Ph \quad 82\%$$

Doklady Chem, <u>245</u>, 140 (1979)

Ph-C≡N→O

+

Ar

$$\xrightarrow[\substack{2) \ CF_3CO_2H \\ CH_2O, \ H_2O}]{} \overset{O \ O}{\underset{\| \ \|}{}} Ph-C-C-Ar \quad \sim60\%$$

Tetr Lett, <u>21</u>, 1747 (1980)

$$2 \ \overset{O}{\underset{\|}{Ph-C-Cl}} \xrightarrow{SmI_2} \overset{O \ O}{\underset{\| \ \|}{}} Ph-C-C-Ph \quad 78\%$$

Tetr Lett, <u>22</u>, 3959 (1981)

JOC, <u>47</u>, 4347 (1982)

Tetr Lett, <u>21</u>, 3479 (1980)

Chem Lett, 779 (1980)

Chem Lett, 305 (1980)

$$\text{Na}_2\text{PdCl}_4 \quad / \quad \underline{t}\text{-BuOOH}$$

59%

Chem Lett, 257 (1980)

$$\text{PdL}_4 \quad / \quad \text{Ph}_2\text{PCH}_2\text{CH}_2\text{PPh}_2$$

62%

JACS, 102, 2095 (1980)

$$\text{P}-\text{CH}_2\text{-(acac)Ni}^{++}$$

73%

Synthesis, 467 (1982)

1) $\text{CH}_3\overset{\text{O}}{\underset{}{\text{C}}}\text{-CN}$, TiCl_4

2) H_2O, $^{\ominus}\text{OH}$

~90%

Tetr Lett, 22, 1171 (1981)

JOC, <u>47</u>, 5099 (1982)

JOC, <u>46</u>, 3771 (1981)

Synth Comm, <u>12</u>, 189 (1982)

Tetr Lett, <u>23</u>, 3073 (1982)

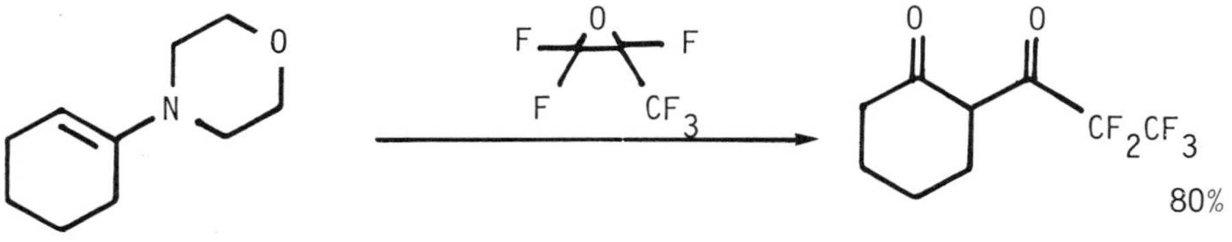

BCS Japan, <u>55</u>, 3345 (1982)

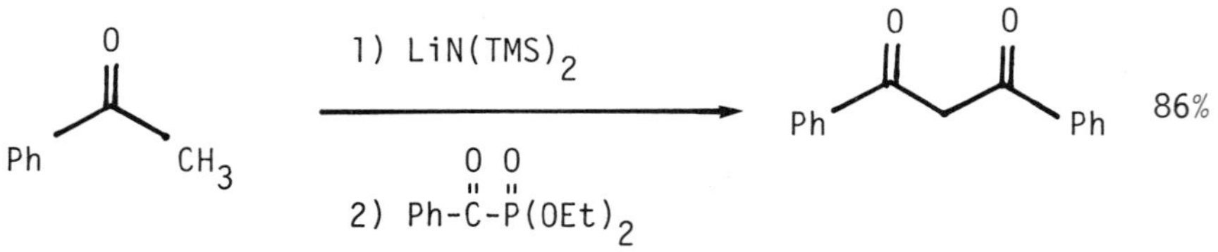

Chem Lett, 1087 (1981)

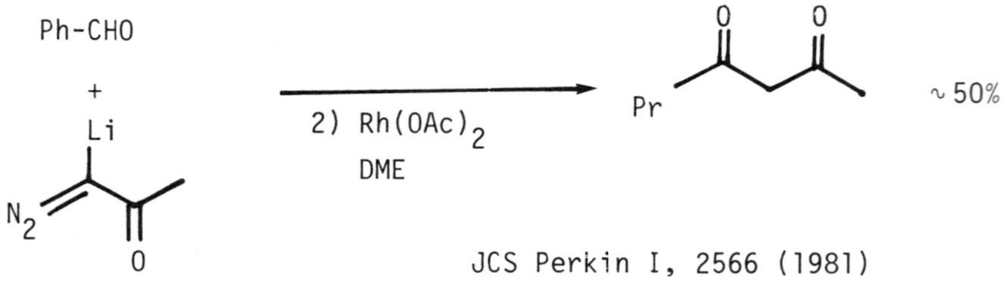

Ph-CHO

+

JCS Perkin I, 2566 (1981)

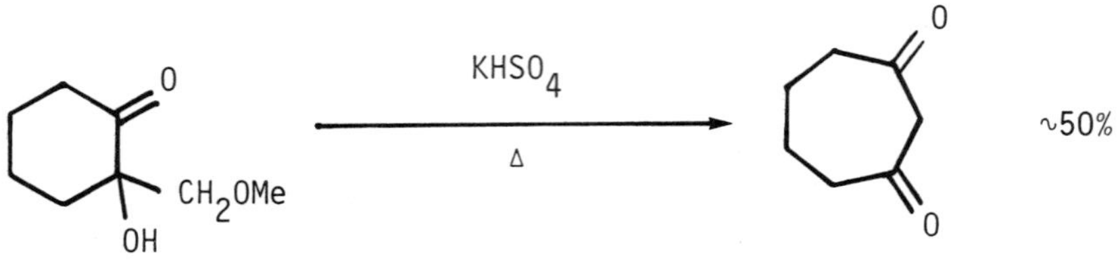

Chem Lett, 551 (1981)

1) s-BuLi

2) cyclohexanone =O

2) [H]

C₅H₁₁ isoxazole

Tetr Lett, 22, 3699 (1981)

PdCl₂, CuCl, O₂

H₂O/dioxane

61%

Chem Lett, 859 (1982)

2 cyclohexanone

FeCl₃, DMF/THF

or 1) LDA   2) Cu(OTf)₂

45-73%

Chem Pharm Bull, 28, 262 (1980)
JOC, 45, 5408 (1980)

JOC, 46, 4631 (1981)

Synthesis, 413 (1980)

Org Syn, 60, 117 (1981)

$$CH_3\text{-}\overset{\overset{\displaystyle O}{\|}}{C}\text{-}CH_3$$

1) CH_3CHO, $Et_2\overset{\oplus}{N}H_2\overset{\ominus}{Cl}$

2) KF, 18-crown-6

3) $CH_3CH_2NO_2$

4) $KMnO_4$-silica gel

$$CH_3\text{-}\overset{\overset{\displaystyle O}{\|}}{C}\text{-}CH_2\text{-}\overset{\overset{\displaystyle CH_3}{|}}{CH}\text{-}\overset{\overset{\displaystyle O}{\|}}{C}\text{-}CH_3$$

80%

JCS Chem Comm, 635 (1982)

CsF

Si(OEt)$_3$

65%

JCS Chem Comm, 122 (1981)

1)

2) H_3O^+

89%

75% ee

Tetr Lett, 23, 3711 (1982)

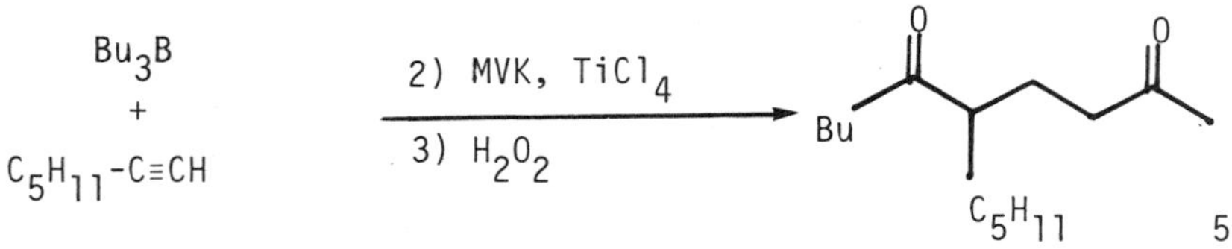
Bu_3B
+
C_5H_{11}-C≡CH
2) MVK, $TiCl_4$
3) H_2O_2
O
O
Bu
C_5H_{11}

58%

Chem Lett, 221 (1980)

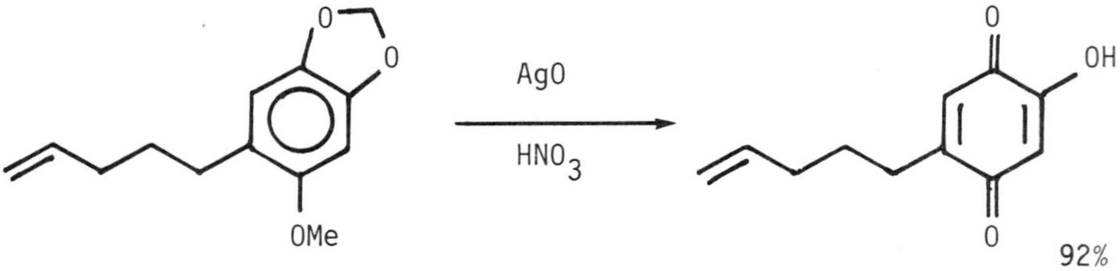
O
O
OMe
AgO
HNO_3
O
OH
O

92%

Synth Comm, 10, 9 (1980)

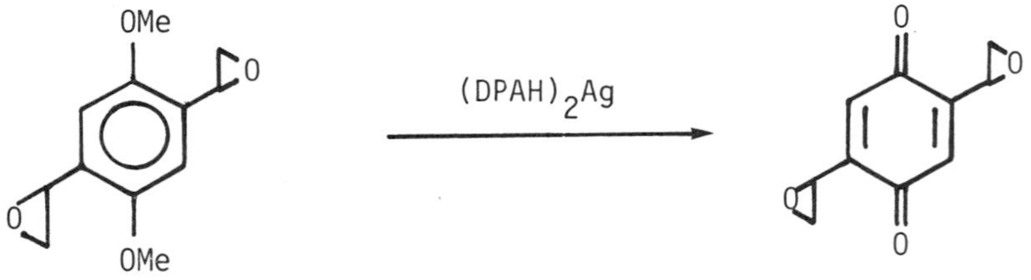
OMe
O
O
OMe
$(DPAH)_2Ag$
O
O
O
O

52%

Chem Lett, 725 (1980)

Section 373 Ketone - Nitrile

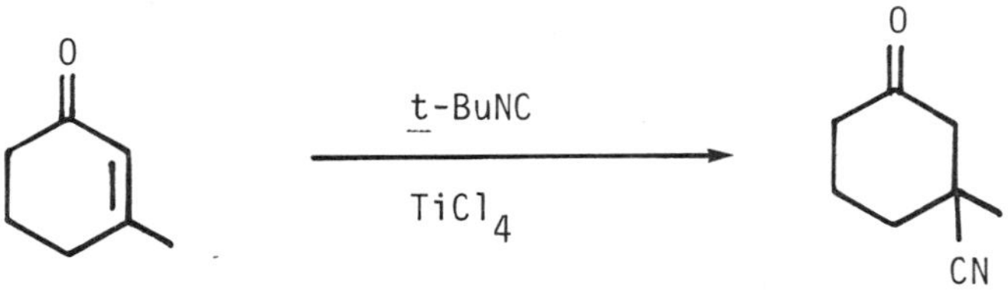
O
t-BuNC
$TiCl_4$
O
CN

84%

JACS, 104, 6449 (1982)

1) LDA
2) PhCH$_2$SCN

70%

Comptes Rendus, 291, 179 (1980)

CH$_2$=CHCN

NaCN, DMF

68%

Org Syn, 59, 53 (1980)

Section 374 Ketone - Olefin

For the oxidation of allylic alcohols to olefinic ketones, see Section 168 (Ketones from Alcohols and Phenols).

For the oxidation of allylic methylene groups (C=C-CH$_2$ → C=C-CO), see Section 170 (Ketones from Alkyls and Methylenes).

For the alkylation of olefinic ketones, see also Section 177 (Ketones from Ketones), and Section 74 (Alkyls from Olefins) for conjugate alkylations.

$$2 \quad \text{[cyclohexenyl-CH=CH-HgCl]} \quad \xrightarrow[\text{[Rh(CO)}_2\text{Cl]}_2,\ \text{THF}]{\text{CO, LiCl}} \quad \text{(}\text{product}\text{)}_2 \text{C=O} \quad 100\%$$

JOC, 45, 3840 (1980)

$$\text{[1-Et-cyclohexene]} \quad \xrightarrow[\text{ZnCl}_2]{\text{Ac}_2\text{O}} \quad \text{[Et-cyclohexene-COCH}_3\text{]} \quad 95\%$$

JACS, 102, 3848 (1980)

$$\text{[cyclohexene]} \quad \xrightarrow[\text{EtAlCl}_2]{\text{CH}_3\text{COCl}} \quad \text{[O=C-CH}_3\text{ cyclohexene]} \quad 73\%$$

JOC, 47, 5393 (1982)

$$\text{[decalone]} \quad \xrightarrow[\text{H}_2\text{O}]{\text{Na}_2\text{O}} \quad \text{[diketone]} \quad 63\%$$

Tetr Lett, 22, 5127 (1981)

Ph-CH=CH-CHCN (with morpholine N-substituent)

1) LDA
2) EtBr

3) H$_2$O
4) Cu(OAc)$_2$, EtOH

Ph-CH=CH-C-Et   78%
        ‖
        O

Chem Lett, 1263 (1982)

1) (pyridine-Se)$_2$ + Br$_2$

2) O$_3$
3) Δ

82%

Tetr Lett, 23, 2105 (1982)

(cyclohexanone with CH$_2$OTs and 2-methylallyl substituents)

DBN or DBU

95%

JOC, 47, 4358 (1982)

(2-acetylcyclohexanone)

1) NaH
2) Se

3) MeI
4) H$_2$O$_2$

83%

Tetr Lett, 22, 3043 (1981)

$$CH_3-\underset{\underset{OBR_2}{|}}{C}=CH-CH_2-\text{(cyclopentyl)} \xrightarrow{\text{PhSeCl}} \text{(enone with cyclopentyl)} \quad 96\%$$

Synth Comm, 10, 667 (1980)

$$\xrightarrow[\text{PhIO}_2]{\text{PhSeSePh}} \quad \sim 80\%$$

JCS Chem Comm, 1044 (1981)

$$\xrightarrow[\text{CH}_3\text{CN}]{\text{Pd(OAc)}_2, \; \underset{\text{PPh}_2}{\overset{\text{CH}_2-\!\!-\text{CH}_2}{|}} \underset{\text{PPh}_2}{|}} \quad 85\%$$

JACS, 104, 5844 (1982)

Tetr Lett, 21, 651 (1980)

Synthesis, 647 (1980)

JCS Chem Comm, 486 (1980)

Synthesis, 60 (1982)

Org Syn, 60, 88 (1981)

Synth Comm, 10, 637 (1980)

JOC, 45, 1046 (1980)

JOC, 45, 3017 (1980)

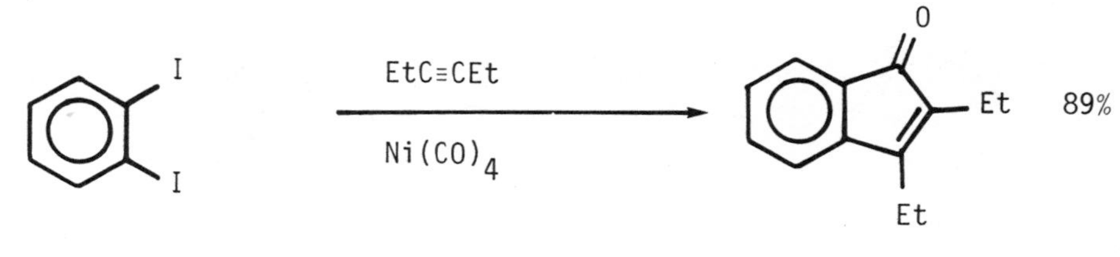
I
I
EtC≡CEt
Ni(CO)4
O
Et
Et
89%

JOC, 45, 5426 (1980)

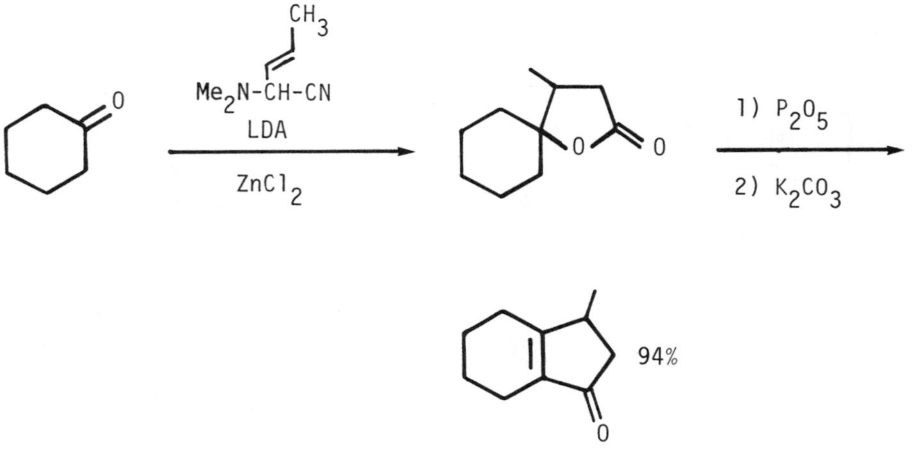
O
CH3
Me2N-CH-CN
LDA
ZnCl2
O
O
1) P2O5
2) K2CO3
94%
O

Tetr Lett, 21, 1205 (1980)

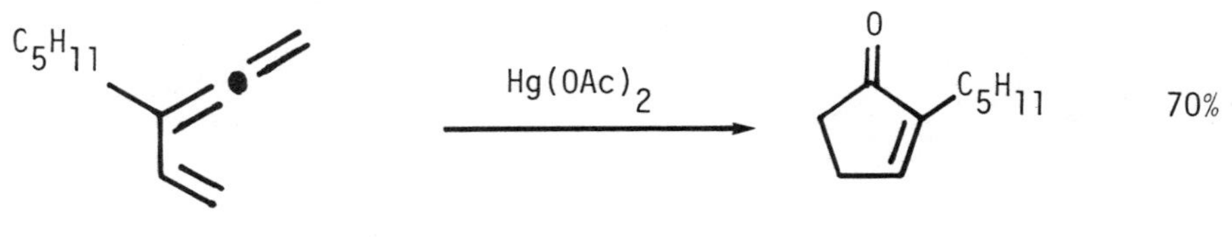
C5H11
Hg(OAc)2
O
C5H11
70%

Tetrahedron, 36, 189 (1980)

Tetr Lett, 23, 4923 (1982)

JACS, 103, 1604 (1981)

JOC, 45, 5399 (1980)

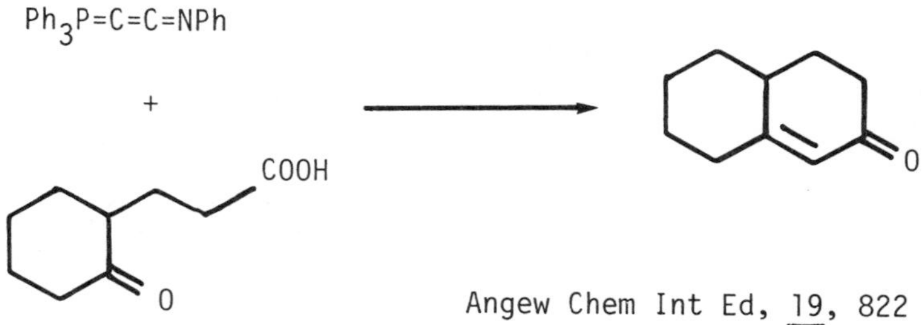
$Ph_3P=C=C=NPh$
+
COOH
O
O

72%

Angew Chem Int Ed, 19, 822 (1980)

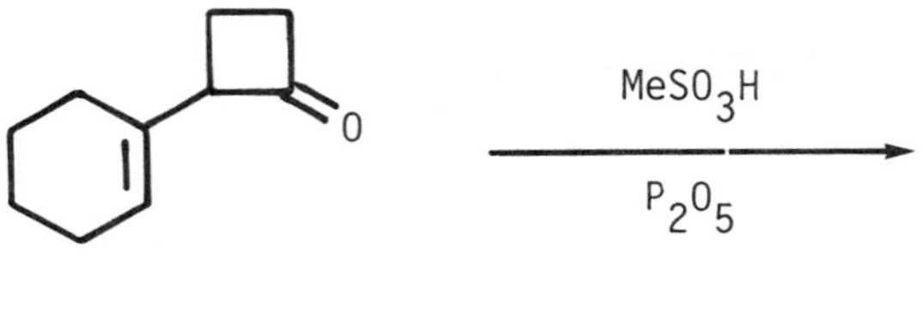
O
$MeSO_3H$
P_2O_5
O

65%

Tetr Lett, 22, 2459 (1981)

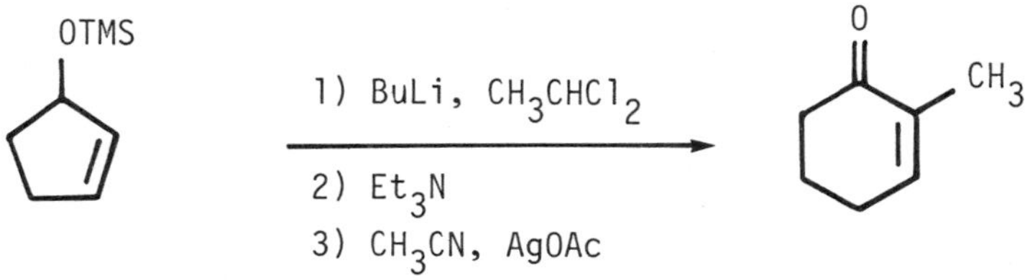
OTMS
1) BuLi, CH_3CHCl_2
2) Et_3N
3) CH_3CN, AgOAc
O
CH_3

66%

Synthesis, 289 and 291 (1981)

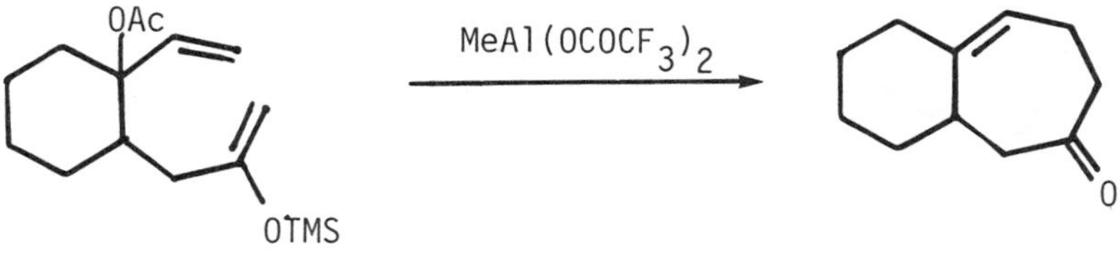
OAc
OTMS
$MeAl(OCOCF_3)_2$
O

65%

Bull Chem Soc Japan, 53, 2050 (1980)

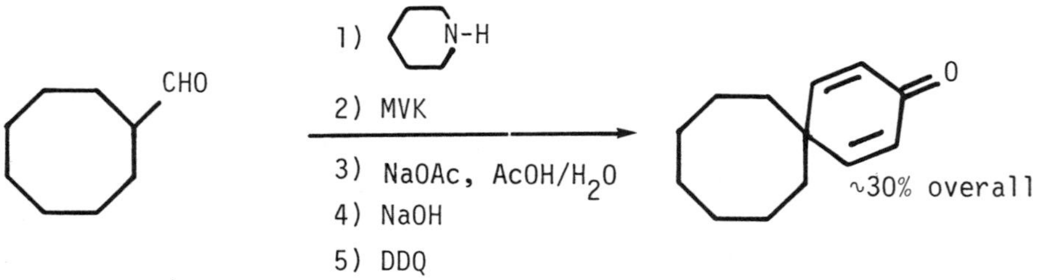

1) ⬡ N-H

2) MVK

3) NaOAc, AcOH/H_2O

4) NaOH

5) DDQ

~30% overall

Org Syn, 61, 129 (1983)

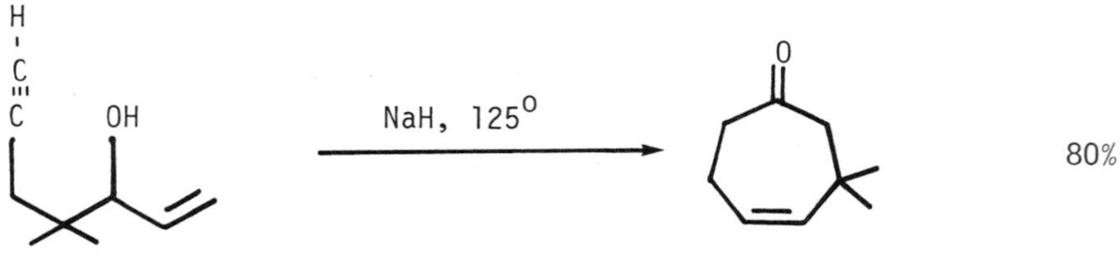

NaH, 125°

80%

Tetr Lett, 21, 2123 (1980)

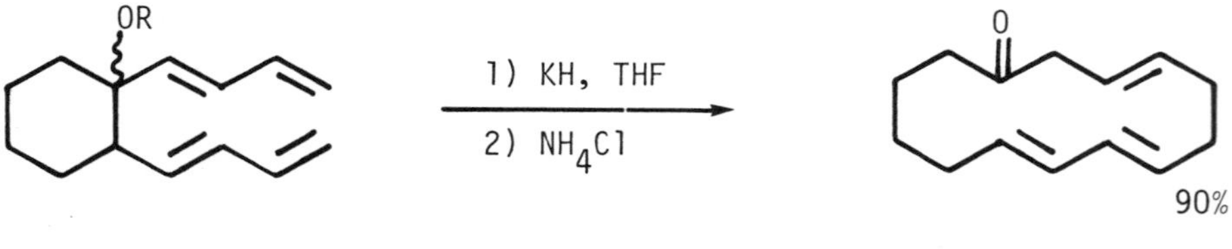

1) KH, THF

2) NH_4Cl

90%

Tetr Lett, 22, 2471 (1981)

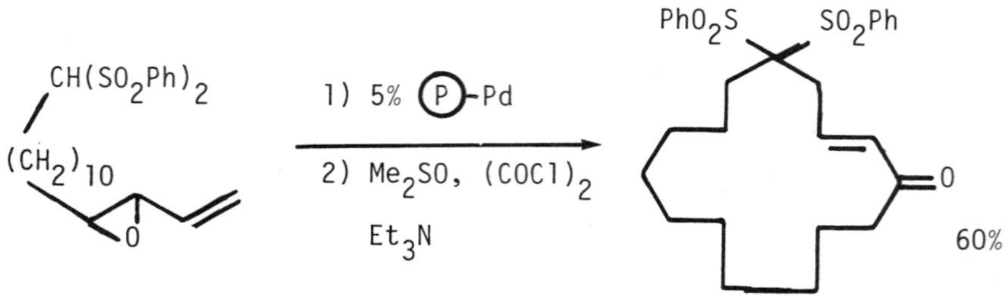

$CH(SO_2Ph)_2$

$(CH_2)_{10}$

1) 5% Ⓟ–Pd

2) Me_2SO, $(COCl)_2$

Et_3N

PhO_2S SO_2Ph

60%

JACS, 104, 6112 (1982)

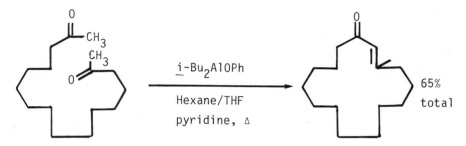

i-Bu₂AlOPh

Hexane/THF

pyridine, Δ

65% total

(mixture of E and Z plus some β,γ-isomer)

Bull Chem Soc Japan, 53, 1417 (1980)

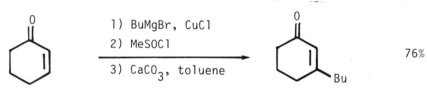

1) BuMgBr, CuCl

2) MeSOCl

3) CaCO₃, toluene

76%

Chem Lett, 1159 (1981)

1) ⟨alkenyl⟩Br, LDA

2) 5% HCl - THF

90%

Chem Lett, 165 (1982)

Nafion-H

H⁺

84%

Synthesis, 473 (1981)

PhSe ⟶ SePh

1) LDA

2) (isobutyl-CH$_2$-Br)

3) H$_2$O$_2$

(product: 6-methyl-2-heptenal aldehyde) H, O   75%

JOC, 47, 1618 (1982)

(tetronic acid / butenolide structure with OH, O, O)   Zn(OAc)$_2$ ⟶   (enone product)   78%

Gazz Chim Ital, 112, 1 (1982)

OMe, COOH (aromatic)

1) Na, NH$_3$

2) C$_7$H$_{15}$Br

3) H$_3$O$^+$

⟶   (cyclohexenone product) O, C$_7$H$_{15}$ ~50%

Org Syn, 61, 59 (1983)

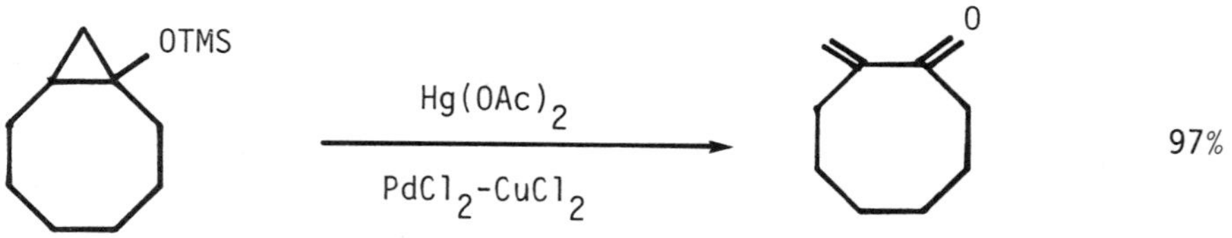
OTMS
Hg(OAc)2
PdCl2-CuCl2
O
97%

Tetr Lett, 21, 4283 (1980)

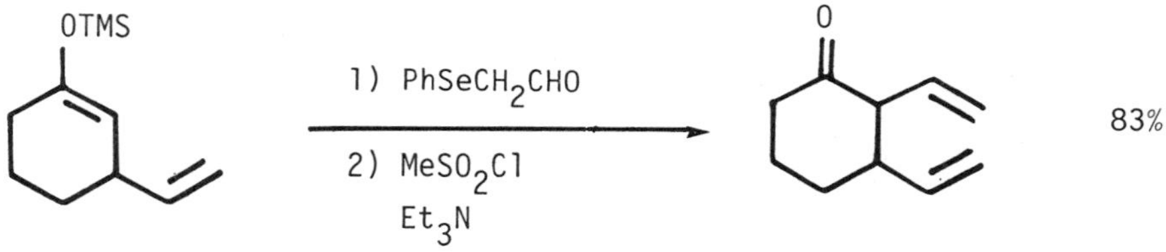
OTMS
1) PhSeCH2CHO
2) MeSO2Cl
Et3N
O
83%

JCS Chem Comm, 434 (1981)

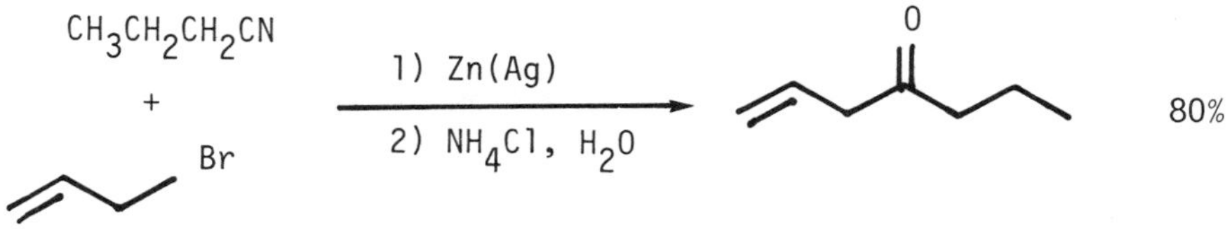
CH3CH2CH2CN
+
Br
1) Zn(Ag)
2) NH4Cl, H2O
O
80%

Tetr Lett, 22, 649 (1981)

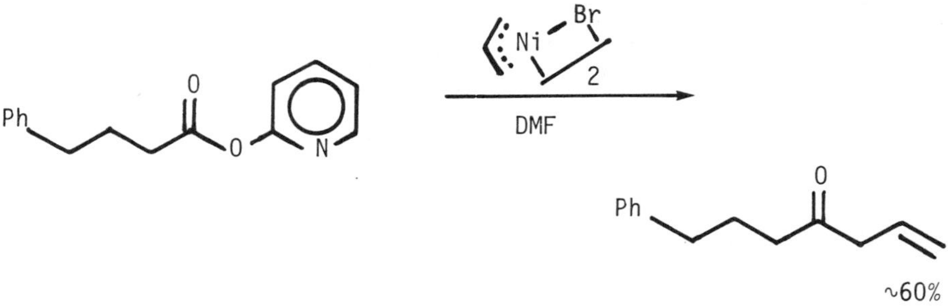
Ph
O
O
N
Ni
Br
2
DMF
Ph
O
~60%

Chem Lett, 1483 (1979)

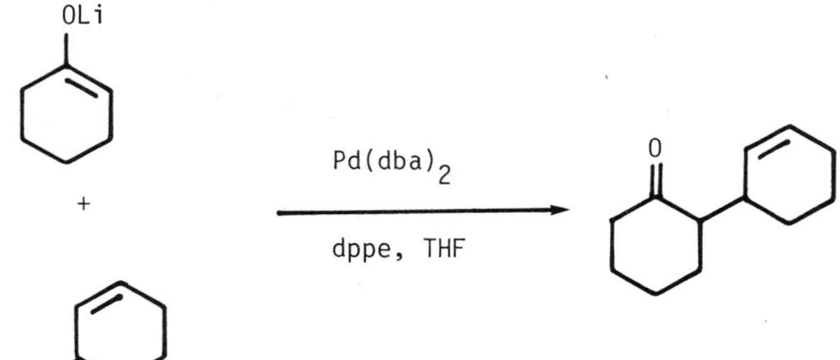

OLi
+
AcO
Pd(dba)$_2$
dppe, THF
O

55%

JCS Chem Comm, 1159 (1981)

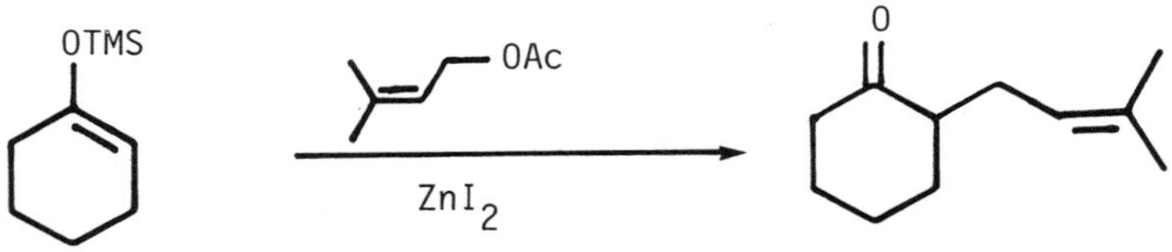

OTMS
OAc
ZnI$_2$
O

71%

Synth Comm, 11, 217 (1981)

C≡CH
1) Cp$_2$ZrClH
2)
O
, Ni(II)
3) H_3O^+
O

78%

JACS, 102, 1334 (1980)

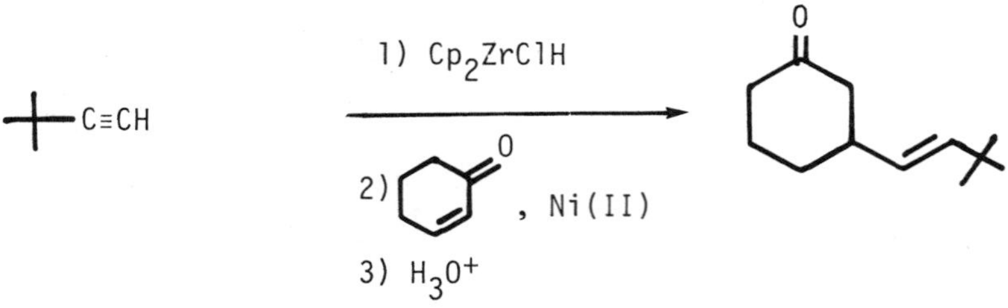

JACS, 102, 6381 (1980)

Tetr Lett, 21, 3199 (1980)

Tetr Lett, 22, 1115 (1981)

J Chem Research(S), 248 (1982)

Section 375 <u>Nitrile - Nitrile</u>

Acta Chem Scand B, <u>34</u>, 289 (1980)

Section 376 <u>Nitrile - Olefin</u>

JACS, <u>104</u>, 1560 (1982)

Chem Lett, 1565 (1982)

JACS, <u>103</u>, 5568 (1981)

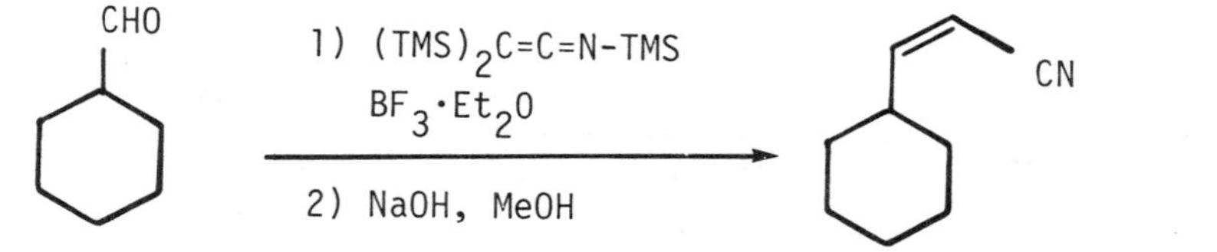
CHO
1) $(TMS)_2C=C=N-TMS$
$BF_3 \cdot Et_2O$
2) NaOH, MeOH
CN

~60%

JCS Chem Comm, 56 (1982)

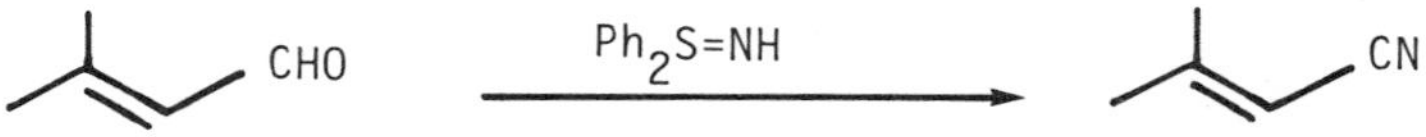
CHO
$Ph_2S=NH$
CN

85%

Synthesis, 1005 (1980)

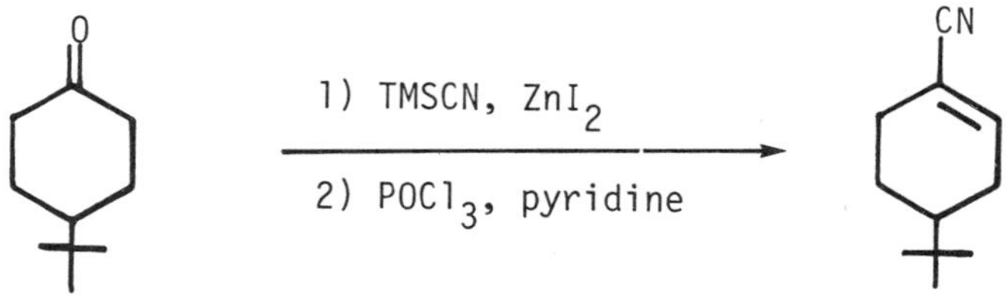
O
1) TMSCN, ZnI_2
2) $POCl_3$, pyridine
CN

82%

Chem Lett, 1427 (1979)

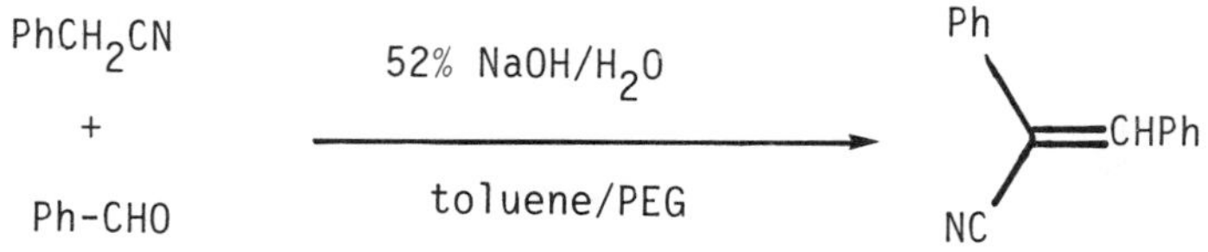
$PhCH_2CN$
+
Ph-CHO
52% $NaOH/H_2O$
toluene/PEG
Ph
CHPh
NC

71%

Synthesis, 913 (1981)

Synth Comm, <u>10</u>, 479 (1980)

JCS Perkin I, 2516 (1980)

Tetr Lett, <u>22</u>, 2573 (1981)

Section 377 Olefin - Olefin

JOC, 47, 3364 (1982)

Tetr Lett, 21, 5019 (1980)

Tetr Lett, 22, 1451 (1981)

JOC, 45, 4536 (1980)

Synth Comm, 12, 739 (1982)

$C_5H_{11}-C\equiv C-CH_2Br$

+

$\xrightarrow{\quad CrCl_2 \quad}$

$H_2C=C=C\underset{HO}{\overset{C_5H_{11}}{\diagup}}$ 78%

Tetrahedron, <u>37</u>, 1359 (1981)

$HC\equiv C$

+

BuMgBr

$\xrightarrow{\quad \text{cat. CuI} \quad}$

$Bu-CH=C=CHCH_2COOH$ 97%

Tetr Lett, <u>22</u>, 2375 (1981)

$Me_2C=C=C\overset{H}{\underset{Li}{\diagup}}$

·+

Ph-I

$\xrightarrow[\text{AlH(i-Bu)}_2]{\quad PdCl_2L_2 \quad}$

$Me_2C=C=CHPh$ 90%

Synthesis, 738 (1982)

$\xrightarrow{\quad Bu_3SnH, \text{ AIBN} \quad}$

100%

Tetr Lett, <u>22</u>, 2675 (1981)

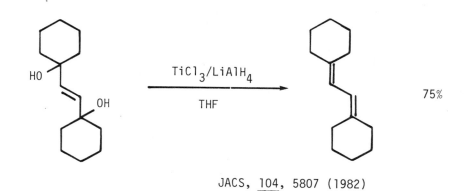

JACS, <u>104</u>, 5807 (1982)

$$P-\overset{Ph}{\underset{Ph}{\overset{|}{\underset{|}{P}}}}-CH_2CH=CH_2$$

+

CHO

Cl

50% NaOH →

HC=CH-CH=CH_2

Cl

78%

Tetr Lett, <u>21</u>, 1375 (1980)

1) ( )_2 BH

2 Bu-C≡CH

2) NaOMe

3) CuBr·SMe_2

→ Bu   Bu

97%

JOC, <u>45</u>, 549 and 550 (1980)

Bu —— B(Sia)$_2$

+

Br —— Hex

$\xrightarrow[\text{NaOEt}]{\text{PdL}_4}$

Bu —— —— Hex   49%

Tetr Lett, <u>22</u>, 127 (1981)

1) $\left(\bigcirc \right)_2$ BCl

2) Me$_2$CHCHO

3) KH

TMS —— $^\ominus$

Me$_2$CH ——

1) Bu$_3$SnCl-BF$_3$

2) C$_9$H$_{19}$CHO

3) KH

C$_9$H$_{19}$ ——

JCS Chem Comm, 1326 (1982)

2   PhSO$_2$ —— (MgBr)

$\xrightarrow[\text{THF}]{2\% \text{ Ni(acac)}}$

80%

Tetr Lett, <u>23</u>, 2457 (1982)

Tetr Lett, 22, 2751 (1981)

Tetr Lett, 23, 1591 (1982)

Tetr Lett, 22, 959 (1981)

OAc

1) $\diagup\!\!\!\diagdown$ $CO_2CH_2CCl_3$

2) PdL_4, Et_3N

87%

JACS, 102, 2841 (1980)

Review: "Palladium-catalyzed Syntheses of Conjugated Dienes"

Pure and Appl Chem, 53, 2323 (1981)

Ph$\diagup\!\!\!\diagdown$$\diagdown$S $\diagdown$ 2-pyridyl

1) BuLi

2) Bu_3SnCH_2I

Ph$\diagup\!\!\!\diagdown$$\diagdown$

73%

Tetr Lett, 23, 2205 (1982)

$H_2C=C=C$ C_5H_{11} $SiMe_3$

BF_3-HOAc

CH_2Cl_2

C_5H_{11}

80%

Synth Comm, 12, 409 (1982)

Tetr Lett, <u>23</u>, 3277 (1982)

JOC, <u>45</u>, 1640 (1980)

Organometallics, <u>1</u>, 259 (1982)

Hx—CH=CH—B(O₂C₆H₄)   →(1) MeLi (2) allyl-Br→   Hx—CH=CH—CH₂—CH=CH₂   89%

Bull Chem Soc Japan, 53, 1471 (1980)

Ph—CH=CH—CH₂—OAc

+

CH₂=C(CH₃)—CH₂—SnBu₃

→ PdL₄ / THF →   Ph—CH=CH—CH₂—CH₂—C(CH₃)=CH₂   69%

Tetr Lett, 21, 2591, 2595, and 2599 (1980)

Bu—CH(CR₂)—CH=CH₂   →(allyl-Br) / CuI→   Bu—CH=CH—CH₂—CH₂—CH=CH₂   43%

Synth Comm, 12, 813 (1982)

CH₃—CH=CH—CH₂—Cl   →⟨cyclopentadienyl Li⊕, Bu₃B⟩→   72%

JACS, 103, 1969 (1981)

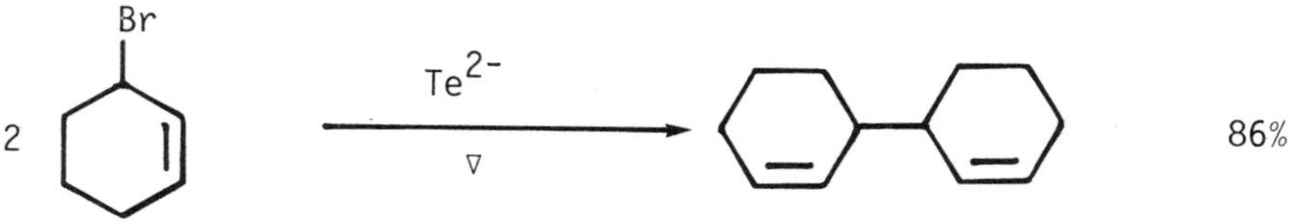

JOC, <u>47</u>, 1641 (1982)

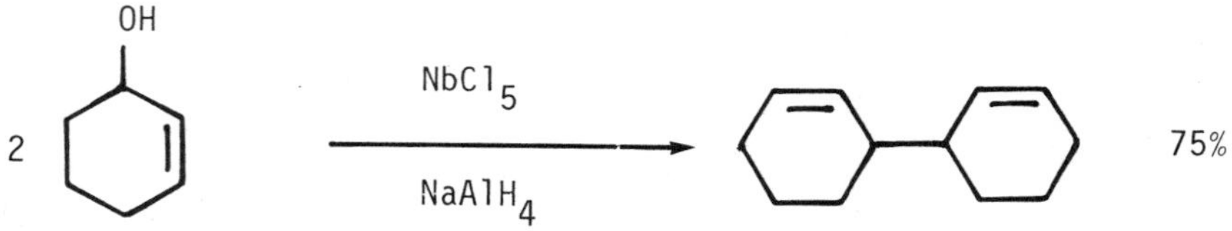

Chem Lett, 157 (1982)

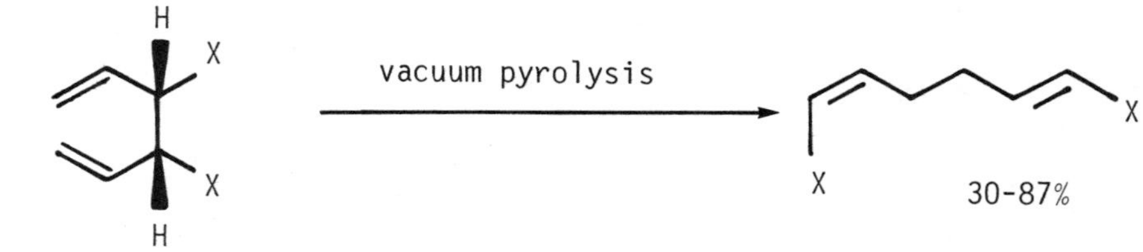

X = -CH$_2$OH, -COOH, -COOMe

JCS Chem Comm, 325 (1982)

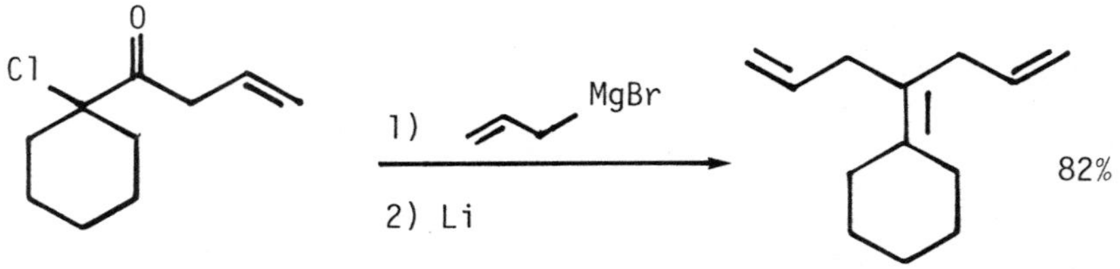

JOC, <u>46</u>, 2721 (1981)

Review: "The Synthesis of Leukotrienes: A New Class of Biolo-
            gically Active Compounds Including SRS-A"

Chem Soc Rev, <u>11</u>, 321 (1982)